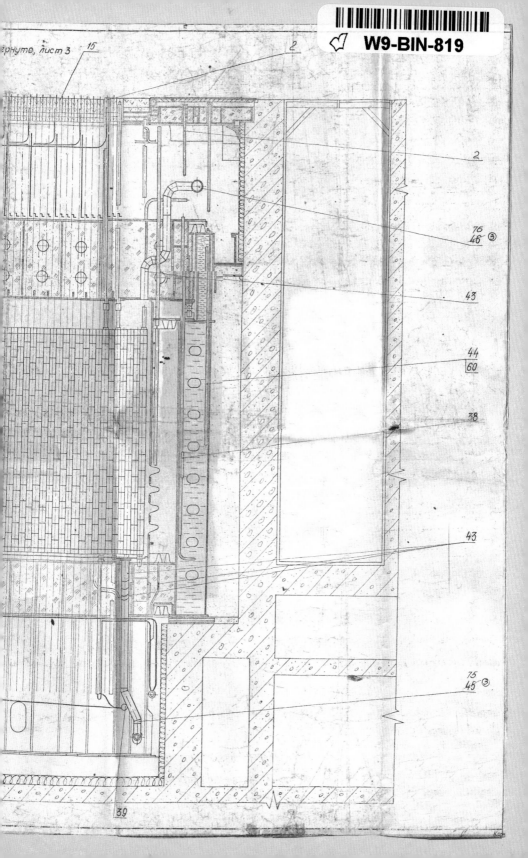

MIDNIGHT

in

CHERNOBYL

The Untold Story of the World's
Greatest Nuclear Disaster

ADAM HIGGINBOTHAM

Simon & Schuster

NEW YORK   LONDON   TORONTO
SYDNEY   NEW DELHI

Simon & Schuster
1230 Avenue of the Americas
New York, NY 10020

First Simon & Schuster hardcover edition February 2019

SIMON & SCHUSTER and colophon are registered trademarks
of Simon & Schuster, Inc.

For information about special discounts for bulk purchases, please
contact Simon & Schuster Special Sales at 1-866-506-1949
or business@simonandschuster.com.

The Simon & Schuster Speakers Bureau can bring authors to your
live event. For more information or to book an event, contact the
Simon & Schuster Speakers Bureau at 1-866-248-3049
or visit our website at www.simonspeakers.com.

Interior design by Paul Dippolito

Manufactured in the United States of America

5   7   9   10   8   6   4

Library of Congress Cataloging-in-Publication Data is available.

ISBN 978-1-5011-3461-6
ISBN 978-1-5011-3464-7 (ebook)

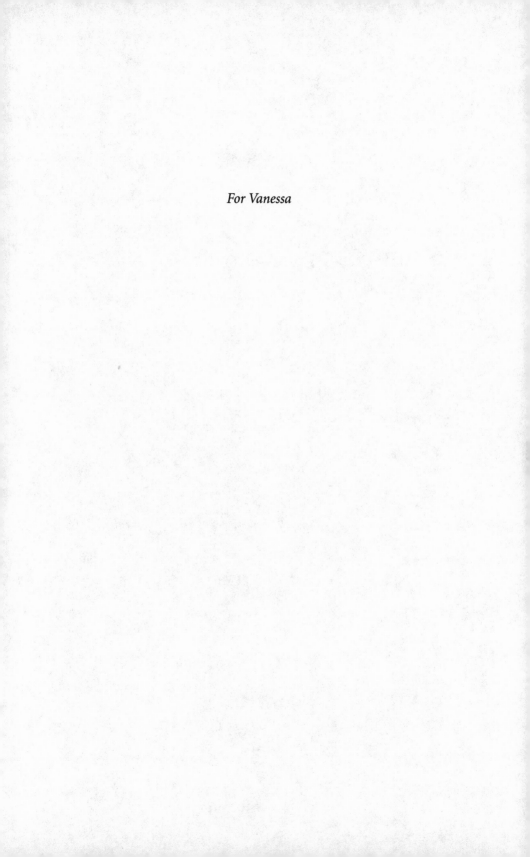

For Vanessa

ARCTIC OCEAN

Bilibino

Arctic Circle

FEDERATIVE SOCIALIST REPUBLIC
(R.S.F.S.R.)

Komsomolsk-
on-Amur

Amur R.

Lake Baikal

CHINA

Vladivostok

MONGOLIA

NORTH
KOREA

JAPAN

0   1000 miles

0   1000 kilometers

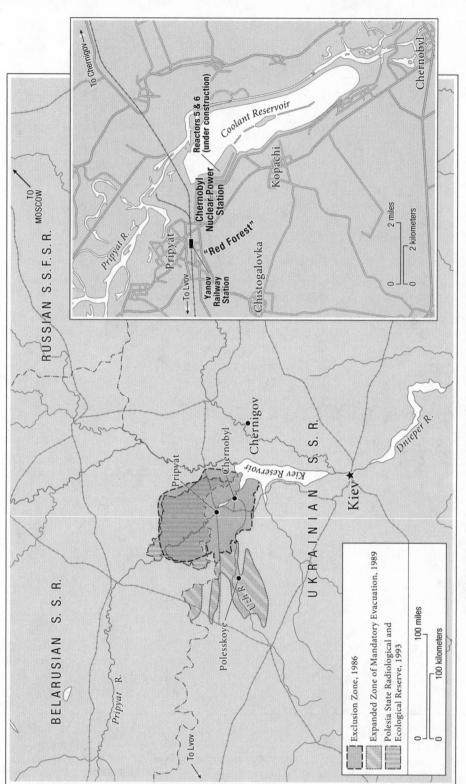

THE CHERNOBYL AREA AND THE EXCLUSION ZONE, 1986–93

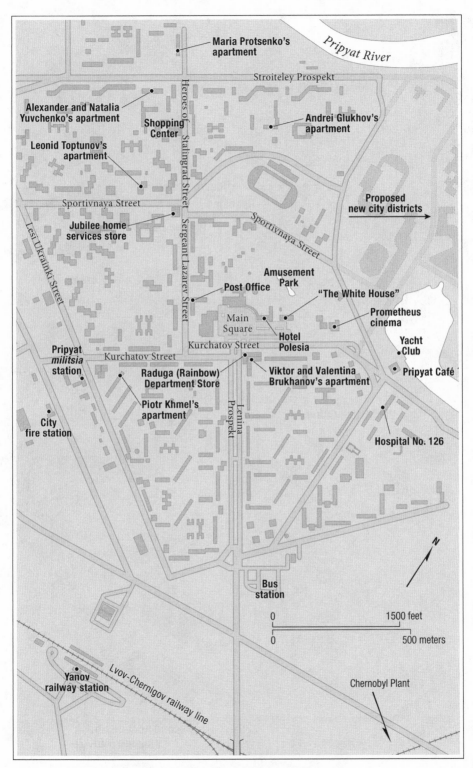

THE CITY OF PRIPYAT IN APRIL 1986

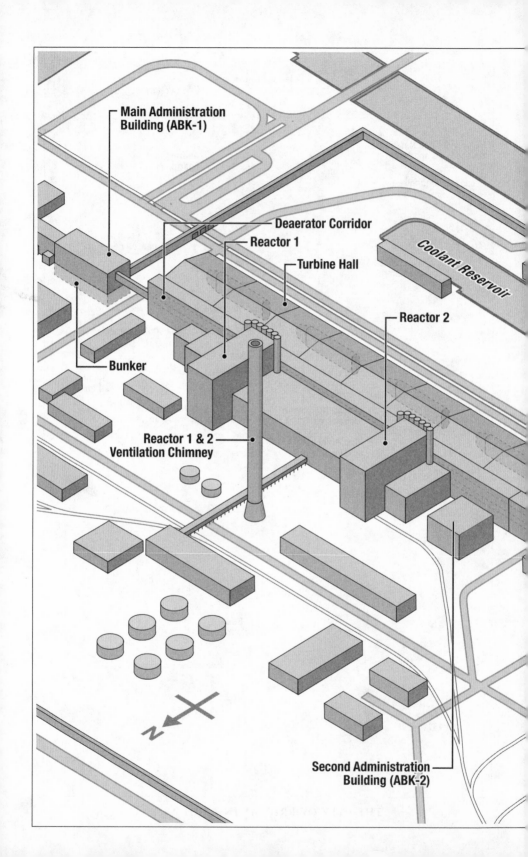

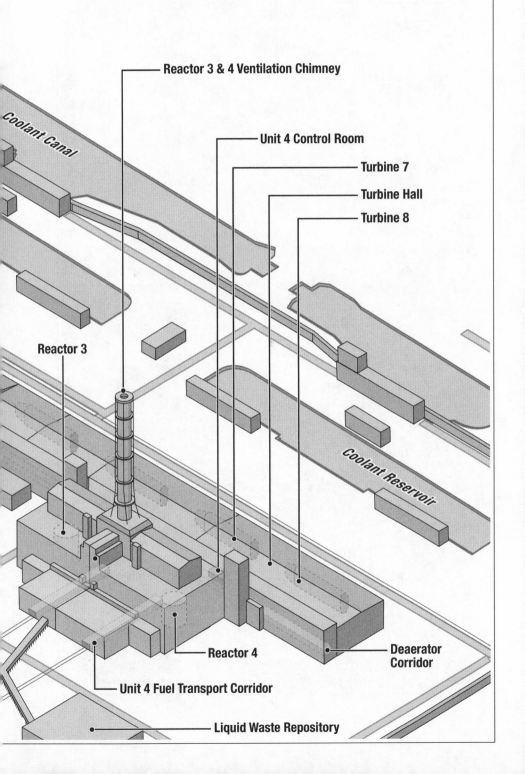

The Chernobyl Nuclear Power Plant in April 1986

Reactor 3 & 4 Ventilation Chimney

Unit 4 Control Room

Turbine 7

Turbine Hall

Turbine 8

Coolant Canal

Reactor 3

Coolant Reservoir

Reactor 4

Deaerator Corridor

Unit 4 Fuel Transport Corridor

Liquid Waste Repository

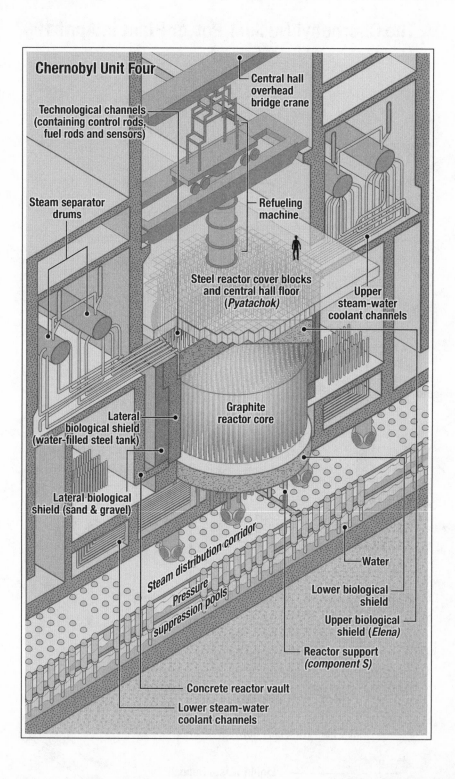

Chernobyl Unit Four

Central hall overhead bridge crane

Technological channels (containing control rods, fuel rods and sensors)

Steam separator drums

Refueling machine

Steel reactor cover blocks and central hall floor (*Pyatachok*)

Upper steam-water coolant channels

Graphite reactor core

Lateral biological shield (water-filled steel tank)

Lateral biological shield (sand & gravel)

Steam distribution corridor

Pressure suppression pools

Water

Lower biological shield

Upper biological shield (*Elena*)

Reactor support (*component S*)

Concrete reactor vault

Lower steam-water coolant channels

Cast of Characters

The Chernobyl Atomic Energy Station and the City of Pripyat

MANAGEMENT

Viktor Brukhanov plant director

Nikolai Fomin chief engineer; deputy to the plant director

Anatoly Dyatlov deputy chief engineer for operations

STAFF

Alexander Akimov foreman, fifth shift of reactor Unit Four

Leonid Toptunov senior reactor control engineer, fifth shift, Unit Four

Boris Stolyarchuk senior unit control engineer, fifth shift, Unit Four

Yuri Tregub senior reactor control engineer, Unit Four

Alexander Yuvchenko senior mechanical engineer, fifth shift, Unit Four

Valery Perevozchenko reactor shop shift foreman, fifth shift, Unit Four

Serafim Vorobyev head of plant civil defense

Veniamin Prianichnikov head of training in plant nuclear safety

FIREFIGHTERS

Major Leonid Telyatnikov chief of Paramilitary Fire Station Number Two (Chernobyl plant)

Lieutenant Vladimir Pravik head of third watch, Paramilitary Fire Station Number Two (Chernobyl plant)

Lieutenant Piotr Khmel head of first watch, Paramilitary Fire Station Number Two (Chernobyl plant)

Lieutenant Viktor Kibenok head of third watch, Paramilitary Fire Station Number Six (Pripyat)

Sergeant Vasily Ignatenko member of third watch, Paramilitary Fire Station Number Six (Pripyat)

PRIPYAT

Alexander Esaulov deputy chairman of the Pripyat *ispolkom*, or city council; the deputy mayor

Maria Protsenko chief architect for the city of Pripyat

Natalia Yuvchenko teacher of Russian language and literature at School Number Four, and wife of Alexander Yuvchenko

The Government

Mikhail Gorbachev General secretary of the Communist Party of the Soviet Union; leader of the USSR

Nikolai Ryzhkov chairman of the Soviet Council of Ministers; prime minister of the USSR

Yegor Ligachev chief of ideology for the Communist Party of the Soviet Union; the second most powerful figure in the Politburo

Viktor Chebrikov chairman of the Committee for State Security (KGB) of the USSR

Vladimir Dolgikh secretary of the Communist Party Central Committee with responsibility for heavy industry, including nuclear power

Vladimir Marin head of the nuclear power sector of the Heavy Industry and Energy Division of the Communist Party Central Committee

Anatoly Mayorets Soviet minister of energy and electrification

Gennadi Shasharin deputy Soviet minister of energy, with specific responsibility for nuclear energy

Vladimir Scherbitsky first secretary of the Communist Party of Ukraine and member of the Soviet Politburo; leader of the Soviet Socialist Republic of Ukraine

Alexander Lyashko chairman of the Council of Ministers of the Soviet Socialist Republic of Ukraine; the Ukrainian prime minister

Vladimir Malomuzh second secretary of the Kiev Oblast Communist Party

Vitali Sklyarov Ukrainian minister of energy and electrification

Boris Scherbina deputy chairman of the Soviet Council of Ministers; first chairman of the government commission in Chernobyl

Ivan Silayev deputy chairman of the Soviet Council of Ministers, responsible for the engineering industry; member of the Central Committee of the Communist Party of the USSR; second chairman of the government commission in Chernobyl

The Nuclear Experts

Anatoly Aleksandrov chairman of the Soviet Academy of Sciences and director of the Kurchatov Institute of Atomic Energy, responsible for the development of nuclear science and technology throughout the USSR

Efim Slavsky minister of medium machine building, in control of all aspects of the Soviet nuclear weapons program

Nikolai Dollezhal director of NIKIET, the Soviet reactor design agency

Valery Legasov first deputy director of the Kurchatov Institute, the immediate deputy to Anatoly Aleksandrov

Evgeny Velikhov deputy director of the Kurchatov Institute; scientific advisor to Mikhail Gorbachev and rival to Valery Legasov

Alexander Meshkov deputy minister of the Ministry of Medium Machine Building

Boris Prushinsky chief engineer of Soyuzatomenergo, the Ministry of Energy's department of nuclear power; leader of OPAS, the ministry's emergency response team for accidents at nuclear power stations

Alexander Borovoi head of the neutrino laboratory at the Kurchatov Institute and scientific leader of the Chernobyl Complex Expedition

Hans Blix director of the International Atomic Energy Agency, based in Vienna, Austria

The Generals

General Boris Ivanov deputy chief of the general staff of the USSR's Civil Defense Forces

General Vladimir Pikalov commander of the Soviet army chemical troops

Major General Nikolai Antoshkin chief of staff of the Seventeenth Airborne Army, Kiev military district

Major General Nikolai Tarakanov deputy commander of the USSR's Civil Defense Forces

The Doctors

Dr. Angelina Guskova head of the clinical department of Hospital Number Six, Moscow

Dr. Alexander Baranov head of hematology, Hospital Number Six, Moscow

Dr. Robert Gale hematology specialist at UCLA Medical Center, Los Angeles

Prologue

SATURDAY, APRIL 26, 1986: 4:16 P.M.

CHERNOBYL ATOMIC ENERGY STATION, UKRAINE

S enior Lieutenant Alexander Logachev loved radiation the way other men loved their wives. Tall and good-looking, twenty-six years old, with close-cropped dark hair and ice-blue eyes, Logachev had joined the Soviet army when he was still a boy. They had trained him well. The instructors from the military academy outside Moscow taught him with lethal poisons and unshielded radiation. He traveled to the testing grounds of Semipalatinsk in Kazakhstan, and to the desolate East Urals Trace, where the fallout from a clandestine radioactive accident still poisoned the landscape; eventually, Logachev's training took him even to the remote and forbidden islands of Novaya Zemlya, high in the Arctic Circle and ground zero for the detonation of the terrible *Tsar Bomba*, the largest thermonuclear device in history.

Now, as the lead radiation reconnaissance officer of the 427th Red Banner Mechanized Regiment of the Kiev District Civil Defense force, Logachev knew how to protect himself and his three-man crew from nerve agents, biological weapons, gamma rays, and hot particles: by doing their work just as the textbooks dictated; by trusting his dosimetry equipment; and, when necessary, reaching for the nuclear, bacterial, and chemical warfare medical kit stored in the cockpit of their armored car. But he also believed that the best protection was psychological. Those men who allowed themselves to fear radiation were most at risk. But those who came to love and appreciate its spectral presence, to understand its caprices, could endure even the most intense gamma bombardment and emerge as healthy as before.

As he sped through the suburbs of Kiev that morning at the head of a column of more than thirty vehicles summoned to an emergency at the Chernobyl nuclear power plant, Logachev had every reason to feel confident. The spring air blowing through the hatches of his armored scout car carried the smell of the trees and freshly cut grass. His men, gathered on the parade ground just the night before for their monthly inspection, were drilled and ready. At his feet, the battery of radiological detection instruments—including a newly installed electronic device twice as sensitive as the old model—murmured softly, revealing nothing unusual in the atmosphere around them.

But as they finally approached the plant later that morning, it became clear that something extraordinary had happened. The alarm on the radiation dosimeter first sounded as they passed the concrete signpost marking the perimeter of the power station grounds, and the lieutenant gave orders to stop the vehicle and log their findings: 51 roentgen per hour. If they waited there for just sixty minutes, they would all absorb the maximum dose of radiation permitted Soviet troops during wartime. They drove on, following the line of high-voltage transmission towers that marched toward the horizon in the direction of the power plant; their readings climbed still further, before falling again.

Then, as the armored car rumbled along the concrete bank of the station's coolant canal, the outline of the Fourth Unit of the Chernobyl nuclear power plant finally became visible, and Logachev and his crew gazed at it in silence. The roof of the twenty-story building had been torn open, its upper levels blackened and collapsed into heaps of rubble. They could see shattered panels of ferroconcrete, tumbled blocks of graphite, and, here and there, the glinting metal casings of fuel assemblies from the core of a nuclear reactor. A cloud of steam drifted from the wreckage into the sunlit sky.

Yet they had orders to conduct a full reconnaissance of the plant. Their armored car crawled counterclockwise around the complex at ten kilometers an hour. Sergeant Vlaskin called out the radiation readings from the new instruments, and Logachev scribbled them down on a map, hand-drawn on a sheet of parchment paper in ballpoint pen and colored marker: 1 roentgen an hour; then 2, then 3. They turned left, and the figures began to rise quickly: 10, 30, 50, 100.

"Two hundred fifty roentgen an hour!" the sergeant shouted. His eyes widened.

"Comrade Lieutenant—" he began, and pointed at the radiometer.

Logachev looked down at the digital readout and felt his scalp prickle with terror: 2,080 roentgen an hour. An impossible number.

Logachev struggled to remain calm and remember the textbook; to conquer his fear. But his training failed him, and the lieutenant heard himself screaming in panic at the driver, petrified that the vehicle would stall.

"Why are you going this way, you son of a bitch? Are you out of your fucking mind?" he yelled. "If this thing dies, we'll all be corpses in fifteen minutes!"

1

The Soviet Prometheus

A t the slow beat of approaching rotor blades, black birds rose into the sky, scattering over the frozen meadows and the pearly knots of creeks and ponds lacing the Pripyat River basin. Far below, standing knee deep in snow, his breath lingering in heavy clouds, Viktor Brukhanov awaited the arrival of the *nomenklatura* from Moscow.

When the helicopter touched down, the delegation of ministers and Communist Party officials trudged together over the icy field. The savage cold gnawed at their heavy woolen coats and nipped beneath their tall fur hats. The head of the Ministry of Energy and Electrification of the USSR and senior Party bosses from the Soviet Socialist Republic of Ukraine joined Brukhanov at the spot where their audacious new project was to begin. Just thirty-four years old, clever and ambitious, a dedicated Party man, Brukhanov had come to western Ukraine with orders to begin building what—if the Soviet central planners had their way—would become the greatest nuclear power station on earth.

As they gathered near the riverbank, the dozen men toasted their plans with shots of cognac. A state photographer posed them between long-handled shovels and a theodolite, the helicopter waiting, squat and awkward, in the background. They stood in the snow and watched as Minister Neporozhny drove a ceremonial stake, centimeter by centimeter, into the iron ground.

It was February 20, 1970. After months of deliberation, the Soviet authorities had at last settled on a name for the new power plant that would one day make the USSR's nuclear engineering famous across the globe. They had considered a few options: the North Kiev, or the Western Ukraine, or, perhaps, the Pripyat Atomic Energy Station. But finally,

Vladimir Scherbitsky, the formidable leader of the Ukrainian Communist Party, signed a decree confirming that the station would take the name of the regional capital: a small but ancient town of two thousand people, fourteen kilometers from where Brukhanov and his bosses stood in the snow-covered field.

The town of Chernobyl had been established in the twelfth century. For the next eight hundred years, it was home to peasants who fished in the rivers, grazed cows in the meadows, and foraged for mushrooms in the dense woods of northwestern Ukraine and southern Belarus. Swept repeatedly by pogrom, purge, famine, and war, by the second half of the twentieth century, Chernobyl was finally at peace. It had evolved into a quiet provincial center, with a handful of factories, a hospital, a library, a Palace of Culture; there was a small shipyard to service the tugs and barges that plied the Pripyat and the Dnieper, the two rivers that met nearby. Water permeated the surrounding countryside, an endlessly flat landscape of peat bogs, marshes, and sodden forests that formed part of the Dnieper River basin, a network of thirty-two thousand rivers and streams that covered almost half of Ukraine. Just fifteen kilometers downstream from the site chosen for the new power station, the rivers joined and flowed onward to the Kiev Sea, a massive hydroelectric reservoir providing fresh water to the two and a half million citizens of the republic's capital, two hours' drive away to the southeast.

Viktor Brukhanov had arrived in Chernobyl earlier that winter. He checked into the town's only hotel: a bleak, single-story building on Sovietskaya Street. Slight but athletic, he had a narrow, anxious face, an olive complexion, and a head of tight, dark curls. The oldest of four children, Brukhanov was born to ethnically Russian parents but raised in Uzbekistan, amid the mountains of Soviet Central Asia. He had an exotic look: when they eventually met, the divisional KGB major thought the young director could be Greek.

He sat down on his hotel bed and unpacked the contents of his briefcase: a notebook, a set of blueprints, and a wooden slide rule. Although now the director and, as yet, sole employee of the Chernobyl Atomic

Energy Station, Brukhanov knew little about nuclear power. Back at the Polytechnic Institute in Tashkent, he had studied electrical engineering. He had risen quickly from lowly jobs in the turbine shop of an Uzbek hydroelectric power plant to overseeing the launch of Ukraine's largest coal-fired station in Slavyansk, in the industrial east of the republic. But at the Ministry of Energy in Moscow, knowledge and experience were regarded as less important qualifications for top management than loyalty and an ability to get things done. Technical matters could be left to the experts.

At the dawn of the 1970s, in a bid to meet its surging need for electricity and to catch up with the West, the USSR embarked upon a crash program of reactor building. Soviet scientists had once claimed to lead the world in nuclear engineering and astonished their capitalist counterparts in 1954 by completing the first reactor to generate commercial electricity. But since then, they had fallen hopelessly behind. In July 1969, as US astronauts made their final preparations to land on the moon, the Soviet minister of energy and electrification called for an aggressive expansion of nuclear construction. He set ambitious targets for a network of new plants across the European part of the Soviet Union, with giant, mass-produced reactors that would be built from the Gulf of Finland to the Caspian Sea.

That winter, as the 1960s came to a close, the energy minister summoned Brukhanov to Moscow and offered him his new assignment. It was a project of enormous prestige. Not only would it be the first atomic power plant in Ukraine, but it was also new territory for the Ministry of Energy and Electrification, which had never before built a nuclear station from scratch. Until this point, every reactor in the USSR had been constructed by the Ministry of Medium Machine Building: the clandestine organization behind the Soviet atom weapons program, so secret that its very name was a cipher, designed to discourage further curiosity. But whatever the challenges, Brukhanov, a true believer, gladly enlisted to carry the banner of the Red Atom.

Sitting alone on his hotel bed, the young engineer confronted his responsibility for conjuring from an empty field a project expected to cost almost 400 million rubles. He drew up lists of the materials to begin building and, using his slide rule, calculated their attendant costs. Then he delivered his estimates to the state bank in Kiev. He traveled to the city almost every day by bus; when there wasn't a bus, he thumbed a ride. As the project had no accountant, there was no payroll, so he received no wages.

Before Brukhanov could start building the station itself, he had to create the infrastructure he'd need to bring materials and equipment to the site: a rail spur from the station in nearby Yanov; a new dock on the river to receive gravel and reinforced concrete. He hired construction workers, and soon a growing army of men and women at the controls of caterpillar-tracked excavators and massive BelAZ dump trucks began to tear pathways through the forest and scrape a plateau from the dun landscape. To house himself, a newly hired bookkeeper, and the handful of workers who lived on the site, Brukhanov organized a temporary village in a forest clearing nearby. A cluster of wooden huts on wheels, each equipped with a small kitchen and a log-burning stove, the settlement was named simply *Lesnoy*—"of the woods"—by its new inhabitants. As the weather warmed, Brukhanov had a schoolhouse built where children could be educated up to fourth grade. In August 1970 he was joined in Lesnoy by his young family: his wife, Valentina, their six-year-old daughter, Lilia, and infant son, Oleg.

Valentina and Viktor Brukhanov had spent the first decade of their lives together helping fulfill the dream of Socialist electrification. Chernobyl was the family's third power plant start-up in six years; Valentina and Viktor had met as young specialists working on the building of the Angren hydroelectric project, a hundred kilometers from the Uzbek capital, Tashkent. Valentina had been the assistant to a turbine engineer, and Viktor, fresh from university, had been a trainee. He was still planning to return to university to finish his master's degree when the head of his department at the plant encouraged him to stay: "Wait," he told him, "you'll meet your future wife here!" Mutual friends introduced Viktor and Valentina in the winter of 1959: "You'll drown in her eyes," they promised. The couple had been dating for barely a year when, in December 1960, they were married in Tashkent; Lilia was born in 1964.

To Valentina, Lesnoy seemed a magical place, with fewer than a dozen families gathered in the huddle of makeshift cottages; at night, when the roar of the bulldozers and excavators subsided, a velvet silence fell on the glade, the darkness pierced by a single lantern and the screeching of owls. Every once in a while, to inspire the workers to help them achieve their construction targets, Moscow sent down Soviet celebrities, including the Gypsy superstar Nikolai Slichenko and his troupe, to perform shows and concerts. The family remained in the forest settlement for another two years, as shock-work brigades excavated the first reactor pit and carved a giant reservoir—an artificial lake 11 kilometers long and 2.5 kilometers wide that would provide the millions of cubic meters of cooling water crucial to operating four massive reactors—from the sandy soil.

Meanwhile, Viktor oversaw the genesis of an entirely new settlement— an *atomgrad*, or "atomic city"—beside the river. The planners designed the settlement, eventually named Pripyat, to house the thousands of staff who would one day run the nuclear complex, along with their families. A handful of dormitories and apartment buildings reached completion in 1972. The new town went up so quickly that at first there were no paved roads and no municipal heating plant to serve the apartment buildings. But its citizens were young and enthusiastic. The first group of specialists to arrive on the site were idealists, pioneers of the nuclear future, keen to transform their homeland with new technology. To them, these problems were trifles: to keep warm at night, they slept with their coats on.

Valentina and Viktor were among the first to move in, taking a three-bedroom apartment in 6 Lenina Prospekt, right at the entrance to the new town, in the winter of 1972. While they waited for the city's first school to be completed, every day their daughter, Lilia, hitched a ride in a truck or car back to Lesnoy, where she attended lessons in the forest schoolhouse.

According to Soviet planning regulations, Pripyat was separated from the plant itself by a "sanitary zone" in which building was prohibited, to ensure that the population would not be exposed to fields of low-level ionizing radiation. But Pripyat remained close enough to the plant to be reached by road in less than ten minutes—just three kilometers as the crow flies. And as the city grew, its residents began to build summer houses in the sanitary zone, each happy to disregard the rules in exchange for a makeshift dacha and a small vegetable garden.

Viktor Brukhanov's initial instructions for the Chernobyl plant called for the construction of a pair of nuclear reactors—a new model known by the acronym RBMK, for *reaktor bolshoy moschnosti kanalnyy,* or high-power channel-type reactor. In keeping with the Soviet weakness for gigantomania, the RBMK was both physically larger and more powerful than almost any reactor yet built in the West, each one theoretically capable of generating 1,000 megawatts of electricity, enough to serve at least a million modern homes. The deadlines set by his bosses in Moscow and Kiev required Brukhanov to work with superhuman dispatch: according to the details of the Ninth Five-Year Plan, the first was due to come online in December 1975, with the second to follow before the end of 1979. Brukhanov quickly realized that this timetable was impossible.

By the time the young director began work in Chernobyl in 1970, the Socialist economic experiment was going into reverse. The USSR was buckling under the strain of decades of central planning, fatuous bureaucracy, massive military spending, and endemic corruption—the start of what would come to be called the Era of Stagnation. Shortages and bottlenecks, theft and embezzlement blighted almost every industry. Nuclear engineering was no exception. From the beginning, Brukhanov lacked construction equipment. Key mechanical parts and building materials often turned up late, or not at all, and those that did were often defective. Steel and zirconium—essential for the miles of tubing and hundreds of fuel assemblies that would be plumbed through the heart of the giant reactors—were both in short supply; pipework and reinforced concrete intended for nuclear use often turned out to be so poorly made it had to be thrown away. The quality of workmanship at all levels of Soviet manufacturing was so poor that building projects throughout the nation's power industry were forced to incorporate an extra stage known as "preinstallation overhaul." Upon delivery from the factory, each piece of new equipment—transformers, turbines, switching gear—was stripped down to the last nut and bolt, checked for faults, repaired, and then reassembled according to the original specifications, as it should have been in the first place. Only then could it be safely installed. Such wasteful duplication of labor added months of delays and millions of rubles in costs to any construction project.

Throughout late 1971 and early 1972, Brukhanov struggled with labor disputes and infighting among his construction managers and faced steady reprimands from his Communist Party bosses in Kiev. The workers complained about food shortages and the lines at the site canteen; he had failed to provide cost estimates and design documents; he missed deadlines and fell pathetically short of the monthly work quotas dictated by Moscow. And still there was more: the new citizens of Pripyat required a bakery, a hospital, a palace of culture, a shopping center. There were hundreds of apartments to be built.

Finally, in July 1972, exhausted and disillusioned, Viktor Brukhanov drove to Kiev for an appointment with his boss from the Ministry of Energy and Electrification. He had been director of the Chernobyl Atomic Energy Station for less than three years, and the plant had not yet emerged from the ground. But now he planned to resign.

Behind all the catastrophic failures of the USSR during the Era of Stagnation—beneath the kleptocratic bungling, the nepotism, the surly inefficiencies, and the ruinous waste of the planned economy—lay the monolithic power of the Communist Party. The Party had originated as a single faction among those grappling for power in Russia following the Revolution of 1917, ostensibly to represent the will of the workers, but quickly establishing control of a single-party state—intended to lead the proletariat toward "True Communism."

Distinct from mere Socialism, True Communism was the Marxist utopia: "a classless society that contains limitless possibilities for human achievement," an egalitarian dream of self-government by the people. As revolution was supplanted by political repression, the deadline for realizing this meritocratic Shangri-la was pushed repeatedly off into the future. Yet the Party clung to its role enforcing the dictates of Marxism-Leninism, ossifying into an ideological apparatus of full-time paid officials—the apparat—nominally separate from the government but that in reality controlled decision-making at every level of society.

Decades later, the Party had established its own rigid hierarchy of personal patronage and held the power of appointment over an entire class of influential positions, known collectively as the *nomenklatura*. There were also Party managers to oversee every workshop, civil or military enterprise, industry, and ministry: the apparatchiks, who formed a shadow bureaucracy of political functionaries throughout the Soviet empire. While officially every one of the fifteen republics of the USSR was run by its own Ministerial Council, led by a prime minister, in practice it was the national leader of each republican Communist Party—the first secretary—who was in control. Above them all, handing down directives from Moscow, sat Leonid Brezhnev, granite-faced general secretary of the Communist Party of the Soviet Union, chairman of the Politburo, and de facto ruler of 242 million people. This institutionalized meddling proved confusing and counterproductive to the smooth running of a modern state, but the Party always had the final word.

Party membership was not open to everyone. It required an exhaustive process of candidacy and approval, the support of existing members, and the payment of regular dues. By 1970, fewer than one in fifteen Soviet citizens had been admitted. But membership brought perks and advantages available only to the elite, including access to restricted stores and foreign journals, a separate class of medical care, and the possibility of travel abroad. Above all, professional advancement in any kind of senior role was difficult without a Party card, and exceptions were rare. By the time Viktor Brukhanov joined in 1966, the Party was everywhere. In the workplace, he answered to two masters: both his immediate managers and the committee of the local Communist Party. When he became director of a nuclear power station, it was no different. He received directives from the Ministry of Energy in Moscow but was also tyrannized by the demands of the regional Party committee in Kiev.

Although by the early seventies many in the Party still believed in the principles of Marxism–Leninism, under the baleful gaze of Brezhnev and his claque of geriatric cronies, ideology had become little more than window dressing. The mass purges and the random executions of the three decades under Stalin were over, but across the USSR, Party leaders and the heads of large enterprises—collective farms and tank factories, power stations and hospitals—governed their staff by bullying and intimidation. These were the thuggish bureaucrats who, according to the novelist and historian Piers Paul Read, "had the face of a truck driver but the hands of a pianist." The humiliation of enduring an expletive-spattered dressing-down delivered at screaming pitch was a ritual repeated daily in offices everywhere. It engendered a top-down culture of toadying yes-men who learned to anticipate the whims of their superiors and agree with whatever they said, while threatening their own underlings. When the boss put his own proposals to the vote, he could reasonably expect them to be carried unanimously every time, a triumph of brute force over reason.

Advancement in many political, economic, and scientific careers was granted only to those who repressed their personal opinions, avoided conflict, and displayed unquestioning obedience to those above them. By the midseventies, this blind conformism had smothered individual decision-making at all levels of the state and Party machine, infecting not just the bureaucracy but technical and economic disciplines, too. Lies and deception were endemic to the system, trafficked in both directions along the chain of management: those lower down passed up reports to their superiors packed with falsified statistics and inflated estimates, of unmet goals triumphantly reached, unfulfilled quotas heroically exceeded. To protect his own position, at every stage, each manager relayed the lies upward or compounded them.

Seated at the top of a teetering pyramid of falsehood, poring over reams of figures that had little basis in reality, were the economic mandarins of the State General Planning Committee—Gosplan—in Moscow. The brain of the "command economy," Gosplan managed the centralized distribution of resources throughout the USSR, from toothbrushes to tractors, reinforced concrete to platform boots. Yet the economists in Moscow had no reliable index of what was going on in the vast empire they notionally maintained; the false accounting was so endemic that at one point the KGB resorted to turning the cameras of its spy satellites onto Soviet Uzbekistan in an attempt to gather accurate information about the state's own cotton harvest.

Shortages and apparently inexplicable gluts of goods and materials were part of the grim routine of daily life, and shopping became a game of chance played with a string *avoska*, or "what-if" bag, carried in the hope of stumbling upon a store recently stocked with anything useful— whether sugar, toilet paper, or canned ratatouille from Czechoslovakia. Eventually the supply problems of the centrally planned economy became so chronic that crops rotted in the fields, and Soviet fishermen watched catches putrefy in their nets, yet the shelves of the Union's grocery stores remained bare.

Soft spoken but sure of himself, Viktor Brukhanov was not like most Soviet managers. He was mild mannered and well liked by many of those beneath him. With his prodigious memory and shrewd financial sense, his excellent grasp of many technical aspects of his job—including chemistry and physics—he impressed his superiors. And at first, he was confident enough in his opinions to openly disagree with them. So when the pressures of the mammoth task he faced in Chernobyl became too much for him, he simply decided to quit.

Yet when Brukhanov arrived in Kiev that day in July 1972, his Party-appointed supervisor from the Energy Ministry took his letter of resignation, tore it up in front of him, and told him to get back to work. After that, the young director recognized that there was no escape. Whatever else his job might require, his most important task was simply to obey the Party—and to implement their plan by any means he could. The next month, construction workers poured the first cubic meter of concrete into the foundations of the plant.

Thirteen years later, on November 7, 1985, Brukhanov stood silently on the reviewing stand in front of the new Pripyat Palace of Culture, where the windows had been hung with hand-painted portraits of state and Party leaders. Power station and construction workers paraded through the square below, carrying banners and placards. And in speeches marking the anniversary of the Great October Revolution, the director was hailed for his illustrious achievements: his successful fulfillment of the Party's plans, his benevolent leadership of the city, and the power plant it served.

Brukhanov had now dedicated the prime of his life to the creation of an empire in white reinforced concrete, encompassing a town of nearly fifty thousand people and four giant 1,000-megawatt reactors. Construction was also well under way on two more reactors, which were scheduled for completion within two years. When Units Five and Six of the Chernobyl station came online in 1988, Brukhanov would preside over the largest nuclear power complex on earth.

Under his direction, the Chernobyl plant—by then formally known as the V. I. Lenin Nuclear power station—had become a prize posting for nuclear specialists from all over the Soviet Union. Many of them came straight from MEPhI–the Moscow Engineering and Physics Institute, the Soviet counterpart to MIT. The USSR, hopelessly backward in developing computer technology, lacked simulators with which to train its nuclear engineers, so the young engineers' work at Chernobyl would be their first practical experience in atomic power.

To trumpet the wonders of the atomic town of Pripyat, the city council—the *ispolkom*—had prepared a glossy book, filled with vivid color photographs of its happy citizens at play. The average age of the population was twenty-six, and more than a third of them were children. The young families had access to five schools, three swimming pools, thirty-five playgrounds, and beaches on the sandy banks of the river. The town planners had taken care to preserve the city's sylvan environment, and each new apartment block was surrounded by trees. The buildings and open spaces were decorated with sculptures and spectacular mosaics celebrating science and technology. For all its modernity and sophistication, the city remained encircled by wilderness, offering a sometimes enchanting proximity to nature. One summer day, Brukhanov's wife, Valentina, watched as a pair of elk swam out of the Pripyat and slouched up the beach before disappearing into the forest, apparently heedless of the bathers gawping from the sand.

As an *atomgrad*, the city and everything in it—from the hospital to the fifteen kindergartens—was considered an extension of the nuclear plant it served, financed directly from Moscow by the Ministry of Energy. It existed in an economic bubble; an oasis of plenty in a desert of shortages and deprivation. The food stores were better stocked than those even in Kiev, with pork and veal, fresh cucumbers and tomatoes, and more than five different types of sausage. In the Raduga—or Rainbow—department store, Austrian-made dining sets and even French perfume were available to shoppers, all without having to spend years on a waiting list. There was a cinema, a music school, a beauty parlor, and a yacht club.

Pripyat was a small place: few of the buildings reached higher than ten stories, and one could cross the whole city in twenty minutes. Everyone knew everyone else, and there was little trade for the *militsia*—the policemen of the Ministry of Internal Affairs—or the city's resident KGB chief, who had an office on the fifth floor of the *ispolkom*. Trouble was confined mostly to petty vandalism and public drunkenness. Each spring, the river gave up another grim harvest, as the thaw revealed the bodies of drunks who had blundered through the ice and drowned in midwinter.

A Western eye may have been drawn to Pripyat's limitations: the yellowing grass bristling between concrete paving slabs or the bleak uniformity of the multistory buildings. But to men and women born in the sour hinterlands of the USSR's factory cities, raised on the parched steppes of Kazakhstan, or among the penal colonies of Siberia, the new *atomgrad* was a true workers' paradise. In home movies and snapshots, the citizens of Pripyat captured one another not as drab victims of the Socialist experiment but as carefree young people: kayaking, sailing, dancing, or posing in new outfits; their children playing on a great steel elephant or a brightly painted toy truck; cheerful optimists in the city of the future.

By the end of December 1985, Viktor and Valentina Brukhanov could look back on a year of triumphs and milestones at home and at work. In August they saw their daughter married and Lilia and her new husband resume their studies at the medical institute in Kiev; soon after, Lilia became pregnant with their first child. In December, the couple celebrated Viktor's fiftieth birthday and their own silver wedding anniversary, with parties in their big corner apartment overlooking Pripyat's main square.

At the same time, Viktor was honored with an invitation to Moscow to join the delegation attending the impending 27th Congress of the Communist Party of the Soviet Union, an important stamp of political approval from above. The Congress promised, too, to be a significant event for the USSR as a whole. It would be the first over which the new general secretary, Mikhail Gorbachev, would preside as leader of the Soviet Union.

Gorbachev had assumed power in March 1985, ending the long succession of zombie apparatchiks whose declining health, drunkenness, and senility had been concealed from the public by squadrons of increasingly desperate minders. At fifty-four, Gorbachev seemed young and dynamic and found an enthusiastic audience in the West. With political opinions formed during the 1960s, he was also the first general secretary to exploit the power of television. Speaking unselfconsciously in his southern accent, plunging into crowds on apparently spontaneous walkabouts finely orchestrated by the KGB, Gorbachev appeared constantly on the nation's flagship TV news show, *Vremya*, watched every night by nearly two hundred million people. He announced plans for economic reorganization—perestroika—and, at the climax of the Party congress in March 1986, talked of the need for glasnost, or open government. A dedicated Socialist, Gorbachev believed that the USSR had lost its way but could be led to the utopia of True Communism by returning to the founding principles of Lenin. It would be a long road. The economy was staggering under the financial burden of the Cold War. Soviet troops were mired in Afghanistan, and in 1983 US president Ronald Reagan had extended the battle into space with the Strategic Defense Initiative, the "Star Wars" program. Annihilation in a nuclear strike seemed as close as ever. And at home, the monolithic old ways—the strangling bureaucracy and corruption of the Era of Stagnation—lingered on.

In the sixteen years that he'd spent building four nuclear reactors and an entire city on an isolated stretch of marshland, Viktor Brukhanov had received a long education in the realities of the system. Hammered on the anvil of the Party, made pliant by the privileges of rank, the well-informed and opinionated young specialist had been transformed into an obedient tool of the *nomenklatura*. He had met his targets and fulfilled the plan and won himself and his men orders of merit and pay bonuses for beating deadlines and exceeding labor quotas. But, like all successful Soviet managers, to do so, Brukhanov had learned how to be expedient and bend limited resources to meet an endless list of unrealistic goals. He had to cut corners, cook the books, and fudge regulations.

When the building materials specified by the architects of the Chernobyl station had proved unavailable, Brukhanov was forced to improvise: the plans called for fireproof cables, but when none could be found, the builders simply did the best they could.

When the Ministry of Energy in Moscow learned that the roof of the plant's turbine hall had been covered with highly flammable bitumen, they ordered him to replace it. But the flame-retardant material specified for reroofing the structure—fifty meters wide and almost a kilometer long—was not even being manufactured in the USSR, so the Ministry granted him an exception, and the bitumen remained. When the district Party secretary instructed him to build an Olympic-length swimming pool in Pripyat, Brukhanov tried to object: such facilities were common only in Soviet cities of more than a million inhabitants. But the secretary insisted: "Go build it!" he said, and Brukhanov obeyed. He found the extra funds to do so by fiddling the city expenses to hoodwink the state bank.

And as the fourth and most advanced reactor of the Chernobyl plant approached completion, a time-consuming safety test on the unit turbines remained outstanding. Brukhanov quietly postponed it, and so met Moscow's deadline for completion on the last day of December 1983.

But, like a spoiled lover, the Soviet Ministry of Energy and Electrification would not be satisfied. At the beginning of the 1980s, the USSR's punishing schedule of nuclear construction had been accelerated further, with breathtaking plans for more and increasingly gigantic stations throughout the western territories of the Union. By the end of the century, Moscow intended Chernobyl to be one part of a dense network of atomic power megacomplexes, each one home to up to a dozen reactors. In 1984, the deadline for completing the fifth reactor was brought forward by a year. Labor and supply problems remained endemic: the concrete was defective; the men lacked power tools. A team of dedicated KGB agents and their network of informants at the plant reported a continuing series of alarming building faults.

In 1985 Brukhanov received instructions for the construction of Chernobyl Two, a separate station of four more RBMK reactors, using a new model fresh from the drawing board and even more Brobdingnagian than the last. This station would be built a few hundred meters away from the existing one, on the other side of the river, along with a new residential area of Pripyat to accommodate the plant workers. A bridge would be required to reach it, and a new ten-story administration building, with an office at the top from which the director could survey his sprawling nuclear fiefdom.

Brukhanov worked around the clock. His superiors could usually expect to find him somewhere in the station at almost any time of the day or night. If something went wrong at the plant—as it often did—the director often forgot to eat, and would subsist for a full twenty-four hours on coffee and cigarettes. In meetings, he withdrew into inscrutable silence, never offering two words when one would do. Isolated and exhausted, he had few friends and confided little, even to his wife.

Brukhanov's staff, too, had changed. The spirited team of young specialists who had first colonized the freezing settlement in the woods all those years ago, and then worked to bring the first reactors online, had moved on. In their place were thousands of new employees, and Brukhanov found maintaining discipline difficult: despite his technical gifts, he lacked the force of personality necessary for management on the Soviet scale. The plant construction chief, a domineering and well-connected Party man whose authority rivaled that of the director, derided him as a "marshmallow."

The Era of Stagnation had fomented a moral decay in the Soviet workplace and a sullen indifference to individual responsibility, even in the nuclear industry. The USSR's economic utopianism did not recognize the existence of unemployment, and overstaffing and absenteeism were chronic problems. As the director of the plant and its company town, Brukhanov was responsible for providing jobs for everyone in Pripyat. The inexorable construction work took care of twenty-five thousand of them, and he had already arranged for the establishment of the Jupiter electronics plant to provide work for more of the women in town. But that wasn't enough. Each shift at the Chernobyl plant now brought hundreds of men and women to the plant by bus from Pripyat, and many of them then sat around with nothing to do. Some were trainee nuclear engineers—aspiring to become a part of the highly qualified technical elite known as *atomshchiki*—who came to watch the experts at work. But others were mechanics and electricians who came from elsewhere in the energy industry—the "power men," or *energetiki*—who harbored complacent assumptions about nuclear plants. They had been told that radiation was so harmless "you could spread it on bread," or that a reactor was "like a samovar . . . more simple than a thermal power plant." At home, some drank from glassware colored with iridescent patterns that, they boasted, were created by having been steeped in the radioactive waters of the plant's used fuel coolant pond. Others listlessly filled out their shifts reading novels and playing cards. Those who actually had important work to do were known—with a bureaucratic frankness that hinted at satire—as the Group of Effective Control. Yet the dead weight of unwanted manpower tugged even at those with urgent responsibilities and infected the plant with inefficiency and a dangerous sense of inertia.

At the top, the experienced team of independent-minded nuclear engineering experts who had overseen the start-up of the station's first four reactors had all left, and senior specialists were in short supply. The chief engineer—Brukhanov's principal deputy, responsible for the day-to-day technical operation of the station—was Nikolai Fomin, the former plant Party secretary and an arrogant, blustering apparatchik of the old school. Balding, barrel chested, with a dazzling smile and a confident baritone voice that rose steeply in pitch when he became excited, Fomin had all the overbearing Soviet charisma Brukhanov lacked. An electrical engineer, his appointment had been pushed through by the Party in Moscow over the objections of the Ministry of Energy. He had no previous experience in atomic power but was ideologically beyond reproach—and did his best to learn nuclear physics through a correspondence course.

By the spring of 1986, Chernobyl was, officially, one of the best-performing nuclear stations in the Soviet Union, and the word was that Brukhanov's loyalty to the Party would soon be rewarded. According to the results of the latest Five-Year Plan, the plant was due to receive the state's highest honor: the Order of Lenin. The staff would win a financial bonus, and Brukhanov would be awarded the star of the Hero of Socialist Labor. At the Ministry of Energy, the decision had already been taken to promote

Brukhanov to Moscow, and Fomin would take his place as plant director. The news would be announced on the May 1 holiday, with a decree of the Presidium of the Supreme Soviet.

Brukhanov had also raised Pripyat from nothing, creating a beautiful model town cherished by its citizens. And despite the appointment of a city council, almost every decision about the *atomgrad*—no matter how trivial—remained subject to his approval. From the outset, the architects had called for the city to be populated with a lush variety of trees and shrubs—birch, elm, and horse chestnut; jasmine, lilac, and barberry. But Brukhanov was especially fond of flowers and ordered them planted everywhere. At a meeting of the *ispolkom* in 1985, he announced a grand gesture. It was his wish that the streets would blossom with fifty thousand rosebushes: one for every man, woman, and child in the city. There were objections, of course. How could they possibly find so many flowers? Yet, by the following spring, thirty thousand good Baltic rosebushes had already been purchased at great expense from Lithuania and Latvia and planted in the long, raised beds beneath the poplar trees on Lenina Prospekt and everywhere around the central square.

Here, on the elevated concrete plaza along Kurchatov Street, at the end of the picturesque promenade into the city, the plans called for Pripyat to have its own statue of Lenin, an architectural necessity for every major town in the USSR. But a permanent monument had not yet been built. The city council had announced a competition for a design, and the plinth where it would stand was occupied by a triangular wooden box, painted with an inspiring portrait, a hammer and sickle, and a slogan: "The name and mission of Lenin will live forever!"

In the meantime, Viktor Brukhanov had given his blessing to a memorial to more ancient gods: a massive realist statue, in front of the city's cinema, six meters tall and cast in bronze. It depicted a Titan, naked beneath the swooping folds of his cloak, holding aloft leaping tongues of flame. This was Prometheus, who had descended from Olympus with the stolen gift of fire. With it, he brought light, warmth, and civilization to mankind—just as the torchbearers of the Red Atom had illuminated the benighted households of the USSR.

But the ancient Greek myth had a dark side: Zeus was so enraged by the theft of the gods' most powerful secret that he chained Prometheus to a rock, where a giant eagle descended to peck out his liver every day for eternity.

Nor did mortal man escape retribution for accepting Prometheus's gift. To him, Zeus sent Pandora, the first woman, bearing a box that, once opened, unleashed evils that could never again be contained.

2

Alpha, Beta, and Gamma

Almost everything in the universe is made of atoms, fragments of stardust that compose all matter. A million times smaller than the width of a human hair, atoms are composed almost entirely of empty space. But at the center of every atom is a nucleus—unimaginably dense, as if six billion cars were crushed together into a small suitcase—and full of latent energy. The nucleus, formed of protons and neutrons, is orbited by a cloud of electrons and bound together by what physicists call "the strong force."

The strong force, like gravity, is one of the four principal forces that bind the universe, and scientists once believed it was so powerful that it made atoms indestructible and indivisible. They also believed that "neither mass nor energy could be created or destroyed." In 1905 Albert Einstein overturned these ideas. He suggested that if atoms could be somehow torn apart, the process would convert their tiny mass into a relatively enormous release of energy. He defined the theory with an equation: the energy released would be equal to the amount of mass lost, multiplied by the speed of light squared. $E=mc^2$.

In 1938 a trio of scientists in Germany discovered that when atoms of the heavy metal uranium are bombarded with neutrons, their nuclei can, in fact, be broken apart, releasing nuclear energy. When the nuclei split, their neutrons could fly away at great speed, smashing into other nearby atoms, causing their nuclei to split in turn, releasing even more energy. If enough uranium atoms were gathered in the correct configuration—forming a critical mass—this process could begin sustaining itself, with one atom's neutrons splitting the nucleus of another, sending more neutrons into a collision course with further nuclei. As it went critical, the resulting chain reaction of splitting atoms—nuclear fission—would liberate unimaginable quantities of energy.

At 8:16 a.m. on August 6, 1945, a fission weapon containing sixty-four kilograms of uranium detonated 580 meters above the Japanese city of Hiroshima, and Einstein's equation proved mercilessly accurate. The bomb itself was extremely inefficient: just one kilogram of the uranium underwent fission, and only seven hundred milligrams of mass—the weight of a butterfly—was converted into energy. But it was enough to obliterate an entire city in a fraction of a second. Some seventy-eight thousand people died instantly, or immediately afterward—vaporized, crushed, or incinerated in the firestorm that followed the blast wave. But by the end of the year, another twenty-five thousand men, women, and children would also sicken and die from their exposure to the radiation liberated by the world's first atom bomb attack.

Radiation is produced by the disintegration of unstable atoms. The atoms of different elements vary by weight, determined by the number of protons and neutrons in each nucleus. Each element has a unique number of protons, which never changes, determining its "atomic number" and its position in the periodic table: hydrogen never has more than one proton; oxygen always has eight; gold has seventy-nine. But atoms of the same element may have varying numbers of neutrons, resulting in different isotopes, ranging anywhere from deuterium (hydrogen with one neutron instead of two) to uranium 235 (uranium metal, with five extra neutrons).

Adding to or removing neutrons from the nucleus of a stable atom results in an unstable isotope. But any unstable isotope will try to regain its equilibrium, throwing off parts of its nucleus in a quest for stability—producing either another isotope or sometimes a different element altogether. For example, plutonium 239 sheds two protons and two neutrons from its nucleus to become uranium 235. This dynamic process of nuclear decay is radioactivity; the energy it releases, as atoms shed neutrons in the form of waves or particles, is radiation.

Radiation is all around us. It emanates from the sun and cosmic rays, bathing cities at high altitude in greater levels of background radiation than those at sea level. Underground deposits of thorium and uranium emit radiation, but so does masonry: stone, brick, and adobe all contain radioisotopes. The granite used to build the US Capitol is so radioactive that the building would fail federal safety codes regulating nuclear power plants. All living tissue is radioactive to some degree: human beings, like bananas, emit radiation because both contain small amounts of the radioisotope potassium 40; muscle contains more potassium 40 than other tissue, so men are generally more radioactive than women. Brazil nuts, with a thousand times the average concentration of radium of any organic product, are the world's most radioactive food.

Radiation is invisible and has neither taste nor smell. Although it's yet to be proved that exposure to any level of radiation is entirely safe, it becomes manifestly dangerous when the particles and waves it gives off are powerful enough to transform or break apart the atoms that make up the tissues of living organisms. This high-energy radiance is ionizing radiation.

Ionizing radiation takes three principal forms: alpha particles, beta particles, and gamma rays. Alpha particles are relatively large, heavy, and slow moving and cannot penetrate the skin; even a sheet of paper could block their path. But if they do manage to find their way inside the body by other means—if swallowed or inhaled—alpha particles can cause massive chromosomal damage and death. Radon 222, which gathers as a gas in unventilated basements, releases alpha particles into the lungs, where it causes cancer. Polonium 210, a powerful alpha emitter, is one of the carcinogens in cigarette smoke. It was also the poison slipped into the cup of tea that killed former FSB agent Alexander Litvinenko in London in 2006.

Beta particles are smaller and faster moving than alpha particles and can penetrate more deeply into living tissue, causing visible burns on the skin and lasting genetic damage. A piece of paper won't provide protection from beta particles, but aluminum foil—or separation by sufficient distance—will. Beyond a range of ten feet, beta particles can cause little damage, but they prove dangerous if ingested in any way. Mistaken by the body for essential elements, beta-emitting radioisotopes can become fatally concentrated in specific organs: strontium 90, a member of the same chemical family as calcium, is retained in the bones; ruthenium is absorbed by the intestine; iodine 131 lodges particularly in the thyroid of children, where it can cause cancer.

Gamma rays—high-frequency electromagnetic waves traveling at the speed of light—are the most energetic of all. They can traverse large distances, penetrate anything short of thick pieces of concrete or lead, and destroy electronics. Gamma rays pass straight through a human being without slowing down, smashing through cells like a fusillade of microscopic bullets.

Severe exposure to all ionizing radiation results in acute radiation syndrome (ARS), in which the fabric of the human body is unpicked, rearranged, and destroyed at the most minute levels. Symptoms include nausea, vomiting, hemorrhaging, and hair loss, followed by a collapse of the immune system, exhaustion of bone marrow, disintegration of internal organs, and, finally, death.

To the atomic pioneers who first explored "radiant matter" at the end of the nineteenth century, the effects of radiation were a bewitching curiosity. Wilhelm Roentgen, who discovered X-rays in 1895, saw the bones of his hand projected on the wall of his laboratory during the course of an experiment and was intrigued. But when he took the world's first X-ray photograph shortly afterward, of his wife's left hand—complete with wedding ring—the result horrified her. "I have seen my own death!" she said. Roentgen later took precautions to shield himself from his discovery, but others were not so careful. In 1896 Thomas Edison devised the fluoroscope, which projected X-rays onto a screen, allowing him to gaze inside solid objects. Edison's experiments required an assistant to place his hands repeatedly on top of a box, where they were exposed to X-rays. When he sustained burns on one hand, the assistant simply switched to using the other. But the burns wouldn't heal. Eventually surgeons amputated the assistant's left arm and four fingers from his right hand. When cancer spread up his right arm, the doctors took that, too. The disease traveled to his chest, and in October 1904 he died, the first known victim of man-made radiation.

Even as the damage caused by external exposure to radiation became apparent, the harmful effects of internal exposure remained little understood. Throughout the early years of the twentieth century, pharmacies sold patent medicines containing radium as a health tonic, drunk by people who believed radioactivity gave them energy. In 1903 Marie and Pierre Curie had won the Nobel Prize for the discovery of polonium and radium—an alpha-particle emitter, roughly a million times more radioactive than uranium—which they extracted from metric tonnes of viscous, tarry ore in their Paris laboratory. Pierre was killed in a road accident, but Marie continued exploring the properties of radioactive compounds until she died in 1934, probably due to radiation-induced bone marrow failure. More than eighty years later, Curie's laboratory notes remain so radioactive that they are kept in a lead-lined box.

Because radium can be mixed with other elements to make them glow in the dark, clock makers used it to create fluorescent numbers on watch faces and hired young women to perform the delicate task of painting them. In the watch factories of New Jersey, Connecticut, and Illinois, the Radium Girls were trained to lick the tips of their brushes into a fine point before dipping them into pots of radium paint. When the jaws and skeletons of the first girls began to rot and disintegrate, their employers suggested they were suffering from syphilis. A successful lawsuit revealed that their managers had understood the risks of working with radium and get done everything they could to conceal the truth from their employees. It was the first time the public learned the hazards of ingesting radioactive material.

The biological effect of radiation on the human body would eventually be measured in rem (roentgen equivalent man) and determined by a complicated combination of factors: the type of radiation; the duration of total exposure; how much of it penetrates the body, and where; and how susceptible those parts of the body are to radiation damage. The parts where cells divide rapidly—bone marrow, skin, and the gastrointestinal tract—are more at risk than other organs such as the heart, liver, and brain. Some radionuclides—such as radium and strontium—are more energetic emitters of radiation, and therefore more dangerous, than others, like cesium and potassium.

The survivors of the atom bomb attacks on Hiroshima and, three days later, Nagasaki provided the first opportunity to study the effects of acute radiation syndrome on a large number of people. They would eventually become the subject of a project spanning more than seventy years, creating a universal database on the long-term effects of ionizing radiation on human beings. Of those who lived through the initial explosion in Nagasaki, thirty-five thousand died within twenty-four hours; those suffering from ARS lost their hair within one or two weeks, and then experienced bloody diarrhea before succumbing to infection and high fever. Another thirty-seven thousand died within three months. A similar number survived for longer but, after another three years, developed leukemia; by the end of the 1940s, the disease would be the first cancer linked to radiation.

The effect of ionizing radiation on both inanimate objects and living beings was explored extensively in the late 1950s by the US Air Force. As part of a government program to develop atomic-powered planes, Lockheed Aircraft built a water-cooled 10-megawatt nuclear reactor in a shielded underground shaft in the woods of North Georgia. At the touch of a button, the reactor could be raised from its shielding to ground level, exposing everything within a three-hundred-meter radius to a lethal dose of radiation. In June 1959 the Radiation Effects Reactor was brought up to full power and unsheathed for the first time, killing almost everything in the vicinity stone dead: bugs fell from the air, and small animals and the bacteria living in and upon them were exterminated, in a phenomenon the technicians called "instant taxidermy." The effect on plants varied: oak trees turned brown, yet crabgrass remained strangely unaffected; pine trees appeared to be the hardest hit of all. The changes in objects caught in the reactor's field seemed equally mysterious: clear Coca-Cola bottles turned brown, hydraulic fluid coagulated into chewing gum, transistorized equipment stopped working, and rubber tires became rock hard.

As profound and terrible as exposure to ionizing radiation might prove for human beings, it's rarely accompanied by any detectable sensation. A person might be bathed in enough gamma rays to be killed a hundred times over without feeling a thing.

On August 21, 1945, two weeks after the bomb was dropped on Hiroshima, Harry K. Daghlian Jr., a twenty-four-year-old physicist on the Manhattan Project, was conducting an after-hours experiment in Los

Alamos, New Mexico, when his hand slipped. The test assembly he had built—a ball of plutonium surrounded by tungsten carbide bricks—went critical. Daghlian saw a momentary blue flash and was struck by a wave of gamma and neutron radiation amounting to more than 500 rem. He quickly disassembled the experiment, walked away, and admitted himself to medical care without visible symptoms. But radiation had killed him as surely as if he'd stepped in front of an oncoming train. Twenty-five days later, Daghlian slipped into a coma from which he never awoke—the first person in history to die accidentally from close exposure to nuclear fission. The *New York Times* attributed his death to burns sustained in "an industrial accident."

From the very beginning, the nuclear power industry has struggled to escape the shadow of its military origins. The first nuclear reactor ever built, assembled by hand beneath the bleachers of the University of Chicago's disused football field in 1942, was the anvil of the Manhattan Project, the essential first step in creating the fissile material needed to forge the world's first atomic weapon. The reactors that followed—built on a remote tract of land along the Columbia River in Hanford, Washington— were constructed solely to manufacture plutonium for use in the United States' growing arsenal of atom bombs. The US Navy was responsible for choosing the reactor design subsequently used in almost every civilian power station in the country. The first nuclear plant constructed for civilian use in the United States was based on blueprints recycled from a planned atomic-powered aircraft carrier.

In the USSR, the pattern was the same. The first Soviet atomic bomb— RDS-1, or "the Article," as it was called by the men who built it—was detonated soon after dawn on August 29, 1949, on a test range 140 kilometers northwest of Semipalatinsk, on the steppes of Kazakhstan. The project, code-named Problem Number One, was led by Igor Kurchatov, a forty-six-year-old physicist with the whiskery, forked beard of a Victorian spiritualist, noted by his secret police minders for his discretion and political cunning. The bomb was a faithful copy of the Fat Man device, which had destroyed Nagasaki almost exactly four years earlier, and con-

tained a core of plutonium produced in a reactor—known as reactor "A,"
or "Annushka"—initially modeled on the ones in Hanford.

Kurchatov had succeeded with the help of a handful of well-placed
spies and information contained in the bestselling book *Atomic Energy for
Military Purposes*—generously published by the US government in 1945
and speedily translated into Russian in Moscow. Nuclear work was the
responsibility of the newly formed First Main Directorate and an "atomic
politburo" overseen by Stalin's sadistic henchman, Lavrenty Beria—head
of the NKVD, forerunner of the KGB. From the start, the Soviet nuclear
project was governed by principles of ruthless expedience and paranoid
secrecy. By 1950, the First Main Directorate would employ seven hundred
thousand people, more than half of whom were forced laborers—including, at one point, fifty thousand prisoners of war—working in uranium
mines. Yet even when their prison sentences were complete, the Directorate packed these men and women into freight cars and shipped them into
exile in the Soviet Far North, to prevent them from telling anyone what
they had witnessed. Many were never seen again. And when Kurchatov's
team succeeded, Beria rewarded them in direct proportion to the punishment he had planned for them in the event of failure. Those that the secret
police chief would have ordered immediately shot—Kurchatov himself
and Nikolai Dollezhal, who designed the Annushka reactor—were instead
awarded the state's highest honor, the title Hero of Socialist Labor, along
with dachas, cars, and cash prizes. Those who would merely have received
maximum prison terms were instead granted the country's next-highest
honor, the Order of Lenin.

By the time the Article exploded, Igor Kurchatov had already decided
to begin work on a reactor dedicated to generating electricity. Development started in 1950 in a newly constructed closed city, Obninsk, two
hours southwest of Moscow. There the same group of physicists who had
built the Annushka reactor were set to work on a new one, this time intended to use the heat of fission to turn water into steam and power a
turbine. Resources were scarce, and some in the nuclear program believed
that a power reactor could never be practical. Only as a concession to
Kurchatov's prestige as the father of the bomb did Beria permit the project to proceed. It was not until the end of 1952 that the government sig-

naled its commitment to nuclear power by naming a new design institute dedicated to creating new reactors: the Scientific Research and Design Institute of Energy Technology, known by its Russian acronym NIKIET.

The following year, the USSR tested its first thermonuclear device—a hydrogen bomb, a thousand times more destructive than the atom bomb—and both emerging superpowers became theoretically capable of wiping out humanity entirely. Even Kurchatov was shaken by the power of the new weapon he had created, which had turned the surface of the earth to glass for five kilometers around ground zero. Less than four months later, US president Dwight D. Eisenhower delivered his "Atoms for Peace" address to the UN General Assembly, part of an attempt to mollify an American public facing a future now menaced by the specter of apocalypse. Eisenhower called for global cooperation to control the incipient arms race and tame the power of the atom for the benefit of mankind. He proposed an international conference to consider the issue. No one was especially surprised when the USSR publicly dismissed the idea as empty propaganda.

But when the UN International Conference on the Peaceful Uses of Atomic Energy finally convened in Geneva, Switzerland, in August 1955, the Soviet delegation arrived in force. It marked the first time in twenty years that scientists from the USSR had been permitted to mix with their foreign counterparts, and they delivered a propaganda coup of their own. They announced that, on June 27 the previous year, they had successfully connected their Obninsk reactor, designated AM-1, to the Moscow grid.

It was the first reactor in the world to use nuclear power for civilian electricity generation, and the scientists christened it Atom Mirny-1— "Peaceful Atom-1." At that moment, the first US nuclear power station, in Shippingport, Pennsylvania, was still more than two years from completion. Housed in a quaint stucco building with a tall chimney that could easily be mistaken for a chocolate factory, AM-1 generated only 5 megawatts—just enough to drive a locomotive—yet symbolized Socialism's superior ability to harness nuclear power for the benefit of mankind. Its launch marked the birth of the Soviet nuclear energy industry and the start of a Cold War technological contest between the superpowers.

Soon after the death of Stalin in 1953, Lavrenty Beria was arrested, imprisoned, and shot. The First Main Directorate was reconstituted and renamed. The new Ministry of Medium Machine Building—Ministerstvo srednego mashinostroyeniya, abbreviated in Russian to MinSredMash, or simply Sredmash—would now supervise everything connected with atomic energy, from uranium mining to bomb testing. Newly appointed Soviet premier Nikita Khrushchev brought an end to the years of Stalinist repression, liberalized the arts, embraced high technology, and promised that True Communism—the workers' Shangri-la of equality and plenty for all—would be achieved by 1980. To help modernize the Soviet economy and also reinforce his hold on power, Khrushchev personally promoted both space travel and nuclear technology.

With the success of Atom Mirny-1, the physicists and their Party bosses glimpsed a panacea that would finally release the Soviet Union from the deprivation of the past and help it on the path to a brighter future. To the Soviet people, still rebuilding amid the devastation of World War II, the Obninsk reactor showed how the USSR could technologically lead the world in a way that benefited ordinary citizens, bringing heat and light into their homes. The physicists who worked on AM-1 received the Lenin Prize, and the power of the atom was hymned in magazine articles, films, and radio programs; the Ministry of Culture introduced elementary school courses teaching children the fundamentals of atomic energy and contrasting the peaceful aims of the Soviet nuclear program with the militaristic intentions of the United States. Alongside the cosmonauts and martyrs of the Great Patriotic War, according to historian Paul Josephson, the nuclear scientists became "near-mythic figures in the pantheon of Soviet heroes."

But the little reactor in Obninsk was not all that it seemed. The principles of its design had not originated with the imperatives of electricity generation but with the need to manufacture plutonium bomb fuel quickly and cheaply. The same team from the Ministry of Medium Machine Building who had built the Annushka reactor oversaw its construction. Its path to completion had been fraught with corrosion, leaks, and instrument failure. And it had first been developed to provide propulsion for nuclear submarines. Only when that proved impractical had the original code name behind its acronym AM—Atom Morskoy, or "Naval Atom"—been revised to suggest more innocent goals.

It was also inherently unstable.

Unlike a nuclear weapon, in which a vast number of uranium atoms fission in a fraction of a second, releasing all their energy in an annihilating flash of heat and light, in a reactor the process must be regulated and delicately sustained for weeks, months, or even years. This requires three components: a moderator, control rods, and a coolant.

The simplest form of nuclear reactor requires no equipment at all. If the right quantity of uranium 235 is gathered in the presence of a neutron moderator—water, for example, or graphite, which slows down the movement of the uranium neutrons so that they can strike one another—a self-sustaining chain reaction will begin, releasing molecular energy as heat. The ideal combination of circumstances required for such an event—a criticality—has even aligned spontaneously in nature: in ancient subterranean deposits of uranium found in the African nation of Gabon, where groundwater acted as a moderator. There, self-sustaining chain reactions began underground two billion years ago, producing modest quantities of heat energy—an average of around 100 kilowatts, or enough to light a thousand lightbulbs—and continued intermittently for as long as a million years, until the available water was finally boiled away by the heat of fission.

But to generate power steadily inside a nuclear reactor, the behavior of the neutrons must be artificially controlled, to ensure that the chain reaction stays constant and the heat of fission can be harnessed to create electricity. Ideally, every single fission reaction should trigger just one more fission in a neighboring atom, so that each successive generation of neutrons contains exactly the same number as the one before, and the reactor remains in the same critical state.

Should each fission fail to create as many neutrons as the one before, the reactor becomes subcritical, the chain reaction slows and eventually ceases, and the reactor shuts down. But if each generation produces more than one fission, the chain reaction could begin to grow too quickly toward a potentially uncontrollable supercriticality and a sudden and massive release of energy similar to that in a nuclear weapon. To maintain a steady state between these two extremes is a delicate task. The first nuclear engineers had to develop tools to master forces perilously close to the limits of man's ability to control.

Infinitesimal and invisible, the scale of subatomic activity inside a nuclear power reactor is hard to comprehend: generating a single watt of electricity requires more than 30 billion fissions every second. Around 99 percent of the neutrons generated in a single fission event are high-energy particles released at enormous speed—"prompt" neutrons that travel at twenty thousand kilometers a second. Prompt neutrons smash into their neighbors, where they cause more fission, continuing the chain reaction, within an average of just ten nanoseconds. This fraction of time—so small that the wits of the Manhattan Project measured it in "shakes," for a "shake of a lamb's tail"—is much too fast to be controlled by any mechanical means. Fortunately, among the remaining 1 percent of neutrons generated in every fission event are a tiny minority released on a timescale more readily perceptible by man, measured in seconds or even minutes. It is only the existence of these delayed neutrons, which emerge slowly enough to respond to human control, that make the operation of a nuclear reactor possible at all.

By inserting electromechanical rods containing neutron-absorbing elements—such as boron or cadmium, which act like atomic sponges, soaking up and trapping delayed neutrons, preventing them from triggering further fission—the growth of the chain reaction can be controlled incrementally. With the rods inserted all the way into the reactor, the core remains in a subcritical state; as they are withdrawn, fission increases slowly until the reactor becomes critical—and can then be maintained in that state and adjusted as necessary. Withdrawing the control rods farther, or in greater numbers, increases reactivity and thus the amount of heat and power generated, while inserting them farther has the opposite effect. But controlling the reactor using only this fraction of less than 1 percent of all neutrons in every fission makes the control process acutely sensitive: if the rods are withdrawn too quickly, too far, in too large a number—or any of the myriad safety systems fail—the reactor may be overwhelmed by the fission of prompt neutrons and become "prompt supercritical." The result is a reactor runaway, a catastrophic scenario accidentally triggering a similar process to the one designed into the heart of an atomic bomb, creating an uncontrollable surge of power that increases until the reactor core either melts down—or explodes.

To generate electricity, the uranium fuel inside a reactor must become hot enough to turn water into steam but not so hot that the fuel itself starts to melt. To prevent this, in addition to control rods and a neutron moderator, the reactor requires a coolant to remove excess heat. The first reactors built in the United Kingdom used graphite as a moderator and air as a coolant; later commercial models in the United States employed boiling water as both a coolant and a moderator. Both designs had distinct hazards and benefits: water does not burn, although when turned to pressurized steam, it can cause an explosion. Graphite couldn't explode, but at extreme temperatures, it could catch fire. The first Soviet reactors, copied from those built for the Manhattan Project, used both graphite and water. It was a risky combination: in graphite, a moderator that burns fiercely at high temperatures and, in water, a potentially explosive coolant.

Three competing teams of physicists produced the initial proposals for what became Atom Mirny-1. These included a graphite-water design, another that used a graphite moderator and helium as a coolant, and a third using beryllium as a moderator. But the Soviet engineers' work on the plutonium production plants meant that they had far more practical experience with graphite-water reactors. These were also cheaper and easier to construct. The more experimental—and potentially safer—concepts never had a chance.

It wasn't until late in the construction of Atom Mirny-1 that the physicists in Obninsk discovered the first major defect with their design: a risk of coolant water leaking onto the hot graphite, which could lead not only to an explosion and radioactive release but also to a reactor runaway. The team repeatedly delayed the launch of the reactor as they devised safety systems to address the problem. But when it finally went critical in June 1954, Atom Mirny-1 retained another profound drawback the scientists never fixed: a phenomenon known as the positive void coefficient.

When working normally, all nuclear reactors cooled by water contain some steam also circulating through the core, which forms bubbles, or "voids" in the liquid. Water is a more efficient neutron moderator than steam, so the volume of steam bubbles in the water affects the reactivity of the core. In reactors that use water as both coolant and moderator, as the volume of steam increases, fewer neutrons are slowed, so reactivity falls. If too much steam forms—or even if the coolant leaks out entirely—the chain reaction stops, and the reactor shuts itself down. This *negative* void coefficient acts like a dead man's handle on the reactor, a safety feature of the water-water designs common in the West.

But in a water-graphite reactor like Atom Mirny-1, the effect is the opposite. As the reactor becomes hotter and more of the water turns to steam, the graphite moderator keeps doing its job just as before. The chain reaction continues to grow, the water heats further, and more of it turns to steam. That steam, in turn, absorbs fewer and fewer neutrons, and the chain reaction accelerates still further, in a feedback loop of growing power and heat. To stop or slow the effect, the operators must rely on inserting the reactor control rods. If they were to fail for any reason, the reactor could run away, melt down, or explode. This *positive* void coefficient remained a fatal defect at the heart of Atom Mirny-1 and overshadowed the operation of every Soviet water-graphite reactor that followed.

On February 20, 1956, Igor Kurchatov materialized before the Soviet public for the first time in more than ten years. The father of the bomb had been enveloped in the secrecy surrounding Problem Number One since 1943, isolated in the clandestine laboratories of Moscow and Obninsk or lost in the vastness of the weapons testing grounds of Kazakhstan. But he now stood before the delegates assembled for the 20th Congress of the Communist Party of the Soviet Union in Moscow, where he revealed a fantastical vision of a new USSR powered by nuclear energy. In a short but galvanizing speech, Kurchatov outlined plans for an ambitious program of experimental reactor technology and a futuristic Communist empire crisscrossed by atomic-propelled ships, trains, and aircraft. He predicted that cheap electricity would soon reach every corner of the Union through a network of giant nuclear power stations. He promised that Soviet nuclear capacity would reach 2 million kilowatts—four hundred times what the Obninsk plant could produce—within just four years.

To realize this audacious vision, Kurchatov—now named head of his own Institute of Atomic Energy—had convinced the chief of Sredmash to let him build four different reactor prototypes, from which he hoped to choose the designs that would prove the basis of the Soviet nuclear industry. But before construction could begin, Kurchatov also had to win over the economics mandarins of Gosplan, who controlled the distribution of all resources throughout the USSR. Gosplan's own Department of Energy and Electrification set targets for everything from how much money could be allocated for building an individual power station to the quantity of electricity it would be expected to produce once complete. And the men and women of Gosplan cared little for ideology, Soviet prestige, or the triumph of Socialist over capitalist technology. They wanted rational economics and tangible results.

Like their counterparts in the West, the Soviet scientists' arguments about how quickly and inexpensively nuclear power could become competitive with conventionally produced electricity were speculative and colored by wishful thinking about electricity "too cheap to meter." But unlike the boosters of the nuclear future in the United States, the Soviets could not rely on golf course sales pitches and entrepreneurial investment from the free market. And economics were not on their side: the capital costs of building any nuclear reactor were colossal, and the USSR was rich in fossil fuels—especially beneath the remote wastes of Siberia, where new oil and gas deposits were being discovered all the time.

Yet the sheer size of the Union, and its poor infrastructure, favored nuclear power. The scientists pointed out that the Siberian deposits were thousands of miles from where they were most needed: in the western part of the Soviet Union, where the majority of its population and industry lay. Moving either raw materials or electricity over these distances was costly and inefficient. Meanwhile, the nuclear plants' closest competition—hydroelectric stations—required flooding huge areas of valuable farmland. Nuclear stations, while expensive to build, had little environmental impact; they were largely independent of natural resources; they could be located close to the sources of demand in major cities; and, if constructed on a large enough scale, they could produce vast amounts of electricity.

Apparently convinced by Kurchatov's promises, Gosplan released the money for two prototype plants: one with a pressurized water reactor of the kind already becoming standard in the United States, and another with a water-graphite channel type—a scaled-up version of Atom Mirny-1. But, just as in they would in the West, construction costs quickly skyrocketed, and Gosplan suspected the scientists had misled them. They scaled back the plans and called a halt to work on the PWR plant, and Kurchatov's vision of the atom-powered future gradually collapsed. He pleaded for the flow of resources to be turned back on, writing to the head of Gosplan to insist that the plants were crucial for determining the future of the Soviet atom. But his pleas went unheeded, and in 1960 Kurchatov died without seeing his dream revived.

In the meantime, the Ministry of Medium Machine Building had completed a new project, hidden inside the clandestine nuclear site known as Combine 816, or Tomsk-7, in Western Siberia. The EI-2, or "Ivan the Second," was a big military water-graphite reactor with a thrifty edge. Its predecessor, Ivan-1, had been a simple model built solely to manufacture plutonium for nuclear warheads. But EI-2 had been adapted to perform two tasks at once. It made weapons-grade plutonium and, as a by-product of the process, also generated 100 megawatts of electricity. And when work on the Soviet civil nuclear program finally restarted two years after Kurchatov's death, by then lagging behind its competition in the United States, it did so with a new emphasis on reactors that were affordable to build and cheap to run. At that moment, it was not the sophisticated experimental reactors of Igor Kurchatov's civilian nuclear program but the stalwart Ivan the Second that stood ready to carry the atomic banner for the Soviet Union.

Less than a year after Igor Kurchatov presented his imperial vision of an atom-powered USSR to the Party congress in Moscow, a toothy, young Queen Elizabeth II made her own ceremonial appearance, outside Calder Hall nuclear power station, on the northwest coast of England. Pulling a lever with elegantly gloved hands, she watched as the needle on an over-sized meter began to spin, showing the first atomic electricity flowing into the British national grid from one of the station's two gas-cooled reactors. It was pronounced the launch of the first commercial-scale nuclear power station in the world, the dawn of a new industrial revolution and a triumph for those who had kept the faith in the peaceful power of the atom when others had feared it would bring only destruction to the world. "For them," a newsreel commentator reported, "this day is a milestone of victory!"

The event was a grand propaganda exercise; the truth was darker. Calder Hall had been constructed to manufacture plutonium for Britain's own nascent atom bomb program. What electricity it did produce was a costly fig leaf. And the military roots of the civilian nuclear industry had entangled not only the technology it relied upon but also the minds of its custodians. Even in the West, nuclear scientists continued to dwell in a culture of secrecy and expedience: an environment in which sometimes reckless experimentation was married with an institutional reluctance to acknowledge when things went wrong.

A year after Calder Hall opened, in October 1957, technicians at the neighboring Windscale breeder reactor faced an almost impossible dead-line to produce the tritium needed to detonate a British hydrogen bomb. Hopelessly understaffed, and working with an incompletely understood technology, they operated in emergency conditions and cut corners on safety. On October 9 the two thousand tons of graphite in Windscale Pile Number One caught fire. It burned for two days, releasing radiation across the United Kingdom and Europe and contaminating local dairy farms with high levels of iodine 131. As a last resort, the plant manager ordered water poured onto the pile, not knowing whether it would douse the blaze or cause an explosion that would render large parts of Great Britain un-inhabitable. A board of inquiry completed a full report soon afterward, but, on the eve of publication, the British prime minister ordered all but two or three existing copies recalled and had the metal type prepared to print it broken up. He then released his own bowdlerized version to the public, edited to place the blame for the fire on the plant operators. The British government would not fully acknowledge the scale of the accident for another thirty years.

Meanwhile, in the USSR, endemic nuclear secrecy had reached fresh extremes. Under Khrushchev, Soviet scientists began to enjoy unprecedented autonomy, and the public—encouraged to trust unquestioningly in the new gods of science and technology—were kept in the dark. In this intoxicating atmosphere, the physicists' early success in taming the power of the peaceful atom made them dangerously overconfident. They began using gamma rays to extend the shelf life of chicken and strawberries, they built mobile nuclear reactors mounted on tank treads or designed to float around the Arctic, and, like their US counterparts, they designed atomic-powered aircraft. But they also used nuclear weapons to put out fires and excavate underground caverns, restricting the size of their explosions only when the seismic shock began to destroy nearby buildings.

Following the death of Igor Kurchatov, the Institute of Atomic Energy had been renamed in his honor, and leadership of Soviet nuclear science had passed to his disciple, Anatoly Aleksandrov. An imposing man with a glisteningly bald head who'd helped build the first plutonium production reactors, Aleksandrov was appointed director of the Kurchatov Institute in 1960. A dedicated Communist who believed entirely in science as an instrument of the Soviet economic dream, he prized monumental projects over cutting-edge research. As the Era of Stagnation began, the Soviet scientific establishment lavished resources on the immediate priorities of the state—space exploration, water diversion, nuclear power—while emergent technologies, including computer science, genetics, and fiber optics, fell behind. Aleksandrov oversaw the design of reactors for nuclear submarines and icebreakers, as well as the prototypes of the new channel-type graphite reactors designed to generate electricity. To reduce the cost of building these, he emphasized economies of scale and insisted on increasing their size to colossal proportions using standardized components and common factory materials. He saw no reason that manufacturing nuclear reactors should be any different from making tanks or combine harvesters. Aleksandrov regarded the serial production of these massive reactors as the key to Soviet economic development, and atomic power as the means to realize Ozymandian dreams of irrigating deserts, bringing tropical oases to the Arctic North, and leveling inconveniently situated mountains with atom bombs—or, as the Russian expression went, "correcting the mistakes of nature."

Despite his breadth of vision and political influence, Aleksandrov did not have sovereignty over Soviet nuclear science. Behind him loomed the sinister, adamantine power of the Ministry of Medium Machine Building and its belligerent chief, the veteran revolutionary Efim Slavsky, variously known as "Big Efim" and "the Ayatollah." Although as young men they had fought on opposite sides in the Russian Civil War—Slavsky on horseback as a political commissar with the Red Cavalry, Aleksandrov with the White Guard—the two atomic magnates were close, and enjoyed reminiscing together over vodka and cognac. But as the Cold War intensified, the military-industrial demands of Sredmash overwhelmed those of the pure scientists at the Kurchatov Institute. In the first few years of its existence, the national priority afforded the atom weapons program had allowed the ministry to consolidate control of a massive nuclear empire, with its own scientists, troops, experimental labs, factories, hospitals, colleges, and testing grounds. Sredmash could call upon almost unlimited resources, from gold mines to power stations, all maintained behind an impenetrable wall of silence.

The very names of Sredmash facilities were classified, and sites that ranged in size from individual institutes in Moscow and Leningrad to entire cities were known by the men and women who worked there as *pochtovye yashchiki*—"post office boxes"—referred to only by code numbers. Led by Slavsky, a cunning political operator with access to the highest levels of government, the Ministry of Medium Machine Building became closed and almost entirely autonomous, a state within a state.

Under the paranoid regime of permanent warfare maintained by Sredmash, any accident—no matter how minor—was regarded as a state secret, policed by the KGB. And even as the USSR's nuclear power industry began to gather momentum in the mid-1960s, the clandestine impulse persisted. In the bureaucratic upheaval that followed the fall of Khrushchev, in 1966 responsibility for operating new atomic stations throughout the USSR was transferred from Sredmash to the civilian Ministry of Energy and Electrification. Yet everything else—the design and technical supervision of the reactors that powered the plants, their prototypes, and every aspect of their fuel cycle—remained in the hands of the Ministry of Medium Machine Building.

As one of the twelve founding members of the International Atomic Energy Agency, since 1957 the USSR had been obliged to report any nuclear accident that took place within its borders. But of the dozens of dangerous incidents that occurred inside Soviet nuclear facilities over the decades that followed, not one was ever mentioned to the IAEA. For almost thirty years, both the Soviet public and the world at large were encouraged to believe that the USSR operated the safest nuclear industry in the world.

The cost of maintaining this illusion had been high.

At 4:20 p.m. on Sunday, September 29, 1957, a massive explosion occurred inside the perimeter of Chelyabinsk-40 in the southern Urals, a Sredmash installation so clandestine that it had never appeared on any civilian map. The forbidden area encompassed both the Mayak Production Association—a cluster of plutonium production reactors and radiochemical factories scraped from the wilderness by forced labor—and Ozersk, the comfortable closed city that housed the privileged technicians who staffed them. It was a warm, sunny afternoon. When they heard the explosion, many of Ozersk's citizens were watching a soccer match in the city stadium. Assuming it was the sound of foundations being dynamited by convicts in the nearby industrial zone, few of the spectators even looked up. The match continued.

But the explosion had taken place inside an underground waste storage tank filled with highly radioactive plutonium processing waste. The blast, which occurred spontaneously after cooling and temperature-monitoring systems had failed, threw the tank's 160-tonne concrete lid twenty meters into the air, blew out the windows of a nearby prisoner barracks, ripped metal gates from the nearby fence, and launched a kilometer-tall pillar of dust and smoke into the sky. Within a few hours, a blanket of gray radioactive ash and debris several centimeters thick had settled over the industrial zone. The soldiers who worked there were soon admitted to the hospital, bleeding and vomiting.

No emergency plans had been prepared for a nuclear accident; at first, no one realized they were facing one. It was hours before the plant managers, away on business, were finally tracked down at a circus show in Moscow. By then, highly radioactive contamination had begun to spread across the Urals—2 million curies of it—falling in a deadly trace six kilometers wide and nearly fifty kilometers long. The next day, light rain and a thick, black snow fell on nearby villages. It took a year to clean up inside the forbidden zone. The so-called "liquidation" of the consequences of the explosion was begun by soldiers who ran into the contaminated areas with shovels and tossed parts of the shattered waste storage container into a nearby swamp. The city leaders of Ozersk, apparently fearing mass panic more than the threat of radiation, attempted to stifle news of what had happened. But as rumors spread throughout the cadres of young engineers and technicians, nearly three thousand workers left the city, preferring to take their chances in what they called the "big world" beyond the perimeter wire rather than remain in their cozy but contaminated homes.

In the remote villages outside the wire, barefoot women and children were instructed to harvest their potatoes and beets, but then dump them into trenches dug by bulldozers, overseen by men wearing protective suits and respirators. Soldiers herded the peasants' cows into open pits and shot them. Eventually ten thousand people were ordered permanently evacuated over the course of two years. Entire settlements were plowed into the ground. Twenty-three villages were wiped from the map, and up to a half million people were exposed to dangerous levels of radioactivity.

Rumors of what had happened in Mayak reached the West, but Chelyabinsk-40 was among the most fiercely guarded military locations in the USSR. The Soviet government refused to acknowledge its very existence, let alone that anything might have happened there. The CIA resorted to sending high-altitude U-2 spy planes to photograph the area. It was on the second of these missions, in May 1960, that Francis Gary Powers's aircraft was shot down by a Soviet SA-2 surface-to-air missile, in what became one of the defining events of the Cold War.

Although it would be decades before the truth finally emerged, the Mayak disaster remained, for many years, the worst nuclear accident in history.

3

Friday, April 25, 5:00 p.m., Pripyat

The afternoon of Friday, April 25, 1986, was beautiful and warm in Pripyat, more like summer than late spring. Almost everyone was looking forward to the long weekend leading into May Day. Technicians were preparing the grand opening of the city's new amusement park, and families were filling their fridges with food for the holiday; some were engaged in the home improvement fad sweeping the city, hanging wallpaper and laying tiles in their apartments. Outside, the scent of apple and cherry blossom lingered in the air. Fresh laundry hung on the balconies on Lenina Prospekt. Beneath their windows, Viktor Brukhanov's roses were in bloom: a palette of pink, red, and fuchsia.

In the distance, the V. I. Lenin Atomic Energy Station, attended by the huge latticed power masts carrying high-tension cables to the switching stations, shone a brilliant white against the skyline. On the roof of the ten-story apartment building on Sergeant Lazarev Street, overlooking the central square, giant, angular white letters spelled out in Ukrainian the mellifluous propaganda jingle of the Ministry of Energy and Electrification: *Hai bude atom robitnikom, a ne soldatom!* "Let the atom be a worker, not a soldier!"

Brukhanov, harried by work as usual, had left for the office at 8:00 a.m. and driven the short distance from the family apartment overlooking Kurchatov Street to the plant in the white Volga he used for official business. Valentina had arranged to take the afternoon off from her job in the plant construction offices to spend time with her daughter and her son-in-law, who had both driven over from Kiev to visit for the weekend. Lilia was already five months pregnant, and the weather was so good that the three of them decided to take a day trip to Narovlia, a riverside town a few kilometers over the border in Belarus.

Alexander Yuvchenko, senior mechanical engineer in the reactor department on the night shift of Chernobyl's Unit Four, spent the day in Pripyat with his two-year-old son, Kirill. Yuvchenko had worked at the station for only three years. Lean and athletic, almost two meters tall, he had built up his towering frame with competitive rowing in high school back in Tiraspol, in the tiny Soviet republic of Moldova. At thirteen, Yuvchenko had been one of the first members of the city rowing club, where the trainer selected only the tallest and strongest boys to test themselves on the fast-flowing waters of the Dniester. At sixteen, he became the Junior League champion in Moldova; his team went on to take second place in the All-Union Youth Competition, competing against teams from across the entire USSR.

But Yuvchenko was also gifted at physics and mathematics, and, at seventeen, he showed such promise as a rower that he faced a painful decision: to go to university or pursue a career in athletics. It was only over the objections of his trainer that he finally chose academia. In 1978 he enrolled to major in nuclear physics at Odessa National Polytechnic University, less than a hundred kilometers away from home, just across the border in Ukraine. He was young and zealous, and had decided to do something futuristic and spectacular: he dreamed of working at a nuclear power station.

Now, at twenty-four, Yuvchenko was deputy secretary of the Chernobyl plant's Komsomol—the Young Communist League, the Party's youth wing. In spite of the long hours he put in at work, he still liked to play ice hockey with his friends from the station, on the rinks poured in the city every winter. In the spring, he and his wife, Natalia, borrowed a neighbor's small motorboat, and together the family took trips down the Pripyat River—idling through the slick, brown water, drifting past forest glades carpeted with sweet-scented lilies of the valley, stopping at empty beaches of fine, white sand, surrounded by towering pines.

Alexander and Natalia had first met as children in Tiraspol, where they were in the same class at school. At twelve, Alexander was already taller than the other boys, gangly and clumsy. Natalia was slight, and spoiled. Her parents were members of the *nomenklatura*: loyal Party members, both with senior management positions in local industry. She wore her dark hair in two braids that dangled to the small of her back. Her blue-gray eyes seemed to change color with her mood and the weather. Alexander noticed her immediately. But if she reciprocated his interest, she didn't show it.

A few years later, Alexander and his family moved into an apartment on Sovietskaya Street, directly opposite the small private house where Natalia lived. They began dating on and off—often breaking up, seeing other people—but always got back together again. Finally, in August 1982, after spending an entire year apart, they married. By then, they were both deep into their studies at the university in Odessa: Natalia was twenty-one; Alexander, just twenty. Kirill was born a year later.

Like all newly qualified Soviet specialists, when Alexander graduated in 1983, he had to choose a posting from a government short list of work assignments. But there was never really any question about which he would take. The Chernobyl nuclear power plant was one of the best and most prestigious nuclear facilities in the entire Soviet Union; it was in Ukraine, close to Kiev, and surrounded by tranquil countryside. Most important, he'd heard that married couples moving to Pripyat could expect to be allocated an apartment in the city. Alexander hoped that he and his new family would have a place of their own within a year, an unthinkable prospect elsewhere in the Soviet Union.

When their son was born, Natalia still had another year of her degree in Russian philology to complete. She stayed in Odessa while Alexander moved into a single men's dormitory in Pripyat and began working at the plant. And when she first visited him there, at the end of December 1983, she saw none of the city's vaunted beauty. In the flat winter light, Pripyat was wan and featureless, stifled between the fallow landscape and grubby, dishwater skies. She was struck only by the concrete monument that marked the entrance to the city in massive, brutalist characters, reading "Pripyat 1970." But the next year, the family was assigned an apartment on the top floor of a big building in a newly completed district of town, on Stroiteley Prospekt—Avenue of the Construction Workers. They moved into it in August, and the two-room flat seemed as big as a palace. From the balcony, the Yuvchenkos had a sweeping view of the Pripyat River and the forest beyond. A fresh breeze blew through the kitchen windows. They hung bright-pink floral wallpaper in the living room and filled it with furniture Natalia's mother had secured through connections at the timber *kombinat* where she worked.

There was little call for the expertise of a philology graduate in the technically focused world of the *atomgrad*, so Natalia went to work as a schoolteacher. School Number Four was enormous, with more than two thousand children: Natalia taught Russian language and literature and supervised a fourth-grade class. She often wondered why she had to spend her time looking after other people's children, while her own son languished in a crèche. By spring 1986, Alexander had been promoted from circulation pump operator to senior mechanical engineer in the Number Four reactor department. At the end of March, he was summoned to the offices of the Pripyat Communist Party. They offered him the job of first secretary of the city Komsomol. Unlike his part-time role in the plant organization, this was a full-time political position, and would mean resigning from the job he loved at Unit Four. Yuvchenko declined; they insisted; he declined again, this time quoting a few lines from Engels. They allowed him to go home, but Alexander knew he couldn't say no forever: nobody could refuse the Party's requests. In the meantime, with two salaries and a place of their own, the Yuvchenkos had all they needed. They began thinking about a second child.

Still, with no family nearby to provide help, life was hardly easy. In the second half of April that year, Kirill became sick with a bad cold. At first, Natalia took time off work to look after him. But the illness lingered, and when she had to go back to her students, the couple began sharing child care. When Alexander was on the night shift at the plant, caring for the boy during the day fell to him. When Natalia returned home from work on the afternoon of April 25, she looked down from the apartment window and spotted her husband on the street below, giving Kirill a ride on the crossbar of his bike. Alexander had worked from midnight until eight the previous night and then spent all day with their son without sleeping. He was due back at the plant again in just a few hours for another shift. Natalia realized how exhausted he must be, and the thought made her uneasy. Despite the bright sunshine and the excited cries of her son drifting up from below, a shadow of apprehension passed over her.

After dinner, Natalia put Kirill to bed and sat down to watch the climactic episode of a TV miniseries, the Soviet blockbuster adaptation of Irwin Shaw's potboiler *Rich Man, Poor Man.* Alexander usually left for the night shift at around 10:30 p.m., but he seemed restless and prepared for work with an odd meticulousness. He spent almost an hour taking a bath. Then he put on a smart new outfit—a pair of slacks and a coveted Finnish-made Windbreaker—as if he were going to a party, not a power station. He poured a cup of coffee, alone in the kitchen. But he wanted company and asked Natalia to join him.

She left the TV, and they spent the next few minutes talking about nothing much until, finally, it was time for him to go.

A few hundred meters away from the Yuvchenkos, in his apartment opposite the big swimming pool on Sportivnaya Street, Sasha Korol was reading on the couch when his friend Leonid Toptunov ambled in. The two nuclear engineers had been close for almost a decade, ever since the early days of their studies at a branch of Moscow Engineering and Physics Institute—MEPhI—in the nuclear city of Obninsk. Now they lived one floor apart in a block of almost identical one-room apartments, occupied by doctors, teachers, and other single, young nuclear engineers. The two men shared keys and let themselves into each other's homes whenever they felt like it.

Korol, the son of a physics teacher, and Toptunov, the only child of a senior army officer attached to the Russian space program, both had science in their blood. They had been born into a world where, in the late 1950s and early 1960s, the startling coups of Soviet engineers regularly humbled the West. Toptunov's father had been at the heart of the USSR's shadow world of clandestine technology, overseeing the construction of rocketry facilities at Baikonur Cosmodrome in Kazakhstan—the site, in

1957, of the surprise launch of Sputnik, which first shattered the United States' complacent assumptions about its technical superiority over an empire of ham-fisted potato farmers.

Toptunov was born within sight of the Baikonur launchpad, in the secret space city of Leninsk, three years later. He grew up surrounded by the hallowed group of men and women who would lead mankind into orbit, idolized not just by the children who lived around the Cosmodrome, but throughout the USSR. Toptunov's father liked to boast that Yuri Gagarin, soon to be the most famous man on earth, used to babysit the infant Leonid. When Gagarin's massive Vostok 1 rocket thundered off the launchpad early one morning in April 1961, Toptunov, just seven months old, was there to witness its blazing exhaust plume vanish into the stratosphere, and a Soviet pilot become the first human being in space.

When Toptunov was thirteen, his father was appointed military attaché at the Dvigatel rocket engine plant in Tallinn, and the family moved to Estonia. Three years later, in July, Toptunov went to Moscow, to sit for the MEPhI entrance exams. Reserved and attentive, he had proved an able student in math. But MEPhI—established with the patronage of Kurchatov, the father of the Soviet atom bomb—was the USSR's most prestigious institute for the study of nuclear engineering and physics. The examination was notoriously difficult, with as many as four students competing for every place; some sat for it again and again before being admitted. While Toptunov toiled over the test papers, his father waited outside the room on a bench. When the young man emerged at last, he was shaking with exhaustion. Toptunov passed on the first attempt, but when he called to give his mother the good news, she pleaded with him not to go. He was her only child, and the idea of nuclear power terrified her; she begged him to remain in Tallinn and study there.

But Leonid had no interest in life in a Baltic backwater. At seventeen, he left home to join the cult of the *atomshchiki*—the disciples of the peaceful atom.

Toptunov met Sasha Korol in 1977, as part of a group of thirty or so first-year students studying atomic power plant engineering at the MEPhI campus in Obninsk. It was a place of thrilling novelties for the aspirant teenaged engineers, a complex surrounded by sixteen other research facilities and with access to two small research reactors. The course work was tough, beginning with general disciplines—mathematics, technical drawing, and chemistry—but also political indoctrination. To excel, the students had to pass courses in historical materialism and "scientific Communism": the study of the history of the Communist Party in the Soviet Union, and the social laws established by Marx and developed by Lenin and Brezhnev, leading to the state of True Communism, then scheduled to arrive in the year 2000.

In their spare time, the young freshmen in Obninsk were like students everywhere. They drank beer and played cards and went to films and shows. Especially popular were the improvised comedy competitions based on the format of the TV show *KVN—Klub vesyolykh i nakhodchivykh*, or the *Club of the Merry and Quick-Witted*—which, although by then long banished from television by Soviet censors, lived on as a cult live spectacle at colleges across the Union. Toptunov, shy looking, with glasses and a lingering hint of puppy fat, was frustrated by his boyish appearance. He grew a mustache that he hoped gave him an air of sophistication. With a charming smile and a thick shock of shaggy, brown hair, he did well with girls.

At MEPhI, Toptunov took up karate—a sport on the long and often inexplicable list of ideas and practices from outside the USSR that were officially forbidden. But information about it was available in samizdat form, and Toptunov learned to kick and punch from illegally circulated homemade manuals. Against the advice of his tutors, who warned their students it could damage their eyesight, and with it their future in the nuclear industry, he also began to box. Although his retinas emerged from the ring intact, he did eventually have his nose broken, leaving him with a chronic nasal dribble. One night after class, Toptunov got into a drunken argument with an overbearing thermodynamics tutor. The dispute escalated, and, in the bathroom, the two men came to blows. Toptunov gave the tutor a black eye. Afterward, the young student was threatened with expulsion but, somehow granted a reprieve, stayed on.

After four years of study at MEPhI, Toptunov and Korol began their diploma projects: Korol focused on a technique to isolate faulty fuel rods, while Toptunov worked on using acoustics to identify irregularities in the performance of the reactor. The diploma research required a six-month internship at a nuclear station somewhere in the USSR, and both chose Chernobyl. They liked it so much there that, upon finally graduating from MEPhI in 1983, they opted to return to Chernobyl for their full-time work assignments. Toptunov and Korol arrived just in time for the completion of Chernobyl's Unit Four, the newest and most advanced of the station's RBMK reactors.

Like all other new engineers, they had to start at the bottom, doing menial work for which they were overqualified—patrolling the plant with an oil can, feeling machinery for hot bearings, mopping up spills—while they learned the practicalities and layout of the station and its equipment. The young specialists learned quickly that it was one thing to understand how the reactor worked in principle, and quite another to understand it in reality. When their working shifts ended, they stayed in the station for hours, putting in extra time to trace the pathway of giant steam pipes and cables by hand, finding the location of huge gate valves in the dark, following myriad connections from room to room and floor to floor. It was common practice, too, for the trainees to return to the plant at all hours of the day and night to observe both routine work and special tests, in the hope of gathering extra knowledge that might speed their advancement.

During the summer and autumn of 1983, as Reactor Number Four was undergoing final assembly, the new trainees were given responsibility for managing quality control. While the giant cylindrical concrete vault, cast to contain the "active zone" of the reactor, was filled slowly with the thousands of tonnes of rectangular graphite blocks that would help moderate its fission, Toptunov, Korol, and the other apprentice operators clambered inside it to check the progress of construction. They compared the work of the assembling teams against the designers' blueprints and looked for leaks and cracks in the graphite pile; they monitored the welding of the scores of water pipelines that would circulate cooling water through the core, a gleaming thicket of narrow-bore stainless steel. Finally, when the vault was filled and the pipework complete, they watched as the reactor was sealed, loaded with fuel, and went critical for the first time, on December 13, 1983.

The work left little time for hobbies, but somehow Toptunov managed to fit them in. When he and Korol had first arrived in Pripyat, Toptunov organized a gym downstairs in the dormitories where they lived—he put up a set of Swedish wall bars for everyone to use—and, later, a class in which he tutored high school students from the city in math and physics. He had a girlfriend who worked as a nurse at the Pripyat hospital, Medical-Sanitary Center Number 126, and he loved to go fishing: the network of artificial canals and the giant cooling reservoir around the station were rich in fish, which thrived in water that had circulated through the plant's reactors as coolant before being flushed, still radioactive but pleasingly warm, toward the river.

After serving his apprenticeship by helping to build the reactor he would one day operate, Toptunov grew closer to qualifying as a senior reactor control engineer. This was perhaps the most demanding job in the entire plant, as the man—even in the nominally egalitarian Soviet Union, it was always a man—who, minute by minute through each eight-hour shift, governed the enormous power of the reactor. The position required rigorous study and practical experience: the operators spoke their own coded language, dense with acronyms and apparently unpronounceable abbreviations that were muddled into a new vocabulary: the ZGIS and the MOVTO, the BShchU, the SIUR, the SIUT, and the SIUB. Then there were thick stacks of manuals and regulations to pore over, followed by a series of examinations in the station's Department of Nuclear Safety. There were also health checks, and security screenings conducted by the KGB. After one of these safety exams, Toptunov sat down with Korol and told him about a strange phenomenon, described deep in the RBMK documentation, indicating that the reactor control rods may—under some circumstances—accelerate reactivity instead of slowing it down.

Only after all of this study and practice was Toptunov permitted to stand behind an existing senior reactor control engineer at the reactor panel in the control room and watch how the work was done. Eventually he was allowed—still under close supervision—to begin touching the buttons and switches of the panel himself.

———

When Leonid Toptunov let himself into Alexander Korol's eighth-floor apartment late on the night of April 25, 1986, he had been qualified as a senior reactor control engineer for just two months. Korol, still an assistant, was behind him but hoping to qualify soon as a senior engineer in the Unit Four reactor department. Toptunov found his old friend lounging on the couch, reading a story about a new medical phenomenon discovered in the United States—AIDS—in a recent Russian edition of *Scientific American.* Leonid told him that there was an electrical test on the turbines scheduled during his shift at Reactor Number Four that night. It would be worth witnessing.

"Let's go together," Toptunov said.

"No, I'll pass," Korol replied. "I've got this interesting article to read."

At a few minutes before eleven at night, Toptunov set off for the bus stop, a few blocks away on Kurchatov Street, where a scheduled service shuttled workers to and from the plant. He walked down to the end of Sportivnaya and took a right at the darkened windows of the Jubilee home services store. Then on past the post office and the technical school, and across the square toward the end of Lenina Prospekt. It was a warm, sultry night; the sky an inky blue, glittering with stars.

On the bus, Toptunov joined his colleagues on the midnight shift in Unit Four. These included the control room staff—Senior Unit Control Engineer Boris Stolyarchuk and Shift Foreman Alexander Akimov—and the engineers from the Reactor Department, among them Leonid's friend Alexander Yuvchenko, wearing his new clothes. It was a short ride. After ten minutes, they were at the steps of the station's main administrative block.

The four-story office building sat, like the bridge of a massive container ship, at the extreme eastern end of the station's four reactors and the turbine hall, which stretched away into the distance, a narrow concrete box almost a kilometer long. Inside the administrative block were the offices of Viktor Brukhanov and his senior staff and one of the plant's two main radiation control points: the sanitary locks that marked the barrier between the plant's "clean" and "dirty"—or potentially radioactive—zones.

Taking the polished marble stairs to the second floor, past stained glass panels depicting life-sized modernist figures in brilliant shades of yellow, scarlet, and cobalt blue, Toptunov and the others arrived at the double doors of the men's sanitary lock. Inside, a narrow bench, painted with the instruction "Take off your shoes!," blocked his path. He sat, removed his footwear, swung his legs over the bench, and walked to the changing room in his socks. He hung his clothes in a narrow steel locker and passed through the door into the "dirty" room wearing just his underpants. Once this door closed behind him, the only way back into the "clean" room was through a radiation monitoring device equipped with sensors to detect alpha- and beta-particle contamination. Toptunov put on freshly laundered white cotton overalls, a white cotton cap like those worn in an operating theater, to protect his hair, and white canvas boots.

The Chernobyl plant was constructed with a utilitarian disregard for high-minded notions of architecture: its form followed function in the most economical ways the station's designers could conceive. The turbine hall housed the station's eight colossal steam turbines in a single row, end to end in a cavernous shed thirty meters high and roofed with corrugated steel. The plant's four reactors were strung out in a line along the length of the turbine hall: giant concrete boxes arranged in the order they had been constructed, from one to four. The first two reactors were housed in separate structures, but—to save time and money—Reactor Number Three and Reactor Number Four had been built together, back-to-back under the same roof, where they shared ventilation and auxiliary systems. Between the turbine hall and the reactors was the spine of the station, which housed the deaerator corridor. Uninterrupted by a single door or dogleg, this seemingly endless hallway ran parallel to the turbine hall, all the way from the main administrative block at one end of the plant to the western end of Reactor Number Four at the other, not quite a kilometer long in total.

The deaerator corridor provided the plant staff with access to every part of the station, including each of the four unit control rooms—one dedicated to each reactor—that lay along it. It was also a key orientation point inside a complex that, with its dark spaces and tang of machine oil, often resembled the dark, roaring voids inside a gargantuan submarine more than an ordinary building. Much of it was navigated along catwalks and clanging steel stairwells, lined with hundreds of kilometers of dense pipework and accessed through heavy steel doors. The layout could be bewildering, and workers found their bearings inside the plant using alphanumeric coordinates, lettered in Russian from *A* to *Ya* along one axis, and along the other by numbers, from 1 to 68. Instead of conventional floors, the levels of the plant were subdivided vertically by "mark" numbers, showing distance in meters from the ground, and painted on the walls of hallways and landings in large red figures. Climbing from mark -5 in the basement to the station's highest point at mark +74.5—the roof of the reactor block—the structure stood more than twenty stories high.

To reach Control Room Number Four, Toptunov, Stolyarchuk, Akimov, and the other men on the night shift had to ascend to mark +10—ten meters above ground level—and then travel almost the entire length of the deaerator corridor: a brisk ten-minute walk from one end of the station to the other. From there, the floor of the Unit Four central reactor hall was higher still: up several flights of stairs, or by elevator, from the control room, to mark +35, or more than ten stories aboveground. Here, accessible through a heavy, airtight door that could be sealed shut against radiation, lay the shining steel lid of Reactor Number Four.

A little more than five hundred meters away from Control Room Number Four, on the other side of the access road that ran beside the plant, the men of the third watch of Paramilitary Fire Brigade Number Two lingered outside their fire station. Their cigarettes glowed in the mellow darkness. It had been a quiet day. As midnight approached, the fourteen firefighters were more than halfway through their twenty-four-hour shift and taking turns to sleep in the ready room. They were not due to be relieved until eight the next morning. The brigade was one of two based close to the Chernobyl plant. Pripyat also had its own paramilitary firefighting crew—Brigade Number Six—who lived beside their station, in a large two-story building near the end of Lesi Ukrainki Street. They had already been called out earlier in the evening, to a blaze reported on the roof of the city bus station. But that had taken the civilian firefighters less than five minutes to put out, and they were soon back at home.

Brigade Number Two was dedicated to protecting the Chernobyl nuclear power plant, but there was never much action. The construction work conducted by shifts of several thousand men, day and night, across the complex, would sometimes cause small fires: welders' sparks might set light to a pile of trash, or a vat of hot bitumen could be knocked over. The fire station, with offices, a canteen, the ready room with a TV set, and the recreation room with a Ping-Pong table, was within easy reach of the plant and the building sites. The red-and-white-striped ventilation stack between Reactors One and Two dominated the view through the large glass doors at the front of the station. Behind the doors sat four fire trucks: compact ZIL-130s and the bigger, six-wheeled ZIL-131 powder trucks that could carry 2,400 liters of water and 150 liters of foam for smothering electrical fires. At the back of the building was a separate garage holding the special equipment, including a Ural mobile fire tanker capable of pumping 40 liters of water a second.

The third watch lacked discipline. It was packed with obstinate old hands who disliked following orders. Many of them were from peasant families, close relatives raised in the surrounding countryside. Among them were two Shavrey brothers, Ivan and Leonid, from just over the border in Belarus, and fifty-year-old "Grandpa" Grigori Khmel, who had two sons who were also firefighters—all of them born in a small village ten kilometers away from where the station now stood. The watch commander, Lieutenant Vladimir Pravik, was just twenty-three, a college graduate who dabbled in photography, drawing, and poetry and was a dedicated member of the Komsomol. His wife taught music at a kindergarten in Pripyat and had given birth to their first child, a daughter, just a few weeks before, at the end of March.

That morning, Pravik had applied to take the day off, offering to switch shifts with his friend Piotr Khmel, chief of the first watch, with whom he'd graduated from the fire safety institute in Cherkasy. Piotr, the younger son of Grandpa Khmel, was a burly, good-natured twenty-four-year-old lieutenant. Khmel had already covered for Pravik after his daughter's birth, and that morning he had been there again with his uniform on, ready to go. But the deputy station commander wouldn't approve the change.

"Major Telyatnikov will get back from his holiday on Monday," he told Pravik. "He'll give you permission."

Khmel went home to get some rest and prepare for work on Saturday, and Pravik once again took command of the troublesome third watch.

Back in Pripyat, Piotr decided to take advantage of his unexpected night off and joined three fellow officers from the fire station for dinner at the restaurant in the city's new shopping center. Despite General Secretary Gorbachev's ongoing Unionwide campaign against alcohol, they had no trouble getting a bottle of vodka. Later, they moved on to *Sovietskoe shampanskoye*—the cheap, mass-produced "people's champagne" originally developed on the orders of Stalin. At around eleven, they went up to Khmel's one-room apartment, in the old low-rise blocks just across the street from the Pripyat fire station. They invited a few girls over to keep the party going. It was well after midnight when Khmel's guests departed, leaving behind a little chocolate and a half-finished bottle of Soviet champagne on the kitchen table.

Tired and drunk, Khmel took a shower and prepared for bed.

Over at the power station, Senior Mechanical Engineer Alexander Yuvchenko was already at his post: a large, windowless space at mark +12.5 in a mezzanine between the reactor halls of Units Three and Four. He had a desk there for his papers and a metal cage stacked with equipment and supplies. Although he hadn't slept in twenty-four hours, he was expecting a quiet night. Earlier in the day, the reactor had been scheduled for a maintenance shutdown, following a series of long-overdue tests on the turbines. By the time he arrived at work, he understood that everything in Unit Four would be powered down. All he and the others on the graveyard shift would be doing was overseeing the cooling of the reactor: easy work.

But down in the control room, there had been a change of plans. The tests were running twelve hours late and were only now beginning in earnest. The impatience of the station's deputy chief engineer was rising. And there was mounting disagreement about how to respond to the troubling data coming in from Reactor Number Four.

4

Secrets of the Peaceful Atom

O n September 29, 1966, the Soviet Council of Ministers, in Moscow, issued a decree approving the construction of the first in a new generation of giant water and graphite nuclear reactors, which would become known by the acronym RBMK, the *reaktor bolshoy moschnosti kanalnyy*, or high-power channel-type reactor. Developed from the Ministry of Medium Machine Building's military workhorse, the plutonium-and-power-producing Ivan the Second, it was a direct descendant of the pioneering Atom Mirny-1 reactor, reimagined on an Olympian scale.

Twelve meters across and seven meters high, the core of the RBMK was a massive cylinder, larger than a two-story house, composed of more than 1,700 tonnes of moderating graphite blocks, and stacked into 2,488 separate columns, each drilled from top to bottom with a circular channel. These channels contained more than 1,600 heat-resistant zirconium-alloy pressure tubes, each of which held a pair of metal assemblies packed with sealed rods of fuel: 190 tonnes of enriched uranium dioxide, compressed into ceramic pellets roughly the diameter of a man's little finger. Once the reactor went critical and the uranium began to heat, releasing the energy of nuclear fission, the fuel assemblies were cooled by water pumped into the core from below. Under enormous pressure—sixty-nine atmospheres, or a thousand pounds per square inch—the water rose to 280 degrees centigrade and turned to a mixture of water and superheated steam, which was then piped out through the top of the reactor to giant separator drums. These directed the steam to turbines to generate electricity, while the remaining water returned to the beginning of the coolant loop to start its journey through the core once more.

The power of the reactor was regulated by 211 boron carbide–filled control rods, most around five meters long, which could be raised or lowered into the reactor core to increase or decrease the rate of the nuclear chain reaction—and thus the level of heat, and energy, it generated. To help protect the plant and its staff from the radiation seething within, the reactor core—the active zone—was surrounded by a massive annular tank filled with water, contained within a steel jacket and surrounded by a giant box packed with sand. All this was further encased in a concrete vault more than eight stories high and crowned with a diadem of metal boxes filled with a mixture of iron shot and the neutron-retarding mineral serpentinite. A biological shield, a shallow stainless steel drum seventeen meters across and three meters deep, known as Structure E—or, more affectionately, Elena—sat on top of the vault like a giant lid. Filled with pebbles and rocks of serpentinite and nitrogen gas, Elena weighed two thousand tonnes—as much as six fully laden jumbo jets—and was held in place almost entirely by gravity. Pierced by ducts providing passage for the fuel channels, and surmounted by hundreds of narrow pipes carrying away steam and water, Elena was concealed beneath two thousand removable steel-clad concrete blocks, which capped the vertical fuel channels and formed the floor of the reactor hall. This tessellated metal circle, the visible face of the reactor during day-to-day operation, was known by the plant staff as the *pyatachok*, or five-kopek piece.

The RBMK was a triumph of Soviet gigantomania, a testament to its creators' unrelenting pursuit of economies of scale: twenty times the size of Western reactors by volume, it was capable of producing 3,200 megawatts of thermal energy, or 1,000 megawatts of electricity, enough to keep the lights on for half the population of Kiev. Soviet scientists proclaimed it the "national" reactor of the USSR—not only technologically unique but the largest in the world. Anatoly Aleksandrov, the bald-headed director of the Kurchatov Institute of Atomic Energy, had personally taken credit for its design, which he filed as a classified invention with the Soviet patent office. In contrast to its principal Soviet competitor, the VVER—a complex piece of engineering derided by its detractors as "the American reactor" because of its similarity to the pressurized water reactors favored in the Unites States—the parts of the RBMK could be made in existing factories and didn't require any specialized tooling. Its modular construction—hundreds of graphite blocks stacked into columns—meant that it could easily be put together on-site and scaled up to become even more powerful if necessary.

Aleksandrov also saved money by dispensing with the containment building, the thick concrete dome built around almost every reactor in the West, intended to prevent radioactive contamination escaping from the plant in the event of a serious accident—but which, because the RBMK was so enormous, would have doubled the cost of building each unit. The subdivision of the reactor into 1,600 pressure tubes was adopted as a less costly solution designed to contain each pair of fuel assemblies in their own thin metal jacket—a feat of Byzantine plumbing that, its inventors argued, made a serious incident exceedingly unlikely. They also devised an accident suppression system that could cope safely with a simultaneous rupture in one or two of these tubes by safely directing the resulting release of high-pressure radioactive steam downward, through a series of valves, and into giant water-filled tanks in the basement beneath the reactor, where it would be cooled and securely contained.

A break in the pressure tubes was one of the worst accidents the designers had ever prepared to encounter with the RBMK—a so-called *maksimal'naya proektnaya avariya*, or maximum design-basis accident. This designation also encompassed other potential calamities, including earthquakes, a plane crashing into the plant—or a complete rupture in one of the large-diameter water pipes in the reactor coolant circuit, which would deprive the core of water and trigger a meltdown. To guard against this last eventuality, the designers devised an emergency cooling system powered by compressed nitrogen gas, and reactor operators at every level of the industry were drilled to maintain a continuous supply of water to the core at all costs.

Worse accidents were theoretically possible, of course: engineering calculations suggested that if more than 2—and as few as 3 or 4—of the 1,600 pressure tubes ruptured simultaneously, the sudden release of high-pressure steam would be enough to lift all two thousand tonnes of Elena and the *pyatachok* off their mounts, severing every one of the remaining steam lines and pressure tubes and resulting in a devastating explosion. Yet the designers saw no need to prepare for such a calamity, which they regarded as outside the realm of reasonable probability. Nonetheless, they granted the scenario its own designation: the beyond design-basis accident.

The Ministry of Medium Machine Building ordered the first-draft plans of the RBMK to be drawn up by a heavy machinery plant in Leningrad, which also built tanks and tractors. But when they received the blueprints, Sredmash dismissed them as technically unsound. One scientist from the Kurchatov Institute warned that the design was too dangerous to be put into civilian operation. Another recognized that the hazards of the positive void coefficient made the new reactor inherently prone to explosion, and—although his superiors attempted to have him dismissed from the institute because of his dissent—he began a letter-writing campaign that eventually reached the Central Committee of the Communist Party and the Soviet Council of Ministers.

But by then, the government—adhering to the rigid needs of central economic planning—had already issued its decree that four of the new behemoth reactors be built. So the designers of NIKIET scrambled to perform a drastic overhaul on the RBMK blueprints, transforming it from a schizoid contraption that could manufacture both plutonium and electricity into a tame generator of power for the civilian grid. Implementing these modifications was difficult and complex work and took far longer than expected: primitive Soviet computing technology made calculating the expected performance of the reactor laborious and produced unreliable results. It was not until 1968 that the new reactor design, now called the RBMK-1000, was complete. So, to save time, Sredmash decided to skip the prototype stage entirely: the quickest way to find out how the new reactors would work in industrial electricity generation would be to put them directly into mass production.

Construction began on the first RBMK reactor in the Soviet Union at a Sredmash installation on the Gulf of Finland, outside Leningrad, in 1970. In the meantime, a pair of technical and economic institutes in Kiev considered possible locations for the first nuclear power station to be built in Ukraine, and quickly narrowed it down to two. When the first proposed site was earmarked for a fossil fuel plant, the Ukrainian Council of

Ministers decreed that the republic's new 2,000-megawatt atomic energy station would be constructed at the other: on a large patch of sandy riverbank near the village of Kopachi, in the Kiev region, fourteen kilometers from the town of Chernobyl.

The first RBMK unit at the Leningrad station started up on December 21, 1973, just one day before *energetiki* across the USSR celebrated their own national holiday, the Day of the Power Engineer. The proud fathers of the RBMK-1000, Anatoly Aleksandrov of the Kurchatov Institute and Nikolai Dollezhal from NIKIET, were both there to witness it come to life. By then, building was also under way on a second unit in Leningrad, and construction workers had broken ground at RBMK stations in Chernobyl and Kursk. But the initial Leningrad reactor had not even reached full power when it became clear that the designers' determination to rush their brainchild from drawing board to full-scale production had come at a steep cost. Serious design faults dogged the RBMK from the outset. Many became apparent immediately; others would take much longer to come to light.

The first problem arose from the positive void coefficient, the drawback that made Soviet graphite-water reactors susceptible to runaway chain reactions in the event of a loss of coolant, and that, in the RBMK, had been exacerbated by attempts to make the reactor cheaper to run. To make it more competitive with fossil energy power stations, the RBMK had been deliberately designed to maximize the electricity output of the uranium fuel it burned up. But it was only when they started up Leningrad Unit One that the designers discovered that the effects of the positive void coefficient grew worse as more of the fuel was burned; the longer it was in operation, the harder the reactor became to control. By the time it reached the end of each of its three-year operational cycles and was shut down for preventive maintenance, the RBMK would be at its most unpredictable. The designers made modifications, but instabilities remained. Yet neither Aleksandrov nor Dollezhal sought to explore these problems further, nor even to fully understand them—and provided no safety analysis of the void coefficient in the manuals accompanying each reactor. The results of the experiments in Leningrad made it obvious that there were important differences between the way they had predicted the reactor would perform in theory and how it worked in practice. But the designers decided not to examine these results too closely. Even as it went into full-scale commercial operation, nobody knew how the RBMK would behave during a major accident.

A second failing of the reactor resulted from its colossal size. The RBMK was so large that reactivity in one area of the core often had only a loose relationship to that in another. The operators had to control it as if it were not a single unit but several separate reactors in one. One specialist compared it to a huge apartment building, where a family in one flat might be celebrating a raucous wedding, while next door another was observing a funeral wake. Isolated hot spots of reactivity might build deep inside the core, where they could prove hard to detect. This problem was especially pronounced during start-up and shutdown, when the reactor was operating at low power—and the systems designed to detect reactivity within the core proved unreliable. During these crucial periods, the engineers at their desks in the control room became almost totally blind to what was happening inside the active zone. Instead of reading their instruments, they were forced to estimate the levels of activity in the core, using "experience and intuition." This made start-up and shutdown the most demanding and treacherous stages of RBMK operation.

A third fault lay in the heart of the reactor's emergency protection system, the last line of defense against an accident. If the operators faced a situation calling for an emergency shutdown—a major coolant leak or a reactor runaway—they could press the "scram" button, activating the ultimate stage of the unit's five-level rapid power reduction system, known in Russian as AZ-5. Pushing this button would drive a special bank of twenty-four neutron-absorbing boron carbide control rods—as well as every one of the remaining 187 manual or automatic control rods that remained withdrawn at the time—simultaneously into the core, quenching the chain reaction throughout the reactor. Yet the AZ-5 mechanism was not designed to bring about an *abrupt* emergency stop. Dollezhal and the technicians of NIKIET believed that suddenly cutting off the electricity generated by the reactor would be disruptive to the operation of the Soviet grid. And they thought that such an immediate shutdown would be necessary only in the extremely unlikely event of a total loss of external power to the plant. So they designed the AZ-5 system to only gradually reduce the reactor's power to zero. Rather than dedicated emergency motors, the system was driven by the same electric servos that moved the manual reactor control rods, used by the operators to manage reactor power during normal operation. Starting from their fully withdrawn position above the reactor, it would take between eighteen and twenty-one seconds for the AZ-5 rods to descend completely into the core; the designers hoped that the rods' slow speed would be compensated for by their great number. But eighteen seconds is a long time in neutron physics—and an eternity in a nuclear reactor with a high positive void coefficient.

Adding to this disquieting list of major design defects, the construction of the reactors also suffered from the shoddy workmanship that plagued Soviet industry. The full start-up of Leningrad's Reactor Number One was delayed for almost a year after fuel assemblies became stuck in their channels and had to be returned to Moscow for repeated testing. The valves and flow meters in other RBMKs, used to regulate the crucial supply of water to each of the more than 1,600 uranium-filled channels, proved so unreliable that the operators in the control room often had no idea to what extent the reactors were being cooled, or if they were being cooled at all. Accidents were inevitable.

On the night of November 30, 1975, just over a year after it had first reached full operating capacity, Unit One of the Leningrad nuclear power plant was being brought back online after scheduled maintenance when it began to run out of control. The AZ-5 emergency protection system was tripped, but before the chain reaction could be stopped, a partial meltdown occurred, destroying or damaging thirty-two fuel assemblies and releasing radiation into the atmosphere over the Gulf of Finland. It was the first major accident involving an RBMK reactor, and the Ministry of Medium Machine Building set up a commission to investigate what had gone wrong. Afterward, the official line was that a manufacturing defect had led to the destruction of a single fuel channel. But the commission knew otherwise: the accident was the result of the design faults inherent in the reactor and caused by an uncontrollable increase in the steam void coefficient.

Sredmash suppressed the commission's findings and covered up the accident. The operators of other RBMK plants were never informed of its true causes. Nevertheless, the commission made several important recommendations, to be applied to all RBMK-1000 reactors: develop new safety regulations to protect them in the event of coolant loss; analyze what would happen in the event of a sharp rise in steam in the core; and devise a faster-acting emergency protection system. Despite their apparent urgency, the reactor designers failed to act on a single one of these directives, and Moscow promptly ordered more of the reactors to be built. The day after the Leningrad meltdown, the Soviet Union's Council of Ministers gave its final approval to construct a second pair of RBMK-1000 units in Chernobyl, expanding the station's projected output to an impressive 4,000 megawatts.

On August 1, 1977, more than seven years after Viktor Brukhanov had watched the first stake being driven into the snow-covered ground beside the Pripyat, and two years later than planned, Reactor Number One of the Chernobyl nuclear power plant at last went critical. The plant's young operators were overwhelmed with pride as they prepared to bring the Ukrainian republic's first nuclear station online. They remained at their posts day and night as the first fuel assemblies were loaded and the reactor was slowly brought up to full power and, finally, connected to the transformers. At 8:10 p.m. on September 27, the scientists and designers from the Kurchatov Institute and NIKIET joined the plant specialists to celebrate as Ukraine's first nuclear electricity ran into the 110- and 330-kilovolt lines and out to the Soviet grid. Together they sang the couplet with which *atomshchiki* across the Union hymned the success of the Soviet Reactor: *A poka, a poka tok dayut RBMK!* "For today, for today, current flows from the RBMK!"

But the Chernobyl operators soon discovered that the reactor on which they had lavished so much attention was an unforgiving mistress. The inherent instabilities of the RBMK made it so difficult to manage that the senior reactor control engineers' work proved not only mentally but also physically demanding. Making dozens of adjustments every minute, they were never off their feet and sweated like laborers digging a ditch. Rumors reached them that up in Leningrad, the Sredmash reactor engineers had doubled up on the control desk, "playing duets" to cope with the complexity of the task. The reactor operators worked the panel so hard that the switches governing the control rods quickly wore out and had to be replaced constantly. When one former nuclear submarine officer first took his seat at the desk in Chernobyl's Unit One, he was horrified by the colossal size of the reactor and how antiquated the instrumentation was.

"How can you possibly control this hulking piece of shit?" he asked. "And what is it doing in civilian use?"

At their first planned maintenance shutdown, the Chernobyl operators found that the serpentine plumbing of the reactor was riddled with faults: the water-steam coolant pipes were corroded, the zirconium-steel joints on the fuel channels had come loose, and the designers had failed to build any safety system to protect the reactor against a failure of its feed-water supply—eventually, the Chernobyl engineers had to design and fabricate their own. Meanwhile, in Moscow, the reactor designers continued to discover further troubling flaws in their creation.

In 1980 NIKIET completed a confidential study that listed nine major design failings and thermohydraulic instabilities which undermined the safety of the RBMK reactor. The report made it clear that accidents were not merely possible under rare and improbable conditions but also *likely* in the course of everyday operation. Yet they took no action to redesign the reactor or even to warn plant personnel of its potential hazards. Instead of engineering new safety systems, NIKIET simply revised the operating instructions for the RBMK-1000. After decades of accident-free operation of military reactors, the atomic chieftains of NIKIET and the Kurchatov Institute apparently believed that a well-written set of manuals would be enough to guarantee nuclear safety. The designers assumed that, so long as they followed the new instructions closely, human beings would act as promptly and unfailingly as any of the plant's electromechanical safety systems.

But the staff of Soviet nuclear power plants, faced with ever-increasing production targets and constantly malfunctioning or inadequate equipment, and answerable to a bewildering and dysfunctional bureaucracy, had long become accustomed to bending or ignoring the rules in order to get their work done. And the updated instructions they received from NIKIET were neither explicit nor explained. One of the new directives stipulated that a minimum number of control rods should henceforth be maintained within the core at all times—but NIKIET did not emphasize that this limit on the operational reactivity margin, or ORM, was a crucial safety precaution intended to prevent a major accident. Deprived of information about why such rules might be important, the operators went on with their work as usual, ignorant of the potentially catastrophic consequences of breaching them.

Meanwhile, every accident that did occur at a nuclear station in the Soviet Union continued to be regarded as a state secret, kept even from the specialists at the installations where they occurred.

Early on the evening of September 9, 1982, Nikolai Steinberg was sitting at the desk in his third-floor office between Chernobyl Units One and Two, overlooking the vent stack shared by the two reactors. Steinberg, a thirty-five-year-old with a short goatee and an easy charm, had worked in Chernobyl since 1971, arriving straight from the Moscow Power Engineering Institute as a graduate in nuclear thermal hydraulics and one of a new breed of bright-eyed *atomshchiki*. He had spent more than two years studying the RBMK at college before the first of the reactors had even been built, watched the first two units of the Chernobyl plant emerge from the ground, and was now chief of the turbine department for Units Three and Four. When Steinberg saw steam rising from the top of the vent stack, he knew it meant trouble: a broken pipe inside the reactor at least, and certainly a release of radiation. He picked up the phone.

But when he got through to the control room of Unit One to warn the operators there to shut down the reactor, the shift supervisor brushed him off. When Steinberg persisted, the supervisor hung up on him. The engineer gathered his staff and waited to be summoned for the emergency. But no call came. Almost six hours later, at midnight, he and his men got into their cars and drove home to Pripyat.

When he returned to work the following morning, Steinberg heard

that there had indeed been a problem in Unit One, but—despite his seniority and experience—could learn nothing further. Director Brukhanov
and the chief engineer of the plant initially insisted that whatever had
happened had caused no radioactive releases, and local KGB officers took
measures "to prevent the spread of panic-mongering, provocative rumors,
and other negative manifestations." In fact, radioactive contamination, carried on the wind and brought down by rain showers, had reached Pripyat
and spread as far as fourteen kilometers from the plant. It included iodine
131, fragments of uranium dioxide fuel, and hot particles containing zinc
65 and zirconium-niobium 95 consistent with partial destruction of the
reactor core. Levels of radiation in the village of Chistogalovka, five kilometers from the station, were hundreds of times higher than normal. But
a team from Soyuzatomenergo—the USSR's atomic energy authority—
contested these findings. Contaminated areas immediately around the
plant were simply sluiced with water and covered with soil and leaves. In
Pripyat, decontamination trucks dispensed foam on the streets, and Lenina Prospekt was discreetly spread with a fresh layer of asphalt.

A subsequent inquiry revealed that there had been a partial meltdown
in Unit One. When the reactor had been brought back online after maintenance, one of the temperamental cooling valves had remained closed;
the uranium fuel in the channel overheated, and it ruptured. No one
was killed, but the damage took eight months to repair. Workers carried
blocks of reactor graphite away in buckets and were exposed to significant
amounts of radiation. The chief engineer took the blame and was de-
moted and reassigned to a job in Bulgaria. The incident was classified top
secret, and those directly involved were forced to sign gag orders by the
KGB. Nikolai Steinberg would wait years before learning the truth about
what had happened.

In the years that followed, there would be even more serious accidents
at nuclear plants elsewhere in the Soviet Union, and all of them would
be covered up. In October 1982 a generator exploded at Reactor Number
One of the Metsamor plant in Armenia; the turbine hall burned down,
and an emergency team had to be airlifted in from the Kola Peninsula,
more than three thousand kilometers away in the Arctic Circle, to help
save the core. Less than three years later, during the start-up of the first re-

actor at the Balakovo plant in Russia, a relief valve burst, and superheated steam at 300 degrees centigrade escaped into the annular compartments surrounding the reactor well. Fourteen men were boiled alive. Both incidents were concealed, and word reached the operators at other stations only through the *atomshchiki* rumor mill and hints in *Pravda*.

Yet the most dangerous suppression of all originated once more within NIKIET, the central nuclear design bureau in Moscow where the RBMK-1000 had been conceived. In 1983, on top of the myriad drawbacks of the reactor that had emerged since it went into operation, the reactor designers learned of one more: a curious design fault in the rods of the AZ-5 emergency protection system. The first conclusive evidence appeared at the end of the year, during the physical start-up of two of the newest RBMK reactors to be added to the Soviet grid: Unit One of the Ignalina nuclear power plant in Lithuania, and Unit Four in Chernobyl, the most advanced of the RBMK-1000 line.

While conducting tests before the two reactors could be brought into normal operation, the start-up teams of nuclear engineers in Ignalina and Chernobyl noticed a small but disturbing glitch. When they used the AZ-5 scram button to shut down the reactor, the control rods began their descent into the core, but instead of completing a smooth shutdown, the rods initially had the opposite effect: for a brief moment, reactor power rose instead of falling. The specialists discovered that the severity of this "positive scram" effect depended on the conditions inside the reactor at the moment the shutdown began—particularly on the ORM, the measurement indicating how many of the 211 control rods were withdrawn from the core. If more than 30 of these rods remained inserted when the scram began, the AZ-5 mechanism worked as intended, and the reactor would shut down quickly and safely. But when the total number of inserted rods fell below 30, the behavior of the reactor at shutdown became increasingly unpredictable, and the AZ-5 system struggled to do its job. With only 15 rods in, the technicians found that initial dampening of fission inside the reactor was marginal; it took six seconds before reactivity began to fall. And under some circumstances—7 rods or fewer—pressing the AZ-5 button might not shut down the reactor at all, but instead trigger a runaway chain reaction. If this happened, the increase in reactor power following an AZ-5 trip might be so great that it would no longer be possible to halt the reaction before the entire reactor was destroyed.

The source of the positive scram effect lay in the design of the control rods themselves, an unintended consequence of NIKIET's desire to "save neutrons" and make the reactor more economical to run. Like all the manual control rods used to manage the reactor during normal operation, the AZ-5 emergency rods contained boron carbide, a neutron poison that gobbles slow neutrons to reduce the chain reaction. But even when fully withdrawn from their water-filled control channels, the tips of the rods were designed to remain at the ready, just inside the active zone of the reactor—where, if they contained boron carbide, they would have a poisoning effect, creating a slight but constant drag on power output. To stop this from happening, the rods were tipped with short lengths of graphite, the neutron moderator that facilitates fission. When a scram shutdown began and the AZ-5 rods began their descent into the control channels, the graphite displaced neutron-absorbing water—with the effect of initially increasing the reactivity of the core. Only when the longer boron-filled part of the rod followed the graphite tip through the channel did it begin dampening reactivity.

It was an absurd and chilling inversion in the role of a safety device, as if the pedals of a car had been wired in reverse, so that hitting the brakes made it accelerate instead of slowing down. With further experiments, the technicians confirmed that the positive scram effect of the rod tips could create local criticality in the lower part of the giant RBMK core—especially if the operators tripped the AZ-5 system when the reactor was running at less than half power.

Alarmed, the director of the nuclear reactor department at the Kurchatov Institute wrote to NIKIET detailing the anomalies in the AZ-5 system and the need to examine them more closely. "It seems likely," he warned, "that a more thorough analysis will reveal other dangerous situations." Nikolai Dollezhal, the chief designer at NIKIET, replied with vaporous assurances: they were aware of the problem, and measures were being taken. But they weren't. Although some partial modifications to the AZ-5 mechanism were approved, even these proved costly and inconvenient, and were executed piecemeal, just one RBMK reactor at a time.

Gradually, Chernobyl's Units One, Two, and Three were approved for the alterations. But Unit Four, already so close to completion, would have to wait for its first scheduled maintenance shutdown, in April 1986.

In the meantime, NIKIET circulated a notification of the positive scram effect to the senior managers of all RBMK plants. But, swept along in a blizzard of bureaucracy, tangled in secrecy, the news never reached the reactor operators. Nonetheless, as far as Anatoly Aleksandrov and the other nuclear chiefs were concerned, the redoubtable RBMK-1000—the Soviet national reactor—had been troubled by nothing more than temporary setbacks. By the time Viktor Brukhanov put his final signature on the paperwork to acknowledge completion of the fourth reactor of the V. I. Lenin nuclear power plant on the last day of 1983, the world still spoke of only one nuclear accident. And that humiliation belonged entirely to the United States.

Early in the morning of March 28, 1979, a handful of water-purifying resin beads smaller than mustard seeds blocked a valve in the secondary coolant loop of Reactor Two of the Three Mile Island nuclear power station near Harrisburg, Pennsylvania. Over the next twenty-four hours, the ensuing cascade of minor equipment malfunctions and human error led to a serious loss of coolant, which partially uncovered the core. The reactor began to melt down and contaminated the containment building with thousands of gallons of radioactive water. The staff had no choice but to vent radioactive gases directly into the atmosphere. Although no one was harmed by the released radiation—contained entirely in a cloud of short-lived isotopes of inert gases that drifted out over the Atlantic Ocean—news of the accident caused widespread panic. The roads of the tristate area were clogged as 135,000 people fled their homes in Pennsylvania. President Jimmy Carter—who had served as a nuclear engineer in the US Navy and knew a disaster when he saw one—attended the scene. The international antinuclear movement, which had slowly been gaining momentum over the previous decade, could not have asked for a more fearsome totem of a dangerous technology slipping its leash. In the United States, the development of the nuclear power industry, already dogged by rising construction costs and growing public apprehension, halted almost overnight.

As bad as it made the United States look, news of Three Mile Island was censored inside the USSR, for fear it could tarnish the ostensibly spotless record of the peaceful atom. Publicly, Soviet officials attributed the accident to the failings of capitalism. Academician Valery Legasov, Aleksandrov's immediate deputy at the Kurchatov Institute, published an article insisting that the events in Three Mile Island were irrelevant to the USSR's nuclear industry because its operators were far better trained and its safety standards higher than those in the United States. Privately, Soviet physicists began to analyze the likelihood of severe accidents at atomic power stations and revised their existing nuclear safety regulations. But neither Sredmash nor NIKIET made any attempt to bring the RBMK reactor into line with these new rules.

In January 1986 the new issue of *Soviet Life*—a glossy, English-language magazine resembling the stalwart American title but published by the Soviet embassy in the United States—featured the Chernobyl nuclear power plant as the centerpiece of a ten-page report on the wonders of nuclear energy. The special section included interviews with the residents of Pripyat, the city "Born of the Atom"; color photographs of the plant; and pictures of smiling station staff. Legasov coauthored another essay in which he boasted, "In the thirty years since the first Soviet nuclear power plant opened, there has not been a single instance when plant personnel or nearby residents have been seriously threatened; not a single disruption in normal operation occurred that would have resulted in the contamination of the air, water, or soil."

In a separate interview, Vitali Sklyarov, the Ukrainian minister of energy and electrification, assured readers that the odds of a meltdown at the plant were "one in 10,000 years."

5

Friday, April 25, 11:55 p.m., Unit Control Room Number Four

Beneath the sickly fluorescent strip lights of Control Room Number Four, a rancid haze of cigarette smoke hung in the air. The midnight shift had only just arrived, but the mood was growing tense. The turbine generator test scheduled to finish that afternoon had not yet begun. The station's deputy chief engineer for operations, Anatoly Dyatlov, was now entering his second day without sleep. He was exhausted and unhappy.

The test was intended to check a key safety system designed to protect Reactor Number Four during an electrical blackout. A total loss of external power providing electricity to the station from the grid was something for which the RBMK designers had planned. It was one scenario of the so-called design-basis accident—in which the plant suddenly lost power, and the giant coolant pumps that kept water circulating through the reactor core came to a rumbling halt. The station had emergency diesel generators, but it would take between forty seconds and three minutes for these to start up and get the pumps going again. This was a dangerous gap— long enough for a core meltdown to begin.

So the reactor designers devised what they called a "rundown unit": a mechanism to use the momentum of the unit's turbines to drive the pumps for those crucial seconds. The rundown unit was a crucial safety feature of Reactor Number Four and was supposed to have been tested before it was approved for use in December 1983. But Director Brukhanov had granted approval to skip the test, to meet his end-of-year deadline. And although similar trials had been attempted since, they had failed each time. By the beginning of 1986, the test was more than two years overdue, but the reactor's first scheduled maintenance shutdown offered an opportunity to conduct the trial in real-world conditions. At two o'clock on Friday afternoon, with new modifications made to one of the unit's two huge turbine generators, Turbine Number Eight, it had—at last—been ready to begin.

But then the central dispatcher of the Kiev electrical grid intervened. Factories and enterprises throughout Ukraine were still in a frenzy of last-minute activity to meet their production quotas and win bonuses before the May Day holiday, and they needed every kilowatt of electricity the Chernobyl nuclear power plant could supply. The dispatcher said that Unit Four could not go off-line to begin the test until after the peak load had passed—9:00 p.m. at the earliest.

By midnight on Friday, the team of electrical engineers waiting to monitor the test were threatening to cancel their contract and return to Donetsk if it didn't start soon. In Control Room Number Four, the staff who had been briefed on the test program had reached the end of their shift and were preparing to go home. And the physicist from the plant's Nuclear Safety Department—expected to be on hand to help the reactor operator through his part in the test—had been told the experiment was already complete. He hadn't shown up at all. Stepping up to the instruments on the senior reactor control engineer's desk, twenty-five-year-old Leonid Toptunov, just two months into his new job, prepared to pilot the capricious reactor through a shutdown for the first time in his life.

But Deputy Chief Engineer Dyatlov was determined to press on. If the test wasn't completed that night, it would have to wait at least another year. And Dyatlov didn't like to wait. At fifty-five, Anatoly Dyatlov looked every inch the austere Soviet technocrat: tall and gaunt, with sharp cheekbones, sparse, gray hair swept straight back from his high forehead, and narrow Siberian eyes that even in photographs seemed to glint with malice. A veteran physicist who had come to Chernobyl after fourteen years working on naval reactors in the Soviet Far East, Dyatlov was one of the three most senior managers at the station with nuclear expertise. He oversaw the operation of both Units Three and Four, with responsibility for hiring and training personnel.

The son of a peasant who had lit the buoys each night on the Yenisei

River near the penal settlements of Krasnoyarsk, Dyatlov had run away from home at fourteen. He went to vocational school and worked as an electrician before winning a place at MEPhI, the Moscow Engineering and Physics Institute. He graduated in 1959 and was posted to a stronghold of the Soviet military-industrial complex: the Lenin Komsomol shipyard in the remote city of Komsomolsk-on-Amur. As the head of the classified Laboratory 23, Dyatlov managed a team of twenty men who installed the reactors of the Yankee- and Victor-class nuclear-powered submarines before they put to sea.

By the time he arrived in Chernobyl in 1973, he had overseen the assembling, testing, and commissioning of more than forty VM reactor cores. These small marine reactors, versions of the VVER, were nothing like the massive graphite-moderated behemoths being constructed in Chernobyl. But Dyatlov, a fanatical specialist, threw himself into learning everything he could about the RBMK-1000. He was there for the commissioning of each one of the four new units in Chernobyl and now worked ten-hour days, six and sometimes seven days a week. He walked to the plant every day from his home in Pripyat—he found that walking helped to keep dark thoughts at bay—and jogged to keep fit. He was rarely in his office but prowled the corridors and gangways of the plant day and night, inspecting the equipment, checking for leaks and errant vibrations, and keeping tabs on his staff. A stickler for detail, Dyatlov knew his job and prided himself on his knowledge of the reactor and its systems—of mathematics, physics, mechanics, thermodynamics, and electrical engineering.

But the manner Dyatlov had developed while running a top-secret military laboratory did not prepare him well for managing the operators and engineers of a civilian nuclear power station. He had no tolerance for shirkers or those who didn't follow his orders to the letter. Even those colleagues he brought with him from Komsomolsk found him hard to work with. He could be high-handed and peremptory, peppering his speech with curses and Soviet navy slang, muttering to himself about the inexperienced technicians he dismissed as *chertov karas*—fucking goldfish. He demanded that any fault he discovered be fixed immediately and carried a notebook in which he recorded the names of those who failed to meet his standards.

The deputy chief engineer believed he was always right and held stubbornly to his own convictions on technical matters, even when overruled from above. And Dyatlov's long experience in the shipyard, and of the chimerical construction targets imposed in Chernobyl, had taught him that the categorical dictates of Soviet bureaucracy and the gray netherworld of Soviet reality were very different things.

Dyatlov had fulfilled every autodidactic expectation of Soviet Man, dedicating himself to his work by day and steeping himself in culture by night; he loved poetry and knew by heart all eight chapters of Pushkin's epic *Eugene Onegin*. Away from work, he could be good company, though he had few close friends. Only long afterward would his secret emerge: before arriving in Chernobyl, Dyatlov had been involved in a reactor accident in Laboratory 23. There was an explosion, and Dyatlov was exposed to 100 rem, a huge dose of radiation. The accident, inevitably, was covered up. Later, one of his two young sons developed leukemia. There could be no certainty that the two events were linked. But the boy was nine when he died, and Dyatlov buried him there, beside the river in Komsomolsk.

Although the specialists who labored under Dyatlov at the Chernobyl plant may have disliked the way he treated them, many admired him, and few doubted his expertise. Eager to learn, they believed that he knew everything there was to know about reactors. And, by crushing dissent and conjuring an air of infallibility, Dyatlov—like the Soviet state itself—expected his underlings to carry out his commands with robotic acquiescence, regardless of their better judgment.

Yet the deputy chief engineer admitted a peculiar reservation about the reactor on which they all worked. For all his hours spent poring over the latest technical revisions and regulations, for all his mastery of thermodynamics and physics, Dyatlov said that there remained something unfathomable about the RBMK-1000: a nuclear enigma even he could never fully understand.

The control room of Unit Four was a large, windowless box, around twenty meters wide and ten meters deep, with a polished stone floor and a low suspended ceiling punctuated by the recessed fluorescent lights and ventilation ducts. Usually it was manned by a team of just four. At the back, the shift foreman had his own desk, from where he could observe the three operators who ran the unit, stationed at three long, gray, steel control panels arranged in a broken arc across the width of the room. On the left sat the senior reactor control engineer, known by the Russian acronym SIUR. On the right was the senior turbine control engineer. And in the center, linking the activity of the other two, the senior unit control engineer, who maintained the water supply that kept it all going—hundreds of thousands of cubic meters of it, flowing around the reactor's primary loop: from the pumps, through the reactor, to the steam separators, out to the turbines, and back again. The three men's control panels were festooned with the hundreds of switches, buttons, gauges, lamps, and annunciator alarms needed to manage the primary processes of spinning electricity from nuclear fission.

Forming a wall in front of the desks was a floor-to-ceiling bank of instrument panels showing the status of all three of the systems on illuminated dials, closed-circuit TV screens, and trembling pen-trace drums, which slowly logged data across scrolling bands of paper. Concealed behind the panel, and in anterooms to the left and right, were thousands of meters of cable leading away into darkness and banks of computer cabinets filled with glowing valves and clicking relays: the complex but antiquated technology that linked the control panels to the reactor itself.

As young Leonid Toptunov took over at the senior reactor control engineer's desk, he faced two giant backlit displays reaching almost to the ceiling and showing the operating conditions within Reactor Number Four. One displayed the status of every one of its 1,659 uranium-filled fuel channels; the other was formed of 211 glowing dials arranged in a circle three meters across. These were the Selsyn monitors, which indicated the position of the boron carbide control rods that could be raised or lowered into the reactor to moderate its chain reaction. Beneath Toptunov's hands lay the panel of switches with which he could select groups of rods, and the joystick that moved them in and out of the core. Nearby, a reactimeter showed the thermal output of the reactor in megawatts, displayed in glowing digital figures. Just behind him stood Shift Foreman Alexander Akimov, who would be responsible for supervising the test under the direction of Deputy Chief Engineer Dyatlov. In the strict technical hierarchy of the power plant, Akimov, an experienced reactor control engineer, was the senior member of the operational staff in the room. Dyatlov's role was administrative: no matter how deep his nuclear expertise, he could no more take the controls of the reactor engineer's desk than an airline executive could step onto the flight deck of one of his passenger jets and fly the plane himself.

Akimov, a gangling thirty-two-year-old with thick glasses, a receding hairline, and a small mustache, was a committed Communist and one of the most knowledgeable technicians at the plant. He and his wife, Luba, were the parents of two young boys, and he spent his spare time reading historical biographies or hunting hare and duck with his Winchester rifle on the Pripyat marshes. Akimov was clever, competent, and well liked, but his colleagues agreed that he was easily pushed around by those above him.

Control Room Number Four had now grown busy. In addition to Toptunov and the other two operators manning the turbine and pump panels, members of the previous shift lingered at their posts, along with others who had come to watch. In an adjacent room, the turbine specialists from Donetsk were on hand to monitor the rundown of Turbine Generator Number Eight. Dyatlov paced the floor.

As soon as the Kiev grid dispatcher gave permission, the operators had resumed the reactor's long, controlled power descent and now held it steady at 720 megawatts—just above the minimum level required to perform the test. But Dyatlov, perhaps assuming that a lower power level would be safer, was adamant that it be conducted at a level of 200 megawatts. Akimov, a copy of the test protocol in his hands, disagreed— vehemently enough for his objections to be noted by those standing nearby, who heard the two men arguing even over the constant hum of the turbines from the machine hall next door. At 200 megawatts, Akimov knew that the reactor could be dangerously unstable and even harder to manage than usual. And the program for the test stipulated it be conducted at not less than 700 megawatts. But Dyatlov insisted he knew better. Akimov, defeated, agreed reluctantly to give the order, and Toptunov began to decrease power further. Then, at twenty-eight minutes past midnight, the young engineer made a mistake.

When Toptunov had assumed responsibility for the reactor at midnight, the unit's computerized regulation system was set to local automatic control, which allowed him to manage regions of the core individually—but was usually switched off when operating the reactor at low power. So Toptunov began the process of transferring the system to global automatic—a form of nuclear autopilot that would help him keep the RBMK on a steady course as the men prepared for the start of the test. Before completing the change, he was supposed to choose a level at which the computer would maintain reactor power in the new operating mode. But, somehow, he skipped this step. The reactor proved as unforgiving as ever. Bereft of fresh instructions, the computer defaulted to the last set point it had been given: near zero.

Now Toptunov watched in dismay as the glowing gray figures on the reactimeter display began to tumble: 500 . . . 400 . . . 300 . . . 200 . . . 100 megawatts. The reactor was slipping away from him.

A series of alarms sounded: "Failure in measuring circuits." "Emergency power increase rate protection on." "Water flow decrease." Akimov saw what was happening. "Maintain power! Maintain power!" he shouted. But Toptunov could not stop the numbers from falling. Within two minutes, the power output of Unit Four had plunged to 30 megawatts—less than 1 percent of its thermal capacity. By 12:30 a.m., the reactimeter display was almost at zero. Yet for at least four more minutes Toptunov took no action. While he waited, neutron-scavenging xenon 135 gas began to build in the core, overwhelming what little reactivity remained. The reactor was being poisoned, plunging into what the operators called a "xenon well." At this point, with the reactor's power stalled at its minimum, and more xenon accumulating all the time, nuclear safety procedures made the operators' course quite clear: they should have aborted the test and shut down the reactor immediately.

But they did not.

Later, there would be conflicting accounts of exactly what happened next. Dyatlov himself would maintain that he was absent from the control room when the power first fell—although he could not always recall exactly why—and issued no instructions to the operators at the senior reactor control engineer's desk during the crucial minutes that followed.

The recollections of others present at the time would be quite different. According to Toptunov, Dyatlov not only witnessed the power fall but also—enraged—told him to withdraw more control rods from the reactor to increase power. Toptunov knew that to do so could certainly increase reactivity but would also leave the core in a dangerously unmanageable state. So Toptunov refused to obey Dyatlov's command.

"I'm not going to raise the power!" he said.

But now Dyatlov threatened the young operator: if he didn't follow orders, the deputy chief engineer would simply find another operator who would. The head of the previous shift, Yuri Tregub—who had stayed behind to watch the test—was well qualified to operate the board and right there at his elbow. And Toptunov knew that such insubordination could mean that his career at one of the most prestigious facilities in the Soviet nuclear industry—and his comfortable life in Pripyat—would be over as soon as it had started.

Meanwhile, the reactor continued to fill with poisonous xenon 135, falling deeper and more inextricably into the well of negative reactivity. Finally, six long minutes after the fall in power had begun, Toptunov, terrified of losing his job, gave in to Dyatlov's demands. The deputy chief engineer withdrew from the console, mopping sweat from his brow, and returned to his position in the center of the room.

But reviving a poisoned reactor is not easy. At first, Toptunov struggled to find the right balance of manual control rods to withdraw. Standing behind him, Tregub noticed the young technician raising them disproportionately from the third and fourth quadrants of the core. The power continued to languish close to zero. "Why are you pulling unevenly?" the veteran engineer asked. "You need to pull from here." Tregub began to prompt him on which rods to choose. As the buttons of the control panel clattered beneath Toptunov's right hand, his left tugged at the joystick. The atmosphere in the control room once more grew tense. Tregub stayed at Toptunov's side for twenty minutes, and together they managed to coax the reactor up to 200 megawatts. But then they could go no further. Xenon poisoning continued to gobble positive neutrons within the core—and they were running out of control rods to extract. More than a hundred of them already stood at their upper-limit stops.

By 1:00 a.m., Toptunov and Tregub had hauled the reactor back from the brink of an accidental shutdown. But to do so they had withdrawn the equivalent of 203 of the unit's 211 control rods from the reactor core. To pull such a large proportion of rods without the authorization of the chief engineer of the plant was forbidden. Yet the engineers knew that the computer system monitoring the number of rods in the core—the operational reactivity margin—wasn't always accurate, and they remained unaware of its importance to the safe running of the reactor. They did not suspect that the simultaneous reinsertion of so many rods into the core could trigger a reactor runaway. At this moment, only a careful stabilization of the reactor, followed by a slowly managed shutdown, might have headed off disaster.

Yet, now, two more of the giant main circulation pumps connected to the reactor came online. Although part of the original test program, the addition of these extra pumps had never been intended for such a low power level. Driving more cooling water into the core, they further upset the delicate balance of reactivity, water pressure, and steam content within the reactor. Running the pump system from his central control desk, twenty-seven-year-old Senior Unit Control Engineer Boris Stolyarchuk struggled to correct the water levels in the steam separator drums, as the pumps thundered toward their maximum capacity, forcing fifteen cubic meters of high-pressure coolant into the reactor every second. The rush of water absorbed an increasing number of neutrons in the core, dampening reactivity, and the reactor's automatic regulation system compensated by withdrawing yet more control rods. A few moments later, the water was moving so fast around the cooling loop that it was entering the core at close to boiling and turning to steam, making the reactor more susceptible to the positive void effect if there was the slightest increase in power.

Now the time to begin the generator rundown had finally come. Some of the operators were clearly nervous. Yet Anatoly Dyatlov felt nothing but calm. The test would proceed, regardless of the fine print of the experiment protocol or the qualms of his subordinates. Ten men now stood ready around the desks and control panels of Control Room Number Four, eyes on their instruments. He turned to Akimov.

"What are you waiting for?" he asked. It was 1:22 a.m.

Simulating the effects of a total power blackout on a single unit of the Chernobyl power plant was a deceptively simple process, and many in the control room mistakenly regarded the rundown test as a matter mostly for the electricians. The reactor's role seemed almost incidental. The test program closely duplicated one conducted on Unit Three in 1984—which, although it had failed to produce the desired results and keep the circulation pumps running, had nonetheless concluded without incident. Nikolai Fomin, the chief engineer, had ordered that test himself and without clearance from above and saw no reason to behave differently this time. He did not notify the State Committee for Nuclear Safety, NIKIET, or the specialists at the Kurchatov Institute in Moscow of his plans. He didn't even bother to tell Director Brukhanov that the test was taking place.

Emboldened by his previous experience, Fomin made two important changes with the new test: this time, all eight of the unit's main circulation pumps would be connected to the reactor, increasing the amount of water driven through the primary circuit during the rundown. But he had also ordered a special piece of equipment, an electrical box designed to be patched in to the control panel circuitry for the test, which would reproduce the effects of the design-basis accident at the touch of a button. The new test program—drafted a month earlier by Gennadi Metlenko, the head of the team of electrical engineers from Donetsk, and approved by Fomin and Dyatlov in April—appeared straightforward.

First, the operators would cut off the supply of steam from the reactor to the turbine, which would begin to spin down. At the same moment, they would press the design-basis accident button. This would send a signal to the reactor's safety systems that all external power to the plant had been lost, triggering the start-up of an emergency diesel generator and connecting the rundown unit of Turbine Number Eight to the main circulation pumps. If all went well, the electricity produced by the coasting turbine would keep the pumps going until the diesel generator could take over. The technicians expected the experiment to take less than a minute. It would begin with an order from Metlenko, who would record the results with an oscilloscope. It would conclude with the operators bringing the reactor to a routine shutdown by tripping the AZ-5 system for a full emergency stop.

By 1:23 a.m., at his desk in the control room, Leonid Toptunov had successfully stabilized the reactor at a power level of 200 megawatts. Dyatlov, Akimov, and Metlenko stood in the center of the room, awaiting the moment to begin. Upstairs at mark +12.5, in the cavernous three-story pump room alongside the reactor vault, Senior Coolant Pump Operator Valery Khodemchuk stood at his post, engulfed in the thunderous roar of all eight main circulation pumps working at once. At the bottom of the reactor core, the pressurized water was now entering the inlet valves at a temperature just a few degrees below boiling. And directly above them, 164 of the 211 control rods had been withdrawn to their upper-limit stops.

The reactor was a pistol with the hammer cocked. All that remained was for someone to pull the trigger. A few seconds later, Metlenko gave the command.

"Oscilloscope on!"

Over on the turbine desk, Senior Turbine Control Operator Igor Kershenbaum closed the turbine steam relief valves. Six seconds later, an engineer pressed the design-basis accident button. Alexander Akimov watched as the needle on the tachometer measuring the speed of Turbine Number Eight dropped, and the four main circulation pumps began to coast down. The control room was calm and quiet; it would soon all be over. Inside the reactor, the cooling water passing through the fuel channels slowed and grew hotter. Deep in the lower part of the core, the amount of coolant turning to steam increased. The steam absorbed fewer neutrons, and reactivity increased further, releasing more heat. Still more of the water turned to steam, absorbing even fewer neutrons and adding more reactivity, more heat. The positive void effect took hold. A deadly feedback loop had begun.

Yet still the instruments on Leonid Toptunov's control panel revealed nothing unusual. For another twenty seconds, the readings from the reactor remained within their normal limits. Akimov and Toptunov talked quietly. Over on the pump desk, Boris Stolyarchuk, absorbed in his tasks, heard no commotion. Behind them all, Deputy Chief Engineer Dyatlov remained silent and impassive. Turbine Generator Number Eight slowed to 2,300 revolutions per minute. It was time to end the test.

"SIUR—shut down the reactor!" Akimov said in a level voice. He waved his hand in the air. "AZ-5!"

Akimov lifted a transparent plastic cover on the control panel. Toptunov pushed his finger through the paper seal and pressed the circular red button beneath. After exactly thirty-six seconds, the test was over.

"The reactor has been shut down!" Toptunov said. High above them in the reactor hall, the rods' electric servomotors whirred. The glowing displays of the 211 Selsyn monitors on the wall showed their slow descent into the reactor. One meter. Two meters—

Inside the core, what happened next took place so fast that it outstripped the recording capacity of the reactor instrumentation.

For one scant second, as the boron carbide–filled upper sections of the rods entered the top of the reactor, overall reactivity fell, just as it was supposed to. But then the graphite tips began to displace the water in the lower part of the core, adding to the positive void effect, generating steam and more reactivity. A local critical mass formed in the bottom of the reactor. After two seconds, the chain reaction began to increase at an unstoppable speed, blooming upward and outward through the core.

In the control room, just as the staff was expecting to relax, the SIUR's annunciator panel suddenly lit up with a frightening succession of alarms. The warning lamps for "power excursion rate emergency increase" and "emergency power protection system" flashed red. Electric buzzers squawked angrily. Toptunov shouted out a warning: *Power surge!*

"Shut down the reactor!" Akimov repeated—yelling, this time.

Standing at the turbine desk twenty meters away, Yuri Tregub heard what he thought was the sound of Turbine Number Eight continuing to decelerate, like a Volga driving at full speed and then beginning to slow down: *wooo-woo-woo-woo*. But then it grew to a roar, and the building started to vibrate ominously around him. He thought it was a side effect of the test. But the reactor was destroying itself. Within three seconds, thermal power leapt to more than a hundred times maximum. In the lower southeast quadrant of the core, a handful of fuel channels over-heated rapidly, and the fuel pellets approached melting point. As the tem-perature climbed toward 3,000 degrees centigrade, the zirconium alloy casing of the assemblies softened, ruptured, and then exploded, dispersing small pieces of metal and uranium dioxide into the surrounding channels, where they instantly evaporated the surrounding water into steam. Then the channels themselves broke apart. The AZ-5 rods jammed at their halfway point. All eight emergency steam release valves of the reactor's protection system snapped open, but the mechanisms were quickly over-whelmed, and disintegrated.

Out on a gantry at mark +50, high above the floor of the central hall, Reactor Shop Shift Foreman Valery Perevozchenko watched in amaze-ment as the eighty-kilogram fuel channel caps in the circular *pyatachok* began bouncing up and down like toy boats on a storm-tossed pond. At Toptunov's control panel, the alarm sounded for *povyshenie davleniya v RP*—"pressure increase in reactor space." The walls of the control room had begun to shake, the oscillations slow but growing in force. At his post on the pump desk, Boris Stolyarchuk heard a rising moan, the protest of a giant beast in anguish. There was a loud bang.

How could this be happening?

As the fuel channels failed, water circulation through the core ceased entirely. The check valves on the massive main circulation pumps closed, and all the remaining water trapped in the core flashed into steam. A neu-tron pulse surged through the dying reactor, and thermal power peaked at more than 12 billion watts. Steam pressure inside the sealed reactor space rose exponentially—eight atmospheres in a second—heaving Elena, the two-thousand-tonne concrete-and-steel upper biological shield, clear of its mountings and shearing the remaining pressure tubes at their welds. The temperature inside the reactor rose to 4,650 degrees centigrade—not quite as hot as the surface of the sun.

On the wall of Control Room Number Four, the lights of the Selsyn dials flared. The needles had stopped dead at a reading of three meters. In desperation, Akimov threw the switch releasing the AZ-5 rods from their clutches, so they could fall under their own weight into the reactor. But the needles remained frozen. It was too late.

At 1:24 a.m., there was a tremendous roar, probably caused as a mixture of hydrogen and oxygen that had formed inside the reactor space suddenly ignited. The entire building shuddered as Reactor Number Four was torn apart by a catastrophic explosion, equivalent to as much as sixty tonnes of TNT. The blast caromed off the walls of the reactor vessel, tore open the hundreds of pipes of the steam and water circuit, and tossed the upper biological shield into the air like a flipped coin; it swatted away the 350-tonne refueling machine, wrenched the high-bay bridge crane from its overhead rails, demolished the upper walls of the reactor hall, and smashed open the concrete roof, revealing the night sky beyond.

In that moment, the core of the reactor was completely destroyed. Almost seven tonnes of uranium fuel, together with pieces of control rods, zirconium channels, and graphite blocks, were pulverized into tiny fragments and sucked high into the atmosphere, forming a mixture of gases and aerosols carrying radioisotopes, including iodine 131, neptunium 239, cesium 137, strontium 90, and plutonium 239—among the most dangerous substances known to man. A further 25 to 30 tonnes of uranium and highly radioactive graphite were launched out of the core and scattered around Unit Four, starting small blazes where they fell. Exposed to the air, 1,300 tonnes of incandescent graphite rubble that remained in the reactor core caught fire immediately.

Inside his workspace on mark +12.5, a few dozen meters away from the control room, Alexander Yuvchenko was talking to a colleague who had come in to collect a can of paint. Yuvchenko heard a thud, and the floor shook beneath his feet. It felt as if something heavy—the refueling crane, perhaps—had fallen to the floor of the reactor hall. Then he heard the explosion. Yuvchenko saw the thick concrete columns and walls of the room buckle like rubber, and the door, blown in by a shock wave carrying a wet, roiling cloud of steam and dust, was torn from its hinges. Debris rained from the ceiling. The lights went out. Yuvchenko's first impulse was to find a safe place to hide. *Finally,* he thought, *the war with the Americans has begun.*

Over in the turbine hall, turbine engineer Yuri Korneyev gazed up in horror as the corrugated steel ceiling panels above Turbine Generator Number Eight began to collapse toward him, tumbling down one after another like a series of massive playing cards, crashing onto the equipment below.

Looking out toward the central hall, former nuclear submariner Anatoly Kurguz saw a dense curtain of steam rolling out toward him. As he was overwhelmed by the searing cloud of radioactive vapor, Kurguz struggled to swing shut the pressurized airlock door, sealing off the hall and saving his colleagues in the reactor shop. It was the last thing he did before losing consciousness.

At his post in the shadow of the main circulation pumps, Valery Khodemchuk was the first to die, vaporized instantly by the explosion or crushed beneath the mass of collapsing concrete and machinery.

Inside Control Room Number Four, tiles and masonry dust fell from the ceiling. Akimov, Toptunov, and Deputy Chief Engineer Dyatlov looked about them in confusion. A gray fog bloomed from the air-conditioning vents, and the lights winked out. When they came back on, Boris Stolyarchuk noticed a sharp, mechanical smell, unlike any he had ever encountered before. On the wall behind them, the indicator lamps monitoring levels of radiation in the room abruptly turned from green to red.

Outside the plant, on the concrete bank of the cooling pond, two off-duty workers had spent the night fishing and still had their lines in the warm outfall from the station's reactors when they heard the first explosion. Turning toward the sound, they glanced back at the plant in time to hear a second report—a thunderous boom, like a plane breaking the sound barrier. The ground trembled, and then both men were struck by a shock wave. Black smoke curled above Unit Four, and sparks and hot debris arced upward into the night. As the smoke dispersed, they could see that the entire height of the 150-meter ventilation stack was now lit from below by a strange, cold glow.

In room 29, on the seventh floor of the Second Administration Building, Alexander Tumanov, an engineer, was working late. From his office window, he had a clear view of the northern side of the plant. At around 1:25 a.m., he heard a roar and felt the building shudder. This was followed by a cracking sound and two heavy thumps. Through the window, he saw a cascade of sparks flying out of Unit Four and what looked to him like fragments of molten metal or burning rags shooting from the unit in all directions. As he watched, larger pieces of blazing debris crashed onto the roofs of Unit Three and the auxiliary reactor equipment building, where they began to burn steadily.

Three kilometers away, the citizens of Pripyat slept on. Inside Viktor Brukhanov's apartment on Lenina Prospekt, the telephone began to ring.

6

Saturday, April 26, 1:28 a.m., Paramilitary Fire Station Number Two

Just after 1:25 a.m., as a purple cone of iridescent flame leapt 150 meters into the air around Chernobyl's candy-cane-striped ventilation stack, the alarm bell sounded at Paramilitary Fire Station Number Two. In the telephone dispatcher's room, the master status board, with its hundreds of red warning bulbs—one for every room in the entire Chernobyl complex—suddenly lit up from top to bottom.

Many of the fourteen men of the third watch had been dozing on their beds in the ready room when a loud thud rattled the station windows and shook the floor, jolting them awake. They were already pulling on their boots as the emergency siren sounded, and ran out onto the concrete apron in front of the station, where the unit's three trucks stood at the ready, keys in the ignition. They heard the dispatcher shout that there was a fire at the nuclear plant and looked over just in time to see a giant mushroom-shaped cloud blossoming into the sky above Units Three and Four, less than five hundred meters away—two minutes by road.

Lieutenant Pravik gave the order to go, and, one by one, the big red-and-white ZIL fire engines bounced off the apron and pulled away. Sergeant Alexander Petrovsky, at twenty-four the second-youngest member of the watch, didn't have time to find his helmet and seized Pravik's service cap instead. It was 1:28 a.m. Behind the wheel of the lead truck was Anatoly Zakharov, a thickset, gregarious thirty-three-year-old who served as the fire station's Party secretary and who worked a second job as a lifeguard in Pripyat, where he had a pair of binoculars and a motorboat to help him pull drunken bathers from the river. Zakharov swung right and followed the plant perimeter fence, heading toward the gate of the plant at top speed. A sharp left, then through the gateway and into the grounds of the plant, flashing past the long, squat shape of the diesel generator station. The radio crackled with questions and instructions: What had happened? What was the damage? Two extra tanker trucks were already coming right behind them; the Pripyat city brigade was also on its way. Lieutenant Pravik called in a number three alarm, the highest-level emergency alert, summoning every available fire brigade in the Kiev region.

Now the giant superstructure of the plant filled the view through Zakharov's windshield. He took the access road to the right, drove between the concrete stilts of an elevated causeway, and sped toward the northern wall of the third reactor. And there, just thirty meters away, he saw what remained of Unit Four.

Up in the Unit Four control room, everyone was talking at once, as Deputy Chief Engineer Anatoly Dyatlov struggled to understand what the instruments were saying. A constellation of warning lamps flashed red and yellow across the consoles of the turbine, reactor, and pump desks, and electric alarm buzzers honked incessantly. The news seemed grave. At Senior Unit Control Engineer Boris Stolyarchuk's desk, the readings showed all eight main safety valves were open, and yet no water remained in the separators. This scenario was the maximum design-basis accident and an *atomshchik*'s worst nightmare: an active zone starved of thousands of gallons of vital coolant, raising the threat of a core meltdown.

And at Senior Reactor Control Engineer Toptunov's control panel, the needles on the dials of the Selsyn gauges were stuck at the four-meter mark, showing the control rods had stopped dead not even halfway through their descent. Toptunov had released the rods from their electromagnetic clutches to let gravity take them all the way to their stops, but somehow they had halted before bringing the reactor to shutdown. The gray LED numbers on the reactimeter—showing activity in the core—fluttered up and down. Something was still going on in there, but Dyatlov and the technicians around him no longer had any means of controlling it.

In desperation, Dyatlov turned to the two trainee reactor control engineers who had come to work that night to observe the test, Viktor Proskuryakov and Alexander Kudryavtsev, and gave them instructions to complete the scram manually. Head up to the reactor hall, he told them, and force the control rods into the core by hand.

The two men obeyed, but almost as soon as they had left the room, Dyatlov realized his mistake: if the rods wouldn't fall under their own weight, they would also be impossible to move manually. He ran into the corridor to call back the trainees, but they had disappeared, swallowed by the clouds of smoke and steam that had filled the halls and stairwells of Unit Four.

Returning to the control room, Dyatlov took command. He gave orders to Shift Foreman Alexander Akimov to dismiss all nonessential personnel still at their posts, including Senior Reactor Control Engineer Leonid Toptunov, who had pressed the AZ-5 scram button. Then he told Akimov to activate the emergency cooling pumps and smoke exhaust fans, and gave instructions to open the gates of the coolant pipe valves. "Lads," he said, "we've got to get water into the reactor."

Upstairs, inside the windowless senior engineers' room on level +12.5, Alexander Yuvchenko was engulfed in dust, steam, and darkness. From beyond the shattered doorway came a terrible hissing sound. He groped along his desk for the telephone connecting him with Control Room Four, but the line was dead. Then someone from Control Room Three rang through with a command: *Bring stretchers immediately.*

Yuvchenko gathered a stretcher and ran downstairs toward mark +10, but before he could reach the control room, he was stopped by a dazed figure, his clothes blackened, his face bloody and unrecognizable. Only when he spoke did Yuvchenko realize it was his friend, the coolant pump operator, Viktor Degtyarenko. He said he had come from near his station, and there were others still there who needed help. Probing the humid blackness with a flashlight, Yuvchenko came upon a second operator on the other side of a pile of wreckage: still able to stand but filthy, wet, and grotesquely scalded by escaping steam. He was quivering with shock but waved Yuvchenko away. "I'm all right," he said. "Help Khodemchuk. He's in the pump room."

Then Yuvchenko saw his colleague Yuri Tregub emerging from the gloom. Tregub had been sent from Control Room Number Four to manually turn on the taps of the emergency high-pressure coolant system and flood the reactor core with water. Knowing this task would require at least two men, Yuvchenko told the injured pump operator where to go to get help and accompanied Tregub toward the coolant tanks. Finding the nearest entrance blocked by rubble, they went down two flights of stairs and immediately found themselves knee-deep in water. The door to the hall was jammed shut, but through a narrow gap, the two men glimpsed inside.

Everything was in ruins. The gigantic steel water tanks had been torn apart like wet cardboard, and above the wreckage, where the walls and ceiling of the hall should have been, they could see only stars. They were staring into empty space; the bowels of the benighted station were lit by moonlight.

The two men turned into the ground-level transport corridor and reeled outside into the night. Standing no more than fifty meters away from the reactor, Tregub and Yuvchenko were among the first to comprehend what had happened to Unit Four. It was a terrifying, apocalyptic sight: the roof of the reactor hall was gone, and the right-hand wall had been almost completely demolished by the force of the explosion. Half of the cooling circuit had simply disappeared: on the left, the water tanks and pipework that had once fed the main circulation pumps dangled in midair. Yuvchenko knew at that moment that Valery Khodemchuk was certainly dead: the spot where he had been standing lay beneath a steaming pile of rubble, lit by flashes from the severed ends of 6,000-volt cables as thick as a man's arm, swaying and shorting on everything they touched, showering the wreckage with sparks.

And from somewhere in the heart of the tangled mass of rebar and shattered concrete—from deep inside the ruins of Unit Four, where the reactor was supposed to be—Alexander Yuvchenko could see something more frightening still: a shimmering pillar of ethereal blue-white light, reaching straight up into the night sky, disappearing into infinity. Delicate and strange and encircled by a flickering spectrum of colors conjured by flames from within the burning building and superheated chunks of metal and machinery, the beautiful phosphorescence transfixed Yuvchenko for a few seconds. Then Tregub yanked him back around the corner and out of immediate danger: the phenomenon that had entranced the young engineer was created by the radioactive ionization of air and was an almost certain sign of an unshielded nuclear reactor open to the atmosphere.

As the three trucks from Fire Station Number Two drew up beside Unit Four, a fire prevention officer from the plant came running out to meet them. He had witnessed the explosion and called in the alarm. Anatoly Zakharov jumped down from his cab and looked around. The ground was littered with blocks of graphite, many of them still glowing red with intense heat. Zakharov had watched the reactor being constructed from the inside out and knew exactly what they were.

"Tolik, what is it?" one of the men asked.

"Lads, it's the guts of the reactor," he said. "If we survive until the morning, we'll live forever."

Pravik told Zakharov to stay by the radio and await instructions. He and squad commander Leonid Shavrey would conduct a reconnaissance to establish the source of the fire. "And then we'll put it out," Pravik said.

With that, the young lieutenant disappeared into the plant.

Inside the turbine hall of Unit Four, the two firemen found a scene of total chaos. Broken glass, concrete, and pieces of metal lay everywhere; a few dazed operators ran here and there through the smoke that rose from the rubble; the walls of the building trembled, and from somewhere above came the roar of escaping steam. The windows along row A had been shattered, and the lights above Turbine Number Seven were blown out; jets of steam and hot water blasted from the ruptured flange of a feed pipeline, and flashes of flame were visible through clouds of steam in the area of the fuel pumps. Some of the roof had caved in, and heavy pieces of debris—hurled out of the reactor building and onto the roof of the hall by the explosion—were still falling from above. At one point, a lead plug that had been used to close a reactor channel tumbled from the ceiling and smashed into the ground within a meter of where one of the turbine operators stood.

Pravik and Shavrey, mere firefighters, had no equipment to measure radiation. Their walkie-talkies weren't working. They found a telephone and tried to call the power station dispatcher to find out more details of the emergency. They couldn't get through. For the next fifteen minutes, the two men ran around inside the plant. But they were unable to establish anything for certain, except that parts of the turbine hall roof had collapsed, and the areas that hadn't seemed to be on fire.

By the time Pravik and Shavrey returned to their men outside Unit Three, the firefighters of the Pripyat city brigade had arrived. By two in the morning, the men of seventeen other fire brigades from all over the Kiev region were racing to the plant, accompanied by search-and-rescue teams, special ladder crews, and tanker trucks. Soon afterward, the chief of the Ministry of Internal Affairs in Kiev established a crisis center dedicated to the emergency and called for updates from the scene every forty minutes.

In his apartment across the street from the Pripyat police station, Piotr Khmel, chief of the first watch of Paramilitary Brigade Number Two, was ready to turn in after his long night of drinking when his doorbell rang. It was Radchenko, a driver from the station house.

"There's a fire in Unit Four," he said. Every man was needed at once. Khmel told him to wait while he put on his uniform, then followed him downstairs to the UAZ jeep waiting on the street. On his way out, the young lieutenant snatched the half-empty bottle of *Sovietskoe shampanskoye* from the kitchen table. As the UAZ yawed into the sharp left-hand bend on Lesi Ukrainki Street, Khmel held tight to the bottle. He drained it to the dregs.

Whatever the emergency, there was no need to waste good Soviet champagne.

In his flat on Lenina Prospekt, plant director Viktor Brukhanov was woken by the telephone within two minutes of the explosion. Beside him in bed, his wife stirred and looked up as the light snapped on. Calls from the station in the middle of the night were not unusual, so she felt no need to be alarmed. But now, as her husband listened in silence to what he was being told, Valentina watched the color drain from his face. Viktor put down the receiver, dressed in a trance, and slipped out into the night without saying another word.

It was not yet 2:00 a.m. when Brukhanov reached the plant. He saw the jagged outline of Unit Four, lit from within by a dim red glow, and knew that the worst had happened.

I'm going to prison, he thought.

Heading into the main administrative building, the director ordered the plant's chief of civil defense to open up the emergency bunker in the basement below. Designed as a refuge for the staff in the event of a nuclear attack, the hardened bunker contained a crisis center with desks and telephones for each of the plant's department heads, decontamination showers, an infirmary for the injured, air filters to scrub poison gas and radionuclides from the atmosphere, a diesel generator, and tanks of fresh water intended to support 1,500 people for a minimum of three days—all sealed behind a steel airlock door. Brukhanov went upstairs to his office on the third floor and tried to raise the chief shift manager of the plant on the phone. There was no reply. He ordered the activation of the automatic telephone alert system, designed to notify all senior personnel of an emergency of the highest degree: a General Radiation Accident. This indicated the release of radioactivity not only within the station but also onto the grounds and into the air surrounding it.

The mayor of Pripyat arrived, accompanied by the plant's resident KGB major and the Party secretaries of both the plant and the city. The apparatchiks had many difficult questions. The director, expected to provide answers, had none.

The bunker was a long, narrow space with a low ceiling, cluttered with tables and chairs, which filled quickly with the department heads summoned by the phone alert. Brukhanov took a seat right beside the door, at a desk equipped with several telephones and a small control panel, and began reporting the news of an accident to his superiors. First, he called Moscow, where he spoke to his boss at the USSR's atomic energy authority, Soyuzatomenergo; then he called the first and second secretaries of the Party in Kiev. "There's been a collapse," he said. "But it's not clear what happened. Dyatlov is looking into it." Then he informed the Ukrainian Energy Ministry and the director of the Kiev region power supply utility.

Soon afterward, the director took the first damage reports from the head of plant radiation safety and the chief shift supervisor: there had been an explosion in Unit Four, but they were attempting to supply cooling water to it. Brukhanov heard that the instrument readings in the control room still showed coolant levels stuck at zero. He feared that they stood on the precipice of the most terrifying catastrophe imaginable: the reactor running dry of water. Nobody suggested to him that the reactor had already been destroyed.

There were soon thirty or forty men in the bunker. The ventilation fans hummed; pandemonium reigned. The hubbub of dozens of simultaneous telephone conversations—the supervisors of every department of the V. I. Lenin nuclear power plant calling their employees, all focused on ensuring a supply of water and getting it pumping to the core of Reactor Number Four—reverberated from the hardened concrete walls. Yet at his desk by the door, Brukhanov seemed stunned: his formerly laconic manner sagged into a dejected stupor, his movements slow and numbed by shock.

After witnessing the horror of the destruction of Unit Four from the outside, Alexander Yuvchenko and Yuri Tregub ran back into the plant to report what they had seen. But before they could reach the control room, they were stopped by Yuvchenko's immediate boss, Valery Perevozchenko, the supervisor of the reactor section on their shift. With him were the two trainees who had been sent by Deputy Chief Engineer Dyatlov to lower the reactor control rods by hand. When Perevozchenko explained their instructions, Yuvchenko tried to tell them that their mission was senseless: the control rods—indeed, the reactor itself—no longer existed. But Perevozchenko insisted. Yuvchenko had only examined the reactor from below, he said; they needed to assess the damage from above.

While Tregub continued on to the control room, Yuvchenko agreed to help find a way to the reactor hall. Orders were orders—and besides, Yuvchenko was the only one with a flashlight. Together the four men picked their way up the stairs from mark +12 toward mark +35. Yuvchenko was last in line as they wound through a labyrinth of collapsed walls and twisted wreckage, until they reached the massive airlock door of the reactor hall. Made of steel and filled with concrete, the door weighed several tonnes, but the crank mechanism used to hold it open had been damaged in the explosion. If they went into the hall and the door swung shut behind them, they would be trapped. So Yuvchenko agreed to stay outside. He braced his shoulder against the door, using all his strength to keep it from closing, while his three colleagues stepped over the threshold.

Inside, there wasn't much room. Perevozchenko stood on a narrow ledge and swept the darkness with Yuvchenko's flashlight. Its yellow beam caught the outlines of the gigantic steel disc of Elena tilted in the air, balanced on the edges of the reactor vault; the hundreds of narrow steam tubes that ran through it had been shorn away in ragged clumps, like the hair of a mutilated doll. The control rods were long gone. As they gazed at the molten crater beneath, the three men realized in horror that they were staring directly into the active zone: the blazing throat of the reactor.

Perevozchenko, Proskuryakov, and Kudryavtsev remained on the ledge for only as long as Yuvchenko held the door: a minute at most. But even that was too long. All three received a fatal dose of radiation in a matter of seconds.

Even as his three colleagues staggered back into the corridor in shock, Yuvchenko wanted to have a look for himself. But Perevozchenko, a veteran of the nuclear submarine fleet, who knew very well what had just happened, shoved the younger man aside. The door slammed shut.

"There's nothing to see here," he said. "Let's go."

In the darkness of the turbine hall, Deputy Section Chief Razim Davlet-bayev battled to contain the chaos sweeping his department. Standing emergency regulations dictated that the plant operators, not the fire brigade, fight any fires inside their part of the station, and the blazes now raging across the multiple levels of the turbine section threatened an even greater catastrophe. The turbine machinery was filled with thousands of liters of highly flammable oil, and the turbine generators with hydrogen—which in normal operation was necessary to cool the generator coils. If either ignited, the resulting fire could spread down the almost one-kilometer length of the turbine hall to engulf the plant's other three reactors or lead to yet another massive explosion inside Unit Four.

Amid the radioactive steam and fountains of boiling water pouring from broken pipes, and sparks showering from severed cables, Davletbayev ordered his men to turn on the sprinklers covering Turbine Number Seven, drain lubricant into emergency tanks, and stem a jet of oil spewing from a broken line on mark +5; a slick was spreading across the floor at level 0 and running into the basement. A team of three engineers fought their way inside the rooms housing the oil feed pump controls, which were flooded with hot water, to turn them off and prevent the fire from spreading. Two machinists extinguished one blaze on level +5, while others fought fires elsewhere. The chief mechanic cut off the pumps from the deaerators, blocking a flow of radioactive water from broken pipes into the turbine room.

Inside the hall, it was hard to breathe, and the humid, steam-saturated air carried the smell of ozone. But the operators gave little thought to radiation, and the panicked dosimetrists who dashed through the unit provided no useful information: the needles of all their monitoring equipment simply ran off the scale. The radiometers capable of taking higher readings remained locked in a safe and could not be released without orders from above. Razim Davletbayev told himself that the distinctive scent filling the turbine hall was caused by the short circuits arcing in the air; later, when he began to feel sick, although he understood that nausea was an early warning of radiation poisoning, he still preferred to put it down to the potassium iodide solution he'd drunk.

Turbine engineer Yuri Korneyev was busy shutting down Turbine Number Eight when Anatoly Baranov, the shift electrician, ran in. Baranov began displacing the hydrogen in both Number Seven and Number Eight turbine generators with nitrogen, averting the possibility of a further explosion. By the time they had finished, an eerie silence had fallen around them and their lifeless machinery. They stepped outside onto a small balcony, to smoke. Only much later did they discover the cost of their cigarette break: the street beneath them was scattered with blocks of reactor graphite, which irradiated them as they rested at the railing.

Elsewhere, the engineers had begun to comb the rubble for the bodies of the dead or injured. The machinists on the floor of the turbine hall had all escaped apparently unscathed from the initial explosion, but after a half hour, Vladimir Shashenok, who had been monitoring the turbine test from compartment 604—the flowmeter room—was still missing. So three men picked their way across piles of debris to the compartment, on an upper landing across from the turbine hall, inside the walls of the reactor building. Their path was strewn with wreckage; they dodged bursts of escaping steam; they waded ankle-deep in water. When they finally reached compartment 604, they discovered that it had been obliterated. Concrete wall panels had been tossed into the street by the explosion. Darkness and swirling dust swallowed the beam of their flashlight. They began calling out to Shashenok in the blackness but heard no response. Eventually they came upon a body: lying unconscious on his side, a bloody foam bubbling from his mouth. They picked Shashenok up by the armpits and carried him out.

Outside, Lieutenant Pravik of Paramilitary Fire Brigade Number Two mounted the fire escape that zigzagged up the northern wall of Unit Three, his boots ringing on the metal treads. A handful of men from the Pripyat brigade accompanied him, including their commander, Lieutenant Viktor Kibenok, and Vasily Ignatenko, a stocky twenty-five-year-old renowned as the champion athlete of Brigade Number Six. Around them they heard only the hum of the plant's remaining three reactors and the crackling of flames.

It was a long way to the top. The flat roofs of Unit Three and its doomed twin were staggered like giant steps. The eight levels of the buildings formed a concrete ziggurat that reached its summit with the top of the ventilation block—twenty stories high and crowned by the red-and-white-striped chimney that towered over the two reactors. From here, the firemen could look straight down into the glowing ruins of the Unit Four reactor hall and glimpse the destruction beyond. Dozens of small fires had broken out on the roofs around them: at the foot of the chimney; on the reactor hall of Unit Three; and, far off in the darkness, on top of the turbine hall. Ignited by fragments of blazing debris thrown from the reactor by the explosion, some of the fires burned fiercely, with flames reaching a meter and a half into the air; others were smaller but strangely incandescent, fizzing and crackling like sparklers. The air was filled with black smoke, but also something else that the firefighters did not recognize: a strange vapor that looked almost like fog but gave off a peculiar smell.

In the darkness around their feet were hundreds of sources of lethal ionizing radiation: lumps of graphite, fragments of fuel assemblies, and pellets of the reactor's uranium dioxide fuel itself, scattered across the rooftops and emitting fields of gamma rays reaching thousands of roentgen an hour.

Yet Pravik and the others were driven on by a more tangible threat: the fires on the roof of Unit Three, directly above the reactor. A breeze was blowing from the west, threatening to spread any one of the small blazes downwind toward reactors Two and One, both of which were still running. If these fires weren't brought under control, the entire station would soon be engulfed in disaster. Pravik moved quickly. Together with Kibenok and his men, they brought hoses to the roof. Pravik ordered his pump trucks connected to the dry standpipes designed to distribute water to the heights of the building through the plant's fire suppression system. But when the pumps were turned on, air whistled through the hoses.

"Give me some pressure!" Pravik yelled over the radio. It was no use: the standpipes had been smashed in the explosion.

For once, even the quarrelsome men of the third watch didn't hesitate to follow orders. Sweating in their heavy canvas uniforms and rubber jackets, they ran out more hoses just as they had been trained to do—five in seventeen seconds. They threw them over their shoulders, dragged them up the staircases, and poured foam onto the roof of Unit Three. Kibenok had a separate line connected to the Pripyat brigade's big Ural fire tanker, which could move forty liters of water a second. Even then, the handful of men on the roof struggled to extinguish even the smallest blazes, caused by materials which seemed to burn more savagely when they poured water on them. These were almost certainly pellets of uranium dioxide, which, superheated to more than 4,000 degrees Celsius before the explosion, had ignited on contact with the air; when hosed with water, the resulting reaction released oxygen, explosive hydrogen, and radioactive steam.

Down on the ground, twenty-four-year-old Sergeant Alexander Petrovsky was ordered to take two men up to the top of the ventilation block and help out. While still a teenager, Petrovsky had been part of a fifteen-man welding crew building Units Three and Four. He had helped construct both reactors and knew every room in the complex, from the basement cable tunnels to the roof. In those days, there had always been some radiation around, and it had never been a problem. He wasn't concerned about catching a bit more.

But Petrovsky had reached only the first level of the roofs—halfway up, at mark +30—when he saw Lieutenant Pravik and the men from the Pripyat brigade coming down toward him. Something terrible had clearly happened to them: staggering and incoherent, the half dozen men were pulling and dragging one another down the staircase, vomiting as they came. Petrovsky told one of his men to get them safely to the ground, while he continued upward with Ivan Shavrey—one of the two Belarusian brothers on the third watch. In his haste to reach the top and help the comrades they imagined still fighting the blazes at mark +71, Shavrey slipped on the steep staircase, and Petrovsky reached out to grab him. As he did so, Petrovsky felt his borrowed service cap slip off his head. He watched helplessly as it tumbled away into the darkness, and went on bareheaded, protected by only his shirt and waterproof jacket.

When they reached the summit, the two firefighters discovered they were alone on the roof. Only a single functioning hose remained. Between them, they began extinguishing what they could. Working around the base of the vent stack, they trained the hose on burning chunks of graphite but found that even when the flames went out, the material gave off an incandescence that no amount of water could douse. After thirty minutes, almost every visible fire around them was out, but one major problem remained: flames were licking from the end of a two-meter ventilation pipe projecting vertically from the roof. There wasn't enough pressure to shoot a jet of water up into it, and Petrovsky was too short to reach directly into the end of the pipe with the hose. But Shavrey was a head taller. It was just as he was passing the heavy cast aluminum nozzle to Shavrey that Sergeant Petrovsky abruptly went blind.

A fatal dose of radiation is estimated at around 500 rem—roentgen equivalent man—or the amount absorbed by the average human body when exposed to a field of 500 roentgen per hour for sixty minutes. In some places on the roof of Unit Three, lumps of uranium fuel and graphite were emitting gamma and neutron radiation at a rate of 3,000 roentgen an hour. In others, levels may have reached more than 8,000 roentgen an hour: there, a man would absorb a lethal dose in less than four minutes.

Petrovsky's loss of vision was sudden and total and endured for only thirty seconds. But it seemed to last an eternity and filled him with terror. When his eyesight returned as unexpectedly as it had left, the sergeant's courage had deserted him. "Fuck this, Vanya!" he shouted to Shavrey. "Let's get the fuck out of here!"

On the other side of the complex, Ivan Shavrey's older brother, Leonid, had been fighting the fires on the roof of the turbine hall. There, down at mark +31.5, flying pieces of debris had smashed gaping holes in the corrugated steel. Some roof panels had collapsed completely into the hall beneath, while others dangled treacherously underfoot, leaving a patchwork of voids almost invisible in the darkness. The heat was so intense that the bitumen surface melted underfoot, tugging at the firemen's boots, making walking difficult. The first men to arrive could not reach all the fires with hoses but inched around the holes to smother them with sand instead.

Descending to the ground for another hose, Leonid Shavrey found that the brigade chief, Major Leonid P. Telyatnikov, had arrived to take command. The major told Shavrey to return to the turbine hall roof and, once all the remaining small blazes had been extinguished, maintain a fire watch until relieved. And it was here that, at a few minutes past three in the morning, Lieutenant Piotr Khmel—still drunk on the People's Champagne—joined Shavrey to maintain a silent lookout for new blazes. Together, amid a field of tangled hoses and radioactive debris, the two men stood and waited for dawn.

In the bunker beneath the administrative block, Director Brukhanov and his senior staff worked the telephones, still struggling to believe what was happening in the world above them. They labored mechanically, punch-drunk from the mounting strain—but could not overcome the power of their faith that a nuclear reactor would never explode. Although by now, many of them had seen for themselves the scale of destruction that surrounded Unit Four, they remained unable—or unwilling—to comprehend the truth. Brukhanov made his own trip to Unit Four, but when he returned to the bunker, he still refused to confront the implications of what he'd seen. He chose to believe instead that the reactor remained intact and that an explosion had taken place elsewhere—in a steam separator drum or perhaps in the turbine oil tank. As long as his men kept feeding water to Reactor Number Four, to head off the possibility of a meltdown, a true catastrophe would be averted.

Yet not everyone had succumbed to wishful thinking and self-delusion. The plant's civil defense chief, Serafim Vorobyev, arrived in the bunker shortly after two in the morning. The first thing he did was remove a powerful DP-5 military radiometer from storage and turn it on. A bulky Bakelite box with a steel detecting wand at the end of a long cable, the DP-5 was designed for use after a nuclear attack, and, unlike the sensitive Geiger counters used by the power station's dosimetrists to monitor workplace safety, it could detect intense gamma radiation fields of up to 200 roentgen per hour. Obliged by regulations to report to the local authorities any accident that resulted in a release of radiation beyond the boundaries of the power station, Vorobyev went up to ground level to take measurements. He'd reached only as far as the bus stop outside the front doors when he took a reading of 150 milliroentgen per hour—more than a hundred times higher than normal. He rushed back to tell Brukhanov to warn the plant staff and the population of Pripyat.

"Viktor Petrovich," he said, "we need to make an announcement."

But the director told him to wait. He wanted more time to think. So Vorobyev went back outside and got into his car to gather more data. As he drove around the plant toward Unit Four, the needle on the DP-5 swung up to 20 roentgen per hour. When he passed the electrical substations, it hit 100 r/h and kept going: 120; 150; 175; finally, passing 200 r/h, the needle went off the scale. Vorobyev now had no idea how high the true levels of radiation around the plant were but knew that they must be enormous. He drove right up to the mountain of debris cascading from the shattered northern wall of the reactor and saw the black trail of graphite leading away into the darkness. Less than a hundred meters away, the first operators were being led out of the plant to a waiting ambulance, oddly excitable, complaining of headaches and nausea or already vomiting.

Vorobyev drove back to the bunker and reported to Brukhanov the most conservative reasonable dosimetry estimate: the station was now surrounded by very high fields of radiation, of up to 200 r/h. It was essential, he said, to warn the people of Pripyat about what had happened. "We need to tell people that there's been a radiation accident, that they should take protective measures: close the windows and stay inside," Vorobyev told the director.

But still Brukhanov stalled. He said he would wait for Korobeynikov, head of the plant's radiation safety team, to make his own assessment. At 3:00 a.m., Brukhanov called his Party boss in Moscow and the Ministry of Internal Affairs in Kiev with a situation report. He described an explosion and a partial collapse of the turbine hall roof. The radiation situation, he said, was being clarified.

It was another hour before the chief of radiation safety arrived. Vorobyev stood by and listened to the man's report in disbelief: his measurements revealed that radiation levels were indeed elevated, but they were a mere 13 *micro*roentgen an hour. He claimed to already have performed a rough analysis and found that the radionuclides in the air were principally noble gases, which would quickly dissipate and therefore posed little threat to the population; there really wasn't much to be concerned about. This assessment was apparently what Brukhanov had been hoping to hear. He stood and, looking around the room, declared darkly, "Some people here understand nothing and are stoking panic." He left no doubt whom he was talking about.

Yet Vorobyev knew that it was simply impossible to approach the station from any direction without passing through radiation fields tens of thousands of times higher than the radiation safety team reported. Every word he'd just heard must therefore be a lie—but his confidence in his own expertise and equipment had been shaken.

Taking the DP-5, Vorobyev went out into the night for a third time to verify his results. Tendrils of amber light were spreading across the sky as he drove toward Pripyat. There, he found a police roadblock, a crowd of people waiting in the open for a bus to Kiev, and hot spots of fallout on the asphalt: levels of gamma radiation rose by thousands of times in the space of a few meters. By the time he returned to the plant from the city, Vorobyev's car and clothes were both so contaminated that the DP-5 could no longer take accurate readings. He clattered down the concrete steps of the bunker on the verge of hysteria, a wild look in his eyes.

"There's no mistake," he told Brukhanov. "We must take action as the plan demands."

But the director cut him off. "Get out," he said and pushed him away. "Your instrument is broken. Get out of here!"

In desperation, Vorobyev picked up the phone to notify the Ukrainian and Belarusian civil defense authorities. But the operator told him he had been forbidden from making long-distance calls. Eventually he managed to get a connection to Kiev on his direct line, which in their haste Brukhanov and his assistants had failed to have cut off. But when Vorobyev delivered his report, the civil defense duty officer who answered refused to believe that he was serious.

When reactor shop supervisor Valery Perevozchenko found his way back to Control Room Number Four, he reported to Deputy Chief Engineer Dyatlov what he had seen on his abortive mission to help lower the control rods by hand: the reactor had been destroyed. Dyatlov assured him that this was impossible. Although he recognized that there had been an explosion somewhere in Unit Four, it didn't occur to him that it might have been in the reactor core itself. Nothing in his decades of experience in the nuclear field—his years overseeing the commissioning of all those submarines in Komsomolsk-on-Amur, the start-ups of Units Three and Four in Chernobyl, the courses and manuals he had studied to keep himself abreast of the latest regulations governing the RBMK-1000—had ever suggested a reactor could explode. As Dyatlov set off down the corridor to examine the unit for himself, he was looking for evidence of gas detonating somewhere in the emergency core cooling system.

Outside in the corridor, he ran into Oleg Genrikh and Anatoly Kurguz, who was covered in terrible burns. The skin hung from his face and hands in scarlet shreds. Dyatlov told them to go immediately to the plant sick bay and then continued down the hall to a window, where he was startled to see that the wall of the unit—all the way from mark +12 to mark +70, or more than seventeen stories in total—had collapsed completely. Heading to the end of the corridor and down the stairs, he walked slowly around the outside of Units Three and Four, taking in the gathered fire trucks, the flames licking the roofs of the buildings, and the debris littering the ground around him.

Running back upstairs to the control room, Dyatlov saw that Leonid Toptunov, although dismissed from his post, had now returned. Dyatlov angrily demanded an explanation for this disobedience and learned that the young operator had left, but then—tugged by his sense of responsibility to the station and his comrades—turned around and came back to help. Dyatlov once again ordered him to leave. But when the deputy chief engineer left the room again a few minutes later, Toptunov stubbornly stayed put. And when a new Unit Four shift foreman arrived to take the place of Alexander Akimov, he, too, remained at his post. The two men were determined to fulfill their orders, to ensure that cooling water reached the reactor by finding the giant gate valves in the feed water complex and opening them—by hand, if necessary.

By now, the levels of radiation in the control room had become dangerously high. Dyatlov's strength, sapped by his repeated sorties through the radioactive debris in and around Unit Four and frequent spasms of vomiting, was fading. Shortly before dawn, he recovered the operating log and gathered the data printouts of the Skala computer—which had been monitoring the reactor in the final moments of its existence—and left Control Room Number Four for the last time.

It was 5:15 a.m. when, weak and retching, radioactive water squelching in his shoes, Dyatlov finally staggered down the steps into the bunker to report to Director Brukhanov. He laid three recordings from the Skala on the desk: two showing the reactor's power levels, and another showing the pressure in the primary cooling circuit. But when Brukhanov and Serhiy Parashyn, the plant's Communist Party secretary, asked him to explain what had happened inside Unit Four, Dyatlov just threw up his hands in bewilderment.

"I don't know. I don't understand any of it," he said.

By five thirty in the morning, the plant was filling with technicians and specialists roused from their beds in Pripyat to help contain the growing catastrophe. Ignoring instructions from above, the shift foreman of Unit Three had ordered the emergency shutdown of his reactor and isolated its control room from the station's ventilation system. At the other end of the plant, Units One and Two remained online, and the operators stood at their posts. But all the alarms were blaring in unison, and the armored doors in the corridors were battened down.

In the hallway outside Control Room Number Four, the aluminum ceiling panels lay scattered on the floor, and contaminated water—much of which had passed through the wreckage of the reactor overhead and was saturated with nuclear fuel—was pouring through from above.

And still Brukhanov's desperate command was relayed from the bunker: "Get that water in!"

Inside the narrow pipeline compartment on mark +27, Alexander Akimov and Leonid Toptunov labored away in darkness at the gate valves controlling the supply of water to the separator drums. The valves were usually opened remotely by electric drives, but the cables had been severed, the power long dead. The two men used all of their ebbing strength to turn the giant wheel—as broad as a man's torso—by hand, one agonizing centimeter at a time. By 7:30 a.m., soaked to the skin and up to their ankles in the radioactive water showering from the ceiling, the two men had managed to open the valves on one coolant pipeline. But both had now been in the extraordinary gamma fields inside Unit Four for more than six hours and were suffering the initial symptoms of acute radiation syndrome. Their white overalls were gray, filthy, and wet, saturated with energetic beta-emitting radionuclides that exposed their skin to hundreds of roentgen an hour. Toptunov vomited constantly; Akimov barely had enough strength to move. No matter how hard they struggled, the final valve wouldn't open. Eventually Akimov was helped from the compartment by his comrades, and the two men stumbled down the stairs toward the Unit Four control room, their path lit by a miner's lamp.

Yet as Toptunov and Akimov entered the plant infirmary, the water they had worked so hard to release gushed uselessly from shattered pipes around the smashed reactor. It spilled through Unit Four from one level to the next, running down corridors and staircases, slowly emptying the shared reserves needed to cool Unit Three, flooding the basement and the cable tunnels that linked them both, and threatening further destruction. Many more hours would pass, and other men would sacrifice themselves to the delusion that Reactor Number Four survived intact, before Director Brukhanov and the men in the bunker acknowledged their terrible mistake.

By 6:35 a.m. on Saturday, thirty-seven fire crews—186 firemen and 81 engines—had been summoned to Chernobyl from all over the Kiev region. Together they had managed to extinguish all the visible fires around the buildings of Reactor Number Four. The deputy fire chief for the Kiev district declared the emergency over. And yet, from inside the remains of the reactor building, wisps of black smoke and something that looked like steam continued to twist upward, drifting away slowly into the bright spring sky.

Picking his way through fallen debris to the end of the deaerator corridor, Senior Unit Engineer Boris Stolyarchuk leaned out through one of the shattered windows of the reserve control room and craned his neck to look down. Dawn had broken. The light was crisp and clear. What Stolyarchuk saw did not frighten him, but he was struck by one thought:

I'm so young, and it's all over.

Reactor Number Four was gone. In its place was a simmering volcano of uranium fuel and graphite—a radioactive blaze that would prove all but impossible to extinguish.

7

Saturday, 1:30 a.m., Kiev

For all the comforts of his state dacha in the southern Kiev suburb of Koncha-Zaspa, where the homes of the Party and ministerial elite lay secluded among fragrant pines, Vitali Sklyarov couldn't sleep. The Ukrainian republic's minister of energy and electrification floundered in bed as midnight came and went, and Friday night turned into Saturday morning. At one thirty, he was staring at the ceiling in frustration when the phone rang.

It was the central load dispatcher for the Ukrainian grid, who monitored the distribution of electricity throughout the republic from an office in Kiev. A call from the dispatcher in the middle of the night meant serious trouble somewhere in Ukraine's vast network of power stations and high-tension lines. For a moment, Sklyarov hoped that—whatever it was this time—there had been no casualties.

At fifty, Sklyarov had been a power man for his entire career. It had taken him sixteen years to work his way up from junior technician at a coal-fired power plant in Lugansk to station director—but after that, his ascent continued, to chief engineer of the Kiev electricity board and, eventually, head of the Energy Ministry itself. He was a lifelong Communist whose work had enabled him to travel far beyond the borders of the USSR but also brought him into frequent contact with the operatives of the dreaded "special services": the KGB. Sklyarov's glimpses of the reality of life beyond the Iron Curtain had only sharpened his cynicism, and his ascent through the ranks of the *nomenklatura* left him with a sense of how to tread carefully amid the minefields of Party politics.

Although the staff of nuclear power plants in Ukraine continued to answer directly to Moscow, the electricity they produced was Sklyarov's responsibility. While deputy energy minister, he had helped commission the first reactor at the Chernobyl station and now dealt directly with the mandarins of the USSR's clandestine nuclear state, Aleksandrov and Slavsky. He had been kept informed of problems at the station, including the September 1982 meltdown in Unit One. Even in conventional stations, over his long career Sklyarov had already witnessed his fair share of accidents—downed lines, total blackouts, cable and oil fires—in which his men had been maimed or killed. But the sudden cascade of problems in Chernobyl now being described by the dispatcher sounded worse than anything even Sklyarov had experienced before.

"There's been a series of operational disturbances at the Chernobyl nuclear plant," the dispatcher said. "Unit Four tripped out at 1:20 a.m. We've had a message about a fire at the station. Fire has broken out in the main hall and turbine hall on Number Four unit. We've lost contact with the plant."

Sklyarov immediately called the Ukrainian prime minister. When Alexander Lyashko heard the news, he told Sklyarov to take it to the top and call the first secretary of the Communist Party of the Ukrainian Soviet Socialist Republic. Vladimir Scherbitsky, leader of the republic and veteran member of the Politburo, was a hardline Party man—a sixty-eight-year-old Brezhnev crony with little interest in the reforming ways of Gorbachev. Scherbitsky had left instructions not to be disturbed for the weekend—he had left Kiev for the countryside, where he kept a coop of his beloved pigeons—and the guard who answered the phone at his dacha refused to wake the boss. Sklyarov called Lyashko back and explained the situation. Five minutes later, Scherbitsky was on the line, still half asleep.

"What happened?" the first secretary mumbled.

The first emergency calls came in to the Union-level ministries in Moscow less than thirty minutes after the explosion. The secure high-frequency lines rang at the USSR Ministry of Energy, the Third Department of the Ministry of Health, and the central command of the Ministry of Defense, and the many tentacles of the centralized state began slowly to twitch with life. From Kiev, the head of the Ukrainian Ministry of Internal Affairs notified the local offices of the KGB, the civil defense, and the public prosecutor, as well as his superiors in Moscow.

Boris Prushinsky—chief engineer of Soyuzatomenergo, the Ministry of Energy's Department of Nuclear Power, and head of OPAS, the emergency response team recently created to respond to accidents at atomic stations—was at home in bed when he was woken by a call from the duty operator. She told him there had been an accident in Unit Four at the Chernobyl power plant. Then she read aloud the code signal used to signify its severity: *Odin, dva, tri, chetyre.* "One, two, three, four." Prushinsky, barely conscious, struggled to recall what the numbers meant: A localized or general accident? A fire? Radiation? With casualties or without? It was no good. He grew impatient.

"Tell me normally," he said. "What happened?" It was 1:50 a.m.

At 2:20 a.m., a call from the central command desk at the Ministry of Defense roused Marshal Sergei Akhromeyev, chief of the general staff of the Soviet army. There had been an explosion at the Chernobyl nuclear plant; it was possible that radionuclides had been released into the atmosphere— no one seemed to know for sure. Akhromeyev told the duty officer to gather more information, and summon a meeting of the general staff. By the time the marshal arrived at his headquarters an hour later, there were no new details. Akhromeyev began issuing orders regardless.

The head of the USSR's civil defense—the branch of the Soviet armed forces responsible for protecting civilians in the event of natural disaster, nuclear war, or chemical attack—was away at a conference in Lvov, in western Ukraine. The marshal found him by telephone and gave instructions to deploy immediately the mobile radiation reconnaissance unit of civil defense stationed in Kiev. He also alerted the special Soviet army brigade dedicated to dealing with radioactive contamination, based east of the Volga River, and arranged for men and equipment to be airlifted to Chernobyl. By the time he left Moscow to lead the operation, Colonel General Boris Ivanov, deputy commander of all civil defense forces in the USSR, understood that he was dealing with an explosion in the Chernobyl plant's gas storage system and a fire in Unit Four. He planned to deploy his troops in accordance with existing plans to protect the workers and nearby population in the event of power station accidents: this was a scenario for which his staff had been specifically trained.

Boris Prushinsky listened as the duty operator gave him the decoded incident report: it was the maximum possible emergency, indicating a General Radiation Accident, with a fire and explosion. He asked the operator to get him a direct line to the power plant. Ten minutes later, a shift supervisor called him from Chernobyl but couldn't provide any details: the technician said that the reactor had been shut down, and cooling water was being supplied to the core; there was no word yet on casualties. While still on the phone, he tried to raise Unit Four on his intercom, but got no reply.

Prushinsky hung up and immediately gave orders to call out all eighteen members of the nuclear emergency response team—for the first time in its existence—and assemble them for a meeting at once. Then he called his friend Georgi Kopchinsky, a physicist who had spent three years as deputy chief engineer in Chernobyl and knew the plant and staff well. He was now in Moscow, serving the Central Committee of the Communist Party as a senior advisor on nuclear power. Prushinsky told him there had been an accident at the plant but that no details were available.

"There was some kind of explosion," he said. "The fourth power unit is burning."

Kopchinsky then telephoned his supervisor, Vladimir Marin, the Communist Party's nuclear industry chief. They agreed that a Party team should convene at the Central Committee offices as soon as possible. Kopchinsky called for a car, dressed, packed a small suitcase, and left for the offices of Soyuzatomenergo on Kitaysky Lane. When he arrived, the director was at his desk. A KGB officer sat silently in the corner. As the members of the emergency nuclear response team arrived from their homes across the city, they made plans to coordinate their response to the accident with the other ministries and departments: the Ministry of Medium Machine Building, the Ministry of Health, and the State Committee for Hydrometeorology, which monitored weather and the environment throughout the Soviet Union.

In the meantime, they tried again and again to reach someone in charge at the Chernobyl Atomic Energy Station. There was no response.

At 3:00 a.m., Vladimir Marin was still at home when his phone rang a second time. It was Viktor Brukhanov himself, calling from the bunker beneath the plant. The director confessed that there had been a terrible accident at the power station—but assured his boss that the reactor itself remained intact. Marin told his wife the news, then quickly dressed and ordered a car to take him to the Central Committee. Before leaving, Marin called his immediate superior, who then passed the message up through the Party hierarchy.

As dawn broke over the Kremlin—and even as an increasing volume of high-frequency traffic flashed across the special telephone lines connecting Moscow, Kiev, and Chernobyl—Brukhanov's reassuring assessment of what had happened began to percolate through the upper echelons of Soviet government.

By 6:00 a.m., news of the accident had reached USSR energy minister Anatoly Mayorets, and he called the Soviet prime minister, Nikolai Ryzhkov, at home. He told Ryzhkov that there had been a fire at the Chernobyl station. One unit was out of commission, but the situation was under control: a team of experts was already on its way to the plant to find out more, and his deputy for nuclear energy—an experienced nuclear specialist—had been recalled from his holiday in the Crimea to lead a government commission on the scene. Ryzhkov told Mayorets to contact his team and call back as soon as he had more information from them.

But over at Soyuzatomenergo, Georgi Kopchinsky and the other nuclear experts had already grasped that the reality might be far worse than anyone imagined. When they reached a shift supervisor at the station by phone, he had sounded incoherent and on the edge of panic. The agency's director instructed him to find someone from the station's senior management and tell them to phone Soyuzatomenergo immediately.

Chernobyl's deputy chief engineer for science was the first to call back. He calmly explained what he knew: Unit Four had been taken off-line for routine maintenance, and some kind of electrical tests were being carried out; exactly what, he couldn't say. During those tests, an accident had occurred.

But when asked about the progress of emergency cooling of the core—the vital work that would ensure Reactor Number Four could soon be repaired and brought back online—the Chernobyl engineer's composure snapped abruptly.

"There's nothing left to be cooled!" he shouted. Then the line went dead.

From his office in Kiev, Ukrainian energy minister Vitali Sklyarov, under instructions to find out what was happening at the stricken plant, had been unable to gather any more concise and accurate information about the accident over the phone, so he dispatched his own deputy minister for nuclear energy to Chernobyl by car. In the two hours it took his emissary to reach the site, Sklyarov called the station again and again and spoke repeatedly with his superiors in Moscow. He gathered an impression of deepening despair. But nobody could tell him anything specific.

At 5:15 a.m., the deputy called from the scene. The plant was still ablaze, he said. Fire crews were struggling to contain it. The roof and two walls of the reactor hall had caved in, the instruments were dead, and the supply of chemically treated water being pumped in to cool the reactor was running out. But when Sklyarov asked the urgent questions that had so far been evaded by everyone else he had spoken to—What was the level of radiation? What was the state of the reactor?—he found that even his own specialist had apparently become incapable of a straight answer.

"Things are very, very bad," was all he would say.

What kind of accident could so confound technical comprehension?

Sklyarov called First Secretary Scherbitsky once again and told him what he had.

"Vitali Fedorovich," Scherbitsky began, and Sklyarov braced himself. It was always a bad sign when the First Secretary used your patronymic. "You need to go there yourself."

Sklyarov, who had almost no interest in seeing a blazing nuclear plant at close quarters, tried to object.

"The station is under the supervision of Moscow. It doesn't belong to us," he said.

"The station might not be Ukrainian," Scherbitsky replied, "but the land and the people are."

In the bunker beneath the power plant, Director Brukhanov sat at his desk in a bewildered stupor, apparently still unable to recognize the true scale of the catastrophe and refusing to accept the radiation readings reported by his plant's own head of civil defense. The chief engineer, Fomin, who had approved the Unit Four turbine test without bothering to notify Brukhanov that it was taking place, appeared to be in shock. His commanding self-assurance had snapped with the finality of a rotten branch, and he sat, repeating the same question again and again, in the small voice of a forlorn child: "What happened? What happened?"

By eight in the morning, samples taken by technicians from the plant's Department of Nuclear Safety revealed the presence of fission products and particles of nuclear fuel on the ground and in water around the station. This provided conclusive evidence that the core of the reactor had been destroyed and that radioactive substances had been released into the atmosphere. By 9:00 a.m., Interior Ministry troops wearing green rubber chemical warfare suits and respirators were blocking access to the plant, and Vladimir Malomuzh, the deputy head of the Party for the entire Kiev region, had arrived to assume overall control of the crisis. Inside Brukhanov's office on the third floor of the main administration building, the Party chief stood by while the director heard reports from his heads of department. The doctor in charge of the plant sick bay provided details of casualties so far. There was one dead and dozens of injured; it was clear that they had been exposed to enormous levels of radioactivity and undeniably exhibited the symptoms of radiation sickness. Yet the chief of the station's external dosimetry, tasked with measuring radiation beyond the plant's limits, insisted there was no need to evacuate Pripyat. Vorobyev, the head of civil defense for the plant, tried to interrupt to say—once again—that they had a duty to inform the city's population of the accident, but this time Malomuzh cut him off.

"Sit down," he snapped. "It's not up to you."

Malomuzh instructed Brukhanov to give him a written situation report, which was then drafted by a handful of staff led by the plant Party
secretary and brought to the director's desk at around 10:00 a.m. The
document was brief—a single typed page—describing an explosion, the
collapse of the roof of the reactor hall, and a fire, which had already been
completely extinguished. Thirty-four people involved in the firefighting
were being examined in the hospital; nine had suffered thermal burns
of varying degrees, and three were in critical condition. One man was
missing, and another had died. There was no mention of any radiation
injuries. The document stated that levels of radiation near Unit Four
had reached 1,000 microroentgen per second—a tolerable 3.6 r/hr. But it
failed to explain that this was the highest reading possible for the equipment used to take the measurements. It concluded with the assurance that
the situation in Pripyat remained normal and that radiation levels there
were being investigated. At the bottom, in the space provided, Brukhanov
signed his name in blue ballpoint.

At around nine in the morning on Saturday, just as a military transport
plane was lifting off from Moscow's Chkalovsky military airport carrying Boris Prushinsky and his emergency nuclear response team toward
Chernobyl, Prime Minister Ryzhkov arrived for work in the Kremlin.
The son of a coal miner, an expert in logistics, Ryzhkov's industrious
management of the Soviet economy had driven his ascent through the
ranks of government; at fifty-six, he was trim and energetic, a moderate
ally to Gorbachev in his campaign for reform. He usually made it into
his office a little late on the weekends, and today was no exception. As
soon as he arrived, he called Anatoly Mayorets for a further update on
the plant fire in Ukraine.

The energy minister's report was bleak. He now believed things were
far more serious than he'd first thought: this was not the usual kind of
accident at all. There had been an explosion in one of the reactors; the
damage was extensive; the consequences were hard to predict; urgent
measures were necessary. Ryzhkov told Mayorets to assemble a second,

more senior team of experts and take a plane to Kiev at once. He sent orders to Aeroflot, to make an aircraft available. Then, according to the normal procedures set down for a major accident, Ryzhkov began organizing yet another, even more senior team—his own handpicked government commission—to travel to the scene and assume responsibility for addressing the accident and its consequences. As chairman, he chose Boris Scherbina, the balding, bulldog-faced deputy prime minister of the USSR in charge of all fuel and energy operations throughout the Union. Ryzhkov located him more than a thousand kilometers away in Orenburg, close to the border with Kazakhstan, where Scherbina was about to make a speech to local oil workers. Ryzhkov told him to wrap it up immediately and get back to Moscow, where a plane would be waiting to take him on to Ukraine.

At 11:00 a.m., with the second team led by Mayorets already in the air, Ryzhkov signed a formal decree to establish the commission. He sent out word to gather its members—including senior figures from the Soviet Academy of Sciences, the Kurchatov Institute, the Prosecutor General's Office of the USSR, the KGB, the Ministry of Health, and the Ukrainian Council of Ministers—as soon as possible.

Academician Valery Legasov, first deputy director of the Kurchatov Institute of Atomic Energy, woke that morning entirely ignorant of what was happening in Ukraine. It was a beautiful day, and he couldn't decide whether to spend it with his wife, Margarita, go in to work at the department he chaired at Moscow State University, or attend a meeting of his Communist Party *aktiv* at the headquarters of the Ministry of Medium Machine Building.

Ever the reliable Communist, Legasov chose the Party meeting. When he arrived, just before 10:00 a.m., a colleague mentioned an unpleasant incident at the Chernobyl plant. The meeting was led by Sredmash chief Efim Slavsky. It was the usual hot air: the old man declaimed at tedious length about the many successes and victories of the ministry and scolded a handful of individuals for their failures. Generally, everything was proceeding as wonderfully as ever: all their plans were being fulfilled, their targets met. In the midst of his traditional hymn to the glories of the nuclear power industry, Slavsky paused, and then mentioned that, apparently, there had been some mishap at an atomic station in Ukraine. But he added quickly that this plant was run by their neighbors at the Ministry of Energy. Whatever the accident was, it wouldn't halt the onward march of Soviet nuclear power.

At noon, they stopped for a break, and Legasov went up to the second floor to chat with a colleague. There, Alexander Meshkov, Slavsky's deputy, gave Legasov some urgent news: he had been selected to serve on a government commission investigating the Chernobyl incident. He was to be at Moscow's Vnukovo Airport at four that afternoon. Legasov immediately called a car to take him over to the Kurchatov Institute. Despite his seniority at the state's leading agency for nuclear research, he was a chemist, not a reactor specialist. He would need some expert advice.

The son of a senior Party ideologue, Valery Legasov was the head of the Communist Party Committee at the Kurchatov Institute. As university students in the 1950s, he and Margarita had joined the Komsomol brigades to raise wheat from the plains of southern Siberia, and he later chose to do his postgraduate work in radiochemistry in the remote Tomsk-7 Chemical Combine rather than accept a cozy position in Moscow. Both an intellectual and a scientist, he believed in the principles of Socialism and an equal society run by an educated elite. Legasov was witty and opinionated, and his privileged background gave him the confidence to speak his mind in a world of cowed apparatchiks. In his spare time, he wrote poetry. Despite his candor, he was beloved by his Party superiors and had risen to power with amazing speed, acquiring state prizes for his work and every accolade available to a Soviet scientist, with the exception of the highest award of all: the Hero of Socialist Labor.

Legasov, thickset but athletic, with dark hair and heavy spectacles, was now approaching the pinnacle of his career and enjoyed the life of privilege accorded to the stars of Soviet science. He played tennis, skied, swam, and traveled widely. He and Margarita occupied a grand villa at Pekhotnaya 26, on a tree-lined street within walking distance of his office—where they entertained visiting friends and colleagues, including his boss, Anatoly Aleksandrov. Now eighty-three, the chairman of the Academy of

Sciences and head of the Kurchatov Institute lived a few doors down from Legasov. Aleksandrov liked to drop round for dinner and play chess; he often remarked that his deputy was always thinking several moves ahead. Still only forty-nine, Legasov seemed set to assume his place as the director of the Kurchatov Institute just as soon as Aleksandrov stepped aside.

Only one man stood in his way: his next-door neighbor, Evgeny Velikhov, a portly and gregarious plasma physicist, descended from a family of inventors and freethinkers. A personal scientific advisor to Gorbachev and the head of his own theoretical research lab on the outskirts of Moscow, Velikhov was also high up in the administration of the Kurchatov Institute—and Legasov's closest rival. He had traveled abroad, was well connected with Western scientists, spoke serviceable English, and liked to wear a Princeton tie. But he was a rare presence in the dining room of Pekhotnaya 26. When Legasov professed to be mystified by his colleague's apparent hostility, his wife had a simple remedy: "Tell him less about your successes."

When he arrived at the institute at lunchtime on Saturday, it took Legasov a while to find the man he wanted: Alexander Kalugin, the resident expert on RBMK reactors. Kalugin had taken the day off, but when he finally learned of Legasov's summons, he brought with him all the technical documents he could find about the reactor and the Chernobyl plant. Legasov then rushed home to tell his wife he was leaving—although what he would be doing and how long he would be gone, he didn't know—and made for the airport. Despite the fine weather, he was still wearing the suit and expensive leather coat he'd put on that morning.

At around 11:00 a.m., a little more than nine hours after the crisis had begun, the first of the planes from Moscow touched down on the tarmac in Kiev. Led by Boris Prushinsky, the Ministry of Energy's nuclear accident emergency response team included scientists from Soyuzatomenergo and the institutes that designed the reactor and the plant itself, members of the KGB, and a quartet of specialists from Moscow's Hospital Number Six—the State Institute of Biophysics' clinic dedicated to treating radiological injuries. When he landed, Prushinsky learned that a government commission was on its way to take control of the situation. But if any further information about the true extent of the accident had reached the Politburo in Moscow, it was not relayed to Prushinsky and his specialists. As they were driven the 140 kilometers to Pripyat by bus, under police escort, their mood was grim—they knew that two men were now dead. But they remained mystified by what might have happened. Perhaps the roof of the reactor hall had collapsed, or some machinery had caught fire. Yet they still believed that the reactor had been shut down safely and was being cooled with water; there would be no more casualties.

So when the bus finally approached the fork in the road separating the city from the nuclear plant, and Prushinsky saw a *militsia* officer wearing a *lepestok* mask with his summer uniform, he was puzzled. The *lepestok*, or "petal," was a Soviet-designed cloth respirator, which filtered radioactive aerosols from the atmosphere, and he couldn't imagine why it might be necessary. But when the team reached Pripyat, someone from the plant was there to meet them and assured him that everything was under control. Relieved, Prushinsky checked into Hotel Polesia, an eight-story concrete building overlooking the central square, and went to lunch in the restaurant downstairs. Afterward, he ambled out onto the sunny hotel terrace and saw Director Brukhanov crossing the plaza toward him.

"What's the problem with the unit?" Prushinsky asked.

Although the shell-shocked director would later continue to give his superiors contradictory information—and for several more hours tell others that Reactor Number Four remained intact—at this moment, Brukhanov acknowledged the truth.

"There is no unit anymore," he said.

Prushinsky was dumbfounded. He knew this man was no nuclear expert. But what he was suggesting was simply inconceivable.

"Have a look for yourself," Brukhanov said in despair. "The separators are visible from the street."

Back in Moscow, the information from Brukhanov's earlier written report was still ascending slowly through the channels of Party bureaucracy. At noon, Deputy Energy Minister Aleksei Makukhin sent a seventeen-line cable to the Central Committee that relayed the director's reassuring prognosis. Although marked "Urgent," it passed from the General Department to the Department of Atomic Energy, and it was not until later on Saturday afternoon that the report was delivered to Gorbachev.

"An explosion occurred in the upper part of the reactor chamber," the cable read. "The roof and parts of the wall panels of the reactor compartment, several roof panels of the machine room . . . were demolished during the explosion and the roofing caught on fire. The fire was extinguished at 0330."

To a government that had developed a strong stomach for industrial accidents, this was familiar territory. Some sort of explosion, yes; a fire, which had already been put out. A serious incident, certainly, but nothing that couldn't be contained. The main thing was that the reactor itself remained undamaged; a potential nuclear catastrophe had been averted.

"The personnel of the AES are taking steps to cool the fuel core of the reactor. In the opinion of the Third Main Directorate of the USSR Ministry of Health," the report stated, "it is not required to take special measures, including the evacuation of the population from the city."

At 2:00 p.m., the second—more senior—wave of government officials, led by Energy Minister Anatoly Mayorets, arrived in Kiev aboard a chartered executive jet from Moscow. Vitali Sklyarov, Mayorets's Ukrainian counterpart, met them on the runway, and together they transferred to a pair of antiquated-looking Antonov An-2 biplanes. Mayorets, new to his job and not a nuclear man, was confident. "You know," he said, "I don't think we're going to be sitting around in Pripyat for too long." He thought they would be on their way home again in forty-eight hours.

"Anatoly Ivanovich," Sklyarov said, "I don't think two days is going to be enough."

"Don't try to frighten us, Comrade Sklyarov. Our main job is to restore the damaged reactor unit as soon as possible and get it back on the grid."

Bumping down on a dirt airstrip outside Chernobyl, they sped into Pripyat, through the dappled shadows of the poplars on Lenina Prospekt. Sklyarov noticed that people were occupied just as they would be on any warm weekend afternoon. Children were playing soccer, fresh laundry hung on balconies, and couples lingered in the central square in front of the new shopping center. He asked someone about the radiation levels and was told that here the readings were around ten times normal background—apparently within permissible limits. Sklyarov, too, grew more optimistic.

The ministers gathered inside the headquarters of the Pripyat Communist Party and *ispolkom*, or city council: a five-story concrete building next to the Hotel Polesia, and called by those who worked in it *belyy dom*, or the "White House." Malomuzh, the Party boss from Kiev, had established his command post there. General Ivanov, the Soviet civil defense chief, had arrived from Moscow and suggested that the Party authorities make a radio broadcast to warn Pripyat's citizens that an accident had taken place. In the meantime, his troops were conducting a radiation reconnaissance of the plant and the city itself.

The assembled ministers and experts now began a fierce debate about how best to cool down Reactor Number Four and clean up the mess left by the accident. But they could take no decisive action before the arrival of the commission chairman, Boris Scherbina, who was still en route from Moscow. Outside, the weather was clear and warm. In the hotel next door, a traditional Ukrainian wedding celebration had begun.

Circling the reactor at low altitude in the helicopter, Boris Prushinsky realized that Director Brukhanov had been right about the fate of Unit Four of the Chernobyl Atomic Energy Station. Yet the head of the nuclear emergency response team still found it hard to believe what he was seeing.

The roof of the central hall had disappeared. Inside was a gaping black crater, where more than ten stories of walls and floors had been carved away as if scooped out from above by a monstrous spoon. The northern wall of the building had collapsed into a shambles of black rubble that tumbled out across the flat roofs of adjoining buildings and toward the station perimeter. Inside the ruins of the hall, he could see the tangled wreckage of the 120-tonne bridge crane, the refueling machine, the main circulation pumps, and the emergency core cooling tanks. The pilot tilted the helicopter to one side, so the plant photographer could shoot through the window. Prushinsky saw that the lid of the reactor—Elena, the two-thousand-tonne disc of concrete and steel designed to shield it from the outside world—was tilted upward to the sky. Beneath it, deep within the reactor vault, and despite the bright sunshine, he could make out the glowing lattice formed by the remaining fuel cells—and a single spot that burned a fierce yellow-red. As the helicopter banked away, Prushinsky forced himself to confront what his mind still refused to accept: Reactor Number Four had ceased to exist.

At a 4:00 p.m. meeting in the Party's conference room in the White House, Chief Engineer Nikolai Fomin finally conceded that the efforts made by his men over the previous twelve hours to keep cooling water circulating through Reactor Number Four had been entirely in vain. He admitted that the reactor had been destroyed and pieces of highly radioactive graphite lay on the ground everywhere. Worse news was to come. That morning, station physicists had entered the contaminated control room of Unit Four and established that the control rods had not descended fully into the reactor before the explosion. They now suspected that the conditions for a new criticality would soon exist in whatever nuclear fuel remained inside the reactor vessel, initiating a new chain reaction—but this time, it would take place in the open air, and they would have no means of controlling it. When the reactor returned to life, it could bring fires and explosions and release waves of deadly gamma and neutron radiation into the atmosphere less than 2,500 meters from the edge of Pripyat. Their estimates suggested they had only three hours to intervene before the new criticality began—sometime after seven in the evening.

Shortly before 5:00 p.m., Senior Lieutenant Alexander Logachev of the Kiev region civil defense unit ran into the White House with the results of his ground radiation survey of the plant. His armored personnel carrier had come down Lenina Prospekt at one hundred kilometers per hour—so fast that the seven-tonne vehicle became airborne as it crested the railway bridge, and then drove right up the steps and across the pedestrian plaza to the main entrance. The map the breathless Logachev presented to Malomuzh showed the radiation reading just beside the plant cafeteria, scribbled hastily in pencil: 2,080 roentgen an hour.

"You mean milliroentgen, son," the Party chief said.

"Roentgen," Logachev said.

Logachev's commander studied the map. He finished one cigarette and then lit another.

"We need to evacuate the city," he said.

The aircraft carrying Scherbina and Academician Valery Legasov landed at Kiev's Zhuliany Airport at 7:20 p.m. on Saturday evening. They were greeted by a delegation of nervous Ukrainian ministers and a glistening line of big black cars, which delivered them to the steps of the Pripyat *ispolkom* in the gathering twilight. As they drove north, Legasov had watched collective farms give way to peasant pasture and boundless levels of marsh, lush water meadows, and dense pine forest. They were wound tight with anxiety about what lay ahead; conversation faltered, and stopped. In the lengthening silence, Legasov willed them on to their destination. But in Pripyat, Scherbina—a veteran of gas pipeline explosions and other industrial catastrophes—emerged from the back of his giant Chaika limousine wearing a confident smile: a command-economy savior come to deliver his underlings from any potentially perilous decision-making.

Sklyarov, the Ukrainian energy minister, had encountered Scherbina often over the years, during the Moscow boss's tours of inspection at the many power plants under construction in the republic. At sixty-six, Scherbina was intelligent, energetic, and hardworking, tough and self-assured—but also emotional and impulsive and always determined to demonstrate that he knew better than anyone else, even the specialists. Small and wiry, what he lacked in stature he made up for with imperious attitude. Some regarded him with respect and admiration. The energy minister had found him all but impossible to work with.

Scherbina introduced himself quietly to each of the gathered experts, until he came to Sklyarov, who, by now, had already driven out to the plant and seen the destruction of the reactor for himself.

"So, are you shitting your pants?" Scherbina asked.

"Not yet," Sklyarov said. "But I think things are going that way."

Upstairs, Boris Prushinsky had just returned from his reconnaissance of the plant. As Scherbina came within earshot, Prushinsky was in the hallway, sharing his alarming discoveries with the Soviet minister for nuclear energy. After the helicopter flight over the reactor, Prushinsky had continued his investigation on the ground, studying the ruins of Unit Four through binoculars. From the station perimeter, he could make out the blocks of graphite scattered on the ground around the plant. It was obvious to him that there had been an explosion inside the reactor and that pieces of nuclear fuel must lie among the debris.

"We have to evacuate the local population," Prushinsky said.

"Why are you being so alarmist?" Scherbina asked.

The first meeting of the government commission began in the third-floor office of the Pripyat Party secretary sometime after 10:00 p.m. As many as thirty government ministers, military officers, and industry specialists took their seats on three rows of chairs beside the door. Scherbina stood in the middle of the room, at a table littered with maps, documents, and ashtrays filled with cigarette butts. It was hot, and the air was thick with smoke; the tension was stifling.

Academician Legasov listened as Scherbina took reports from Malomuzh—the regional Party boss—and Mayorets, the Soviet energy minister. But they provided no detailed information about the situation at either the plant or in the city and had no plan on how to deal with the consequences of the accident. They said only that during a turbine rundown experiment in Unit Four, there had been two explosions in quick succession, and the reactor hall had been destroyed. There were hundreds of injuries among the staff: two men were dead, and the rest were in the city hospital. The radiation situation in Unit Four was complex, and although levels in Pripyat had departed significantly from normal, they posed no threat to human health.

Scherbina divided the members of the commission into groups. One, headed by Meshkov, the deputy head of the Ministry of Medium Machine Building, would begin investigating the causes of the accident. A second would take further dosimetry readings. General Ivanov of the civil defense and General Gennadi Berdov of the Ukrainian Interior Ministry would prepare for a possible evacuation. Evgeny Vorobyev, deputy health minister for the entire Soviet Union, would take care of all medical matters. Finally, Valery Legasov himself would oversee the team containing the effects of the disaster.

Like the station's own physicists, Legasov's first concern was the possibility of a new chain reaction in the remains of Reactor Number Four. The plant operators had already tried to drench the nuclear fuel by pouring bags of boric acid powder—which contained neutron-absorbing boron—into the water tanks of the cooling system. But the chemical solution had disappeared into the maze of broken piping tangled in the reactor hall. They couldn't be certain where it had gone, and now supplies were running low. Ukrainian energy minister Sklyarov ordered another ten tonnes of the powder to be sent from Rovno nuclear power station, more than three hundred kilometers away to the west. But the Rovno station director was reluctant to part with it—what if *he* had an emergency? And when it was finally dispatched by truck, the vehicle broke down. It would not arrive in Chernobyl until the following day.

At the same time, Legasov realized that the heroic but doomed efforts of the plant operators to cool the shattered reactor core with water had resulted only in flooding the basement spaces of Units Three and Four with contaminated water and sending clouds of radioactive steam billowing into the atmosphere. In addition, there was the poisonous torrent of radioactive aerosols being carried into the air from the crater of Reactor Number Four—where the glowing lattice of fuel cells and the ominous spot of incandescence that Prushinsky had glimpsed suggested strongly that something was on fire. Somehow the blaze had to be put out and the reactor sealed.

But the debris thrown from inside the core also made the station and its grounds a radioactive minefield. Unit Four was now potentially lethal to approach for anything but the shortest time. Getting close enough to cover the reactor, or even fighting the fire using conventional methods like foam—or water, as the British had done in Windscale nearly thirty years earlier—was impossible. Yet no one on the commission could suggest how the burning reactor might be smothered. Legasov looked around him in consternation: the politicians were ignorant of nuclear physics, and the scientists and technicians were too paralyzed by indecision to commit to a solution. Everyone knew that *something* must be done—but what?

And as dense clouds of radionuclides continued to roil into the sky above Reactor Number Four, the experts assembled in the White House still could not agree on whether to evacuate Pripyat. The civil defense radiation scouts had been taking hourly readings on the streets of the city since noon, and they found the figures alarming: on Lesi Ukrainki Street, less than three kilometers from the reactor, by midafternoon, they had recorded readings of 0.5 roentgen an hour; by nightfall, it was up to 1.8 roentgen. This reading was tens of thousands of times higher than normal background radiation, but the Soviet deputy minister of health insisted that it posed no immediate threat to the population. He pointed out indignantly that even after the still-undisclosed 1957 disaster in Mayak, the population of the secret city had not been told to leave. "They never evacuated people there!" he said. "Why do it here?"

Indeed, the threshold officially stipulated by the Soviet authorities for evacuation in the event of a nuclear accident lay far off. According to the state document "Criteria for Making a Decision on Protection of the Population in the Event of an Atomic Reactor Accident," only if citizens were forecast to acquire a lifetime dose of 75 rem—fifteen times the annual level deemed safe for workers inside a nuclear power station—did evacuation become mandatory. Even the regulations governing when it was necessary to tell the population that a radiation leak had taken place were contradictory, and it was unclear who had the final say in authorizing evacuation. Scherbina may have feared creating panic in Pripyat. But at that point, he had little reason to believe that Soviet citizens—long hardened to news of misfortune and distrustful of official information—would really lose their heads if warned of an accident; more urgent was the state's compulsion for secrecy. By daybreak on Saturday, the men of the *militsia* had sealed off the entire area with roadblocks, and the KGB then cut off the city's long-distance telephone lines. By nightfall, the local lines were dead, too, and there had still been no radio broadcast to notify the citizens of Pripyat of the accident, let alone warn them to stay indoors or close their windows. Even so, in the event of an evacuation, Scherbina knew there would be no way of concealing the exodus of the fifty thousand residents of an entire *atomgrad*.

And yet, the civil defense commanders and the physicists disagreed with the health minister's sanguine forecast: even if the radiation situation in the town seemed tolerable in the short term, it was unlikely to improve. So far, the plume of vapor from the reactor had been drifting north-northwest—away from Pripyat and Kiev, toward Belarus; by noon on Saturday, chemical troops had registered external dosages of radiation along its path measuring a life-threatening 30 roentgen per hour, as far as fifty kilometers from the plant. But the wind could change at any moment, and there were already thunderstorms to the southeast. If even the smallest amount of rain fell on Pripyat, it would bring down radioactive fallout, with terrible consequences for the population. From Kiev, the Ukrainian prime minister had already unilaterally given orders to arrange transport—more than a thousand buses and trucks—for a possible evacuation of the city. Yet nothing would move without sanction from the top. And Scherbina wanted more information before making a decision. He resolved to wait until morning.

In the meantime, something had begun stirring inside the yawning vault of Unit Four. At around eight o'clock on Saturday evening, the plant's deputy chief engineer for science noticed a ruby glow shimmering within the ruins. This was followed by a series of small explosions and brilliant white flashes that leapt from the wreckage of the central hall like geysers of light, illuminating the full height of the 150-meter vent stack. Two hours later, a team led by a member of the Ministry of Energy's nuclear research institute, VNIIAES, was taking samples from the coolant canal when the walls of Unit Four were shaken by a thunderous roar. The technicians took shelter beneath a box-girder bridge as incandescent fragments showered from the sky, and the needles of their dosimetry equipment ran off the end of their dials.

Back in Pripyat, the meetings of the government commission wound on. An air of unreality still prevailed: at one point, the chairman's assistants drafted an action plan for repairing Reactor Number Four and reconnecting it to the Soviet electrical grid, although by then it was obvious that such a feat was impossible. And, according to Vitali Sklyarov's account, at sometime before midnight, a functionary interrupted a meeting to tell Scherbina that General Secretary Gorbachev would be calling him shortly for a situation report. The deputy minister ordered the room cleared. As Sklyarov rose to leave, Scherbina stopped him.

"No, no. Sit down," he said. "Listen to what I'm going to say. Then you're going to tell your superiors exactly the same thing."

The VCh—the scrambled high-frequency line, from Moscow—rang, and Scherbina answered.

"There's been an accident," the deputy minister told Gorbachev. "Panic is total. Neither the Party organs, the secretary of the oblast, nor the rayon committees are here at present. I'm going to demand that the minister of energy restart all units. We're going to take all measures to liquidate the accident."

Scherbina was silent for a few moments as Gorbachev spoke.

At last, Scherbina said, "Okay," and replaced the receiver in its cradle. He turned to Sklyarov.

"Did you hear all that?"

He had; he was appalled. "You can't restore the reactor, because there is no reactor," he said. "It no longer exists."

"You're a panicker."

"I've seen it with my own eyes."

A few minutes later, the special telephone rang again. This time it was Scherbitsky, the Ukrainian Communist Party chief.

Scherbina repeated to Scherbitsky what he'd just told Gorbachev: a can-do shock-work action plan of fantasy and denial. Then he handed the phone to Sklyarov.

"He wants to talk to you. Just say what I was saying."

"I don't agree with what Comrade Boris Evdokimovich is saying," Sklyarov said. "We need to evacuate everyone."

Scherbina snatched the phone from the energy minister's hand.

"He's a panicker!" he yelled at Scherbitsky. "How are you going to evacuate all these people? We'll be humiliated in front of the whole world!"

8

Saturday, 6:15 a.m., Pripyat

It was sometime after 3:00 a.m. when Alexander Esaulov was jangled awake by the telephone. *Shit!,* he thought as he fumbled for the receiver. *Another weekend ruined.*

With his wife and children off with the in-laws for a few weeks, he'd been looking forward to enjoying a few days to himself: perhaps squeezing in a little fishing. Having the two children at home—a five-year-old daughter and a son about to turn six months—there was always plenty of work to do, even without his job. And as the deputy chairman of the Pripyat city *ispolkom*—the equivalent of deputy mayor—Esaulov spent his days reeling from one administrative headache to another.

He had come to Pripyat from Kiev, where he had worked in the city's municipal financial planning department. It was a nice step up for the thirty-three-year-old accountant and his family: out of the rotting communal apartment with a queue for the bathroom every morning, into the clean air of the countryside, and a prestigious job with his own secretary and the use of a car—dilapidated but serviceable—for work. Still, Esaulov found his new responsibilities onerous. He had to manage Pripyat's city budget, expenses, and income—but also served as the head of the planning commission and oversaw transportation, health care, communication, road and street cleaning, the employment bureau, and the distribution of building materials. There was always something going wrong, and the citizens of Pripyat were never hesitant to complain when it did.

On the phone was Maria Boyarchuk, the secretary from the *ispolkom*. She'd just been woken by a neighbor who had come from the nuclear power plant. There had been an accident: a fire, maybe an explosion.

By 3:50 a.m., Esaulov was at his desk in the *ispolkom* offices, on the second floor of the White House. The chairman—the city mayor—had left for the plant to find out what was going on. Esaulov telephoned the head of Pripyat's civil defense, who sprang from his bed and raced over to the office. But neither of them had any idea what to do. The plant had its own civil defense staff, and the city had never been involved in any of its exercises. There had been accidents at the station before, but they had always been cleaned up with a minimum of fuss.

Now they phoned every number they had at the station, but no one would tell them anything. They considered driving there, but didn't have a car. All they could do was sit and wait. Outside the window, the streetlamps threw pools of amber light across the square; the apartments along Kurchatov Street remained dark and silent.

But as dawn approached, Esaulov watched from behind his desk as a single ambulance raced down Lenina Prospekt from the direction of the plant. Its emergency lights flashed, but the siren remained silent. The driver took a sharp right at the Rainbow department store, tore along the southern side of the square, and then swung away in the direction of the hospital. A few moments later, a second ambulance followed, and it, too, disappeared around the corner.

The blue lights faded into the distance, and the city streets were still once more. But then another ambulance sped past. And another. Esaulov began to suspect that there might be something different about this accident, after all.

As dawn broke, word began to spread among those with friends and relatives on the midnight shift at the plant that there had been some kind of accident. But nobody could say exactly what.

At around 7:00 a.m., Andrei Glukhov, who worked in the reactor physics laboratory at the plant, was in his flat on Stroiteley Prospekt when the phone rang. It was a friend from the Instrumentation and Control Department. He, too, was at home and had heard that something had gone wrong at the station, but he didn't know any details. As a member of the Nuclear Safety Department, Glukhov had the authority to make calls directly to the control rooms of every reactor at the plant. Would he mind making a few enquiries?

Glukhov hung up and phoned his friend Leonid Toptunov, on the senior reactor control engineer's desk of Unit Four. But nobody answered. *Strange*, he thought. *Maybe he's busy.* He tried Control Room Number Two, where the senior reactor control engineer picked up immediately.

"Good morning, Boris," Glukhov said. "How is everything going?"

"Okay," the engineer said. "We're raising power on Unit Two. Parameters are normal. Nothing special to report."

"Okay. How about Unit Four?"

There was a long silence on the line.

"We've been instructed not to talk about it. You'd better look out of the window."

Glukhov went out to the balcony. The apartment was on the fifth floor, and from his position just behind the new Ferris wheel, he had a decent view of the plant. But he couldn't see much out of the ordinary. Some smoke hung in the air above Reactor Four. Glukhov had a cup of coffee and told his wife he would head to Kurchatov Street to meet the bus bringing the night shift back from the plant. They'd be able to tell him what was going on.

He waited at the bus stop, but the men from the shift never arrived. Instead, a truck filled with policemen pulled up. Glukhov asked what had happened. "It's not clear," the policeman said. "The wall of the reactor hall has collapsed."

"*What?*"

"The wall of the reactor hall has collapsed."

It was an unbelievable idea. But Toptunov would surely have an explanation.

Perhaps I just missed the bus, Glukhov thought. *Leonid might already be at home.*

It was less than a fifteen-minute walk from the bus stop to Toptunov's apartment building. Glukhov climbed to the top floor, turned right at the head of the stairs, and walked to the door at the end of the hall: number 88, handsomely upholstered in red imitation leather. He pushed the buzzer. He pushed it again. There was no reply.

The Pripyat hospital, Medical-Sanitary Center Number 126, was a small complex of biscuit-colored buildings behind a low iron fence on the eastern edge of the city. It was well equipped to serve the growing town and its young population, with more than 400 beds, 1,200 staff, and a large maternity ward. But it hadn't been set up to cope with a catastrophic radiation accident, and when the first ambulances began to pull up outside in the early hours of Saturday morning, the staff was quickly overwhelmed. It was the weekend, so it was hard to find doctors, and, at first, no one understood what they were dealing with: the uniformed young men being brought from the station had been fighting a fire and complained of headaches, dry throats, and dizziness. The faces of some were a terrible purple; others, a deathly white. Soon all of them were retching and vomiting, filling wash basins and buckets until they had emptied their stomachs, and even then unable to stop. The triage nurse began to cry.

By 6:00 a.m., the director of the hospital had formally diagnosed radiation sickness and notified the Institute of Biophysics in Moscow. The men and women arriving from the plant were told to strip and surrender their personal effects: watches, money, Party cards. It was all contaminated. Existing patients were sent home, some still in their pajamas, and the nurses broke open emergency packages designed for use in case of a radiation accident, containing drugs and disposable intravenous equipment. By morning, ninety patients had been admitted. Among them were the men from Control Room Number Four: Senior Reactor Control Engineer Leonid Toptunov, Shift Foreman Alexander Akimov, and their dictatorial boss, Deputy Chief Engineer Anatoly Dyatlov.

At first, Dyatlov refused treatment and said he only wanted to sleep. But a nurse insisted on putting in an IV line, and he began to feel better. Others, too, seemed not to have been too badly injured. Alexander Yuvchenko felt dizzy and excited but soon fell asleep, woken only when a nurse came to attach a drip. He recognized her as a neighbor from his apartment building and asked her to find his wife when she finished her shift, to reassure her that he would be home soon. In the meantime, Yuvchenko and his friends tried to estimate how much radiation they had received: they thought 20 rem, or perhaps 50. But one, a navy veteran once involved in an accident on a nuclear submarine, spoke from experience: "You don't vomit at fifty," he said.

Vladimir Shashenok, rescued from the wreckage of compartment 604 by his colleagues, had been one of the first to arrive. Burns and blisters covered his body, his rib cage was caved in, and his back appeared to be broken. And yet, as he was carried in, the nurse could see his lips moving; he was trying to speak. She leaned closer. "Get away from me—I'm from the reactor compartment," he said.

The nurses cut the shreds of filthy clothing from his skin and found him a bed in intensive care, but there was little they could do. By 6:00 a.m., Shashenok was dead.

It was not yet eight o'clock when Natalia Yuvchenko heard her doorbell ring. She had awoken early, tired and anxious. Her son's cold had kept him awake and crying throughout the night, and the apprehension Natalia had felt the evening before only deepened. But the city schools, like those throughout the Soviet Union, had classes on Saturday morning, and she was due to begin teaching at eight thirty. So she washed, dressed, and waited for Alexander to get back from the plant. As the night shift finished at eight, if he hurried to the bus, he'd make it home just in time to take over with Kirill before Natalia had to leave.

But instead of her husband, a stranger was at the door: a woman whose face looked familiar but whom, at first, she couldn't place. It was a neighbor who worked at the hospital.

"Natalia," she said, "your husband asked me to tell you that you shouldn't go to work. He's in the hospital. There's been an accident at the station."

Around the corner, at home on Heroes of Stalingrad Street, Maria Protsenko heard a commotion in the apartment downstairs. Just as she did whenever she wanted to telegraph the neighbors below with important news, or to share something on the stove that was particularly tasty, Protsenko rapped on the kitchen radiator with a spoon. The response came clanging right back: *Come down!*

Protsenko was a small but formidable forty-year-old who wore her dark, curly hair cropped sensibly short, born in China to Sino-Russian parents, yet forged in the crucible of the USSR. Her grandfather had been arrested and disappeared into the Gulag during Stalin's purges; when she was a baby, both her older brothers died from diphtheria because they were kept from seeing a doctor by the curfew in the Chinese border town where they lived. After that, her grief-stricken father sank into opium addiction, and her mother fled into Soviet Kazakhstan, where she raised Maria alone. A graduate in architecture from the Institute of Roads and Transport in Ust-Kamenogorsk, Protsenko had been Pripyat's chief architect for seven years, with her own office on the second floor of the city *ispolkom*. It was from there that she oversaw the execution of Pripyat's new construction projects, with a profoundly un-Soviet eye for detail. Barred from Party membership by her Chinese birth, she brought an outsider's zeal to her work. She roamed the streets with a ruler, checking on the quality of the concrete paneling in new apartment buildings. She chastised the construction workers for shoddy sidewalks: "Children will break their legs, and then how will you feel?" When persuasion wouldn't work, she lashed them with invective. More than a few of the men were afraid of her.

Many of Pripyat's apartments and major buildings—the Palace of Culture, the hotel, the *ispolkom* itself—were erected from standardized blueprints produced in Moscow and designed to be reproduced identically in every town in every corner of the USSR. But Protsenko did what she could to make them unique. In spite of the prevailing state doctrine calling for gloomy "proletarian aesthetics"—rejecting decadent Western notions of individuality in the interests of economy—she wanted them to be beautiful. Protsenko worked frugally with small supplies of hardwood, ceramic tiles, or granite to decorate the interiors of Pripyat's public buildings, designing parquet flooring and wrought-iron screens in a botanical pattern for the restaurant, or inlaying small sections of marble in the walls of the Palace of Culture. She watched as the city grew from just two microdistricts to three, then four. She helped choose the names of the new streets as they were added, and attended to the finer points of each of the city's new amenities. The library, the swimming pool, the shopping center, the sports stadium—all passed through her hands.

As she left her apartment that morning, Protsenko was still expecting to spend the day at the office, busy with preparations for another expansion of the city. Only the day before, she had received a delegation from the urban design institute in Kiev. Together they were planning the infrastructure of Pripyat's sixth new district, to be constructed on reclaimed land beside the river to accommodate the workers who would operate the first reactors of Director Brukhanov's massive new Chernobyl Two plant. Dredging was under way, bringing up sand from the river bottom to provide the foundations of the extra neighborhoods. When they were complete, Pripyat would be home to as many as two hundred thousand people.

By the time Protsenko made her way to the apartment downstairs, it was past eight o'clock on Saturday morning. Her fifteen-year-old daughter had already left for school; her husband, who worked as a mechanic for the city, was still asleep in bed. She found the neighbors—her close friend Svetlana and her husband, Viktor—sitting at their kitchen table. Despite the early hour, they were drinking shots of moonshine vodka, or *samogon*. Svetlana explained that her brother had called from the plant. There had been an explosion.

"We're going to chase away the *shitiki*!" Viktor said, raising a glass. Like many construction and energy workers at the plant, he believed that radiation created contaminated particles in the blood—*shitiki*—against which vodka was a useful prophylactic. Just as Protsenko was telling him that she didn't think she could handle *samogon*, whatever the need, her own husband appeared in the doorway: "You've got a phone call."

It was the secretary from the *ispolkom*. "I'm coming over," Protsenko said.

By 9:00 a.m., hundreds of members of the *militsia* had been mobilized on the streets of Pripyat, and all roads into the town had been cut off by police roadblocks. But as the city's leaders—including Protsenko, Deputy

Mayor Esaulov, the Pripyat civil defense chief, and the directors of schools and enterprises—gathered for an emergency meeting at the White House, elsewhere in Pripyat the day began exactly as it would on any warm Saturday morning.

Across the city's five schools and in the Goldfish and Little Sunshine kindergartens, thousands of children started their lessons. Beneath the trees outside, mothers walked babies in their strollers. People took to the beach to sunbathe, fish, and swim in the river. In the grocery stores, shoppers stocked up on fresh produce, sausage, beer, and vodka for the May Day holiday. Others headed off to their dachas and vegetable gardens on the outskirts of the city. Outside the cafe beside the river jetty, last-minute preparations were under way for an alfresco party to celebrate a wedding, and at the stadium, the city soccer team was warming up for an afternoon match.

Inside the fourth-floor conference hall of the White House, Vladimir Malomuzh, the Party's second secretary for the Kiev region, took the stage. Malomuzh had arrived from Kiev just an hour or two earlier and, since the Party took precedence over government in emergency situations, was now in charge. Beside him stood the two most powerful men in the city: plant director Viktor Brukhanov and construction boss Vasily Kizima.

"There has been an accident," Malomuzh said, but offered no further information. "The conditions are being evaluated right now. When we know more details, we'll let you know."

In the meantime, he explained, everything in Pripyat should carry on as usual. Children should stay at school; stores should remain open; the weddings planned that day should continue.

Naturally, there were questions. Members of the Young Pioneers of School Number Three—1,500 children in all, part of the Communist equivalent of the Scouts—were due to assemble in the Palace of Culture that day. Could they proceed with the meeting? There was a children's health run scheduled for the following day through the streets of the city. Should that go ahead, too? Malomuzh assured the school director that there was no need for a change of plans; everything should continue as normal.

"And please do not panic," he said. "Under no circumstances should you panic."

———

At 10:15 a.m., a solitary armored car—the lead radiation reconnaissance scout vehicle of the 427th Red Banner Mechanized Regiment of the Civil Defense Forces of the USSR—swung slowly left off the Kiev road toward Pripyat. Hatches sealed and dosimetry instruments on, its engine whined as it rose over the bridge across the railway line. The city came into view through the crew's thick bulletproof windows. Everything seemed normal.

Trailing behind at a distance of eight hundred meters, as combat protocol required, the remaining vehicles of the reconnaissance column caught up with the scout car on the square outside the White House. The civil defense troops had been instructed to conduct a radiation survey of the city and its surroundings but lacked any detailed maps of the plant, or Pripyat. Up on the second floor of the White House, a detachment of men found Maria Protsenko, who had maps of the city but no means to copy them. As photocopiers could be used to create samizdat materials, access to the few in the USSR was controlled tightly by the KGB. Protsenko sat down at her drawing board and began turning out schematics of the city as quickly as she could by hand.

At noon, as the radiation reconnaissance troops divided into groups and set off to take dosimetric readings across the city, an Mi-8 helicopter of the Soviet Armed Forces 225th Composite Air Squadron headed toward Pripyat from the south. In the pilot's seat was Captain Sergei Volodin, who, along with his two-man crew, had been on standby that morning at the military airport in Borispol. The duty was part of a regular rotation, which required one helicopter crew to be ready for emergencies anywhere in the Kiev military district. Volodin and his men were otherwise accustomed to the more comfortable task of ferrying Soviet dignitaries around the republic—their helicopter was specially modified for the purpose, its carpeted cabin furnished with comfortable armchairs, a toilet, and even a bar. Although they had received mandatory training to prepare them to fly combat missions in the mountains of Afghanistan, the call-up had never come.

At around nine that morning, Volodin had received orders to conduct an airborne radiation survey around the Chernobyl plant. On the way there, he was to collect a senior officer of the civil defense who could supply the necessary details. After filing his flight plan, Volodin went to the duty officer to collect personal dosimeters for himself and his crew. But the instruments' batteries were corroded. Only the squadron's chemical services officer could replace them, and he was on the other side of the airfield, building a garage for the base commander. Volodin decided he could do without personal dosimeters. And although he and his crew had been issued respirators and rubber chemical protection suits for the mission, flying in such an outfit seemed impossible. The weather was still warm, and it was hot inside the cockpit, even in their summer uniforms. At around 10:00 a.m., the flight engineer started the engines, and Volodin set off in his shirtsleeves. He collected the civil defense officer—a major equipped with his own military radiation detection equipment—and continued toward Pripyat to gather further instructions.

Volodin knew Chernobyl well. He often ferried the squadron's helicopters to the big military aircraft plant in Kaunas, Lithuania, for annual maintenance, and the trips would take him past the gleaming white boxes of the power station. Sometimes, just out of curiosity, he'd turn on the DP-3 battlefield radiometer installed behind his seat in the cockpit. Intended for use after a nuclear attack, the DP-3 could be adjusted across four ranges of sensitivity: measuring up the scale from 10 to 100, 250, and, finally, 500 roentgen an hour. But he'd never seen the needle even flicker.

Now, as the pilot approached the station at a height of two hundred meters, he could see white smoke drifting above the buildings. He told the engineer to turn on the cockpit radiometer. His navigator prepared to make the calculations necessary to estimate from their airborne readings what the radiation dosage might be on the ground. Volodin caught sight of a yellow Ikarus bus driving between the incomplete Fifth and Sixth Units of the power station. *Well*, he thought, *if people are still at work down there, everything must be okay.*

Then he saw that the western end of the plant had collapsed. Inside, something was burning.

"Eighteen roentgen an hour," the flight engineer reported. "Climbing rapidly." The civil defense major opened the door of the cockpit to report that his hand was radioactive. He'd opened a cabin window to take his own readings outside: 20 roentgen an hour.

Leaving the plant behind, Volodin prepared to set down the helicopter in Pripyat so that the major could gather detailed instructions for the survey flight. He circled the city to land against the wind. He noticed how many people were out on the streets, fishing on the riverbank, planting potatoes in their allotments. The sky was a clear blue, the forest a vivid green. A flock of white gulls wheeled overhead.

Volodin brought down the Mi-8 near a playground on the southwestern edge of the city, where he hoped to avoid creating too much disturbance. But the machine always attracted attention whenever it landed near civilians. It was quickly surrounded by a crowd of adults and children. The adults wanted to know what was going on at the station and how soon they'd be able to get back to work there. The children wanted to see inside the helicopter. While the civil defense major went into the city, Volodin allowed them in, six or seven at a time, to look around.

Back at the plant, the staff summoned during the night by emergency telephone calls had been joined by members of the regular morning shift, who arrived for work as usual at eight o'clock. At the construction headquarters, just four hundred meters from Unit Four, the daily briefing began as it always did, but it was broken up by news of an accident at the plant, and everyone was sent home. Yet there was no great sense of alarm. Some construction workers took advantage of an unexpected day off, going to their dachas or swimming at the beach. There were mishaps all the time at the plant, and radiation had never seemed to hurt anyone. The last time something like this happened, trucks had appeared in Pripyat to spray the streets, and children had played barefoot in the decontaminant foam when they passed.

From her desk in the White House, Maria Protsenko phoned home to tell her husband to vacuum and wash the floors of the apartment and make sure that when their fifteen-year-old daughter returned home from school she changed her clothes and showered. Yet when she called back two hours later, she found them both unflustered by her warnings. They were watching a movie together on TV, and her daughter hadn't even bothered to wash: "When the movie finishes, I'll go," she said.

Even those who had seen the developing catastrophe firsthand found it hard to reconcile the destruction at the plant with the carefree atmosphere on the streets of Pripyat. A manager working on the Fifth and Sixth Units had seen the blaze for himself as he returned late at night by road from a trip to Minsk. Just an hour after the explosion, he stopped his car less than a hundred meters from the shattered reactor hall of Unit Four and watched, transfixed and terrified, as firemen on the roof struggled to contain the flames. Yet when he awoke at home in Pripyat at ten the following morning, everything seemed so normal. He felt determined to enjoy the day with his family.

Elsewhere, however, there were signs that not everything in the city was quite as it should be. The technician's next-door neighbor, an electrical assembly man, spurned the beach that morning in favor of the roof of his apartment building, where he lay down on a rubber mat to sunbathe. He stayed up there for a while, and noticed that he began to tan right away. Almost immediately, his skin gave off a burning smell. At one point, he came down for a break, and his neighbor found him oddly excited and good humored, as if he'd been drinking. When no one else seemed interested in joining him up on the roof, the man returned there alone and continued to work on his accelerated tan.

But at the plant, the nuclear engineers on the morning shift recognized all too clearly the danger the city was facing and tried to warn their families. Some managed to reach them by telephone and told them to stay indoors. Knowing that the KGB was monitoring the calls, one tried to use coded language to prepare his wife to escape the city. Another persuaded Director Brukhanov to let him go home for lunch and then packed his family into the car to take them to safety, only to be turned back at the end of Lenina Prospekt by an armed *militsia* officer manning a roadblock. The city had been sealed off. No one would be permitted to leave without official clearance.

Arriving at Yanov station at around 11:00 a.m., Veniamin Prianichnikov, the director of the plant's technical training programs, had missed all the drama of the preceding twelve hours. He had been away on a business trip in Lvov. On his way home on the train that morning, he overheard other passengers discussing rumors of a major accident. Prianichnikov, an experienced nuclear physicist whose expertise had taken him from the plutonium factories of Krasnoyarsk-26 to the atomic testing grounds of Kazakhstan, had worked on the Chernobyl project at its inception and was proud of his position at the plant. He knew the reactors well and refused to believe the tittle-tattle: an explosion in the reactor core was impossible under any conditions he could imagine. He argued so vociferously with his fellow passengers that they almost came to blows.

But when he arrived in Pripyat, he saw the tanker trucks of the 427th Mechanized Regiment of civil defense spraying the streets with a detergent that left a white foam in the gutters. The physicist recognized it as desorbent solution, designed to absorb and contain radionuclides when they settled on the ground. And there were *militsia* officers everywhere. Prianichnikov ran back to his apartment to warn his wife and daughter but found nobody home.

From the apartment, he tried calling the station, but the line was dead. Taking to his bicycle, he found his wife a few kilometers outside of town at their dacha, tending her flowers. She refused to believe anything was amiss. Only when he showed her the dark specks of graphite on the leaves of her strawberry plants did she agree to return home.

Prianichnikov suspected the accident was a catastrophic failure of the reactor, but without a dosimeter, he found it hard to convince his neighbors of such a heretical idea. He couldn't make them listen, and—as someone whose father and grandfather had both died at the hands of the Party—he knew that it could be dangerous to try too hard.

When the civil defense major returned to Captain Volodin's helicopter, he brought news that the damage they'd seen at the plant had been caused by an explosion. A government commission was on its way from Moscow: when they arrived, they would need a full report on the current situation. The major said he would accompany Volodin and his men as they flew a triangular route around the city to locate areas of potentially high radioactive contamination. Before they took off again, Volodin told everyone nearby to take their children indoors and close the windows.

At around 1:30 p.m., the pilot took the helicopter up to a hundred meters, flew north, over the first of three villages close to Pripyat, and then turned to the west. The cockpit dosimeter remained at zero. Volodin dropped to fifty meters and continued toward the next village; nothing. He brought the helicopter down farther, to just twenty-five meters, but the needle of the radiometer didn't move. Volodin suspected that it just wasn't sensitive enough to get a reading. Passing the final turning point on the survey flight plan, Volodin began following the railway line, in the direction of the Chernobyl plant itself.

On his right, he could see the village of Chistogalovka, where more people were tilling their gardens. The wind was now blowing southwest, carrying a thin trail of white smoke—or steam, perhaps—from the direction of the plant and the railway station, toward the village.

Chistogalovka wasn't a part of the survey flight plan, but Volodin decided to take some readings anyway. What if the smoke was radioactive? This stuff could be falling right on people's heads. As he passed the railway station, he pulled the control yoke over. The helicopter banked right.

Large beads of liquid began to form on the canopy. At first, Volodin thought it was rain. But then he noticed that it wasn't breaking over the glass like water: instead, it was strange, heavy, and viscous. It flowed slowly like jelly and then evaporated, leaving a salty-looking residue. And the sky remained clear. He bent over the control panel, and looked up: directly above him, the same whitish smoke was blowing overhead, thin in some places, dense in others. Almost like a cloud.

"Captain, it's maxed out!" the flight engineer shouted.

"What's maxed out?"

"The DP-3. The needle's stuck."

"Then switch to a higher range," Volodin said, and turned to check the dial himself. But the radiometer was already calibrated to its most extreme setting. The needle was glued to the far end of the range, at 500 roentgen per hour. And Volodin knew that the device took its readings from a receiver in the back of his seat. It seemed impossible: the level of radiation *inside* the cockpit had risen beyond the worst expected in a nuclear war. Whatever it was, he had to get them away from the cloud immediately.

Volodin threw the yoke forward. The nose of the helicopter dipped down and to the left. The treetops flashed beneath them, a smear of green. He pushed the machine to its maximum speed, away from the railway station and toward Pripyat. Then the cockpit door flew open, framing the terrified civil defense major, his own radiometer in his hand.

"What have you done?" the officer screamed against the howl of the engines. "You've killed us all!"

Natalia Yuvchenko had spent all morning trying to discover what had happened to her husband, Alexander. First, she had gone downstairs to the public phone and called the hospital, but they wouldn't tell her anything. Then she heard that the KGB was there, and no one was being allowed in. But she couldn't stay at home, knowing nothing. And Alexander wasn't the only one who hadn't returned from work as expected. Her close friend Masha came up from her apartment downstairs to say that her husband, who worked in Unit Three, hadn't come home either.

So Natalia left her son, Kirill, in the care of a neighbor, and together the two women went from door to door, apartment to apartment, building to building, down the street and across the courtyards—running up one echoing concrete staircase and then another, ringing doorbells, searching for someone from the plant who could tell them what had happened. She tried to send a telegram to her parents, but the post office was closed. Masha picked up the phone to call her mother and father in Odessa. But the line had been cut off.

Eventually Masha's husband came home, apparently unhurt but confirming news of an accident. He explained that he had helped get Alexander to the hospital before dawn that morning. Then another neighbor said that he had seen him in the hospital. He was all in one piece, and the neighbor knew where Natalia could find him: on the second or third floor, at the back. She might not be able to get inside, but she could certainly call up to him through the window.

It was already late in the afternoon when Natalia found her way to

Hospital Number 126. Alexander appeared in the window, bare chested, wearing pajama bottoms. He leaned out and asked his wife if she had left the windows of the apartment open the night before.

Natalia was relieved. He looked normal and unhurt, although his arm and shoulder were bright red, as if from an angry sunburn. And, more troublingly, the hair at his temples appeared to have turned completely white.

"Of course!" she replied. "It was so hot and stuffy."

Behind her husband, Natalia could see other people moving around inside the hospital ward: more patients, perhaps. She couldn't really tell. None of them came near the windows. She was terrified someone would notice her there, and she'd be taken away.

"Natasha," Alexander said, "close all the windows. Throw away all the food that has been out. And wash everything in the apartment."

He couldn't say much else. The KGB was there, interrogating everybody. But the couple agreed to meet the same way the following day. By now, other women had managed to smuggle vodka, cigarettes, and folk remedies to their husbands, some even passing them up through the hospital windows, tying bags to the end of a rope. Alexander said there were a few things he'd like Natalia to bring him: a towel, a toothbrush, toothpaste—and something to read. These requests were the normal things anyone might want to have in the hospital. It seemed the panic was over. Natalia felt quite sure that now, as soon as whatever had gone wrong at the plant was resolved, everything would be okay. She went home and did everything just as her husband had said.

At 4:00 p.m., the members of the OPAS medical team began to triage the patients. Alexander Esaulov, deputy mayor of Pripyat, stood by as the doctor in charge produced a worn notebook and began reading a list of symptoms down the phone to someone at the Institute of Biophysics in Moscow.

"Many are in grave condition," he said in a hollow voice. "The burns are bad. Some have severe vomiting, and a large number of burns on the extremities. The patients' condition is exacerbated by thermal burns. They should be urgently evacuated to Moscow." But when he explained that there were as many as twenty-five people in need of an emergency airlift, there were protests on the line. The specialist's voice turned harsh.

"Well, then organize it," he said.

Still more patients were arriving all the time, displaying the symptoms of radiation sickness. After some debate, the director of the hospital made the decision to distribute to everyone in Pripyat stable iodine—a prophylactic against iodine 131, the radioisotope that represents a particular threat to children. But there weren't enough iodine pills in the dispensary, and it was imperative that the crisis remain secret. So Esaulov used his Party contacts in the neighboring districts of Chernobyl and Polesia to quietly appeal for help. By nightfall, a total of twenty-three thousand doses of potassium iodide had arrived, and preparations were under way to deliver them door-to-door to apartments across the city.

At 8:00 p.m., Second Secretary Malomuzh summoned Esaulov back to the White House. The deputy mayor found the building surrounded by parked vehicles of all kinds: Volgas, Moskvitches, *militsia* patrol and escort cars, military jeeps, and the new black sedans of senior Party officials. Inside, on the third floor, a group of colonels and generals in uniform waited outside the office where the government commission was meeting. Malomuzh instructed Esaulov to transfer the most seriously injured patients from the Pripyat hospital to Borispol Airport outside Kiev. From there, a military jet provided by General Ivanov, the civil defense chief, would fly them to Moscow.

From the office window, Esaulov could see a large crowd leaving an evening screening at the Prometheus Cinema and mothers strolling with their children toward the cafe on the pier. The sound of tinkling glasses drifted up from a wedding celebration in the restaurant below. He heard shouts of "Kiss!" and then, in slow chorus, "One! Tw-oooo! Thre-eeeee!"

By nightfall on Saturday, the telephone lines and the hardwired radio speaker boxes in every apartment in Pripyat had fallen silent. The boxes—*radio-tochki*, or "radio points"—hung on the walls of homes throughout the Soviet Union, piping in propaganda just like gas and electricity, over three channels: all-Union, republic, and city. Broadcasts began every morning at

six with the Soviet anthem and the cheerless greeting *Govorit Moskva*—
"Moscow speaking." Many people left the radio on constantly—at one time,
switching it off was regarded with suspicion—a susurrating trickle of Party
enlightenment in every kitchen. When the boxes were silenced and the
phones went dead, even those in Pripyat who had spent the afternoon soaking up the sunshine began to realize something unusual was happening.

And then officials from local housing bureaus, or *zheks*, came around,
telling people to mop their stairwells, and young women from the Komsomol began knocking on doors, handing out stable iodine tablets. Word
spread that all the remaining reactors at the plant had been shut down.
There were rumors of a citywide evacuation. Some people even packed
their bags and went down into the street, expecting to be taken away at
any minute. But no official word came.

Alexander Korol had spent much of the morning sitting in Leonid
Toptunov's apartment, waiting for his old friend to come home and explain what had happened inside Unit Four. He'd heard that there had been
a maximum design-basis accident at the plant. But he refused to believe
it. Eventually Toptunov's girlfriend, the nurse, arrived and explained that
everybody from the midnight shift was in Hospital Number 126. Some of
them were being flown to a special clinic in Moscow that evening.

It was past 9:00 p.m. when Korol found his way to the hospital, clutching a towel, toothpaste, and Toptunov's toothbrush. When he got there,
two red Ikarus buses were drawn up at the front steps. One was being
loaded with injured firemen and his friends from the night shift on Unit
Four. They were all still wearing their hospital pajamas, and many appeared perfectly healthy. Korol climbed aboard and found Toptunov: Leonid looked just as he always did. But his friend could see that the seats
and walls of the bus had been covered with sheets of plastic, and when
Toptunov spoke, he seemed bewildered and disoriented.

Korol asked him what had happened. "I don't know," the young operator said. "The rods came halfway, and then stopped."

Korol asked no more questions. He knew that few other people in Pripyat would have any idea that the men were being taken from the city, or
where they were going. He began moving through the bus with a pen and
paper, scribbling down the names and addresses of his friends' relatives—

so at least he could tell them that their loved ones were being transported to Moscow. As he did so, two more men were brought aboard the bus, on stretchers.

One of them looked up. "Hi, Korol!" he said brightly.

But Korol had no idea who the stricken specialist was. His face had turned a bright red and was so swollen that he was completely unrecognizable. And when Korol saw the man on the second stretcher, his body seared by 30 percent burns, he realized that, whatever had happened with the control rods, this was no minor accident. By then, his time was up: his friends were leaving. Korol climbed down from the bus and watched as it pulled away from Hospital Number 126.

That night, Korol and other senior engineers from the plant gathered in groups in one another's homes, drinking beer and discussing what might have caused the accident. There were many theories, but no answers. They turned on the TV hoping for some news, but there was no mention of the plant or an accident.

In their big corner apartment at the end of Lenina Prospekt, Valentina Brukhanov had waited in vain all day for news of her husband, whom she had last seen leaving silently before dawn. It was long after midnight when the station director returned home, bringing a permit that would allow their pregnant daughter and son-in-law to take the family car, slip through the *militsia* cordon, and escape the city. He stopped for only a few minutes. He said he had to get back to the plant. "You know the captain is always the last one off the ship. From now on," he told Valentina, "you'll be responsible for the family."

When Veniamin Prianichnikov at last got through to his boss at the station by phone, he had been told that they were conducting an exercise, and he should mind his own business. That night, Prianichnikov shut his wife and daughter inside the apartment. He instructed them to pack their suitcases and be ready to leave town on the first train the next morning. The family was just preparing for bed when they heard strange sounds coming from the direction of the station. From their sixth-floor balcony, they watched as yellow and green flames flared a hundred meters into the sky above the torn ruins of Reactor Number Four.

In the small hours of Sunday morning, General Ivanov's plane lifted off from the tarmac at Borispol Airport, carrying twenty-six men suffering the initial symptoms of acute radiation syndrome. They included Leonid Toptunov; his shift foreman, Alexander Akimov; Deputy Chief Engineer Dyatlov; Alexander Yuvchenko; and the firefighters who had fought the blazes on the roof of the reactor hall. Most had no idea where they were being taken, or why. They worried about the fate of their families—and about what had happened at the plant. The flight to Moscow took less than two hours. Those who were still conscious vomited all the way.

As the new day began in the *militsia* station in Pripyat, where the Department of Internal Affairs had established an emergency headquarters, the duty officer recorded a series of notes in the official log. At 7:07 a.m., he wrote: "People are resting. At 8:00 a.m., the staff office will start working. The situation is normal. The radiation level is rising."

9

Sunday, April 27, Pripyat

The first of the big transport helicopters came in soon after dawn, dropping low over the rooftops surrounding the town square. The concrete facades of the White House and the apartment buildings on Kurchatov Street rang with engine noise from its twin turbines, the air whirled with dust, and rotor wash tore the petals from the flower beds. Major General Nikolai Antoshkin, the baby-faced forty-three-year-old chief of staff of the Soviet Air Defense Forces Seventeenth Airborne Army, stood below, signaling to the pilot with his uniform cap until the machine came to rest in the street outside the Hotel Polesia.

Dispatched from the central command post of the Kiev military district the previous night, General Antoshkin had reached Pripyat by car after midnight on Saturday, accompanied by an air force chemical warfare expert. He had only the vaguest idea of what was happening at the power plant, and had no instructions, staff, or equipment—not even the two-way radios needed to communicate directly with his pilots. As soon as he arrived in Pripyat, he went to the White House to report to Boris Scherbina. The chairman of the government commission was brief: "We need helicopters," he said.

Using a telephone in one of the offices now annexed by generals and admirals of the army, nuclear navy, and civil defense, Antoshkin roused his deputy in Kiev from bed to scramble the heavy helicopter regiments. Flying by night, in rain and low clouds, and menaced by thunderstorms, the first machines came into the nearby military air base in Chernigov from across Ukraine and Belarus. Antoshkin used the emergency powers granted to the government commission to summon test pilots from the helicopter training school at Torzhok, north of Moscow, and had more fliers transferred from bases a thousand kilometers away, on the border of Kazakhstan.

By sunrise on Sunday, the general was in command of a disaster-response aviation task force of eighty helicopters, awaiting orders at four different airfields around the plant, with still more being redirected toward Chernobyl from airfields across the Soviet Union. He had also been awake for more than twenty-four hours.

Inside the hotel, the din of the arriving aircraft shook Boris Scherbina, Academician Legasov, and the other commission members from their beds. They had remained in session deep into the night, attempting to disentangle the tightening knot of problems surrounding the wreckage of Unit Four: the threat of a new chain reaction in the reactor; the fire, and the need to smother the invisible plume of radionuclides streaming into the atmosphere; but also the questions of whether to begin evacuating the city and how to solve the mystery of what had caused the accident in the first place.

Legasov estimated that the reactor had contained 2,500 tonnes of graphite blocks, which had caught fire and already reached a temperature of more than 1,000 degrees centigrade. The intense heat could soon melt both the zirconium cladding of the fuel cassettes remaining in the core and the uranium dioxide pellets they contained, adding yet more radioactive particles to the cloud escaping from the shattered core. Legasov determined that the graphite would burn at a rate of around one tonne an hour. Even accounting for the material that had been thrown from the core by the explosion, if his calculations were correct and what remained burned unchecked, the blaze could roar on for more than two months—releasing a column of radionuclides into the air that would spread contamination across the USSR and circle the globe for years to come.

But the problem was one of unprecedented complexity. Ordinary firefighting techniques were useless. The graphite and nuclear fuel were burning at such high temperatures that neither water nor foam could possibly quench them: so hot that water would not only evaporate instantly into steam, further distributing radioactive aerosols in a cloud of toxic vapor, but also could separate into its constituent elements, oxygen and hydrogen, creating the possibility of a further explosion. And besides, the colossal fields of gamma radiation surrounding the reactor made it impossible to approach by land or water for prolonged periods.

Legasov and the other increasingly exhausted nuclear specialists debated for hours, throwing out every idea they could think of, scavenging information from books and manuals, and via phone and teletype from Moscow. Firefighting chiefs from the Ministry of the Interior and experts from the Energy Ministry sought help from their counterparts in the capital. One physicist, unable to find the answer to a simple question in the technical materials held at the power station, phoned his wife and asked her to look it up at home. In his office at the Kurchatov Institute, the eighty-three-year-old director himself, Anatoly Aleksandrov—chairman of the Soviet Academy of Sciences, holder of the patent on the RBMK reactor, and Legasov's mentor—sat beside a scrambler telephone, offering advice to the scientists in Pripyat on how to regain control of Reactor Number Four. Legasov suggested covering it with the iron shot stockpiled at the plant's construction site for making radiation-resistant "heavy" concrete; Scherbina wanted to sail fireboats up the Pripyat and shoot water into the reactor hall with high-pressure hoses. But the iron shot sat in a warehouse that lay directly in the path of the fallout drifting from Unit Four and was already too contaminated to approach; pouring more water into the reactor would be dangerous and pointless.

The arguments continued through the night. Meanwhile, the team from the Ministry of Energy's nuclear research institute, VNIIAES, returned from their reconnaissance at the plant, where they had witnessed the terrible light show blossoming from the ruins of Unit Four. They reported to Scherbina that the radiation situation was perilous.

At 2:00 a.m., Scherbina telephoned his Party boss in Moscow, Vladimir Dolgikh—the Central Committee secretary responsible for heavy industry and energy—and requested permission to abandon the city. By the time the scientists finally crawled into their beds in the hotel, a few hours before daybreak, Scherbina had also reached a decision about the burning reactor: to smother it by bombarding it from the air, using Antoshkin's helicopters.

But the members of the government commission had still reached no conclusions about what combination of materials would work, or how, exactly, such an operation might be accomplished.

At around 7:00 a.m. on Sunday, Boris Scherbina walked into the White House office now occupied by the Soviet Union's leading military authorities on radiation: General Boris Ivanov, deputy chief of Soviet civil defense forces, and Colonel General Vladimir Pikalov, leader of the Chemical Warfare Troops of the USSR. He announced himself ready to address the issue of evacuation.

"I've made my decision," Scherbina said. "What's your opinion?"

Ivanov gave him the radiation report. Far from falling, as the Ministry of Health officials had hoped, the level of contamination on the streets of Pripyat was increasing. There was no question in the minds of the civil defense chief and his regional deputy: the population of the city was in danger not only from the radionuclides continuing to drift from the reactor but also from the fallout already accumulated on the ground.

They must be evacuated. The officers' view was backed up by a separate report from the director of Hospital Number 126. Only Pikalov, the imposing, beetle-browed commander of the chemical troops, a decorated veteran of the Great Patriotic War, suggested that there was still no hurry to get the people of Pripyat to safety.

Scherbina told them that he had made up his mind: the evacuation should begin that afternoon. But he still held back from giving the order. First, he wanted to see Reactor Number Four for himself.

Soon after 8:00 a.m., still wearing the same business suits in which they had embarked from Moscow the day before, Scherbina and Valery Legasov climbed aboard a Mi-8 helicopter parked in the middle of the city's soccer pitch. They were joined by Generals Pikalov and Antoshkin and two men from the Kiev prosecutor's office carrying a brand-new video camera with which to record the scene. It was less than a two-minute flight from Pripyat to the power station, and as the aircraft banked around the western end of the long turbine hall, the six men gazed through the circular portholes of the cabin at the dreadful spectacle below.

To even the most recalcitrant Soviet eye, it was clear that Unit Four of the Chernobyl nuclear power plant would never again generate a single watt of electricity. In the crisp light of the new day, it was obvious that the reactor had been completely destroyed. The roof and upper walls of the reactor hall had gone, and inside Legasov recognized the upper lid of the reactor, thrown aside by what must have been an enormous explosion and resting at a steep angle on top of the vault. He could see graphite blocks and large pieces of the fuel assemblies scattered across the roof of the machine hall and on the ground beyond. A white pillar of vapor—most likely the product of the graphite fire, Legasov thought—floated from the crater, reaching several hundred meters into the sky. And, ominously, deep within the dark ruins of the building, the scientist could see spots of deep crimson incandescence where something—he didn't know what—appeared to be burning fiercely.

As the helicopter headed back to Pripyat, Legasov knew conclusively that he was dealing with not merely yet another regrettable failure of Soviet engineering but also a disaster on a global scale, one that would affect the world for generations to come. And now it was up to him to contain it.

At ten o'clock on Sunday morning, a full thirty-two hours after the catastrophe had begun, Boris Scherbina gathered the Soviet and local Party staff in the *gorkom* offices of the White House. At last, he gave the order to evacuate Pripyat.

At 1:10 p.m., the *radio-tochki* in kitchens across the city finally broke their silence. In a strident, confident voice, a young woman read aloud the announcement drafted that morning by a gaggle of senior officials and approved by Scherbina:

> Attention! Attention! Dear comrades! The City Council of People's
> Deputies would like to inform you that, due to an accident at the
> Chernobyl nuclear power plant in the city of Pripyat, adverse ra-

diation conditions are developing. Necessary measures are being taken by the Party and Soviet organizations and the armed forces. However, in order to ensure complete safety for the people—and, most importantly, the children—conducting a temporary evacuation of city residents to nearby localities in the Kiev region has become necessary. . . . We ask that you remain calm, be organized, and maintain order during this temporary evacuation.

The emergency proclamation was worded carefully: it didn't tell the citizens how long they would endure their enforced absence, but deliberately led them to believe it would be a short time. They were told to pack only their important documents and enough clothing and food for two or three days. They should close their windows and turn off their gas and electricity. Municipal workers would remain behind to maintain city utilities and infrastructure. The empty homes would be guarded by police patrols. Some, fearing what would happen in their absence, packed only their most valuable possessions—party dresses, jewelry, flatware. Others packed winter clothes and prepared for the worst.

Earlier that morning, Natalia Yuvchenko—carrying a towel, a toothbrush, and the other things her injured husband had asked her to bring to him—had returned expectantly to the grounds of Medical-Sanitary Center Number 126. But when she reached the spot beneath the window where she and Alexander had talked the day before, she found no sign of him or any of the other men and women from the plant. The windows of the building stood open, but the entire wing of the hospital—which just a few hours earlier had been filled with patients—was now completely deserted. She looked around for someone to ask about where they might have gone but found no one.

When Yuvchenko got back to her apartment on Stroiteley Prospekt, her neighbors told her about the announcement of a citywide evacuation: they would be gone for three days; buses would be coming to collect everyone; until then, keep all children inside; wait. There was no time for fear or panic. There were too many questions to be answered: Where are my friends? Where will we be going? When will we return?

Yuvchenko focused on immediate necessities: First, she had to make sure she had all of the family documents. She gathered their internal passports, their university diplomas, vaccination certificates, and the papers for the apartment. Then: Where would she find the milk she needed to feed Kirill for three days? All of the shops were closed. Above all, she needed to find her husband.

But it wasn't long before she discovered why Alexander had vanished so suddenly. Soon afterward, Sasha Korol—working through the list of addresses he had scribbled down on the bus the previous evening—arrived at her door and explained what he had witnessed of the medical airlift to Moscow. Unbidden, he also brought Natalia money—100 rubles, almost a month's salary—and a carton of milk for the baby.

Setting down the milk on the seat of Alexander's bicycle in the hallway, she went into the bedroom to pack. She filled a small suitcase—clothes for the boy, a couple of dresses, some shoes—and went downstairs to wait.

On the second floor of the White House, as military officers, scientists, and members of the government commission came and went, Maria Protsenko had remained at her desk throughout Saturday night. There was much to do and few people to do it: most of the technical staff of the *ispolkom* had been sent home.

Regardless of the crisis at the plant, Protsenko had felt determined to take care of the mountain of paperwork required for the planned development and expansion of the city. She remained confident it would proceed on schedule.

Yet every hour the chemical troops of the civil defense kept returning and asked her to draw another set of maps, as they continued to log the rising levels of radiation in Pripyat and beyond. And at eight o'clock on Saturday evening, the mayor had told her to prepare the city for a possible evacuation. Nothing was confirmed yet. But if the order came, he said she should be ready to get the entire population of Pripyat out at short notice—by bus and by train.

In the meeting room across the hall from her office, Protsenko joined a group of twenty members of the city administration to make the arrangements. The architect laid out the maps and counted every apartment building in the city, while the chiefs of the internal passport department and the district *zheks* added up the number of families in each complex, and how many children and elderly each family included. Together with the city's head of civil defense, Protsenko then calculated the number of buses that would be needed to collect them all from each of the city's six microdistricts.

In all, there were some 51,300 men, women, and children in Pripyat, of whom more than 4,000 were station operators and construction workers expected to remain behind to take care of essential services in the city and work at the power plant itself. To get all the families out safely would require more than a thousand buses—plus two river vessels and three diesel trains, which would be directed to Yanov railway station and used to ship out the single men who filled the city's dormitory complexes.

At the same time, in Kiev, the Ukrainian Ministry of Transport began requisitioning buses from motor transport enterprises all over the city and the surrounding towns and suburbs, summoning their drivers to work late on Saturday night and preparing them to travel toward Pripyat under police escort. At 11:25 p.m. they received the Ukrainian Council of Ministers' order to move. By 3:50 a.m., five hundred vehicles had already arrived at the city limits, with five hundred more reaching Chernobyl town less than a half hour later. Before dawn, a column of vehicles some twelve kilometers long had halted on the road toward Pripyat, while the drivers awaited further instructions and ate from a mobile canteen. The entire operation was conducted beneath a blanket of secrecy. By midmorning on Sunday, the bus stops of Kiev were crowded with frustrated passengers waiting in vain for scheduled services that would never show up.

By lunchtime on Sunday, the citizens of Pripyat had begun to gather outside their buildings to await their departure from the city, clutching small shopping bags of belongings, supplies of food—boiled potatoes, bread, lard— and handfuls of documents. There was no sign of panic. Despite warnings to stay inside, parents found it hard to control young children, who ran around and played in the dusty streets. Some families set off on foot.

At the same time, the crews of two helicopters of the Fifty-First Guards Helicopter Regiment prepared to begin an airborne assault on Reactor Number Four. The operation, approved by Boris Scherbina at eight that morning, began in a frenzy of improvisation. Not only takeoff and landing sites, but flight plans, speed, trajectory, radiation conditions—all had to be worked out by General Antoshkin and his men, and all were subject to the short-tempered demands of the chairman of the government commission. As the pilots began their reconnaissance sorties establishing flight paths over the reactor, Scherbina's attention turned to locating the thousands of tonnes of material they intended to drop into it.

Valery Legasov and the other scientists would eventually formulate a complex cocktail of substances to dump inside the ruins of Unit Four—including clay, lead, and dolomite—which they hoped would quench the graphite fire, cool the incandescent nuclear fuel, and block the release of further radionuclides into the atmosphere. Old man Aleksandrov and the nuclear physicists from the Kurchatov Institute recommended the lead, as well as the dolomite—a naturally occurring mineral containing calcium and magnesium carbonate. With its low melting point, the scientists believed that the lead would be liquefied in the heat of the blaze, helping to cool the temperature of the nuclear fuel and trapping radionuclides released from the ruined core. They also hoped that it would flow into the bottom of the reactor vessel, where it would solidify to form a barrier against gamma radiation. They intended the dolomite to cool the fuel and also to decompose chemically in the heat of the blaze, releasing carbon dioxide, which would deprive the graphite fire of oxygen. Aleksandrov suggested adding clay, too, which might seal up the reactor and help absorb radionuclides.

But none of these substances could be found at the power plant. Lead, in particular, was among the many raw materials in short supply throughout the USSR. And it was imperative that the operation begin at once. Scherbina instructed the pilots to begin bombing the reactor with powdered boron—the neutron moderator that would head off the possibility of further chain reactions in the remaining uranium—which had finally arrived by truck from the nuclear power plant in Rovno. Legasov had ridden out into the shadow of the reactor in an armored personnel carrier to take neutron radiation readings himself, and his data seemed to confirm that the chain reaction had now ceased amid the debris. But the physicists wanted to be certain it couldn't start again.

In the meantime, Scherbina sent General Antoshkin and two deputy ministers of the USSR—both nuclear specialists—down to the bank of the Pripyat, where they personally began filling sacks with sand. Academician Legasov argued that sand would stifle the fire and create a filtering layer on top of the burning reactor, to trap the escaping hot particles and radioactive gases. It was also cheap and plentiful. Maria Protsenko's preparations to expand the city had already involved dredging the river for tonnes of it, which now lay piled on the bank beside the city's waterfront cafe—just two blocks from the square in front of the Hotel Polesia, where the helicopters landed. This was just as well, as the quantities required were enormous: the scientists suggested that the reactor should be covered with a layer of absorbents at least a meter thick. According to their calculations, around fifty thousand bags would do it.

It was hot on the riverbank, and the general and the two ministers—still wearing their suits and city shoes—were soon bathed in sweat. And if the sun was bad, the radiation was worse. They had neither respirators nor dosimeters. One of the ministers sought help from the manager of a group of nuclear assembly specialists, who demanded that his men be paid a bonus for working in a contaminated area. But even with their help, the scale of the task was overwhelming. Two of the specialists drove to a nearby collective farm—named Druzhba, or "Friendship"—where they found *kolkhoz* workers out in the fields, busy with spring sowing. The farmworkers, relaxed and happy in the sunshine, didn't believe what they were told about the accident, the need to stifle the burning reactor—or that the soil they were tilling was already poisoned with radiation. Only after the farm director and Party secretary arrived and repeated the explanation of the crisis again and again did the workers agree to help. Eventually between 100 and 150 men and women from the *kolkhoz* volunteered to join the effort on the riverbank, reinforced by troops from the Kiev civil defense detachment.

But Boris Scherbina remained implacable. Back inside the White House, he drove the ministers and generals to work harder and more quickly and reserved a furious contempt for the representatives of the nuclear ministries. They seemed to be gifted at blowing up reactors, he roared, but pathetic at filling sandbags.

If he was aware of the rising level of contamination in the air all around them, Scherbina didn't show it. The chairman seemed to regard the dangers of radiation with the haughty disdain of a cavalry officer striding across a battlefield bursting with cannon fire. And almost everyone else on the commission followed his lead: mentioning the radioactivity surrounding them seemed almost tactless. Among the ministers, an air of Soviet bravado prevailed.

At last, early on Sunday afternoon, the first ten bags of sand—each weighing sixty kilos or more—were carried up to the square and loaded aboard one of General Antoshkin's helicopters.

There were 1,225 buses in all, painted in a kaleidoscope of colors, representing more than a dozen different Soviet transport enterprises: some red, some yellow, some green, and some blue; some half red, half white; others with a stripe—plus 250 trucks and other vehicles in support, including ambulances from the civil defense, repair trucks, and fuel tankers. At 2:00 p.m., a full day and a half after the pall of radionuclides had first begun drifting into the atmosphere, the motley caravan of vehicles waiting at the Pripyat city limits at last began to move.

Maria Protsenko was waiting for them on the railway bridge at the entrance to the city. With a map of Pripyat folded under her arm, she was dressed for the hot weather in a blouse, skirt, and summer sandals. She was joined by a major from the army and another from the police. They shook hands. They all understood what they had to do. Nobody said much.

As the first bus approached, the police major waved it to a halt, and Protsenko climbed aboard. She showed the driver the map and gave him his instructions: the buses were to proceed in groups of five; she told him which microdistrict they should go to, how to get there, the building they should stop at, and the route they should take out of the city. Then she climbed down, the policeman waved forward another group of buses, and Protsenko showed the next driver her map.

Gradually, five by five, hour by hour, she watched as each of the one thousand vehicles drew away down the scenic incline of Lenina Prospekt and, with a sharp turn and a tart belch of low-octane gasoline fumes, disappeared into the city.

Outside the 540 separate entrances of its 160 apartment buildings, the citizens of Pripyat mounted the steps of their buses, and the doors banged shut behind them.

At around 3:00 p.m., Colonel Boris Nesterov, deputy commander of the Air Forces of the Kiev Military District, a helicopter pilot with twenty years' experience who had served in Syria and seen combat in the mountains of northern Afghanistan, saw his target come into view ahead. Bringing the powerful Mi-8 transport in from the west at an altitude of two hundred meters, he prepared to cut his speed as he closed in on the red-and-white-striped vent stack of Unit Four. Behind him in the cargo compartment, the flight engineer had already slid open the side door and clipped his harness to the airframe; the pile of ten sandbags stood ready at his feet.

Nesterov slowed to a hundred kilometers per hour, and gave the command: "Prepare to drop!"

The ruins of Reactor Number Four came up fast. The colonel's headphones filled with static, the cockpit thermometer spiked abruptly from 10 to 65 degrees centigrade, and the radiometer housed in the back of his seat ran off the scale. Through the cockpit glass between his foot pedals, Nesterov saw the pillar of white vapor and the edges of the reactor glowing red, like a blast furnace during smelting.

The helicopter was equipped with no bombsights or targeting mechanisms that could help them here. To drop the sandbags into the reactor vault, the flight engineer had to aim as best he could by eye, estimate a trajectory, and shove them through the door one at a time. As he leaned out over the reactor, he was enveloped in clouds of toxic gas and blasted by waves of gamma and neutron radiation. He had no protection apart from his flight suit. The intense heat rising from below made it impossible for Nesterov to hover: if the helicopter lost forward momentum, it would be caught in the column of superheated air, its rotor blades would encounter a calamitous drop in torque, and the machine would fall abruptly out of the sky.

The colonel throttled back to sixty kilometers per hour. He fought to hold the helicopter steady and hoped the flight engineer could keep his footing. "Drop!" he shouted. The engineer hefted the first of the sandbags out into the sky above Unit Four, then another; and another. "The cargo has been dropped!"

Nesterov swung away to starboard and prepared to come around again.

At 5:00 p.m., Maria Protsenko folded her map, flagged down one last bus, and, climbing aboard, rode it down Lenina Prospekt, a lone passenger entering a deserted city. She directed the driver from one side of Pripyat to the other, stopping in each district to check on the results of her work. At six thirty, Protsenko returned to the *ispolkom* to tell the mayor that her task was complete.

"Vladimir Pavlovich, that's it. Everybody has been evacuated," she said.

With the exception of the maintenance staff and the men left behind to tend to the surviving reactors of the power station, the city was empty.

As her report was passed up the chain to the chairman of the government commission, Protsenko felt no sorrow, only the satisfaction of an important job well done. It was just as they had it in the Young Pioneers: "the Party speaks—and the Komsomol says, It's done!"

It wasn't until later that night that she began to feel sick: her throat was raw, and she developed a blinding headache; her feet and ankles burned and itched. She made no conscious connection to radiation, partly because she knew nothing about the effects of the radioactive alpha and beta particles in the dust that had blown around her bare legs during her hours on the railway bridge, but also because she preferred not to think about it. When the diarrhea began, Protsenko told herself she had eaten some bad cucumbers; the headaches and the sore throat—well, she had gone without sleep for two days. She soothed the itching by lifting her legs into the bathroom sink and running her feet under cold water. But it soon returned.

Protsenko went back to her desk, where she once again took to drawing maps for the chemical troops, who were now conducting radiation surveys of the area every sixty minutes. They began sweeping the inside of the *ispolkom* itself for radiation and warned her that the hallways were all contaminated. The janitor was long gone, so Protsenko took a wet cloth and mopped the linoleum herself; there were no gloves, so she used her bare hands.

As the multicolored convoy of buses wound its way through the narrow roads surrounding the city, few of the departing citizens knew where they were going. Nobody had told them anything. But they were confident they would soon return. Part of the convoy was well beyond the city limits when someone realized that the vehicles were carrying dangerous levels of radioactive dust on their wheels and had to double back to Pripyat for decontamination. One member of the power station staff traveled more than fifty kilometers away from the city, on a bus with his wife and children, before telling them to go on alone while he returned to the plant to help his colleagues; the driver dropped him in the city of Ivankov, where he had to talk the local *militsia* commander into allowing him to go back. Some evacuees persuaded their drivers to take them all the way to Kiev, but the Interior Ministry's plan called for everyone from Pripyat to be distributed throughout the small towns and villages of rural Polesia, where they would be taken in—one family at a time—by farmers and *kolkhoz* workers.

Viktor Brukhanov's wife, Valentina, wept as she was driven away from the city. On Natalia Yuvchenko's bus, the passengers whispered anxiously about where they might end up. They scoured the roadsides for the names of villages as they drove on through one settlement after another and saw the pity on the leathery faces of the peasants who stood in their yards to watch them pass.

On the third floor of the White House, the meetings of the government commission continued. Downstairs, Maria Protsenko remained at her desk. It was around 8:00 p.m. when she glanced out of the window and noticed a woman walking across the square into town. She was alone and carrying a suitcase. Protsenko couldn't understand it. Every woman and child in the city was supposed to have been taken to safety hours ago. She dispatched a duty officer to go down and investigate and watched from her office as he stopped the woman and questioned her. They talked, the woman nodded and then carried on as before, taking her suitcase with her. When the guard returned, Protsenko discovered that word of the emergency in Pripyat had apparently not yet prevented the trains passing through the railway station from stopping there on their normal timetable. The woman was returning from a weekend away and alighted from the train from Khmelnitsky, three hundred kilometers away to the southwest—with no reason to believe that anything had changed in her absence.

When the security guard explained to her what had happened, she seemed neither frightened nor panicked. Of course she would agree to be evacuated, she told him. "But first, I'm going home."

Yet, as the woman carried her suitcase back to her apartment building, she found Pripyat eerily transformed. In the space of just a few hours, Viktor Brukhanov's beloved city of the future had become a ghost town. Abandoned laundry flapped in the breeze on the balconies of Lenina Prospekt. The beaches were deserted, the restaurants empty, the playgrounds silent.

Now the streets echoed with new sounds: the barking of bewildered pet dogs, their fur so contaminated with poisonous dust that their owners had been forced to leave them behind; the whine of civil defense reconnaissance vehicles; and the relentless throb of helicopter engines, as the pilots and engineers of the Fifty-First Guards Helicopter Regiment returned again and again to fling bags of boron and sand into the mouth of the radioactive volcano.

PART 2

DEATH OF AN EMPIRE

10

The Cloud

Lifted skyward on a pillar of fierce heat from the shattered core, convoyed by obliging winds, the invisible cloud of radiation had traveled thousands of kilometers since its escape from the carcass of Unit Four.

Unleashed in the violence of the explosion, it had soared aloft into the still night air, until it reached an altitude of around 1,500 meters, where it was snatched by powerful wind currents blowing from the south and southeast, pulled away at speeds of between fifty and a hundred kilometers an hour, and flew northwest across the USSR toward the Baltic Sea. The cloud carried gaseous xenon 133, microscopic fragments of irradiated graphite, and particles composed of pure radioactive isotopes, including iodine 131 and cesium 137—which generated such heat that they warmed the air around them and took flight like hundreds of thousands of tiny hot air balloons. At its heart, it pulsed with some 20 million curies of radioactivity. By the time Soviet scientists finally began regular airborne monitoring near the accident site on Sunday, April 27—a full day after the accident occurred—the invisible monster had slipped away, leaving them ignorant of its size and intensity. Their measurements revealed only its tail. Within twenty-four hours, it had reached Scandinavia.

At midday on Sunday, an automatic monitoring device at the Risø National Laboratory north of Roskilde silently logged the cloud's arrival in Denmark. But because it was a Sunday, the readings went unnoticed. That evening, a soldier at the Finnish National Defense Forces' measuring station in Kajaani, southern Finland, recorded an abnormal increase in background radiation. He reported it to the operational center in Helsinki, but no further action was taken. Late that night, the plume encountered rain clouds over Sweden, and the moisture in them began to scavenge and concentrate the contaminants it contained.

When the rain finally fell from the clouds, around the city of Gävle, two hours' drive north of Stockholm, it had become heavily radioactive.

Shortly before seven o'clock on Monday morning, April 28, Cliff Robinson was eating breakfast in a coffee room of the Forsmark nuclear power station, sixty-five kilometers southeast of Gävle on the Gulf of Bothnia. Robinson, a twenty-nine-year-old Anglo-Swedish technician in the plant's radiochemistry lab, commuted to work each morning on a bus bringing construction workers to Forsmark, where they were building a large underground repository for nuclear waste.

When he had finished his coffee, Robinson stepped into the locker room to brush his teeth. On his way back, he passed through a radiation monitoring point, and the alarm bell rang. Still half asleep, the technician was puzzled. He had only just arrived and hadn't yet entered the reactor block: he couldn't possibly be contaminated. But the alarm brought down a member of the plant radiation protection staff, to whom Robinson explained what had happened. He walked through the detector again. Once more, the bell sounded. But on a third attempt, the monitor remained silent. Baffled, the two men decided that the equipment was faulty. Perhaps the alarm threshold had been miscalibrated. The dosimetrist told Robinson to go back to work. The machine could be fixed later.

Coincidentally, Robinson's job in the lab was to measure radioactivity at Forsmark-1, within the station building and in what it expelled into the environment. The reactor was only six years old but had been plagued by minor technical faults, and leaking fuel rods had already led to several small radioactive releases that winter. His Monday-morning routine took him first to the upper levels of the plant to gather air samples from the vent stack and then to the lab to analyze them. This took time. At around 9:00 a.m., he went back downstairs for another coffee. But when he approached the radiation monitoring point, he saw his path blocked by a long line of plant workers, each of whom was setting off the alarm bell. Now more perplexed than ever, Robinson took a shoe from one of the men, placed it in a plastic bag to prevent cross-contamination, and returned to the lab. He put the shoe on the germanium detector, a sensitive tool for measuring gamma rays, and prepared to wait.

But the results returned with terrible speed, exploding in steep, green peaks across the computer screen. Robinson's heart froze. He had never seen anything like it before. The shoe was intensely contaminated with the entire spectrum of fission products usually found inside the core at Forsmark-1: cesium 137, cesium 134, and short-lived iodine isotopes—but also a number of other elements, including cobalt 60 and neptunium 239. These, he realized, could have originated only with nuclear fuel that had been exposed to the atmosphere. Robinson immediately telephoned his boss, who, fearing the worst, told him to return to the chimney stack and take a fresh set of air samples.

At 9:30 a.m., the plant manager, Karl Erik Sandstedt, was alerted to the contamination. But the senior staff of Forsmark remained as confused by it as Robinson had been. They couldn't trace it backward to a source within the plant, and yet, given the weather conditions, radiation levels on the ground outside conformed to what they would expect of a major leak from one of the Forsmark reactors. At ten thirty, Sandstedt ordered approaches to the station sealed off. Local authorities issued a precautionary alert: a warning was broadcast on the radio instructing the population to keep its distance from Forsmark, and police set up roadblocks. Thirty minutes later, Robinson was still in the lab, at work on his new batch of samples, when he heard sirens sound throughout the building: the entire plant was being evacuated.

But by then, state nuclear and defense agencies in Stockholm had received reports of similarly high levels of contamination at a research facility in Studsvik, two hundred kilometers away from Forsmark. Air samples taken in Stockholm also showed elevated radiation and an isotope composition containing graphite particles, suggesting a catastrophic accident in a civilian nuclear reactor, but one of a very different type from those at Forsmark. By 1:00 p.m., using meteorological calculations developed to help monitor the Partial Test Ban Treaty on nuclear weapon trials, the Swedish National Defense Research Institute had also modeled prevailing weather patterns across the Baltic. These established beyond doubt that the radioactive contamination hadn't originated in Forsmark at all. It had come from somewhere outside Sweden. And the wind was blowing from the southeast.

At around eleven in the morning Moscow time, Heydar Aliyev was in his office in the Kremlin when the phone rang, summoning him to an emergency meeting of the Politburo. As the deputy prime minister of the USSR, Aliyev was one of the most powerful men in the Soviet Union. Once the head of the Azerbaijani KGB, and one of only twelve voting members of the Politburo, he held joint responsibility for making the most profound decisions affecting the course of the empire. Yet by Monday morning, even Aliyev knew only the vaguest details about a nuclear accident in Ukraine. Not one word about Chernobyl had appeared in the Soviet press or been reported on radio or television. Authorities in Kiev, without prompting from Moscow, had already acted to suppress awareness of the situation by scientists. On Saturday, after instruments at the Kiev Institute of Botany registered a sharp increase in radiation, KGB officers arrived and sealed the devices "to avoid panic and the spreading of provocative rumors." Even so, by the time General Secretary Gorbachev assembled the emergency meeting to discuss what had happened, Aliyev realized that radiation would soon be detected far beyond the borders of the USSR.

The dozen men—including Aliyev; Prime Minister Ryzhkov; propaganda chief Alexander Yakovlev; Gorbachev's emerging conservative opponent, Yegor Ligachev; and Viktor Chebrikov, the head of the KGB—convened not in the usual Politburo conference room but in Secretary Gorbachev's gloomy office on the third floor of the Kremlin. In spite of recent renovations, elaborately patterned carpets, and a domed ceiling hung with crystal chandeliers, the room was cavernous and uncomfortable. Everyone was nervous.

Gorbachev asked simply, "What happened?"

Vladimir Dolgikh, the Central Committee secretary in charge of the Soviet energy sector, began by explaining what he knew from his telephone conversations with Scherbina and the experts in Pripyat. He described an explosion, the destruction of the reactor, and the evacuation of the city. The air force was using helicopters to bury the ruined unit in sand, clay, and lead. A cloud of radiation was moving south and west and had already been detected in Lithuania. Information was still scant and conflicting: the armed forces said one thing, scientists another. Now they needed to decide what—or whether—to tell the Soviet people about the accident.

For Gorbachev, this was a sudden and unexpected test of the new openness and transparent government he had promised the Party conference just a month earlier; since then, glasnost had been nothing more than a slogan. "We should make a statement as soon as possible," he said. "We can't procrastinate."

Yet the traditional reflexes of secrecy and paranoia were deeply ingrained. The truth about incidents of any kind that might undermine Soviet prestige or provoke public panic had always been suppressed: even three decades after it had happened, the 1957 explosion in Mayak had still, officially, never taken place; when a Soviet air force pilot mistakenly shot down a Korean Air jumbo jet in 1983, killing all 269 people on board, the USSR initially denied any knowledge of the incident. And Gorbachev's grip on power remained tenuous, vulnerable to the kind of reactionary revolt that had destroyed Khrushchev and his program of liberalization. He had to be careful.

Although the official record of the meeting would later appear to show broad agreement on the need to make a public statement about the accident, Heydar Aliyev insisted it was misleading. According to the deputy prime minister's account, he advocated for immediate and total honesty: all of Europe would soon know that something terrible had happened, and this disaster was simply too big to hide. What was the point in trying to conceal what was already public? But before he could finish, Yegor Ligachev, widely regarded as the second most powerful man in the Kremlin, cut him off. "What do you want?" he asked truculently. "What information do you want to release?"

"Come off it!" replied Aliyev. "We can't conceal this!"

Others at the table argued that they didn't have enough information yet to tell the public and feared causing panic. If they released any news at all, it had to be strictly circumscribed. "The statement should be for-

mulated in a way that avoids causing excessive alarm and panic," said Andrei Gromyko, chairman of the Presidium of the Supreme Soviet. And by
the time they took a vote, Ligachev had apparently prevailed: the Politburo resolved to take the traditional approach. The assembled Party elders drafted an unrevealing twenty-three-word statement to be issued by
the state news agency, TASS—and designed to combat what the Central
Committee's official spokesman called "bourgeois falsification . . . propaganda and inventions."

Whatever Gorbachev's intentions, it seemed that the old ways were
best after all.

By 2:00 p.m. in Stockholm, Swedish state authorities were in unanimous
agreement: the country had been contaminated as the result of a major
nuclear accident abroad. Just over an hour later, the country's Foreign
Ministry approached the governments of East Germany, Poland, and the
USSR to ask if such an incident had taken place on their territory. Soon
afterward, the Swedes sent an identical communiqué to their representatives at the International Atomic Energy Agency. By that time, both Finnish and Danish governments had confirmed that they, too, had detected
radioactive contamination inside their borders.

Back in the town of Chernobyl, the single tiny hotel where Viktor
Brukhanov had once sat on a bed and sketched plans for his nuclear future was filling up with exhausted apparatchiks sent from Moscow. Radionuclides continued to boil from the smoldering remains of Reactor
Number Four, as the helicopter pilots of the Seventeenth Airborne Army
attempted to cover the reactor and bring the graphite fire raging beneath
them under control. Yet the Soviet authorities assured the Swedes that
they had no information about any kind of nuclear accident within the
USSR.

That afternoon in Moscow, the Swedish embassy's science attaché
contacted the State Committee for the Utilization of Atomic Energy—
the kindly public face of Sredmash, the Ministry of Medium Machine
Building. But the committee would neither confirm nor deny a problem
with any of their reactors. Early in the evening, at a cocktail party at the

Swedish embassy, Ambassador Torsten Örn buttonholed an official of the
Soviet Foreign Ministry and asked him directly if he knew of a recent nuclear accident within the USSR.

The official told Örn he would make a note of the inquiry but provided no further comment.

Finally, at 8:00 p.m. on Monday, April 28, almost three days after the
toxic cloud first rolled into the night sky above Unit Four, Radio Moscow broadcast the TASS statement agreed upon in Gorbachev's office.
"An accident has taken place at the Chernobyl nuclear power plant," the
announcer read. "One of the atomic reactors has been damaged. Measures are being taken to eliminate the consequences of the accident. Aid
is being given to those affected. A government commission has been set
up." In its brevity and frugality with the truth, the bulletin was typical
of Soviet news reports, a continuation of the way the state had covered
conventional industrial accidents for decades. An hour later, Radio Moscow's World Service repeated the announcement in English, for foreign
listeners, and followed it by cataloging the long record of nuclear accidents in the West. Both statements avoided mentioning when, exactly, the
Ukrainian accident might have taken place.

At 9:25 p.m. Moscow time, *Vremya*, the flagship nightly TV news
show broadcast throughout the Soviet Union, carried the same twenty-
three-word statement, read in the name of the Council of Ministers of
the USSR. It was the twenty-first item of news. There were no pictures.
Only the grave expression on the face of the anchor and the mention of
the Council of Ministers suggested that something extraordinary might
have happened.

The following morning, Tuesday, April 29, the press in Moscow remained entirely silent about the accident. In Ukraine, the daily papers in
Kiev reported the news, but their editors did their best to keep it quiet:
Pravda Ukrainy printed a short report at the bottom of page three, beneath an article recounting the story of two pensioners struggling to have
telephones installed in their homes. *Robitnycha Hazeta*—the Ukrainian
workers' daily—took care to bury its Chernobyl story below the Soviet
soccer league tables and coverage of a chess tournament.

Back in the Kremlin, General Secretary Gorbachev had convened the second extraordinary meeting of the Politburo in two days, once more at ten thirty in the morning. Now he was concerned that the initial response to the developing catastrophe had been flat-footed: radiation was still spreading, elevated levels had already been reported in Scandinavia, and the Poles were asking awkward questions. Was it possible that contamination could reach Leningrad—or Moscow?

Vladimir Dolgikh gave his colleagues the latest news: the plume of radioisotopes drifting from Chernobyl had divided into three traces, heading north, south, and west, and the Ministry of the Interior had cordoned off an area of at least ten kilometers around the plant, but the levels of radiation released from the reactor were falling. Chebrikov, the KGB chief, disagreed: his sources saw no indication that the radiation situation was improving. In fact, they were facing disaster. Further evacuations in the area were already under way, there were almost two hundred victims of the accident hospitalized in Moscow, and Vladimir Scherbitsky, the Ukrainian premier, had reported outbreaks of panic in his republic.

Everyone present agreed that they must seal up the reactor completely as quickly as possible. To take control of the situation, they moved to set up a special seven-person operations group, headed by Prime Minister Ryzhkov and including Dolgikh, Chebrikov, the minister of the interior, and the minister of defense. The group, granted emergency powers to command all Party and ministerial authorities throughout the Union, would coordinate the disaster response from Moscow, placing all the resources of the centralized state at the disposal of the government commission in Chernobyl.

Discussion turned once more to what to tell the world about what had happened. "The more honest we are, the better," Gorbachev said, suggesting that they should at least provide specific information to the governments of the Soviet satellite states, to Washington, DC, and London. "You're right," said Anatoly Dobrynin, recently appointed to the Central Committee after twenty years as the Soviet ambassador to the United States. "After all, I'm sure the photos are already on Reagan's desk." They agreed to cable statements to their ambassadors in world capitals, including Havana, Warsaw, Bonn, and Rome.

"Should we give information to our people?" asked Aliyev.

"Perhaps," Ligachev replied.

On Tuesday evening, *Vremya* broadcast a new statement issued in the name of the Soviet Council of Ministers. This conceded that two people had been killed as a result of an explosion at the Chernobyl plant, that a section of the reactor building had been destroyed, and that Pripyat had been evacuated. There was no mention of a radioactive release. This time the report was relegated to sixth place, behind the latest encouraging news about the mighty Soviet economy.

By then, the rest of the world's media had the scent of a spectacular catastrophe behind the Iron Curtain, and Chernobyl had become headline news in the West. Newspapers and TV networks threw their correspondents into a hunt for more details, no matter how vaporous their sources. Denied permission to travel to Ukraine and faced with a bureaucratic wall of silence, the foreign press corps in Moscow did what they could with scant material. Luther Whittington of the wire service United Press International, only recently arrived in the USSR, had bumped into a Ukrainian woman in Red Square just a few weeks earlier, who he believed had contacts in the emergency services. Whittington telephoned the woman at home in Kiev, and understood her to say that eighty people had been killed immediately and an additional two thousand had died on the way to the hospital as a result of the explosion. No independent confirmation could be found for these claims, and one of Whittington's Moscow colleagues, Nicholas Daniloff of *U.S. News & World Report*, would later be convinced that the UPI man's Russian was so primitive that he might have misunderstood what his source had said. Nonetheless, it was a sensational story that flashed immediately across the international wires, with predictable results.

"'2,000 DIE' IN NUKEMARE; Soviets Appeal for Help as N-Plant Burns out of Control," screamed the front page of the *New York Post* on Tuesday—the same morning that news of the accident was being modestly tucked away beneath the sports results in Kiev. In London the following day, the *Daily Mail* went with "'2,000 DEAD' IN ATOM HORROR."

That night, this lurid death toll became the lead story on TV news across the United States; a Pentagon source told NBC that satellite images of the plant had revealed such devastation that thousands of deaths were inevitable, and two thousand "seemed about right, since four thousand worked at the plant." Soon afterward, US Secretary of State George Shultz received a secret intelligence assessment describing the Soviet claim of only two fatalities as "preposterous."

Meanwhile, the radioactive cloud had continued north and spread west to envelop all of Scandinavia—before the weather stagnated and the contamination drifted south over Poland, forming a wedge that moved down into Germany. Heavy rain then deposited a dense band of radiation that reached all the way from Czechoslovakia into southeastern France. The West German and Swedish governments lodged furious complaints with Moscow over its failure to promptly notify them of the accident and requested more information about what had happened, but to no avail. Instead, Soviet embassy officials contacted scientists in both Bonn and Stockholm seeking advice on nuclear firefighting, especially tips on how to extinguish burning graphite. The widening rumors, experts' public speculation about a reactor meltdown, and now—more terrifying still— the possibility of a radioactive fire that could not be put out spread panic throughout Europe.

In Denmark, pharmacies quickly sold out of potassium iodide tablets. In Sweden, imports of food from the USSR and five Eastern European countries were banned, a radioactive particle was reportedly discovered in a nursing mother's breast milk, and government switchboards jammed with calls from people asking if it was safe to drink water or even go outside. In Communist Poland, where state television assured the public that they were not in danger, authorities nonetheless distributed stable iodine to children and restricted the sale of dairy products. In Holland, a Dutch radio ham reported listening in to a shortwave exchange in which someone in the Kiev area claimed that not one but two reactors in Chernobyl were melting down. "The world has no idea of the catastrophe," the Ukrainian pleaded through the static. "Help us."

Soviet spokesmen dismissed these stories as opportunistic Western propaganda but, confounded by the state's reflexive secrecy, had few facts with which to fight back. On Wednesday evening, Radio Moscow admitted that in addition to the two deaths, another 197 people had been hospitalized as a result of the accident, but 49 of them had already been released after checkups. The "radiation situation," the announcer added obliquely, was "improving." A single photograph of the plant, said to have been taken soon after the explosion, was also broadcast on television, demonstrating that it had not been destroyed entirely; at the same time, Radio Kiev announced its intention to stifle "Western rumors" of thousands of fatalities.

Meanwhile, KGB chief Chebrikov notified his superiors that he was battling the bourgeois conspiracy at its source. He was, he told the Party Central Committee, undertaking "measures to control the activities of foreign diplomats and correspondents, limit their ability to gather information on the accident at the Chernobyl nuclear power plant, and foil attempts to use it for stoking the anti-Soviet propaganda campaign in the West."

That same day in Moscow, Nicholas Daniloff found that he was unable to telex his dispatches back to *U.S. News & World Report* headquarters in Washington, DC. He made arrangements instead to use the machine at the UPI office, twenty minutes' drive across town. As he was getting ready to leave, Daniloff looked out of the window to see several men tugging at the cables beneath an open manhole in the courtyard—apparently attempting to sever his remaining communications with the outside world. It didn't help: if anything, the Soviet attempts to hamper Western reporting made the rumors worse. By the end of the week, the *New York Post* was going to press with "unconfirmed reports" from Ukraine that fifteen thousand people had been killed in the accident, their bodies buried in mass graves excavated in a nuclear waste disposal site.

In the headquarters of the government commission in Pripyat, the emergency had continued to escalate. On Sunday evening, the young air force commander, General Antoshkin, went to the White House to report the progress of his pilots' bombing operation to Chairman Boris Scherbina. Using just three helicopters, they had managed to dump ten tonnes of boron and eighty tonnes of sand into Unit Four by nightfall. It was a heroic effort, in terrible conditions, but Scherbina was unimpressed. Such quantities were a pathetic show in the face of a planetary catastrophe: "Eighty tonnes of sand for a reactor like this is like hitting an elephant with a BB gun!" he said. They would have to do better.

Antoshkin ordered in heavy-lift helicopters, including the giant Mi-26, the most powerful helicopter in the world, nicknamed *korova*—"the flying cow"—and capable of carrying up to twenty tonnes in a single load. He stayed up that night trying to devise ways to increase the effectiveness of the operation. The next day, the commission drafted the population of several towns and villages in the nearby countryside into the effort to fill bags with sand. Antoshkin improvised an air traffic control system to speed up loading and requested decommissioned braking parachutes from MiG-23 fighter jets for use as improvised cargo nets. In the meantime, the flight engineers continued to heft the bags into Unit Four by hand and remained almost entirely unprotected from the gamma rays rising from the ruined building.

The crews flew from dawn to dusk every day and at night returned to their airfield in Chernigov to decontaminate their machines, discard their uniforms, and scrub radioactive dust from their bodies in a sauna. But it proved almost impossible to entirely remove the radiation from the helicopters, and when they returned each morning to begin a new mission, the airmen found the grass beneath their parked aircraft had turned yellow overnight. Most crews flew a total of ten to fifteen sorties over the reactor, making two or three bombing runs each time, but the first pilots did even more: one returned seventy-six times to Reactor Number Four in the first three days. According to Antoshkin, when they touched down in the landing zones after the second or third sortie, some flight engineers leapt from their machines to vomit into the bushes on the riverbank.

By the morning of Tuesday, April 29, the work of Antoshkin's crews seemed to be having an effect: radioactivity escaping from the reactor began to fall, and the temperature dropped from more than 1,000 degrees to 500 degrees centigrade. But radiation levels in the deserted streets of Pripyat had now become so dangerous that the government commission was forced to withdraw to a new headquarters nineteen kilometers away in Chernobyl town. The territory immediately surrounding the plant—an area roughly one and a half kilometers in diameter, which officials soon began calling the *osobaya zona*, or Special Zone—was highly contaminated with both debris and nuclear fallout. Scientists, specialists, and those operators who had stayed behind to tend to the needs of the remaining three reactors at the station now approached it only in armored personnel carriers.

In Moscow, Prime Minister Ryzhkov chaired the first meeting of the Politburo Operations Group that afternoon, and the hammer of the centrally planned economy immediately began to swing. Academician Legasov and the other scientists calculated that a total of 2,000 tonnes of lead would eventually be needed to help smother the graphite fire, but Legasov was afraid to ask for such an enormous quantity of a scarce resource on short notice. Scherbina—an old hand in the ways of the system—went ahead and ordered 6,000 tonnes just in case, and Ryzhkov had every train carrying lead on the Soviet rail network redirected toward Chernobyl. The first 2,500 tonnes arrived the next morning.

By the time the light failed on Tuesday night, Antoshkin's helicopter crews had dropped another 190 tonnes of sand and clay inside the walls of Unit Four. But the fire blazed on, and radionuclides continued to pour from the wreckage of the reactor. A scientific report revealed to the Party in Kiev that levels of background radiation in the Ukrainian cities of Rovno and Zhitomir, more than a hundred kilometers to the west and southwest of the plant, respectively, had already increased almost twentyfold. The regional civil defense leaders had made preparations for evacuating the settlements within a ten-kilometer radius of the plant—ten thousand people in all—and pleaded with Scherbina to give permission for the operation to proceed. But, to their consternation, he refused.

The next morning, a shipment of parachutes was delivered to Chernobyl by helicopter: not surplus equipment, as General Antoshkin had suggested, but fourteen thousand brand-new paratroop canopies, requisitioned from airborne detachments all over the USSR. When Antoshkin conducted tests, he discovered that each chute could carry as much as 1.5 tonnes without breaking. By sunset, his crews had managed to hit the reactor with another 1,000 tonnes of absorbents. When the general delivered his report that night, he saw Scherbina's face brighten for the first time since the bombing operation had begun.

On Wednesday, April 30—the eve of the annual May Day celebrations, which would fill the streets of towns and cities throughout the USSR with parades and rallies—the wind blowing over the Chernobyl plant shifted once more. This time it turned almost due south, carrying energetic alpha and beta contamination directly toward Kiev, along with dangerously high levels of gamma radiation, in the form of iodine 131—the radioisotope that concentrates dangerously in the thyroid gland, particularly in children. At exactly one o'clock that afternoon, radiation levels on the streets of the city began to rise abruptly. By nightfall, recorded radioactivity on Nauki Prospekt, near the heart of Kiev on the eastern bank of the Dnieper, had reached as high as 2.2 milliroentgen per hour (or 22 microsieverts per hour)—hundreds of times higher than normal. The progress of the radioactive cloud had been tracked by the weather-monitoring equipment of the USSR State Committee for Hydrometeorology, which that day sent classified reports of its findings to Prime Minister Ryzhkov in Moscow, but also to the leaders of the Ukrainian Communist Party in Kiev, including First Secretary Scherbitsky.

Inside the Ukrainian Ministry of Health, the republic's senior doctors, normally orderly and restrained, started to panic. They discussed the need to take precautions against radioactive aerosols and to broadcast a warning to the population. But they took no action. Valentin Zgursky, the chairman of the Kiev City Executive Committee—the city mayor, also responsible for coordinating its civil defense—had worked at a factory manufacturing gamma measurement tools and was familiar with the dangers of radiation. He tried to convince Scherbitsky to cancel the big May Day parade they had scheduled to pass through the center of the city the following morning. But the first secretary told him that the orders had come down from Moscow. Not only would the parade take place, but they were all expected to attend and to bring their families with them—to demonstrate that there was no reason for anyone in Kiev to panic.

The following morning, preparations for the parade got under way as usual. Party organizers had hung banners, and spectators filled the streets. At ten o'clock, the first secretary was scheduled to open the parade from his position in the center of the tall reviewing stand overlooking the Square of the October Revolution. But with only ten minutes to go, he was nowhere to be seen; his central spot on the platform remained empty. The members of the Ukrainian Politburo, the city mayor, and the other assembled dignitaries became agitated: no one else but the first secretary had the right to start the procession, and he had never been late for the May Day parade in all his years in office. At last, his Chaika tore down the hill toward the rostrum and pulled up sharply. Scherbitsky climbed from the backseat, red faced and cursing.

"I told him that we can't possibly conduct the parade on Khreschatyk," he muttered to the assembled dignitaries. "It's not like Red Square. This is a ravine: there's radiation pooling here. . . . He told me: You'll put your Party card on the table if you bungle the parade."

The furious first secretary left no one in any doubt about whom he had been speaking to. The only man in the Union with the authority to threaten him with expulsion from the Party was Gorbachev himself.

"To hell with it," Scherbitsky said. "Let's start the parade."

Soon after ten o'clock, the cheering crowds began their march down the broad avenue of Khreschatyk. The sun was warm, the atmosphere festive. There were seas of red banners and biers of brightly colored peonies in spring shades of magenta, yellow, and blue; ranks of Party veterans clad in gray suits and scarlet sashes; the girls of the Young Pioneers—the Communist Girl Scouts—in white uniforms and red neckerchiefs, waving boughs of cherry blossom; and young dancers in the embroidered blouses and billowing *sharovary* pants of traditional Ukrainian Cossack dress, swinging arm in arm in extended lines or spinning in tight circles.

Carrying their children or leading them by the hand, the citizens of Kiev marched in dense blocks beneath the boulevard's famous chestnut trees, clutching balloons and bearing placards depicting Soviet and Ukrainian Party leaders, past the fountains of the Square of the October Revolution, where six-story portraits of Marx, Engels, and Lenin gazed impassively from the grimy facades of the apartment buildings.

Scherbitsky, who waved and smiled paternally from the rostrum as his subjects passed, had made some concessions to the perils of the fall-out carried on the breeze: instead of four hours, the parade lasted only two; instead of between four thousand and five thousand people from every district, Party organizers gathered only two thousand. But the first secretary ensured that his grandson Vladimir participated in the march. Kiev mayor Zgursky brought his three sons and two grandchildren. Some on the rostrum that morning had armed themselves with dosimeters and consulted them discreetly, but constantly. Others simply stole occasional glances at the sky.

Later, when the wind changed direction again, threatening to carry the plume of radionuclides north toward Moscow, Soviet pilots flew repeated missions to seed the clouds with silver iodide, designed to precipitate moisture from the air. The capital was spared. But three hundred kilometers to the south, peasants watched as hundreds of square kilometers of fertile farmland in Belarus were lashed with black rain.

In Moscow, the May Day procession swept through Red Square just as it did every year, and a carnival atmosphere filled the city. Workers from all over the Soviet Union paraded nine abreast past Lenin's red granite mausoleum, waving paper carnations and scarlet banners as Gorbachev and the other members of the Politburo looked on. But afterward, Prime Minister Ryzhkov convened another urgent meeting of the Chernobyl Operations Group, attended by more than a dozen senior ministers, including those representing health, defense, and foreign affairs, as well as Ligachev and Chebrikov, the ideological and KGB chiefs of the USSR. Leading the nuclear scientists at the table was Anatoly Aleksandrov, head of the Kurchatov Institute.

As the festivities continued in the streets outside, the group confronted the emergencies spiraling from Chernobyl. First, they heard how the scale of the crisis had overwhelmed the Soviet Health Ministry. Ryzhkov relieved the ministry's chief of his responsibility for the medical response and handed control to the deputy health minister, Oleg Schepin. Ryzhkov instructed him to provide the group with daily reports of the number of victims of the accident hospitalized throughout the USSR and how many of them had been diagnosed with radiation sickness. Despite their relocation from Pripyat, Scherbina, Legasov, and the other members of the government commission had already been exposed to dangerous levels of radiation and had to be replaced as soon as possible. They would need to order medical supplies from the West and pay for them in hard currency; the group also had to give serious consideration to foreign doctors offering to come to Moscow and assist in treating radiation victims.

Marshal Akhromeyev, the chief of the General Staff of the USSR, reported that Ministry of Defense troops had begun decontamination work in the area immediately surrounding the Chernobyl plant, but more men were urgently needed, as the radiation continued to spread. It was also clear that attempts to conceal the truth of what had happened from the international community seemed only to have made matters worse. Western diplomats and reporters were deluging Moscow with protests and questions about the nature and proportions of the accident. Ryzhkov resolved to organize a press conference for foreign journalists and delegated the details to Scherbina, Aleksandrov, and Andranik Petrosyants, chairman of the Soviet State Committee for the Utilization of Atomic Energy.

Ryzhkov now assembled the members of the replacement government commission and gave them orders to fly to Chernobyl the following day. This new team would be led by Ivan Silayev, the fifty-five-year-old former Soviet minister of aviation industry. Its chief scientist, intended to replace Valery Legasov, was Evgeny Velikhov—Legasov's next-door neighbor and rival to succeed old man Aleksandrov as head of the Kurchatov Institute.

When the meeting concluded, Ryzhkov went to see Gorbachev in his office. The prime minister and Ligachev, the conservative Party boss, had decided that the time had come for them to visit the scene of the accident. Ryzhkov told the general secretary of his intentions and waited for Gorbachev to say he would join them. But apparently nothing could have been further from the Soviet leader's mind. The next day, Ryzhkov and Ligachev—along with the head of the KGB—flew to Kiev without him.

Accompanied by Scherbitsky, the Ukrainian first secretary, and the prime minister of Ukraine, the men from the Politburo flew by helicopter to the Chernobyl plant. They stopped in a village where evacuees from Pripyat had been temporarily resettled. Ryzhkov found the people he met to be strangely calm and suspected that they were still ignorant of the magnitude of the disaster enfolding them. When they asked when they would be allowed to return home, the ministers couldn't say. They were told to wait and be patient.

At 2:00 p.m., Ryzhkov and Ligachev sat for a briefing in the regional headquarters of the Party, in Chernobyl town. Talk turned to evacuating the settlements surrounding the plant. Scherbina told Ryzhkov that evacuation of the ten-kilometer zone had just begun.

On the table were reconnaissance maps of the area around the nuclear plant, the result of separate military, meteorological, and geological surveys, all top secret and marked to show the spread of the radioactive fallout from the plant. Overlaid upon one another, they showed a ragged blot, centered on Pripyat and extending to the southeast, where it covered Chernobyl; it crossed the border into Belarus to the north, and a long trace of heavy fallout extended due west, forming a forked tongue reaching out toward the towns of Vilcha and Polesskoye. The contamination had spread far beyond the boundaries of the ten-kilometer zone and now imperiled tens of thousands of people across a vast area—in some places, as far as thirty kilometers from the plant.

Prime Minister Ryzhkov examined the maps carefully. Some locations seemed safe for now, others were clearly not, and in some villages, the fallout was patchy and varied from street to street. Clearly something had to be done, but it was hard to know exactly what. Everyone in the room was awaiting his decision.

"We will evacuate the population from the thirty-kilometer zone," Ryzhkov said at last.

"From all of it?" someone asked.

"From all of it," the prime minister replied and circled the area on the map for emphasis. "And start immediately."

11

The China Syndrome

From high up on the roof of the Hotel Polesia, Colonel Lubomir Mimka had an excellent view: the center of Pripyat spread out before him in a bright panorama, from the glinting constructivist sculpture above the entrance to the music school on his left, to the row of colored pennants fluttering above the plaza to his right. He and his radio operator had the city to themselves. The hotel was empty, and even the birds had left: the twittering of sparrows that had once danced through the branches of the poplar and acacia trees on the streets below had long since ceased. Exploring the banquet hall downstairs, the two men had discovered a mysterious black carpet stretching from wall to wall, and it was only when the radioman had begun to cross it, sweltering in his rubber chemical protection suit, his boots crunching beneath him, that they realized the floor was covered with thousands of dozing flies, apparently intoxicated by radiation. Unspooling a fire hose from the wall, they washed the floor clean, and moved down there when the heat of the sun and the toll of the radioactivity upstairs became too much to bear. But from the eighth-story rooftop, framed by the heavy concrete supports of the outdoor terrace, Colonel Mimka's sighting position on Reactor Number Four—just over three kilometers away but clearly visible on the horizon—was perfect.

Almost from the beginning of the bombing operation, the air force had been using the hotel as an improvised control tower, but now General Antoshkin had also devised a system that enabled his pilots to drop hundreds of tonnes of material into the burning reactor every day. As one helicopter after another clattered into its run over Unit Four, Mimka watched from the roof and issued final instructions to the pilots over the radio, estimating distance and trajectory by eye. There were now dozens of machines involved—the medium-heavy Mi-8, the heavy Mi-6, and the superheavy Mi-26—taking off in a continuous carousel from three separate landing zones. Each carried at least one upturned parachute slung beneath its belly, filled with bags of sand or clay by civil defense troops and work gangs drawn from the local settlements. Lifting off from the field in a whirlwind of dust, they approached the reactor at a hundred kilometers per hour.

Mimka waited until they were three hundred meters from their target, judging their position against the pylons rising from the transformer yards around the station. He gave the order "Ready!" and the pilot moved his finger to the release button. Two or three seconds passed. Mimka said, "Drop!" The pilot released the payload, and the helicopter, suddenly unburdened, turned quickly away—before returning to collect another cargo from the landing zone.

Mimka rose at four o'clock each morning and was given a blood test for radiation exposure with his breakfast. He was in the air at six and took a reconnaissance flight over the reactor before being dropped in the square outside the hotel. He remained on the roof until the light gave out sometime after 9:00 p.m., and took the last flight of the day himself to log a second set of radiation and temperature readings over Reactor Number Four. Then there was decontamination and dinner at ten, followed by debriefing. Sleep came finally at midnight—and four hours later, a cadet was shaking him awake again.

General Antoshkin had imposed an exposure limit of 22 rem on his men, although many routinely underreported it in order to keep flying longer. They were issued both bitter potassium iodide tablets and a sweet medicinal paste—shipped in from a pharmaceutical plant in Leningrad, intended to help combat radiation—which they called *pastila*. When the first shipments of lead arrived—in ingots and sheets and ten-kilogram bags of hunters' shotgun pellets delivered from stores with the price tags still on them—the pilots improvised their own protection. They lined the floor of the cabins with the four- and five-millimeter-thick sheets and filled the wells of their seats—designed to accommodate parachute packs—from the bags of shot. They even had a rhyme about it: *Yesli hochesh byt' otsom, zakrivay yaitso svintsom.* "If you want to be a dad, cover your balls in lead."

Even as the bombardment continued, Academician Legasov and the other scientists sent down from the Kurchatov Institute of Atomic Energy and the Ministry of Medium Machine Building in Moscow still had little idea what was going on inside the burning reactor. The pilots were now targeting a red glow they could see inside Unit Four, but no one could be certain exactly what was causing it. Back in the capital, physicists were hauled into their offices at the Kurchatov Institute late at night to help calculate how much uranium fuel might still be left inside the ruins of Reactor Number Four. And five or six times every day, scientists joined the helicopter crews flying over the accident site to monitor radiation levels in the air and estimate the heat of the burning core by analyzing the radioactive isotopes in the atmosphere. They took surface temperature readings of the reactor, using a Swedish-made thermal imaging camera. They watched as the pilots' payloads plunged toward their mark, and saw shaggy mushroom-shaped clouds of black radioactive smoke and dust rise a hundred meters into the air, hanging there before being snatched by the breeze and swept away across the countryside. As dusk fell, a beautiful crimson halo rose above the building. Looking down at the incandescent mass on his reconnaissance flights above the reactor in the evening gloaming, Colonel Mimka was reminded of the lava he had seen among the volcanoes of Kamchatka in the Soviet Far East.

At the very start, one member of the Kurchatov group in Chernobyl—the RBMK reactor specialist Konstantin Fedulenko—tried to tell Legasov that the whole helicopter operation might be misguided. He had seen for himself that each of the cargo drops into the shattered building was hurling heavy radioactive particles into the atmosphere. And, given the small size of the target—partially concealed by the tilted concrete lid of Elena—and the speed of the pilots' approach, there seemed little chance that any of the sand or lead was making it into the eye of the reactor vault itself.

But Legasov disagreed. He told Fedulenko that it was too late to change course. "The decision has been made," he said. The two scientists argued for a few minutes, until Fedulenko eventually admitted the full extent of his fears: that all their efforts to quench the graphite fire were a total waste of time. He said they should just let the radioactive blaze burn itself out.

Legasov didn't want to listen. He insisted that they had to take immediate action—whether it was effective or not.

"People won't understand if we do nothing," Legasov said. "We have to be seen to be doing something."

Day after day, the volume of material dumped into the reactor mounted: on Monday, April 28, the helicopter crews flew 93 sorties and dropped a total of 300 tonnes; the next day, they managed 186 sorties and 750 tonnes. On the morning of Wednesday, April 30, they began dropping lead, and that day blanketed Unit Four with more than 1,000 tonnes of absorbents, including sand, clay, and dolomite. In the landing zones, the men of Special Battalion 731—a hastily created new formation of army reservists mustered overnight across the Kiev region—worked for sixteen hours a day beneath the whirling blades of the helicopters, piling sacks into parachute canopies and securing them to the cargo hard points of the aircraft. The hot weather and the rotor wash created an almost constant tornado of radioactive dust reaching thirty meters into the air. The soldiers wore no protective clothing—not even petal respirators. The dust filled their eyes and mouths and caked beneath their clothes. At night, they slept fitfully in their irradiated uniforms, in tents beside the Pripyat. At dawn, they rose to start again.

As the airborne attack went on, the levels of radionuclides issuing from the reactor continued their downward trajectory: from 6 million curies of radiation on Sunday to 5 million on Monday, 4 million on Tuesday, and 3 million on Wednesday. By the end of the day, the incandescent spot the airmen had been targeting seemed to have been extinguished. On the evening of the next day, Thursday, May 1, General Antoshkin reported to Boris Scherbina that his pilots had bombarded Reactor Number Four with more than 1,200 tonnes of lead, sand, and other materials. Some members of the government commission got to their feet and applauded. Scherbina graced the general with a rare smile. Then he set a new target for the following day: 1,500 tonnes.

But by the next evening, Valery Legasov and the team of scientists analyzing the latest data from Unit Four had made a horrifying and apparently inexplicable discovery. Instead of continuing to fall, the radioactive releases from the reactor had now suddenly begun to increase again, doubling from 3 million to 6 million curies overnight. The temperature of the burning core, too, was rising rapidly. By Thursday night, Legasov's estimates suggested that it was already approaching 1,700 degrees centigrade.

The academicians now feared that the uranium dioxide fuel and zirconium cladding remaining inside the vault of Reactor Number Four had become so hot that they had started to fuse into a mass of radioactive lava in what amounted to a total core meltdown. Worse still, the 4,600 tonnes of sand, lead, and dolomite that had been flung into the damaged building from two hundred meters in the air, combined with the impact of the initial explosions a week before, might have fatally compromised the foundations of the reactor. If the temperature of the molten fuel reached 2,800 degrees centigrade, they suspected it could begin burning through the reinforced concrete floor of the reactor vault. Pressed down from above, it could eat its way through the bottom of the vault, into the basement of the building, and deep into the earth below. This was the doomsday scenario of reactor accidents: the China Syndrome.

Although first envisioned by US nuclear engineers, the China Syndrome had been made infamous as the title of the hit Hollywood movie released less than a month before the accident at Three Mile Island. It starred Jane Fonda as an intrepid TV reporter, aghast to discover how a mass of melted uranium fuel might burn through the base of a faulty reactor in California and continue, inexorably, until it reached the other side of the world, in China. And while this hypothetical nightmare defied the laws of physics, geology, and geography, if a core meltdown had begun in Chernobyl, the China Syndrome posed two real threats. The first and most obvious was to the local environment. The power station sat just a few meters above the water table of the Pripyat River, and if the melted fuel penetrated that far, the consequences would be catastrophic. A whole range of toxic radionuclides would poison the drinking water not only supplied to Kiev but also that of everyone in Ukraine who drew from the waters of the Dnieper River basin—some thirty million people in all—and, beyond that, flow into the Black Sea itself.

But the second threat was even more immediate and frightening to contemplate than the poisoning of the water table. The molten fuel would reach the Pripyat and the Dnieper only if it escaped the foundations of the building. Before that happened, it would have to pass through the steam suppression pools, the flooded safety compartments beneath Reactor Number Four. And some of the scientists feared that if the white-hot fuel made contact with the thousands of cubic meters of water held in the sealed compartments there, it would bring about a new steam explosion orders of magnitude larger than the first. This blast could destroy not only what remained of Unit Four but also the other three reactors, which had survived the accident intact.

Amounting to a gargantuan dirty bomb formed of more than five thousand tonnes of intensely radioactive graphite and five hundred tonnes of nuclear fuel, such an explosion could exterminate whatever remained alive inside the Special Zone—and hurl enough fallout into the atmosphere to render a large swath of Europe uninhabitable for a hundred years.

On Friday, May 2, the new team ordered to replace Boris Scherbina and the members of his government commission, led by Ivan Silayev and including Legasov's old rival Evgeny Velikhov, arrived in Chernobyl from Moscow.

By now, Scherbina and his group were exhausted, and—after five days of often reckless disdain for the intangible dangers surrounding them—thoroughly irradiated. The commission members had not been given iodine tablets or dosimeters until they had already been in the accident zone for twenty-four hours, and not everyone had bothered to use them. Now their eyes and throats were red and raw from exposure to radioactive dust; some noticed their voices becoming high and squeaky—a strange side effect of alpha contamination. Others felt sick, their heads swam, and they became so agitated they could barely concentrate. On Sunday, May 4, when they eventually returned to Moscow, Scherbina and the others were hospitalized and examined for symptoms of radiation sickness. They surrendered their clothes and expensive foreign-made watches, too contaminated to be saved, for burial. One of Scherbina's assistants was showered eighteen times in an attempt to remove the radioactive particles from his skin. Nurses shaved the heads of everyone, except for Scherbina, who declared that such treatment was beneath the dignity of a member of the Council of Ministers of the USSR and assented only to a short trim.

Yet despite his mounting radiation dose and the departure of his colleagues, Valery Legasov chose to stay behind in Chernobyl. By the end of Sunday, emissions from the reactor had reached 7 million curies, even higher than the day the helicopter operation had first begun. And now Legasov found himself at odds with Evgeny Velikhov over how to respond.

Like Legasov, Velikhov had no direct experience with nuclear power reactors and arrived at the scene planning to learn on the job. His manner didn't impress the generals, who preferred the athletic and decisive Legasov—an avowed Socialist, a Soviet leader in the traditional mold—to the portly academic with his Western friends and loud check shirts. But Velikhov could count on his long-standing relationship with Gorbachev to guarantee a direct line to the general secretary—who had already taken a personal dislike to Legasov, begun to suspect he wasn't being told the whole truth about the accident, and needed someone in Chernobyl he could trust.

Now, in addition to their different personalities, the two scientists were divided by their approach to the threat of meltdown in Unit Four. Velikhov had recently seen *The China Syndrome*—screened to a restricted audience in the Physics Department of Moscow State University just over a year before—and feared the worst. However, Legasov and other nuclear specialists at the scene were unswayed by the Hollywood version of events. They believed the chances of a full meltdown were insignificantly small.

The scientists still had little real idea what might be happening deep in the bowels of Unit Four. They had no reliable data from inside the burning reactor, and even their measurements of radionuclides escaping into the atmosphere had a 50 percent margin of error. They knew nothing about the state of the graphite and lacked a full inventory of the fission products being released by the fuel; they couldn't be certain if the zirconium was burning, or how any of these elements might be interacting with the thousands of tonnes of different materials dropped from the helicopters. They did not know how hot nuclear fuel might react with an enclosed body of water. Nor did they have any hypothetical models to help them.

In the West, scientists had been simulating the worst-case scenarios of reactor meltdowns for fifteen years, in ongoing research that had only intensified after the disaster in Three Mile Island. But Soviet physicists had been so confident of the safety of their own reactors that they had never bothered indulging in the heretical theorizing of beyond design-basis accidents. And appealing directly to Western specialists for help at this stage seemed unthinkable. Despite the growing atmosphere of alarm among the physicists at the burning reactor, the government commission and the Politburo remained determined to conceal the news of a possible meltdown from the world beyond the thirty-kilometer zone.

Velikhov contacted the head of his research lab on the outskirts of Moscow and summoned his team to work over the May Day holiday weekend. The dozen scientists weren't given any details over the phone, and even when they reached the lab were told about the accident only in the most general terms. They were required to find out all they could about the potential speed of a reactor core meltdown—but they were all theoretical physicists, experts in the study of esoteric phenomena related to the interaction of lasers and solids, plasma physics, and inertial fusion. None of them knew anything about nuclear reactors, and the first thing they had to do was learn all they could about the RBMK-1000. They plundered the library for reference books on the properties of different radioisotopes, decay heat, and thermal conductivity and commandeered the lab's suite of Soviet-built mainframe computers to begin making calculations.

Meanwhile, as Velikhov and Legasov disagreed about the risks of a meltdown, the graphite blazed on, and the temperature inside Reactor Number Four continued to rise. Velikhov called Gorbachev in Moscow. What was happening in Chernobyl was so secret he wouldn't be allowed to call his wife for six weeks. But when he needed to talk to the general secretary, he could reach him on the car phone in his limousine right away. "Should we evacuate Kiev?" Gorbachev asked.

Velikhov admitted that he just couldn't be sure.

The new leader of the government commission, Ivan Silayev, a Hero of Socialist Labor and holder of two Orders of Lenin, a long-serving technocrat with a direct manner and a steep crest of silver hair swooping back from his forehead, was less volatile than Boris Scherbina. But he faced an even more dire situation than his predecessor had: the fire, the escaping radiation, the core meltdown, and now the possible explosion. He began demanding updates from the scene every thirty minutes. The members of the commission started work at 8:00 a.m. and finished at 1:00 a.m. Many slept for only two or three hours a night.

In his headquarters in Chernobyl town, Silayev adopted a typically Soviet approach to the crisis: rather than choose a single course of action to halt the possible meltdown, he gave orders for dynamic action and patriotic sacrifice on all fronts at once. He issued instructions to the plant staff to find a way of piping nitrogen gas into the reactor vault to blanket the melting core and starve the graphite fire of oxygen. He summoned subway construction engineers from Kiev to begin drilling into the ground below Unit Four to freeze the sandy soil with liquid nitrogen or ammonia and protect the water table from the melting fuel. And he sent out word to find men brave enough to enter the darkened basement rooms directly beneath the reactor itself, open the valves of the steam suppression pool, and pump out the five thousand cubic meters of highly radioactive water they would find there.

In the meantime, General Antoshkin's helicopter assault on Reactor Number Four continued.

At 1:00 a.m. on Saturday, May 3, Captain Piotr Zborovsky of the 427th Red Banner Mechanized Regiment of the civil defense had just finished washing in the field bathhouse, in his camp thirty kilometers south of the plant. He was toweling himself dry when he got word that someone was looking for him. A colonel and a major general approached. He had never seen either of them before.

"Get ready," the general said. "The head of the government commission wants to see you."

Zborovsky, at thirty-six, was a sixteen-year veteran of disaster remediation, nicknamed *Los*—"Moose"—on account of his physical strength. He had so far spent three days with his men in clouds of whipping dust and downdraft, heaving sacks of sand and clay into the parachutes beneath Antoshkin's helicopters. He hadn't eaten since breakfast the previ-

ous morning and was looking forward to a medicinal dose of 100 grams
of vodka.

"I'm not going anywhere until I have dinner," Zborovsky said.

"We'll wait," the general replied.

The steam suppression pools lay deep in the warren of claustrophobic
spaces beneath Reactor Number Four. They were contained in a single,
massive concrete tank with a volume of seven thousand cubic meters, split
into two stories, cluttered with a forest of thick pipework, partitioned into
corridors and compartments, and half filled with water. The pools had
been part of the main reactor safety system intended to prevent a steam
explosion in the event of a pressure channel rupture inside the core. In
an emergency, the escaping steam was supposed to be released through
safety valves and directed down into the pools, where it would bubble up
through the water, condensing harmlessly into liquid as it did so.

But on April 26, the condensation system had been quickly overwhelmed and failed during the final destruction of Reactor Number Four.
Now neither the station staff nor the scientists knew how much water the
tanks contained, or even if they remained intact. Technicians from the plant
had opened one valve connected to the system and heard only the whis-
tling of air rushing in. But the scientists nevertheless suspected there was
still water in the tanks. An order went out to find a good spot for blasting a
hole through the wall—almost two meters thick and sheathed in stainless
steel—using explosives. When this command reached a shift foreman at
Unit Three, he suggested that there might be a less potentially catastrophic
method of accomplishing the task. Examining the plant blueprints, he
identified a pair of valves intended to drain the tanks for maintenance—
located deep in the subterranean maze beneath the reactor—and scouted
the route with a flashlight and a DP-5 military dosimeter.

Before the accident, opening the valves would have been a simple task:
walk down a staircase to level -3, three meters below ground level, to corridor 001, a long concrete passageway connecting Units Three and Four;
find the valve compartment; and turn the wheels on valve numbers 4GT-
21 and 4GT-22. But now radioactive water flooded corridor 001. In the

valve compartment, it was one and a half meters deep. When he reached that point, the shift foreman's DP-5 ran off the scale, and there was no telling how much radiation lay inside. The valves could not be opened until the corridor had been cleared.

It was still the small hours of the morning when Moose Zborovsky was shown in to the meeting room on the second floor of the government commission headquarters in Chernobyl. Deputy Minister Silayev came out from behind his desk and stood to attention, his thumbs on the seams of his pants.

"Comrade Captain, you have an order from the government: to pump water out from under Unit Four."

Zborovsky didn't have a chance to think. "Yes, sir!"

"You'll be given details by military headquarters," Silayev said. "Be ready at 0900."

Only on his way back down the stairs did the captain consider the latest conditions reported near Unit Four: 2,800 roentgen an hour beside the outer wall of the reactor. He had been taught in military technical school that 700 rem was a lethal dose. At the wall, he'd get that in fifteen minutes. How much radiation would there be beneath the core itself?

Zborovsky drove the 120 kilometers back to the civil defense base in Kiev to gather men and equipment, stopping at home on the way. Knowing that his clothes were highly contaminated, he stripped in the hallway before going inside his apartment. He kissed his sleeping twelve-year-old son and said good-bye to his wife; he didn't tell her where he was going.

Reporting back to Silayev at nine o'clock on Saturday morning, the captain learned that the operation had to be planned from scratch. Even the elementary questions of how to get into the basement of Unit Four to get the water out, and where it could go once removed, had not been considered. At the meeting of the government commission that morning, the experts couldn't agree on a safe location for the storage of five thousand cubic meters—enough to fill two Olympic swimming pools—of highly radioactive effluent. While he waited for a decision, Zborovsky scouted the site in an armored personnel carrier and found a place to break through the wall into the basement service tunnel. Once again fearing the consequences of using explosives so close to the damaged reactor, Zborovsky asked for volunteers from his company to go in with a sledgehammer. Five men stepped forward. The radiation was high; the captain estimated each of them could work for a maximum of twelve minutes. When they finally broke through, Zborovsky went into the basement with a rope tied around his waist, like a deep-sea diver. He walked on through the blackness until his footsteps began to squelch. Gradually, the water level rose to a depth of more than four meters. It felt warm—45 degrees centigrade, hot as bathwater—and reeked of hydrogen sulfide.

Back in Moscow, Evgeny Velikhov's team of theoreticians began conducting experiments to investigate the behavior of melting nuclear fuel. In the absence of real data from the plant, Velikhov had arranged for boxes of papers on the phenomenon to be flown over from his contacts in the West, but the scientists were in too much of a hurry to read and synthesize the mountain of material. They decided it was quicker to do the research themselves. They worked around the clock and slept on the chairs in the office. In the lab, they heated metal cylinders and uranium fuel pellets with carbon dioxide lasers, laid them on pieces of concrete, and recorded the results. They sent samples to Kiev, where a specialist examined the interaction between uranium dioxide, molten heavy concrete, and sand. They quickly confirmed Velikhov's worst fears: a mass of fuel weighing only ten kilograms could generate enough heat to melt right through the reinforced concrete floor of the reactor vessel and then keep going, traveling downward as quickly as 2.5 meters a day. But they also found that the incandescent uranium could fuse with and absorb pieces of debris, metal, and sand to form entirely new substances—highly radioactive and possessed of as yet unknown characteristics.

In Chernobyl, the commission remained paralyzed over where to send the radioactive water filling the suppression pools, and, in the meantime, the temperature recorded in the reactor above it continued to rise. Silayev held meeting after meeting. Moose Zborovsky tried to sleep when he could, a few minutes at a time, as the arguments went on into the night—the academicians, generals, and politicians shouting over one another.

In the middle of it all, Gorbachev phoned from Moscow, his voice loud enough to be heard by everyone in the room:

"Well? Have you made up your minds?"

Meanwhile, the plant physicists, consumed with fear, wandered around like zombies: seized not by the long-term terrors of radiation but by the imminent threat of the explosion that might kill them—and everyone for hundreds of meters in every direction—at any moment.

Finally, after two days of deliberations, Zborovsky himself thought to ask the advice of one of the plant's senior engineers about the water. The technician described two open-air pools perfect for the purpose, just outside Pripyat. To reach them from the basement of Unit Four would require one and a half kilometers of hoses, but each pool had a capacity of at least twenty thousand cubic meters. Ominously, the temperature of the water in the basement had begun rising. It was now 80 degrees. And by 6:00 p.m. on Sunday, Legasov's readings from the reactor touched 2,000 degrees centigrade. Something was happening. They had to act quickly.

12

The Battle of Chernobyl

Shortly after 8:00 p.m. on the evening of Friday, May 2, President Ronald Reagan landed at Tokyo's Haneda Airport aboard Air Force One, at the climax of a ten-day tour through Asia and the Pacific. He had arrived in Japan to attend the first-ever meeting of the G7 nations—including the leaders of Great Britain, France, Germany, and Canada—but from the beginning, the trip had been overshadowed by the nuclear disaster unfolding on the other side of the world.

The first reports of the radiation detected over Sweden had reached Reagan aboard the presidential plane as he was leaving Hawaii on Monday, and his planned day off in Bali on Wednesday had been interrupted by a briefing on what US intelligence knew so far about events at the Chernobyl plant. Since then, Soviet dissembling about the accident had metastasized into a global diplomatic and environmental crisis. From high-resolution spy satellite photographs taken over Ukraine, in which they could make out details as small as individual fire hoses laid in the direction of the reactor cooling canals around the plant, CIA analysts knew that the scale of the catastrophe was far greater than Moscow acknowledged. And officials at the US Nuclear Regulatory Commission had begun to suspect that at least one of the other reactors at Chernobyl was now threatened by the ongoing crisis at Unit Four. Yet Moscow had rebuffed Reagan's public offer of medical and technical assistance, and American nuclear experts could only speculate about what was really happening at the crippled plant.

At the same time, Soviet attempts to suppress further details of the accident were unraveling. In a classified report to Gorbachev on May 3, Soviet foreign minister Eduard Shevardnadze warned that continued secrecy was counterproductive and had already bred distrust not only in

Western Europe but also with friendly nations planning to embrace Soviet nuclear technology, including India and Cuba. Shevardnadze wrote that taking the traditional approach to the accident was also endangering Gorbachev's dream of brokering a historic nuclear disarmament initiative with the United States. Western newspapers were asking how a nation that couldn't tell the truth about a nuclear accident could be trusted to be honest about how many nuclear missiles it had.

On the morning of Sunday, May 4, President Reagan broadcast his weekly radio address to the United States from his suite in the Hotel Okura. He talked about his summits in Southeast Asia, the need for an expansion in free trade, and the problems of international terrorism—alluding to the recent strike on Colonel Gadhafi's compound in Tripoli by F-111s of the US Air Force, in retaliation for the Libyan-sponsored bombing of a disco frequented by American soldiers in Berlin.

Reagan reiterated his sympathy for the victims of the accident and his offer of assistance, but then his tone hardened. He contrasted the openness of "free nations" with the "secrecy and stubborn refusal" of the Soviet government to inform the international community of the risks they shared from the disaster. "A nuclear accident that results in contaminating a number of countries with radioactive material is not simply an internal matter," Reagan said in his folksy rasp. "The Soviets owe the world an explanation."

That day, radioactive rain fell on Japan before being carried eastward by the jet stream—out across the Pacific at an altitude of 9,000 meters and a speed of 160 kilometers an hour—toward the coasts of Alaska and California. The following afternoon, Monday, May 5, a delegation from the International Atomic Energy Agency landed in Moscow, at the invitation of the Soviet government. The team, led by IAEA director general Hans Blix, had been promised a full and honest accounting of what had been going on at the Chernobyl nuclear power plant.

In the hours before their arrival, the Politburo gathered once more inside the Kremlin to discuss the crisis. Among the two dozen men around the table this time were Boris Scherbina, the aged nuclear chieftain academician Anatoly Aleksandrov, and Efim Slavsky, the pugnacious head of the Ministry of Medium Machine Building. Valery Legasov had flown in from Chernobyl to present his report in person.

There was much to discuss: the outlook was grim.

Prime Minister Nikolai Ryzhkov took the floor with a detailed analysis and described what he had seen on his visit to the accident zone two days earlier. The helicopter operation to extinguish the fire was proceeding successfully, he said, and the danger of a renewed chain reaction inside the wreckage had been averted so far. But the response of Soviet and local authorities to the accident had been marred by failure and incompetence. "The extreme conditions revealed, in practice, a high degree of organization in some, and the absolute helplessness of others," he said.

The evacuation of a zone extending to thirty kilometers around the plant was still under way, with a hundred thousand people already removed from the area, including two districts in Belarus. But the results of the initial operation had been chaotic: "Five or six thousand people are simply lost," Ryzhkov said. "Where they are now is unknown."

The civil defense and the Ministry of Health had failed utterly in their responsibilities. There had been no clarity or plan. People leaving the evacuation zone had not even received blood tests for radiation exposure. The fiasco made a mockery of the USSR's decades of preparation for the consequences of nuclear war. "I can only imagine what would have happened here had something more serious occurred," the prime minister said in disgust.

So far, more than 1,800 people, including 445 children, had been hospitalized; more were expected. High levels of radioactivity now covered the western Soviet Union, from Crimea in the south to Leningrad in the north, exceeding the natural background levels by five or ten times in most places. The chief of chemical troops of the USSR had already gathered two thousand men inside the evacuated zone and had been ordered to develop a decontamination plan. Ryzhkov had given instructions for a thirty-kilometer dam to be constructed around the accident site to prevent spring rains from flushing contamination from the surface of the ground in the ten-kilometer zone into the Pripyat and Dnieper rivers. He suggested military engineers be given just forty-eight hours to complete the task.

And now, the prime minister explained to his comrades, they must face the greatest threat of all: the reactor meltdown. The scientists had presented him with two possible prognoses for the molten fuel currently scorching its way toward the basement of Unit Four. The first was that the heat of radioactive decay could gradually dissipate on its own; according to their calculations, that might take months.

The second scenario, presented by Academician Legasov and old man Aleksandrov, was much darker. In addition to Velikhov's fear that, reaching 2,800 degrees, the collision of the molten fuel with the water of the suppression pool could result in a blast of steam that would destroy the remains of Unit Four and obliterate Unit Three, the academicians now warned Ryzhkov to consider a further possibility: "a nuclear explosion, with even more disastrous consequences."

Next, Scherbina took the floor, and Legasov presented a technical breakdown of the obstacles they faced: the radiation releases, the burning graphite, the rising temperature of the melting core, and the need to act quickly. Aleksandrov chimed in. There was disagreement and bickering.

"Don't get carried away," Gorbachev's conservative deputy, Ligachev, told Scherbina.

"You've got roentgen and milliroentgen mixed up," Scherbitsky, the Ukrainian Communist Party head, said to the deputy minister of hydro-meteorology.

Akhromeyev, chief of the Soviet General Staff, argued that they should blast through the wall of the suppression pools with a hollow-charge shell. Schadov, the minister of coal, said that such an approach was too dangerous. He suggested that, if the water could be pumped out, his men would stabilize the spaces by filling them with concrete. "If necessary," he said, "we'll dig a mining tunnel under the building."

Legasov concurred: they should excavate beneath the reactor to pump in nitrogen gas and cool it from below. He assured Gorbachev that there was no need yet to send out an emergency appeal for help from the West. If the worst should happen, the maximum evacuation zone would have to be extended no farther than 250 kilometers from the station.

But Gorbachev had already spoken to Velikhov, who had stayed behind in Chernobyl, and the general secretary now believed that they were approaching a terrible denouement: in the event of another explosion, they might need to expand the exclusion zone to a radius of 500 kilometers. That would mean evacuating a vast area of one of the most densely populated regions of the USSR, relocating everyone living in the largest cities in both Belarus and Ukraine, including Minsk and Lvov. In Kiev—a city of more than two million people, the third largest in the USSR—the republican authorities had begun quietly drawing up an evacuation plan but were terrified of having to enact it. They foresaw mass panic and the looting of stores, apartments, and museums. Hundreds of people would be crushed in stampedes at the railway stations and airports.

"We've got to pick up our pace and work around the clock," Gorbachev said. They had to act not merely as if they were at war, he explained, but as if they were under nuclear attack. "Time," he said, "is slipping away."

They were still discussing what to do next when Scherbina received a message from beneath Unit Four: Captain Zborovsky's water-pumping operation had begun.

Zborovsky had set out for the plant with twenty men, recruited from civil defense companies and fire stations across the region. When they arrived on the scene, they found it eerily silent, deserted except for the skeleton staff of operators tending to Units One, Two, and Three. Abandoned equipment surrounded the chaos of debris near Reactor Number Four: the fire trucks left behind by their colleagues more than a week before, too irradiated to be recovered, were now dented and smashed, hit by mistimed loads of lead and sand released from Antoshkin's helicopters. Although the air offensive had been halted temporarily, a thin column of smoke—or vapor—rose into the air from the rubble. Pieces of graphite littered the ground, still lying where they had been thrown by the explosion, shimmering in the hot sun.

Back at their base in Kiev, the fire crews had tried deploying hoses onto the ground from a helicopter to reduce the time the firemen would have to spend in the high-radiation zone near the reactor. But these experiments had failed. So the men would have to string out the one and a half kilometers of hoses by hand. They drilled again and again, practicing their routes and shaving seconds off the time it took to put them together and attach them to the special ZIL fire trucks, equipped with powerful pumps that could move 110 liters of water a second.

The city of Pripyat in the early 80s, with the Chernobyl Atomic Energy Station visible on the horizon. Reactor Unit Four of the plant lay just three kilometers from the southeastern edge of the town.

A view of Pripyat down the length of Lenina Prospekt, the poplar-lined boulevard into the city.

The Raduga—or Rainbow—department store on the corner of Kurchatov Street and the end of Lenina Prospekt. Plant director Viktor Brukhanov, like other senior members of the station staff, lived in an apartment above the store. The letters on the rooftop read "Glory to Lenin!" and "Glory to the Party!"

4

Pripyat was surrounded by forest and white-sand beaches; the daily *Raketa* hydrofoil service provided a cheap and swift connection to Kiev, two hours to the south along the Dnieper River.

Viktor Brukhanov with his wife, Valentina, and their son, Oleg, collecting mushrooms in the woods near Pripyat, 1980.

Alexander Yuvchenko, senior mechanical engineer of the fifth shift of Unit Four, and his wife, Natalia, posing in borrowed hats on the night of his twenty-fourth birthday, October 25, 1985.

Natalia with their son, Kyrill, then two years old, at home in Pripyat for New Year's 1985.

Viktor Brukhanov (*center, wearing sunglasses*), Chernobyl plant Communist Party secretary Serhiy Parashyn (*on the director's left*), and other Party chiefs from the power station and Pripyat lead the city's Victory Day parade on May 9, 1985—celebrating the fortieth anniversary of the Soviet defeat of Germany in the Great Patriotic War.

Anatoly Aleksandrov, the octogenarian director of the Kurchatov Institute and head of the Soviet Academy of Sciences, lecturing with pictures of the nuclear-powered icebreakers he helped pioneer. Aleksandrov personally endorsed the breakneck expansion of nuclear power in the USSR and took credit for the invention of the RBMK reactor.

The central hall of Unit Three of the Chernobyl plant, showing the reactor's 1600 fuel channels with their top covers removed. The RBMK-1000 model reactors of Units Three and Four, constructed back-to-back, were almost identical.

Unit Four, the newest and most advanced of the reactor blocks of the Chernobyl station, photographed soon after its completion in 1983.

Leonid Toptunov (*left*), senior reactor operator on the midnight shift of April 25, with his friend Alexander "Sasha" Korol (*center*) in 1981, on a trip with an unidentified friend two years before they graduated from the Moscow Engineering Physics Institute.

Lieutenant Vladimir Pravik, the twenty-three-year-old head of the third watch on duty at the Chernobyl plant fire station on the night of April 25.

Alexander Akimov, foreman of the midnight shift in the control room of Reactor Number Four, on April 25–26.

The first photograph of Unit Four after the accident, shot from a helicopter by Chernobyl plant photographer Anatoly Rasskazov, at approximately 3:00 p.m. on April 26, 1986.

A sketch made by a member of the team sent to open the giant electrically operated valves feeding coolant to the stricken reactor in the early hours of April 26: Alexander Akimov (*bottom left*) and Leonid Toptunov (*bottom right*) are shown up to their ankles in radioactive water; Akimov can no longer stand unsupported; Toptunov is seized by vomiting.

18

Senior Lieutenant Alexander Logachev, lead radiation reconnaissance scout of the 427th Red Banner Mechanized Regiment of the Kiev region civil defense, with his daughter in 1984.

At first, Captain Zborovsky wasn't afraid of what lay ahead. After all, he thought, his commanders would never have given him an assignment that they knew was certain to kill him. Only when he had entered the plant did he begin to comprehend the threat he faced. The staff there had already seen many of their friends flown out for treatment in the special clinic in Moscow, and they looked at him with the pity reserved for a condemned man.

The specialists and management from the plant who had remained behind to man the station were still nominally led by Director Viktor Brukhanov and his chief engineer, the once-bombastic Nikolai Fomin. The two men continued to sit by their phones in the dimly lit bunker beneath the plant, awaiting instructions from the government commission. But they were wrung out by exhaustion, radiation exposure, and shock. Fomin had remained in the bunker for five days, curling up to sleep beside the humming equipment in the ventilation room. Since the final evacuation of Pripyat, Brukhanov and the other operators had been sent to live in a Pioneer camp thirty kilometers from the plant site, named Skazochny, or "Fairy Tale."

A summer camp where the children of the nuclear workers could spend part of their long school vacations, Skazochny was a settlement of small brick and timber dormitories built deep in the forest, decorated with whimsical sculptures of dragons, sea creatures, and characters from Slavic myth. Now the woods and the fields nearby were crammed with ambulances, cars, fire engines, and military vehicles. A dosimetric checkpoint stood at the entrance gate. All over the camp, hanging on the fences and plastered across the windows of the canteen, was a sea of notes: messages written by plant workers seeking their wives and children, displaced families from Pripyat announcing the names of the villages where they could be found, and pleas for information about relatives who had been lost amid the frenzy of evacuation.

While Captain Zborovsky and his men prepared their pumping operation, the other parallel efforts to arrest the meltdown had also begun. First, the subway engineers arrived from Kiev and dug a large pit in the ground beside Reactor Number Three. Using specialized Japanese-built drilling equipment, they began to bore horizontally toward Unit Four, with the intention of creating a series of parallel holes 140 meters long running beneath its foundation. The engineers hoped that these would then carry narrow pipes bearing liquid nitrogen, which would freeze the ground, halting the progress of any melting nuclear fuel before it reached the water table.

At the same time, technicians from the power station embarked on Legasov's plan to extinguish the burning reactor with nitrogen gas. The idea was to use the plant's existing pipe network—which before the accident had distributed the various gases used in plant maintenance—to help direct the nitrogen through the basement and into the ruins of the reactor hall. From the outset, the members of the plant staff executing this scheme recognized that it was pointless: the pipework in the area beneath the reactor was almost certainly damaged, and even if it reached the reactor hall, the nitrogen couldn't hope to starve the fire of oxygen, because the building had no roof; instead of concentrating around the burning graphite and displacing air from the flames, the gas would just drift uselessly into the atmosphere. But orders were orders.

Silayev's government commission sent out instructions for all the available liquid nitrogen in Ukraine to be rerouted to Chernobyl by truck and by rail. The two massive vaporizers needed to turn the liquid into gas were located at the Cryogenmash plant in Odessa and flown to the airport in Chernigov, while a special shed to accommodate them was constructed near the administrative block of the power station. When they arrived, carried by a pair of Antoshkin's giant Mi-26 "flying cow" helicopters, the machines proved too large to fit through the door of the shed. The operators had to smash a larger opening with hammers. At 8:00 p.m., the technicians reported to Silayev that pumping could begin as soon as the nitrogen arrived. It was due that night, but the next morning there was still no sign of it. The operators waited all day. At 11:00 p.m., Director Brukhanov received a phone call from Silayev.

"Find the nitrogen," the commission chairman said, "or you'll be shot."

Accompanied by a detachment of troops, Brukhanov managed to locate the convoy of tanker trucks sixty kilometers away in Ivankov. The drivers, apparently terrified by the spectral horrors of radiation, had stopped in their tracks and refused to go any farther. Soldiers with machine guns took up positions at each end of the convoy, and the drivers were at last persuaded to deliver their cargo, at gunpoint.

It was eight in the evening on Tuesday, May 6, when Moose Zborovsky's men finally pulled on military respirators and L-1 chemical protection suits—the heavy, rubberized overalls designed for combat during nuclear war—and drove out toward Reactor Number Four. Zborovsky had conducted his own radiation survey and calculated where they could go and for how long. The gamma fields varied wildly, from 50 roentgen in places near Unit One, to the most dangerous areas—no more than 250 meters from Unit Four—where exposure reached 800 roentgen. The men stopped the trucks inside the transport corridor—the large passageway beneath the reactor through which railcars delivered fresh fuel for the plant. They ran out the hoses in just five minutes—a third of the regulation time—and started the pumps. Leaving the engines running, they closed the gates of the transport corridor behind them and sprinted to a nearby bunker. At last, the water level in the basement began falling. From their posts beneath the station, Brukhanov and Fomin telephoned Silayev, who relayed the news to Moscow.

Every few hours, three men raced out to top off the trucks with gas and oil; two others were sent to take readings of the radiation and the water temperature every sixty minutes. At three in the morning on Wednesday, a pair of firemen ran into the bunker to report that the hoses were ruptured. A team of chemical troops conducting radiation reconnaissance in the dark had driven over them in an armored personnel carrier, cutting them in twenty places and crushing the gaskets connecting them together. Radioactive water was gushing onto the ground just fifty meters from the reactor. Two sergeants sprinted out to repair the breaks: they needed twenty new sections of hose; each section took two minutes to replace. They worked on their knees, in a widening pond of gamma-emitting water. The two-fingered rubber mittens of their L-1 suits were clumsy and hot; they threw them off and used their bare hands. An hour later, the task complete, the men retired, exhausted, with an odd taste of sour apples in their mouths.

The pumping went on all night and into the next day. After fourteen hours of continuous operation, one truck's engine coughed to a halt. It had to be replaced. Zborovsky's men were all frightened: one was sent back to the Chernobyl fire station to fetch a case of medicinal vodka but lost his nerve on the way and never returned. Another began ranting incoherently and was taken to the hospital, vomiting. When it was Moose's turn again to take the radiation readings, he instructed a firefighting captain to come with him—in case he passed out or lost his way inside the building. The officer refused.

"Don't bring out the beast in me, you bastard!" Zborovsky roared. "Or I'll have my troops tie you up and throw you out beside Unit Four. Fifteen minutes out there, and you won't be able to utter another word."

The officer climbed into a rubber suit and did as he was told.

One hundred and forty kilometers away, details of what had happened at the plant had begun seeping into Kiev. The news spread by word of mouth and via the warbling "enemy voices"—the Russian-language radio programming broadcast into the Soviet Union by the BBC, Radio Sweden, and Voice of America—or at least those that KGB jamming could not disrupt. Waves of rumor and anxiety rippled through the city. The surveillance department of the Ministry of Internal Affairs—the MVD—reported on wild speculation about the number of victims of the accident and the contamination of air and water. One informant overheard a taxi driver describing how Pripyat had been evacuated amid chaos and looting, which even government troops had been unable to control; that a government minister was among those killed; that pregnant women were being induced to have abortions; and that the Dnieper was already completely radioactive.

Soviet authorities were still assuring the public that the danger from the plant was confined to the thirty-kilometer zone. But the streets of Kiev had been emitting gamma radiation for days, as hot particles carried in fallout from the reactor melted slowly into the surface of the asphalt. Scherbitsky, the Ukrainian Communist Party chief, knew that the radiation dose rates in the city had increased steeply. And radioactive iodine in the water of the Dnieper River basin had indeed reached levels a thousand times higher than normal.

Meanwhile, the head of the Ukrainian KGB warned that the numbers of casualties ascribed to the accident being broadcast on TV by Moscow and by Kiev were sharply contradictory. But his colleagues procrastinated about what—and when—to tell the people.

Finally, on Tuesday, May 6—ten days after the crisis began—the Ukrainian health minister appeared on local radio and TV to warn Kievans to take precautions against radiation: to remain indoors, close their windows, and protect themselves against drafts. By then, word had gone around that senior Party members had quietly sent their children and grandchildren to the safety of Pioneer camps and sanatoria in the south. Days earlier, at a central Kiev pharmacy favored by members of the Ukrainian Central Committee, the doctor and writer Iurii Shcherbak had discovered a long queue of well-to-do pensioners waiting patiently to buy stable iodine. Worse still, rumors had also leaked of the possibility of a devastating second explosion at the plant and the government's secret contingency plan for the total evacuation of the city. Many people recognized the state's reassuring official statements as hollow propaganda.

That evening, crowds gathered at the railway station as thousands of people attempted to escape the city. Men and women spent the night sleeping on the concourse to keep their places in line for tickets. The Soviet internal passport system prevented most citizens from leaving their areas of registration without good reason, so many workers hurriedly applied for vacations; some who were refused simply quit their jobs in desperation. Fleets of orange street-cleaning trucks soon appeared, to start what would become an incessant effort to wash hot fallout from the city streets. By then, crowds had begun to form outside the city's banks, some of which were forced to close just a few hours after opening; others limited withdrawals to 100 rubles per person. By afternoon, many banks had run out of money. When the pharmacies sold out of stable iodine pills, people resorted to drinking tincture of iodine, intended for use as an external antiseptic, burning their throats. Lines outside liquor stores quadrupled in length as people sought protection from radioactivity with red wine and vodka, forcing the Ukrainian deputy minister of health to announce, "There is no truth to the rumor that alcohol is useful against radiation."

By Wednesday, crowds of frantic Kievans were fighting for tickets out of the city, struggling to flee in numbers not seen since the German blitzkrieg swept east in 1941. At the railway station, men and women thrust fistfuls of rubles directly into the hands of the coach attendants, pressed into four-seat compartments ten at a time, and climbed into the luggage racks. Others attempted to make their escape by road, and traffic choked the southern routes from the city: almost twenty thousand people left by car or bus in one day alone. The government added extra flights at the airport and doubled the number of trains leaving Kiev for Moscow, where Western reporters witnessed railcars arrive packed with unaccompanied children, eyes wide and noses flattened against the windows, their relatives waiting anxiously on the platform.

Fearing mass panic, aware of the simmering crisis at the plant, the Ukrainian prime minister began to consider an organized evacuation of every child in the city. But the government commission in Chernobyl had provided no directives on the matter. And nobody in the republican apparat really wanted to bear responsibility for taking such a drastic step, which—quite impossible to conceal or suppress—would telegraph to the outside world just how terrifying the situation had become. The prime minister needed expert advice. He asked for the Kremlin's respective mandarins of radiation medicine and meteorology—Leonid Ilyin and Yuri Izrael—to be sent to Kiev for an urgent consultation.

In Moscow, the Western fact-finding team from the International Atomic Energy Agency—the director general, former Swedish diplomat Hans Blix, and the American Morris Rosen, director of nuclear safety—had been granted permission to see the plant in person and become the first officials from outside the Soviet Union to visit the scene. They were due to fly to Kiev on Thursday, May 8. When Evgeny Velikhov heard the news, he was horrified. The academician asked Deputy Minister Silayev to call Gorbachev with a message: "Tell him that our outhouse is overflowing, and they'll have to climb a mountain of shit."

It wasn't until around four in the morning on Thursday that the taps in the valve compartment began to emerge from beneath the contaminated water in corridor 001. Deputy Minister Silayev insisted that men be sent in to open them immediately. But the basement was filled with kilometers of pipework, and all the valves looked the same. It was pitch black. Only someone with intimate knowledge of the network of narrow, darkened rooms could hope to navigate the task and emerge safely. Three men from the Chernobyl station staff were selected for the job—two to open the valves and one to escort them in case anything went wrong—and issued with wet suits, personally driven over to the plant by a Ukrainian deputy minister. Clutching wrenches and flashlights, with pencil dosimeters clipped to their chests and to their ankles at water level, they stepped into the basement shared between Units Three and Four.

Boris Baranov, chief shift manager for the plant, went first, followed by two engineers, Alexey Ananenko and Valery Bespalov. As they descended the stairwell toward level -3, Baranov stopped to take a reading in the corridor leading beneath Unit Four. He extended the telescopic arm of his DP-5 to its maximum and held the sensor out into the darkness. The dosimeter immediately ran off the scale on every one of its ranges. There was nothing else for it: "Move quickly!" Baranov said, and the three men took off at a sprint. As he ran, one of the engineers couldn't help himself, and looked back. He glimpsed a giant cone of something black and crumbling, mixed with fragments of concrete—material that had spilled into the passageway from the shattered building above. His tongue tingled with the metallic taste of liquid radiolysis.

The route down to the entrance of corridor 001 had been surveyed by a dosimetrist with a DP-5 radiometer, who took his final measurement directly above the surface of the water in the corridor. Beyond that, the basement remained a dangerous mystery. No one knew how much water it contained or how radioactive it became as they moved deeper inside. Exposure increased with every moment spent in the tunnel: each second counted.

Baranov kept watch as the two engineers went in. The space was eerily silent. The slosh of the water parting around their feet resounded from the low ceiling; their ears filled with the sound of their own ragged breathing, stifled by their damp petal respirators. But the men discovered that the water was now only ankle deep and found a large-diameter pipe running along the floor, wide enough to walk on. The valves themselves were intact and labeled clearly: numbers 4GT-21 and 4GT-22 opened easily. A few moments later, Ananenko recognized the sound of the water gurgling from the suppression pools above their heads.

By dawn on May 8, the imminent threat of a second catastrophic explosion from beneath the reactor had been averted. Soon afterward, an official in civilian clothes sought out Moose Zborovsky at his post in the bunker and handed him an envelope sent from the government commission. Inside he found 1,000 rubles in cash.

The academicians' relief at the emptying of the suppression pools was brief. While the efforts of the soldiers and engineers headed off the possibility of a devastating steam explosion, the threat to the water table remained, and the scientists' fear of the China Syndrome only intensified. Some estimates now suggested that if it melted through the foundations of Unit Four, an incandescent mass of fuel might sink as far as three kilometers into the earth before stopping. The metro builders from Kiev had already begun drilling toward the reactor, hoping to freeze the soil with liquid nitrogen, but their efforts were hampered by rain, dust, and highly radioactive debris. They were repeatedly stopped by massive underground obstacles not shown on the plant blueprints, including the foundation plates of the cranes used during the station's construction. Precious drill bits snapped, and they had to start again at ever-greater depths.

At the same time, Silayev gave orders to begin pumping gaseous nitrogen into the steam suppression pools, while more men were directed into the basement as part of a plan to fill them with liquid concrete as soon as they had been emptied of water. And by the end of the week, the Politburo had granted permission for the most desperate measures yet: Soviet diplomats were reported to have approached the German Atom Forum, West Germany's leading nuclear industry group, requesting foreign help. The Soviet emissaries did not provide any specific details of the problem at hand but said they urgently required guidance on "how to handle something extremely hot that may have melted through the nuclear plant floor."

In their lab outside Moscow, the scientists at Velikhov's laboratory continued their round-the-clock research into the properties of molten uranium dioxide, with instructions from the Politburo to reach the most conservative possible prognosis of the meltdown. The physicists worked alongside two separate groups of mathematicians who sat at their computers day and night to test their theories. Running an entire cycle of a single testing algorithm took between ten and fourteen hours, so a colleague sat beside each mathematician to correct his mistakes when he flagged or wake him when he fell asleep. Only when the two groups' results matched could they be confident of their conclusions.

They were aghast at the results. If the molten fuel spread out over a large enough area—forming a layer no more than ten centimeters thick—it would begin cooling faster than it could melt soil or concrete, and eventually stop moving and solidify on its own. But they also found that the new substance believed to be oozing from the melting reactor core—a slurry of uranium dioxide mixed with sand, zirconium, and lead, forming a man-made radioactive lava, or corium—could behave in unexpected ways. If it was covered from above—for example, by several thousand cubic meters of liquid concrete—the heat of radioactive decay would be trapped, and the corium would melt down even more quickly. And while, in theory, using a series of pipes to freeze the earth beneath the melting fuel might stop its progress, the computer model revealed that it would do so only within strict limits. If the cooling pipes were any more than four centimeters apart, the corium could simply divide into separate tongues and burn through the spaces between them—before coalescing on the other side into a single mass, like some primitive but resourceful new life-form, to continue its relentless downward path. The scientists realized that the efforts of the metro engineers were doomed to failure, and the attempt to fill the suppression pools with concrete must be stopped.

The scientists no longer saw themselves as cloistered academics working in the esoterica of pure physics, but as the only men standing between the ignorant fools in Chernobyl and a global disaster. Folding a dot-matrix printout of their computer simulation into his luggage, Vyacheslav Pismenny, the head of the lab, took the next available flight, aboard a Yak-40 executive jet, to Kiev.

On the morning of Thursday, May 8, only hours after water had begun to empty from the steam suppression pools beneath Reactor Number Four,

Hans Blix and Morris Rosen of the IAEA embarked from Moscow for their visit to the Chernobyl station. They were met at the airport in Kiev by Evgeny Velikhov, and together they flew northwest by helicopter.

It was hot inside the aircraft, and they were all sweating in their green overalls. The plant grew steadily closer. Rosen, a veteran administrator from the US nuclear industry, asked Velikhov what range he should set on his dosimeter.

"About a hundred," Velikhov replied.

"Milliroentgen?"

"No. Roentgen."

Rosen looked queasy. His device wasn't designed for such high radiation exposures. But Velikhov assured him that everything would be fine. His own Soviet-made meter could comfortably measure in that range—and, besides, he made this flight himself every day.

What the academician did not share with his American counterpart was how little he understood about the radiation levels around the plant. Velikhov was particularly puzzled by why they did not fall as he expected when he moved away from Unit Four, but declined more slowly than the inverse-square law would suggest. Only later did he discover that on each flight they took, he and his fellow scientists were being exposed to powerful gamma fields not just from the reactor below but also from dozens of fuel fragments scattered across the platforms of the vent stack.

Still, Velikhov could afford—at last—some optimism. While the desperate work had continued to battle the meltdown beneath the reactor, the levels of radionuclides escaping into the air above it had suddenly started to fall—as steeply and inexplicably as they had begun rising five days before.

As Reactor Number Four came into view, Rosen and Blix could see a light trail of smoke drifting from the ruins, but the level of radioactive release, while still significant, was approaching zero—and the graphite fire was apparently all but extinguished. The temperature on the surface of the reactor had plunged, from 2,000 degrees centigrade to just 300 degrees. Although the Soviet scientists were at a loss to comprehend exactly why, it seemed that—thirteen days after it had begun—the emergency might be over at last. Even so, Rosen didn't want to take any chances. When the helicopter was still eight hundred meters out, Velikhov asked if he'd like to get closer.

"No," the American said. "I can see perfectly from here."

At a press conference in Moscow the following day, Rosen told reporters that the graphite fire was out and that measurements taken during their helicopter flight revealed that "there is relatively little radioactivity now." He felt confident there was no longer any risk of a meltdown. "The situation appears to be stabilizing," he said. "I can say that a competent—a very competent—group of Soviet experts is working at the site. They have many very sensible ideas and are carrying out this work now, at this very moment."

That Sunday, May 11, Moscow Central TV broadcast its first report from inside the thirty-kilometer Chernobyl exclusion zone, including footage of masked policemen stopping traffic at roadblocks, deserted houses, and a well sealed with plastic. Upstairs in the headquarters of the government commission in the center of town, Velikhov and Deputy Minister Silayev gave interviews. Sitting beneath a portrait of Lenin, at a table in an echoing conference room, surrounded by white-suited technicians conferring over maps and notebooks, Silayev appeared pale but jubilant. "We have come to the conclusion today that the primary, most important threat has been eliminated," he said, and sorted through a folder of aerial photographs of the reactor until he found one taken that day. "This is the latest," he said. "As you can see, this shows a completely calm state. You can see no smoke here, and certainly no glowing spots.

"This is, of course, a historic event. What was predicted by the world—and particularly by the bourgeois newspapers in the West, which shouted from the rooftops that an enormous catastrophe was imminent—is no longer a threat. We are firmly convinced the danger has passed."

Back in Moscow, the theoretical physicists continued to insist that the molten corium still moving somewhere deep inside Reactor Number Four remained a terrible threat. But there was fierce disagreement about their findings. The atomic specialists from the Kurchatov Institute and Sredmash dismissed them as the opinions of academic interlopers who lacked any practical experience in nuclear reactors. They argued that it was almost certain that the corium would stop melting through the basement levels of Unit Four long before it breached the deepest foundations of the building. And the theoreticians agreed that this scenario was indeed the most likely—but it was by no means guaranteed. They calculated that the chances of a ball of radioactive lava searing through all four of the 1.8-meter-thick reinforced concrete floors beneath the reactor and reaching the water table of the fourth-longest river in Europe were as high as one in ten.

In their formal report, the theoreticians advised that the only guaranteed defense against the China Syndrome was an audacious construction project to be launched in the most perilous circumstances imaginable. They recommended the excavation of a chamber deep beneath Unit Four, about five meters high and thirty meters square, designed to house a massive, purpose-built water-cooled heat exchanger, which would chill the earth and stop the molten corium in its tracks. To illustrate the nature of the threat they were facing, Pismenny, the head of the lab, arrived at a meeting at Sredmash headquarters in Moscow carrying a large chunk of concrete—melted during their experiments, a deformed pellet of uranium dioxide still embedded within it.

The Sredmash construction chief apparently needed no more convincing. "Build it," he said.

13

Inside Hospital Number Six

"Two steps back! Two steps back, otherwise I will not talk to anyone! Two steps back!"

The chief economist of the Pripyat city council climbed onto a stool and surveyed the mob that packed the small room and the hallway beyond, snaking down the stairs and into the street outside. Normally a sweet-natured woman with an easy laugh, Svetlana Kirichenko had now spent days marooned in Polesskoye—a small town with rutted streets, a modest square, and a monument to Lenin, about fifty kilometers west of the Chernobyl station. She and the handful of remaining staff from the Pripyat *ispolkom* had set up an office in the Polesskoye town hall and were now facing the wrath and confusion of their exiled citizenry. The frustrated crowd pressed forward, demanding to see the mayor; they sat bawling children on her desk; they asked what they could do with sick grandparents and when they would receive their salaries; above all, they wanted to know when they could return home.

By the time night fell on Sunday, April 27, at least twenty-one thousand people had been swept from their modern apartments in Pripyat and dropped off by bus in more than fifty small towns and villages scattered across the sodden flatlands of northwestern Ukraine. Told they would need to prepare for only three days away from home, the uprooted families soon ran out of food, money, and clean clothes, and then discovered that even what they had thought was clean was not. When a dosimetrist set up an improvised monitoring station at a desk in the street outside the Polesskoye town hospital, a line of evacuees formed in front of him. The queue moved quickly but never seemed to shorten. Touching his monitoring device to the clothes, hair, and shoes of one person after another, the dosimetrist chanted a slow mantra in a tired, flat voice: "Clean ... Contaminated ... Contaminated .... Clean ... Shake out your clothes downwind ... Clean ... Contaminated ... Contaminated ... Contaminated ..."

At first, many of the peasant families who took in the evacuees were kind and hospitable, and they made the best of it. Viktor Brukhanov's wife, Valentina, a trained engineer, was lodged with a laboratory chief on a collective farm in the village of Rozvazhev, where she took to milking cows. But Valentina had been separated from both her pregnant daughter and her mother during the evacuation and had no idea what had befallen her husband or where any of them might be—and had no way of finding out.

Thirty kilometers away, Natalia Yuvchenko and her two-year-old son, Kirill, were among the 1,200 refugees who had been billeted among the clay-and-thatch homes of Lugoviki, a rural settlement on the River Uzh without a single telephone. The last time she had seen her husband, Alexander, he had been waving to her from inside his hospital ward in Pripyat and telling her to go home and close the windows. Since then, she had received no information about where he had been taken or what condition he might be in. With two other families from her building in Pripyat, Yuvchenko and her son were taken in by an elderly peasant couple who gave up the bedroom of their small house for the newcomers. Yuvchenko and the others with young children shared the bed; everyone else slept on the floor. On Monday, the old man took the children fishing, but Kirill was still sick, and the house was damp.

By Tuesday, there was no longer enough food to feed three families, and Yuvchenko was almost out of money. She appealed to her former neighbor: "Sergei, let's get out of here," she said, and together they gathered enough cash for bus tickets to Kiev. When they arrived, she took Kirill to the airport and boarded a flight for Moldova, where her parents and Alexander's parents still lived directly across the street from one another. From there, Yuvchenko set out, once more, to find out what had happened to her husband.

By Wednesday, the official information blackout about the accident remained in force, concealing the news from even those at other atomic plants. But some details had begun to leak, and the two families worked their connections to discover what they could. Through an uncle in Moscow with military contacts, Natalia Yuvchenko established that the most badly injured men from the plant had been taken to a special hospital in the city, part of the Third Directorate of the health care system, reserved for workers in the Soviet nuclear industry. Yuvchenko and her mother-in-law flew to Moscow that morning and found a city apparently unaware of the crisis in Ukraine, bustling with preparations for the following day's May Day celebrations.

The two women immediately disagreed about where they could find Alexander. Natalia had been given an address of a hospital in a highly restricted area, on the grounds of the Soviet Institute of Biophysics. But Alexander's mother had learned of another location—a cancer research center on Kashirskoye Shosse, in an entirely different part of the city, and she insisted her sources were correct; Natalia didn't want to argue. When the staff at the oncology center told them that they had no patient by the name of Alexander Yuvchenko, the two women hailed a cab and told the driver to take them all the way across town, to Hospital Number Six.

It was midafternoon by the time they reached their destination, but when she saw it, Yuvchenko knew immediately that she had come to the right place. Nine stories of austere brown brick surrounded by a lawn and a cast-iron fence, Hospital Number Six itself looked unremarkable, but the scene surrounding it was not: the entrances were closely guarded, and technicians with radiation monitoring equipment were checking the shoes and pants of everyone entering and leaving the building.

A crowd had gathered directly outside the main entry checkpoint. Among those clustered there were many faces Yuvchenko recognized from Pripyat. All were as bewildered and frightened as she was, but none was being allowed inside the hospital. Instead, as Yuvchenko stood watching, a doctor emerged from the front door and began reading aloud from a list of names of the patients from the Chernobyl plant and their current condition. The crowd was noisy and anxious, pushing, jostling, and shouting questions; when some couldn't hear what the physician said, he had to repeat himself again and again. Even so, straining to separate his words from the hubbub, Yuvchenko heard no mention whatever of her husband. Finally, she elbowed her way through to the front of the huddle.

"What about Alexander Yuvchenko?" she asked. The doctor looked up from the list.

"You," he said. "Come inside with me."

The first patients from the plant had landed in Moscow soon after dawn on Sunday, April 27. They had been met at Vnukovo Airport by doctors clad in PVC aprons and protective suits, and buses with seats sheathed in polyethylene. The specialists of Hospital Number Six—a six-hundred-bed facility reserved for treatment of the nuclear workers of the Ministry of Medium Machine Building and home to two floors dedicated to radiation medicine—had cleared the entire department in preparation for their arrival. Some were still in the same clothes they had been wearing at the moment of the explosion; many were covered with radioactive dust; and once they had been admitted to the hospital, their transport proved beyond the limits of practical decontamination. The aircraft that delivered the first wave of patients was dismantled, and one bus was sent to the campus of the Kurchatov Institute, where it was driven into a pit and buried.

By evening on Sunday, a total of 207 men and women, mostly plant operators and firemen—but also security guards who had remained at their posts beside the burning unit, construction workers who had waited at a bus stop beneath the plume of fallout, and the anglers from beside the inlet channel—had been admitted to the wards of the hospital. One hundred fifteen of them were initially diagnosed with acute radiation syndrome. Ten had received such massive doses of radiation that the doctors immediately regarded their survival as impossible.

The head of the Clinical Department of Hospital Number Six was sixty-two-year-old Dr. Angelina Guskova. She had begun her career in radiation medicine more than three decades before, at the genesis of the Soviet nuclear weapons program. In 1949 she had just qualified as a neurologist when she was ordered to Chelyabinsk-40, the closed city in the southern Urals, to treat soldiers and Gulag prisoners working in the plutonium factories of the Mayak Production Association. Sent to one of the most sensitive and secret locations in the USSR, even professionals like Guskova often had little idea where they were going, and once they arrived, they were forbidden from leaving or communicating with the outside world. When Guskova failed to return from Mayak after two years, her mother assumed she had been arrested, swallowed in the dungeons of the KGB. But while her mother wrote letters petitioning for her release by the secret police, the young doctor was forging a new career on the harsh frontiers of biophysics.

In Mayak, Guskova encountered the first victims of acute radiation sickness she had ever seen: thirteen Gulag prisoners who arrived in her clinic suffering from nausea and vomiting. Failing to understand their symptoms, the doctor treated them for food poisoning and sent them back to work. It was only when the men returned, complaining of fever and internal bleeding, that she discovered that they had been exposed to terrible fields of radiation while digging trenches in the soil near Radio-chemical Factory Number 25, which had been heavily contaminated with radionuclides. By that time, at least one unfortunate prisoner had already received what was regarded as a lethal dose: 600 rem.

Later, the young women who worked at the benches inside the factory began to suffer from another mysterious malady, which rendered them weak and dizzy and led to an aching so severe it made one victim want to "climb the walls." Guskova would be among the first physicians in history to record the symptoms of this new disease—chronic radiation sickness, or CRS—caused by long-term low-level exposure to radioactive isotopes. She devised methods of screening and treating it, produced studies that suggested to her bosses in Sredmash that workers' exposure to radiation caused little harm if managed carefully, and was quickly promoted. She traveled to the secret weapon proving grounds of Semipalatinsk—the hundreds of thousands of square kilometers of Kazakh steppe known as "the Polygon"—to witness the first Soviet atom tests and treated the cameramen who rushed into the blast area immediately after the explosions to recover their film. Guskova became personal physician to the father of the bomb, Igor Kurchatov himself, and in September 1957 she was in Mayak to give emergency aid to the victims of the USSR's first nuclear disaster, following the explosion of Waste Tank Number Fourteen. That same year, at the age of thirty-three, she was appointed to the new radiation medicine clinic being established at the Institute of Biophysics in Moscow.

Over the next thirty years, the nuclear empire of the newly formulated Ministry of Medium Machine Building expanded with ferocious speed, in a gallop toward Armageddon that spared little time for safety. The price of progress was paid by unfortunate reactor technicians and irradiated submariners who one by one fell in their traces before receiving clandestine burials or being sent for examination in Moscow at Guskova's department in Hospital Number Six. The accidents themselves remained secret, and, afterward, those patients who survived were forbidden from disclosing the true cause of the illnesses that would dog them for the rest of their lives. But Guskova and her colleagues gathered an awful bounty of clinical information about the impact of radioactivity on human beings. Alarmed by the refusal of Sredmash to acknowledge the dangers inherent in the breakneck development of the atomic power industry, in 1970 she completed a book that described the possible consequences of a serious accident at a civilian nuclear power plant. But when she presented the manuscript to the deputy health minister of the USSR, he threw it furiously across his office and forbade her from publishing it. The following year, she codified her clinical findings from her years of treatment in the book *Radiation Sickness in Man*, for which she was awarded the Lenin Prize.

By 1986, Guskova had spent more than ten years presiding over the largest radioactive injury clinic in the USSR. She had treated more than a thousand victims of severe radiation exposure and knew perhaps more than any other physician in the world about nuclear accidents. A committed Communist, and one of few women in the upper echelons of Soviet medical administration, she was tough minded and widely feared by her staff but remained proud of the work she had done to protect the people and security of the USSR. She lived alone in an apartment on the grounds of Hospital Number Six, the phone at her bedside ready to alert her of the next nuclear emergency.

It took only a few moments for Natalia Yuvchenko to pass through the checkpoint, mount the five stone steps, and cross the threshold of Hospital Number Six. But the time stretched into an eternity of numb horror. *This is the end*, she thought.

Only once the massive wooden doors of the hospital closed behind her did Yuvchenko discover the truth. She had been picked from the crowd not to be told that she was a widow but because of the privileged status granted by her family connections.

Through his contacts with the Ministry of Medium Machine Building, Natalia's uncle had arranged for a special pass that would admit her to the hospital. He had spent hours waiting for her inside earlier that morning, bewildered by why she was taking so long to arrive.

Yuvchenko climbed into a narrow, cramped elevator—just large enough to hold two people and the operator. The hospital was dim and crumbling, with parquet floors and high ceilings. Here and there, loose wires dangled from holes in the walls. Everyone working there, from the soldiers mopping the hallways, to the doctors and technicians, was identically gowned in white or blue, with caps and masks covering mouths and noses. Damp cloths lay folded across the threshold of each room, to keep radioactive dust at bay. When the elevator shuddered to a stop at the eighth floor, Yuvchenko opened the door and turned left, into room 801. And there, sharing the space with a man she didn't recognize—a firefighter named Pravik—was Alexander. His thick, unruly hair had been clipped down to the scalp.

"Damn!" he said. "Look how ridiculous I am! Look at this head!"

After the days of fears and uncertainty since she'd last seen him, Natalia felt only joy. Regardless of whatever had happened to him that night at the plant, here was the same Sasha she had always known: he didn't look like someone who belonged in a specialist hospital.

By the time they awoke in their hospital beds on Monday morning, Yuvchenko and the other operators from the plant—including Deputy Chief Engineer Dyatlov, Shift Foreman Alexander Akimov, and the young senior reactor control engineer, Leonid Toptunov—no longer felt the acute effects of radiation sickness. The dizziness and vomiting that had seized them in the early hours of Saturday had passed. The firefighters—big, healthy young men who had gone to work that night filled with strength and vitality—were once again boisterous and cheerful and sat playing cards on their beds. Some felt so well that it was all the doctors could do to prevent them from discharging themselves. The remaining symptoms of their ordeal now seemed mild: some men were left with searing headaches, loss of appetite, and a dryness in the mouth that no amount of drinking seemed to slake. Others noticed a reddening of their skin and mild swelling where they had been exposed to gamma rays or radioactive water had splashed on them or soaked through their clothes.

Alexander Yuvchenko's head had been shaved by a nurse when he arrived—as part of a protocol devised following the Mayak disaster, when heavily exposed victims had been profoundly shocked to find their hair falling out in clumps, weeks after the accident. The radioactivity in some of the Chernobyl operators' hair now registered a thousand times higher than normal, and once cut, it was gathered in a plastic bag for burial. But Sasha seemed happy enough to joke about his baldness and otherwise looked fine. What could be wrong?

He told Natalia he didn't want to talk in the room. "Let's go and have a smoke," he said.

As befits a disease created unwittingly by mankind, acute radiation syndrome is a cruel, complex, and poorly understood affliction that tests modern medicine to its limits. The radiation exposure responsible for causing ARS may be over in a few seconds and unaccompanied by any initial reaction. But its destructive effects begin immediately, as the high-energy rays and particles of alpha, beta, and gamma radiation snap strands of DNA, and the exposed cells start to die. Nausea and vomiting set in, with a speed and intensity contingent on the dose, and the skin may redden. But the nausea eventually passes, the discoloration of all but the most severe burns fades within eighteen hours, and the patient enters a comfortable latency period. Depending on the severity of their exposure, this deceptive period of apparent well-being can last for days or even weeks, and only afterward will the further symptoms of ARS develop. The lower the dose, the longer the latency and the greater the likelihood of recovery—given the right treatment.

The patients arriving from Chernobyl had been exposed to radiation in a terrible variety of ways: the firefighters who had climbed to the roofs of Unit Three had breathed in alpha- and beta-emitting smoke, been dusted with fallout and shot through with gamma rays from the fuel pellets and core fragments surrounding them. Their doses depended on where they stood. A few meters here or there would make the difference between life and death. The operators struggling to contain the damage inside Unit Four had been mantled in dust and radioactive steam from the explosion and from broken pipes, had been drenched in water heavy with beta-emitting particles, and had searched through ruins littered with debris from the reactor core. Some had breathed in radioactive xenon, krypton, and argon, short-lived but intensely radioactive gases that would sear the soft tissue of their mouths and airways. Others would suffer widespread skin burns from gamma rays or from the beta particles that had settled on their skin or soaked into their clothes. Some had been exposed for minutes, others for much longer. Alexander Akimov, who—along with Toptunov—had labored ankle-deep in radioactive water in the fruitless attempt to cool the shattered reactor, came off the plane in Moscow in the same filthy overalls he had worn throughout that night. They continued to irradiate his skin for more than twenty-four hours, until they were finally removed by the triage nurses at Hospital Number Six.

Yet by the time they reached Moscow, a full day after the accident began, only the most gravely affected of the 207 patients exhibited any outward signs of sickness.

A half dozen of the firefighters—led by Lieutenant Pravik from the Chernobyl station brigade, reinforced by men from the Pripyat town squad, and afforded no protection against gamma waves by their canvas uniforms—had absorbed such acute doses that when they arrived in Hospital Number Six, their complexions had already faded from red to a waxy gray, the outer layer of their skin killed by radiation. Internal damage was just as hard to discern but would eventually prove equally grievous, affecting the parts of the body where cells naturally multiply the fastest, especially the lungs and airways, intestines, and bone marrow. The treatment available for the affected organs was limited to blood transfusions, antibiotics to fight infections, and—for the worst cases—a bone marrow transplant, a risky procedure fraught with complications and side effects that could themselves prove fatal.

As Dr. Guskova and her team knew, by the time the outward symptoms of ARS appeared—including swelling, skin burns and necrosis, bloody diarrhea and hemorrhaging, the decimation of bone marrow, corrosion of the airways and digestive system—it would be too late for them to intervene. And without detailed knowledge of the circumstances of a victim's exposure, an accurate picture of their dose—and the appropriate treatment—was hard to ascertain. Even in the smallest and closely defined nuclear accidents, triage was almost entirely a matter of estimation and guesswork. In the chaos that followed the explosion of Reactor Number Four, few of the accident victims had been aware of how or where they had been exposed to radiation. The station's own monitoring staff had been overwhelmed; the firefighters had been issued no radiometric equipment at all, and the operators had been wearing only crude personal badge dosimeters designed for daily use inside the plant and measuring only up to 2 rem. Those that had been recovered from the overalls of hospitalized staff had been carefully bagged and flown to Moscow, only to be inadvertently destroyed during decontamination.

But Guskova's decades of work in radiation pathology had helped her pioneer a method of biological dosimetry, gauging exposure based on interviews and tests. These included the time taken for the initial onset of vomiting and a count of white blood cells, or leukocytes. Manufactured in the bone marrow, these cells are the foundation of the body's immune system and among the most reliable biological markers of the effects of ARS. By measuring the patient's leukocyte count and the rate at which it was falling, the doctors could provide a corresponding estimate of the dose each had received. It was a laborious process. Lacking automatic blood cell counters of a kind common in the offices of Western hematologists, the clinicians had to conduct the counts by eye, under a microscope; instead of twenty seconds, each one took a half hour.

The white cell test was part of a battery of analyses used to provide a likely prognosis for each patient, and the patients quickly grew used to having blood drawn daily, either from a fingertip or a vein. The doctors also took samples to measure the levels of strontium and cesium contaminating their skin, and examined their urine for evidence of sodium 24—which would indicate exposure to nuclear fission and can render the body itself radioactive. But it was the blood test that remained the crucial bellwether of who would survive and who—almost certainly—would not.

When Natalia Yuvchenko went to ask the doctors about Alexander's condition, they said she would simply have to wait.

"Within the first three weeks, we'll know," they told her. "Just be prepared for the worst."

By May 1, Guskova and her staff had completed the work of identifying the patients who were less badly injured and moving those who required intensive care into separate rooms, to prevent cross-infection. When a doctor came to Piotr Khmel's room to discuss his test results, he seemed puzzled that the young firefighter's counts revealed relatively little damage, despite the initial reddening of his skin. He asked Khmel if he had recently taken a holiday anywhere sunny. The physician seemed to think a vacation was a more likely explanation for his patient's tan than having been exposed to the gamma radiation of the burning reactor. There were only two reasons his white cell count could be so healthy.

"Either you weren't there, or you'd been drinking," the doctor said. "Tell me the truth."

Khmel, wary of what would happen if the hospital reported that he'd been drunk on duty, sheepishly admitted he'd been out that night. There had been a lot of vodka. "It was Officers' Day," he said.

The doctor smiled and clapped him on the shoulder. "Nice job, Lieutenant. Now we'll make you better."

By now, the relatives of the victims had begun to reach the hospital not just from Pripyat and Kiev but also from elsewhere in the Soviet Union. Lieutenant Pravik's mother was one of the first to arrive and barely left her son's side from that moment on. The doctors suggested to wives and parents that they bring food to help keep up their loved one's strength and recommended making goose or chicken broth. From his bed, Pravik sent a cheerful letter to his young wife and their month-old daughter, in which he apologized for his poor handwriting and his absence from home.

"Greetings, my darlings!" he wrote. "A big hello to you from the holiday-maker and moocher. . . . I'm slacking from my responsibilities in raising

Natashka, our little one. Things are good here. They have settled us in the medical clinic for observation. As you know, everyone who was there before is now here, so I'm enjoying having my entire entourage around. We go out for walks, in the evenings we take in the sights of Moscow at night. The one downside is we have to take it all in from our window. And probably for the next one or two months. Sadly, those are the rules. Until they complete their assessment, they can't discharge us.

"Nadya, you are reading this letter and crying. Don't—dry your eyes. Everything turned out okay. We will live until we're a hundred. And our beloved little daughter will outgrow us three times over. I miss you both very much. . . . Mama is here with me now. She hotfooted it over here. She will call you and let you know how I'm feeling. And I'm feeling just fine."

The parents of Senior Reactor Control Engineer Leonid Toptunov had been at their dacha outside Tallinn when they heard there had been an accident at the power station where their son worked, and they rushed home. On Tuesday they received a telegram from Leonid: "MAMA I AM IN HOSPITAL IN MOSCOW I FEEL OK," he wrote and added the address where they could find him. Vera Toptunova and her husband took the first available flight from Estonia. When they arrived at Hospital Number Six the next day, they were taken upstairs and led down a narrow corridor, where Leonid came out of his room to meet them. Wearing short white pajamas and a matching cap, he seemed well. He could walk by himself and insisted he felt good. "Everything is fine! Don't get upset, Mama," he said, and smiled. "Everything is okay."

But when she looked down, Vera could see that he wasn't fine at all. Where Leonid's pajama pants ended, they revealed that something terrible had already begun happening to his skin: it was the ugly damson color of a day-old black eye, as if the surface of his legs and feet had been bruised all over, or dipped in something corrosive.

Dr. Robert Gale was a man of regular habits. Each morning, he rose early, while his wife and three children were still asleep, to shave and then take a swim in the pool of their house in Bel-Air, in the fragrant foothills of the Santa Monica Mountains. Afterward, he would begin making calls to colleagues in New York and Europe, where the working day had already begun. On April 29, he was still in the bathroom, listening to the radio, when he first heard news of the accident. But it wasn't until later that morning, when he learned there had been casualties at the Chernobyl plant, that it occurred to him he might be able to help.

At forty, Gale was a hematologist at the UCLA Medical Center and a specialist in bone marrow transplants. Favoring wooden clogs he had custom-made on Melrose Avenue and broad ties decorated with images of whales or sheep, he was an avid jogger who ate frozen yogurt for lunch every day and an ardent self-publicist who enjoyed his reputation as a maverick. He was also the committee chairman of the international research-sharing registry for bone marrow transplants and recognized that its resources might be vital in helping to save the lives of anyone struck by acute radiation syndrome. Gale knew that the USSR had already formally rejected offers of medical assistance from the US State Department, but he planned a different approach: through his friend and patron Armand Hammer. At around nine thirty that morning, he picked up the telephone.

The chairman of American oil company Occidental Petroleum, Armand Hammer was a renowned philanthropist and art collector. Born in New York to committed Communists, he had first traveled to the USSR in 1921 after having broken off his medical studies, ostensibly to look after the Soviet interests of his father's drug company. In Moscow, he met Lenin, who granted Hammer trade concessions that became the basis of a business fortune—and opened a direct line to Soviet leaders that would endure for almost seventy years. Although he would eventually be unmasked as one of history's great charlatans—a knowing tool of the Soviet secret police, a con man, and a traitor—at eighty-seven, Hammer was still burnishing his reputation as a globe-trotting humanitarian, described by Walter Cronkite as "an almost unique bridge between Communism and capitalism."

Gale had met Hammer while visiting the USSR in 1978, to attend a medical conference at Moscow State University, and later got to know him well through Hammer's initiative to find a cure for cancer. He could think of no better conduit for an offer of help to the casualties from Chernobyl.

Gale tracked down Hammer at a hotel in Washington, DC, and explained the potential importance of bone marrow transplants in saving the victims of radiation exposure. Later that same day, Hammer addressed a letter to Mikhail Gorbachev, petitioning on Gale's behalf, and delivered it by telex to the Kremlin. By Thursday afternoon, the doctor was striding through Los Angeles International Airport—tickets in hand and attended by a retinue of press photographers—on his way to Moscow.

On the landings of Hospital Number Six, the operators from the plant gathered to talk, smoke, and discuss the mystery that preoccupied them all: the cause of the accident that had put them there. KGB officers and investigators from the Soviet prosecutor's office went from room to room to conduct bedside interrogations, and the firemen and engineers made wild guesses, but nobody really knew how the explosion could have happened. Those trained in nuclear engineering and reactor physics—among them Deputy Chief Engineer Dyatlov, Alexander Akimov, Leonid Toptunov, and Sasha Yuvchenko—were still at a loss to understand it.

"We are open to any suggestion, lads," Dyatlov told the young technicians who had followed his orders that night. "Don't be afraid to come out with even the most far-fetched ideas."

Even when their condition began to worsen, they never discussed who was to blame. At their son's bedside, the parents of Leonid Toptunov—who had pressed the AZ-5 button, which had triggered the explosion—were both afraid to bring up the subject of the accident. But eventually Vera felt bold enough to ask him about it directly.

"Lionechka," she said. "What happened—how could it happen?"

"Mama, I did everything right," he said. "I did everything according to the regulations."

Then a doctor interrupted, signaling her not to disturb her son any further. They never discussed the accident again.

On the morning of Thursday, May 1, Ludmilla Ignatenko was summoned to Angelina Guskova's sixth-floor office, and the doctor explained that it would be necessary to conduct a bone marrow transplant on her husband. Vasily Ignatenko, a sergeant in the Pripyat town fire brigade and the most accomplished athlete in the unit, had fought the fires on the roof of Unit Three alongside Lieutenant Pravik. A marrow donor from his immediate family was now required to save his life. Guskova explained that his closest relatives were already en route to Moscow for this purpose.

Six days had now passed since the accident, and the initial latency period of ARS was ending for the most seriously irradiated patients. Vasily had been placed on an IV and was given constant injections. That night, he surprised Ludmilla with a bouquet of flowers he had asked his nurse to help smuggle in, and together the couple watched the May Day fireworks from his room, high on the eighth floor of the hospital. Vasily was still able to stand, and he wrapped his arms around her as they gazed from the window. But his condition had already declined so badly that he could no longer drink the broth she brought him. The doctors suggested trying raw eggs, but he couldn't keep those down, either.

Identifying marrow donors for the most acutely exposed patients was also becoming difficult: their white blood cell counts were falling so quickly that it was hard to find enough of them to complete the tissue-typing analysis. For those relatives shown by the tests to be good potential donors, the donation process was its own ordeal. Among the first to undergo the procedure was Vera Toptunova, then fifty years old. After administering a general anesthetic, doctors made two incisions in her buttocks and used heavy-duty, six-inch-long needles to puncture her hip bones and draw out the marrow, a teaspoon at a time. It took around ninety minutes to make the two hundred insertions necessary to fill a beaker with a quart of reddish-pink fluid. Technicians strained this to remove pieces of fat and bone, processed it in a centrifuge, placed it in a bag, and transfused it into a vein in her son's arm. Then the wait began for the marrow cells to reach the cavities in his bones and begin manufacturing healthy new blood cells.

When Vasily Ignatenko heard that his younger sister Natalia was the best candidate for donation, he refused permission for the doctors to proceed. "I won't accept bone marrow from Natasha!" he said. "I'd rather die!" Even when his wife explained that it would cause her no long-term harm, Ignatenko resisted. Eventually his older sister Lyuda submitted to the procedure instead.

By the end of the first week, Hospital Number Six's head of hematology, Dr. Alexander Baranov, had overseen three bone marrow transplants on some of the worst-affected patients, including Toptunov and Akimov. But three more patients had been so irradiated that there were no leukocytes left in their bodies with which to make a match. On these men, the Soviet doctors instead tried a new, experimental transplant technique using cells from the livers of stillborn or aborted fetuses. This treatment stood even less chance of success than the marrow graft, but Guskova's staff knew there was little else they could do: these patients were already beyond help.

By then, the limitations of biological dosimetry were becoming apparent. Guskova's initial calculations had suggested that some men had received only low doses of radiation—less than some cancer patients receive as part of standard radiotherapy treatment. Yet this analysis could reveal only the effect of gamma rays on bone marrow and took no account of the damage wrought by internal irradiation, caused by breathing radioactive smoke, dust, and steam or ingesting radioactive particles. And as the visible signs of beta burns slowly became more apparent on the victims' skin, the doctors were astonished by their scale and severity. On May 2 Dr. Baranov estimated that ten of his patients wouldn't make it out of Hospital Number Six alive. Before long, he would increase the number to thirty-seven.

Even so, the patients and their families had high hopes for the imminent arrival of the American doctor they had heard about and the vaunted expertise and life-saving foreign medicines he would bring.

After checking in to the Sovietskaya Hotel near Red Square on Friday evening, Robert Gale rose early the next morning, pulled on a singlet emblazoned with the letters "USA," and went for an eight-mile run through the streets of Moscow. Afterward, he met Alexander Baranov for breakfast at the hotel. Gaunt and bald, Baranov was a pioneering Soviet surgeon responsible for the first bone marrow transplant ever performed in the

USSR, but he had the haunted look of a man who had watched many of his patients die in agony. He chain-smoked constantly and was in the habit of improvising ashtrays from scraps of paper, which he crumpled into the trash each time he finished a cigarette. After breakfast, the two men were driven to Hospital Number Six, where Baranov introduced Gale to Angelina Guskova. She was cordial but disappointed that the boyish-looking American surgeon arrived carrying only a small bag and not the expensive Western equipment she had been led to expect. Afterward, Baranov took him to visit the patients on the eighth floor.

Up here was the hospital's sterile unit, where the transplant recipients recovered after their operations. Until the transplanted marrow cells became sufficiently well established to begin producing blood components—a process that could take two weeks or a month—the patients' immune systems would be all but useless, leaving them susceptible to hemorrhages, minor infections, and even pathogenic attack from the bacteria in their own intestines, any of which might prove fatal.

In the sterile unit, Gale found four patients sealed inside "life islands"—plastic bubbles designed to provide a vital line of defense in the doctors' battle to keep the men alive long enough for the marrow cells to engraft. The patients breathed air that had been filtered or passed through a duct where it was sterilized by ultraviolet light. To further isolate them from infection, they could be reached only by staff whose hands and clothing had been sterilized or through portholes in the plastic fitted with gloves. Because the hospital had far fewer life islands than it needed, their use was rationed. To Gale, who had never seen a beta burn before in his life, the four men he examined that afternoon appeared sick, but not alarmingly so. He participated in his first transplant procedure, assisting Baranov in drawing marrow from a donor, soon afterward.

After he had received the transfusion from his sister, Vasily Ignatenko was transferred to the eighth floor and placed in a life island. The staff tried to keep his wife out, but Ludmilla got in anyway, reaching into the bubble to moisturize his lips. Young soldiers now came to his room instead of nurses, wearing gloves to give him injections and disposing of blood and plasma. No one wanted to be in the room anymore—perhaps, Ludmilla thought, for fear of contamination. Some of the staff, especially the younger ones, had become irrationally terrified of the patients and believed that the radiation disease was somehow infectious, like the plague.

Ignatenko recovered quickly from the transplant procedure. But his overall condition had already begun a sudden and frightening slide. His appearance changed from minute to minute: his skin changed color, his body became distended. He had difficulty sleeping, so they gave him tranquilizers, adding to the dozens of pills he had to take every day. His hair began to fall out, and he became angry. "What's all this about?" he asked. "They said I'd be bad for two weeks! Look how long it's been!"

Gradually it became harder and harder for him to breathe. Cracks covered his arms; his legs swelled and turned blue. Eventually the painkillers stopped working. By Sunday, May 4, he could no longer stand.

The worst-affected patients in Hospital Number Six were attacked from both without and within. As their white blood cell counts collapsed, infection crawled across the skin of the young operators and firemen: Thick black blisters of herpes simplex encrusted their lips and the inside of their mouths. Candida rendered their gums red and lacy, and the skin peeled back, leaving them the color of raw meat. Painful ulcers developed on their arms, legs, and torsos, where they had been burned by beta particles. Unlike thermal burns caused by heat alone, which heal slowly over time, radiation burns grow gradually worse—so their external beta burns expanded outward in waves from wherever radioactive material had touched them and ate into the tissue below. The men's body hair and eyebrows fell out, and their skin darkened—first red, then purple, before finally it became a papery brown-black and curled away in sheets.

Inside their bodies, the gamma radiation ate away the lining of their intestines and corroded their lungs. Anatoly Kurguz, who had fought to close the airlock door to the reactor hall in the moments after the explosion and was enveloped in steam and dust, had so much cesium inside his body that he became a dangerous source of radiation. He began having hysterical fits, and one of the doctors, the burns specialist Dr. Anzhelika Barabanova, had to physically lie down on top of him, using her bodyweight to make him remain in bed. Radiation readings around Kurguz's room eventually became so high that the head of the department had to move from her office next door to elsewhere in the hospital; the parquet flooring in the hall outside it was so contaminated that it was taken up and replaced.

Within the first twelve days after the accident, Alexander Baranov and Robert Gale conducted fourteen bone marrow operations, and Armand Hammer and the Sandoz Corporation arranged for hundreds of thousands of dollars' worth of medicines and equipment to be air-freighted into Moscow from the West; Gale received Soviet approval to bring in more colleagues from New York and Los Angeles. But the doctors knew that much of their effort was probably futile: afterward, Gale announced at a press conference in Moscow that as many as three-quarters of their transplant patients would probably die.

For Shift Foreman Alexander Akimov, who had spent hours bombarded from all sides by energetic sources of gamma rays and wading in contaminated water, the transfusion of marrow from his twin brother could do nothing to arrest his metabolic collapse. His contaminated overalls alone had exposed him to 10 grays—equivalent to 1,000 rem of radiation, causing a beta burn that covered almost his entire body, with the exception of a thick band of unexposed skin around his waist, where the overalls had been fastened with a heavy military belt. But Akimov had also received a separate dose of 10 grays to the lungs, resulting in acute pneumonia. His temperature rose; his intestines disintegrated and oozed from his body in bloody diarrhea. On one visit, his wife, Luba, looked back from the window and noticed her husband had begun pulling out his mustache in tufts.

"Don't worry," he told her. "It doesn't hurt."

Akimov knew that he might not leave the hospital alive, but while he remained well enough to speak, he told a friend that if he lived, he'd like to pursue his love of hunting and become a gamekeeper. Luba suggested that they could live on a river with their two sons, tending buoys and regulating navigation, just as Deputy Chief Engineer Dyatlov's father had done. Whatever happened, Akimov was sure of one thing: "I'll never go back to work in the nuclear field," he said. "I'll do anything. . . . I'll start my life from scratch, but I'll never go back to reactors."

By the time Sergei Yankovsky, the chief investigator of the Kiev prosecutor's office, came to Akimov's room to question him about the accident, the engineer's body had become grossly swollen. He could barely speak. The doctors had little time for the investigators and now demanded to know why Yankovsky was torturing a dying man. They told him that Akimov wouldn't last more than a few more days. The detective's attempts at interrogation proved useless.

Before he left, Yankovsky bent close to the stricken nuclear engineer's bed. "If you remember anything," he said, "just write it down."

On May 6 Akimov celebrated his thirty-third birthday. Soon afterward, he fell into a coma.

On the evening of Friday, May 9—Victory Day, marking the Soviet triumph over the Nazis in the Second World War—the patients watched from the windows of the hospital as another fireworks display lit the sky. But this time they felt little happiness. Vasily Ignatenko had begun to shed his skin, and his body bled. He coughed and gasped for air. Blood trickled from his mouth. Lying alone in his room, Piotr Khmel received an encouraging note from his friend Pravik, delivered by one of the doctors: "Congratulations on the holiday! See you soon!" Khmel hadn't seen his old classmate since they had arrived in the hospital together twelve days earlier and now had no idea where in the building he might be. But he scribbled his own note in reply, reciprocating the greeting.

The deaths began the next day. The first was a firefighter from the Chernobyl power plant brigade, Sergeant Vladimir Tishura, who had climbed to the roof with Pravik minutes after the explosion. On May 11 Pravik and Kibenok—the commander of the Pripyat brigade—both succumbed to their injuries. Grotesque rumors would later reach Pravik's men back in Ukraine: that he had been exposed to such intense radiation that the color of his eyes had changed from brown to blue, and doctors found blisters on his heart. That same day, Alexander Akimov became the first of the plant operators to slip away: he died with his eyes open, his skin black.

Dr. Guskova now forbade further communication among the patients, confining them to their rooms. Outside the windows, the trees were in full bloom; the weather was perfect. Beyond the fence on Marshal Novikov Street, Moscow went about its business as usual. Those men and women who survived lay alone in their beds, attached to IVs or blood exchange machines for hour after hour, often with only the nurses for company. News of further losses among their friends and colleagues came in whispers from relatives and the disquieting sound of loaded gurneys rolling down the long corridors of the hospital.

While the first of his comrades from Unit Four were taken to the cemetery, Alexander Yuvchenko's ordeal was just beginning. As the doctors had warned, the beta burns on his body were slow to reveal themselves. At first, small red spots appeared on the back of his neck. Then more lesions arose on his left shoulder blade, hip, and calf, where he had braced himself against the massive door of the reactor hall, and the slime of beta- and gamma-emitting radionuclides covering it had soaked through his wet overalls.

Yuvchenko moved into intensive care, one of only four gravely ill men held in separate rooms on a single floor. Next door was his boss, reactor shift foreman Valery Perevozchenko. The former marine had taken a huge dose of gamma rays when he entered the reactor hall and gazed into the burning core—but had kept Yuvchenko from looking inside, sparing him the worst of the radiation. Still the burns on Yuvchenko's body darkened and spread, and his skin blackened and peeled away, revealing tender, baby-pink flesh beneath. And what had begun on his shoulder blade as something like a sunburn gradually blistered and necrotized, turning yellow and waxy as the radiation ate down toward the bone. The pain became almost unbearable, and the nurses administered morphine. The doctors began discussing the need to amputate and summoned special equipment from Leningrad to determine if his arm could be saved.

On Tuesday, May 13, Ludmilla Ignatenko took the bus out to Mitino cemetery in the northwest suburbs of Moscow with her friends Nadia Pravik and Tanya Kibenok, whose husbands had died two days earlier. She watched as the men's bodies were lowered into the ground. Ludmilla had left the hospital at nine that morning to prepare for the journey but gave instructions to the nurses to tell Vasily that she was merely taking a rest. By the time she returned to Hospital Number Six that afternoon, her husband, too, was dead. The morticians who came to prepare him for burial found his body so swollen they couldn't get him into his uniform. When he was eventually laid to rest beside his comrades in Mitino, the young fireman's corpse was sealed inside two thick plastic bags, a wooden coffin, and a zinc box, nested together like an irradiated *matryoshka* doll.

That same day, Valery Perevozchenko succumbed to his injuries. Natalia Yuvchenko tried to keep the news from her husband, but, from his bed, Alexander heard the beeping of the machines in the room next door fall silent. On May 14 three more operators from Unit Four died, including Leonid Toptunov, whose parents stayed at his bedside until the end. With 90 percent of his skin covered by beta burns, and his lungs destroyed by gamma radiation, the young man awoke late that night gasping for breath. He suffocated before his bone marrow transplant could take effect. In the end, the doctors calculated that he had absorbed 1,300 rem of radiation, more than three times the lethal dose. Viktor Proskuryakov, one of the two trainees who had ventured out onto the gantry with Perevozchenko and gazed into the burning reactor, was covered in terrible burns, especially on his hands, where he had held Yuvchenko's flashlight. He hung on for three more days but died on the night of May 17.

By the end of the third week in May, the death toll from the accident had reached twenty, and Alexander Yuvchenko grew frightened. His white blood cell counts plunged to zero, and his remaining hair fell out. *When will it be my turn?* he wondered. Alone in their rooms, the most seriously injured survivors began to fear the darkness, and the lights in some wards were kept on almost constantly.

A good Communist, Yuvchenko wasn't religious and knew no prayers. Yet each evening, he lay awake and pleaded with God to let him live through one more night.

14

The Liquidators

On Wednesday, May 14, 1986, more than two and a half weeks after the explosion inside Unit Four, Mikhail Gorbachev finally appeared on TV to address the accident in public for the first time. Reading from a prepared statement on *Vremya* to its audience of two hundred million people distributed across thirteen time zones and broadcast live simultaneously on CNN, the most telegenic leader in the history of the Soviet Union now looked wan, lost, and haunted. The accident at Chernobyl "has sharply affected the Soviet people and raised the concern of international society," he said—in a speech that sometimes veered into the defensive, often flared with anger, and eventually wound on for twenty-six minutes.

Gorbachev railed against the "mountain of lies" told by the United States and its NATO allies about the accident, part of what he called a "vile" campaign to distract attention from their failure to engage with his recent proposals for nuclear disarmament. He thanked Robert Gale and Hans Blix and expressed his sympathy for the families of the dead and injured. "The Soviet government will take care of the families of those killed and of those who have suffered," he said. He assured viewers that the worst was behind them, but warned that the task was far from over: "This is the first time that we have truly encountered a force as terrible as nuclear energy escaping human control. . . . We are working twenty-four hours a day. The whole country's economic, technical, and scientific resources have been mobilized."

Forty-eight hours earlier, the Soviet defense minister—Hero of the Soviet Union Marshal Sergey Sokolov—had arrived in Chernobyl accompanied by a team of senior staff officers and medical administrators from his ministry. A military task force, spearheaded by the radiation specialists of the chemical warfare troops and the civil defense, had been arriving in the thirty-kilometer zone since the beginning of the month. Young men in Kiev, Minsk, and Tallinn had been summoned from their workplaces— or roused by a knock on the door in the middle of the night—and taken to be issued uniforms, sworn under oath, and told they should consider themselves to be at war. They learned their final destination only once they arrived in the zone. Now Marshal Sokolov, who had sent the Soviet armed forces across the border into Afghanistan in 1979, had come to lead his men on one more heroic military campaign to protect the motherland, which would become formally known as "the Liquidation of the Consequences of the Chernobyl Accident."

From every republic of the USSR, men, women, and equipment poured into Chernobyl, as the full might of the centralized state—and the largest army on earth—ground into desperate action. Soldiers and heavy equipment flew in aboard huge Ilyushin-76 military transport planes. Scientists, engineers, and other civilian workers arrived from everywhere between Riga and Vladivostok. Bureaucratic strictures, plan targets, and financial priorities were abandoned. With a single telephone call, any necessary resources could be rushed to the station from almost anywhere in the Union: tunneling experts and rolled lead sheets from Kazakhstan; spot welding machines from Leningrad; graphite blocks from Chelyabinsk; fishing net from Murmansk; 325 submersible pumps and 30,000 sets of cotton overalls from Moldova.

The spirit of patriotic mass mobilization was boosted by the first detailed coverage of the accident appearing in the Soviet press, as the propaganda experts in the Kremlin finally found an angle on the catastrophe. *Izvestia* and *Pravda* published awestruck blow-by-blow accounts of the courageous sacrifice of the firefighters who helped battle the initial blaze, alongside portraits of the miners and subway workers busy tunneling beneath the ruins. Although these stories appeared to bear the hallmarks of glasnost—frank descriptions of the dangers of radiation and visits to the injured men in Hospital Number Six—the openness was circumscribed. There was no room for talk of confusion, incompetence, or lack of safety precautions; each of the firemen apparently strode selflessly into danger fully aware of the risks he faced, ready to take his place in the pantheon of Soviet heroes. The causes of the accident were not explored. Elsewhere, it was made clear that the emergency would soon be over. According to the weekly newspaper *Literaturna Ukraina*, the atom "had temporarily got out of control." Soviet scientists "had a solid handle on everything taking place in and around the reactor." The residents of the evacuated territory, the newspapers reported, would be able to return to their homes just as soon as decontamination work had been completed.

The first cleanup efforts began at the station itself, even as the struggle continued to contain the radiation still leaking from the smoldering husk of Reactor Number Four. The contaminated area was divided into three concentric regions: the outermost thirty-kilometer zone, the ten-kilometer zone, and, within those, the most toxic region of all, the Special Zone, immediately surrounding the plant. The work fell to military engineers and civil defense and chemical warfare troops under the command of the Soviet General Staff, many of them young conscripts. It was chaotic.

No formal plans, either civilian or military, had ever been devised to clean up after a nuclear disaster on such a scale. Even by the middle of May, there still weren't enough plant specialists available to supervise an improvised operation, and there was disagreement over setting the maximum dose of radiation that workers could receive safely. Naval doctors, whose expertise had been earned the hard way, through decades of accidents in the close quarters of nuclear submarines, insisted on the Ministry of Defense standard of 25 rem. But both the Soviet Health Ministry and the head of the chemical troops, General Vladimir Pikalov, wanted twice that: 50 rem, the level prescribed for his soldiers in the event of war. Three weeks went by before they finally agreed to the lower limit, by which time many men had been dangerously overexposed. Even then, the 25 rem maximum proved difficult to monitor and was often deliberately disregarded by unit commanders.

The civilian nuclear industry specialists arriving from other atomic plants across the USSR to help in the cleanup were horrified by the lack of preparation around them. They found too few trained dosimetrists to effectively monitor their radiation exposure. No comprehensive survey had yet been made of the area, and the volume of radionuclides belching from the reactor changed constantly, so reliable radiation information was almost impossible to obtain. There was a chronic shortage of dosimeters. A platoon of thirty soldiers often had to share a single monitoring device: the dose registered by the man wearing it was simply attributed to the others equally, regardless of where they had been or what work they had done.

The task of clearing the largest and most heavily radioactive pieces of debris from around the reactor fell to soldiers driving massive IMR-2 combat engineering vehicles. Designed to clear the way for troops advancing through minefields or in the chaos following a nuclear attack, these were battle tanks equipped with bulldozer blades and telescopic crane booms instead of gun turrets, bearing hydraulic pincers large enough to pluck fallen telephone poles or tree trunks from their path and move them aside. To minimize their radiation exposure, the drivers' compartments had been lined with lead, and each man was permitted to work for only a few minutes before being replaced at the controls. Yet one of the first machines to enter the tangle of wreckage around Unit Four quickly got into trouble. The driver, unable to see properly through his narrow armored viewing slots, drove into a maze of wreckage and became trapped on all sides by debris. His commander couldn't reach him by radio, and his allotted time in the high-radiation zone was ebbing away. Finally, the colonel drove over and, leaning from the open hatch of his own armored vehicle, shouted directions to the stricken driver until he found his way to safety. The soldier was saved, but the few moments in the open were too much for the colonel: the next day, he was sent to a military hospital, suffering from radiation sickness.

By May 4, the first two colossal radio-controlled bulldozers—one built in Chelyabinsk and the other imported from Finland—were flown to the plant for use in clearing the radioactive rubble and soil from around Unit Four. This area remained the most deadly section of the Special Zone, where gamma exposure from the mountain of pulverized wreckage cascading from the northern wall of the reactor building reached thousands of roentgen an hour. Men could work there unprotected for only a few seconds. After shielding their sensitive remote-control units with sheets of lead, technicians began their first experiments with the bulldozers. Operating from the relative safety of a specially protected nuclear and chemical reconnaissance vehicle parked a hundred meters away, they used them to shunt scattered fragments of nuclear fuel back toward Unit Four. But the Scandinavian machine failed quickly, unable to climb the steep escarpment of radioactive debris, and its nineteen-tonne Soviet counterpart lasted only a little longer before it stopped dead in the shadow of the reactor, where it proved impossible to revive. By September, several of the bright-yellow machines could be found abandoned in a nearby field.

While the Ministry of Energy urgently sought additional remote-controlled equipment abroad, and Prime Minister Ryzhkov's Politburo Operations Group made plans to lay a blanket of latex solution across the top of the reactor, the members of the government commission applied an enduring Soviet remedy to the mountain of radioactive rubble at the north wall: they ordered it covered with concrete. Energy Ministry construction teams fed the gray slurry through a pipeline eight hundred meters long; it boiled as it smothered fuel cassettes thrown from the reactor by the explosion, and soon geysers of hot radioactive cement leapt into the air. At the same time, the civil defense reservists of Special Battalion 731 began work on removing the topsoil around the reactor by hand. Even as other troops moved through the high-radiation zone by armored personnel carrier, these men started work in the open wearing regular army uniforms, protected only by cotton petal respirators. They excavated the soil near the reactor walls with ordinary shovels and placed it in metal containers for transport and burial in the partially finished radioactive waste storage vaults under construction for the Fifth and Sixth Units. Their shifts lasted as little as fifteen minutes, but the weather was hot and the radiation relentless. Their throats itched, they felt dizzy, and there was never enough water to drink. Some men suffered nosebleeds, while others began vomiting. Called to help clear chunks of reactor graphite lying on the ground near Unit Three, a detachment of chemical warfare troops arrived in a truck and began to pick them up using their hands.

Such tasks exposed the liquidators to their maximum allowed annual doses of radiation—often in a matter of seconds. In the high-radiation areas of the Special Zone, a single task that anywhere else could be completed by a single man working for an hour might now require thirty men to work for two minutes each. And the new regulations dictated that as soon as they each reached their 25 rem limit, they should be sent from the zone, never to return. Each job had to be measured not only in time but also in the number of individuals who would have to be "burned" to accomplish it. In the end, some commanders resolved that it was better to keep deploying the men they had—men who had already been exposed to their limits—than to burn fresh troops in the danger zone.

Meanwhile, underground, the battle against the China Syndrome had intensified. Plant physicist Veniamin Prianichnikov—having successfully evacuated his family from Pripyat by train—had finally been permitted to return to the station at the beginning of May, only to discover everything in his office buried beneath two centimeters of radioactive ash. On May 16 he received orders to gather temperature and radiation measurements from directly beneath the reactor, in the hope of establishing exactly how imminently the melting core might burn through its concrete base. Although the scientists now believed the heat generated by the decaying fuel was falling, they estimated it was still as high as 600 degrees centigrade. It was up to Prianichnikov and his team to gather accurate data on what might happen next: whether the molten corium was still moving, and whether the water of the Pripyat and Dnieper rivers was still under threat of catastrophic contamination.

Using a plasma torch powered through a huge electrical transformer brought in from Moscow, it took a team of soldiers eighteen hours to cut their way through the thick concrete wall into the subreactor space. Inside the darkened compartment, with the hundreds of tonnes of molten nuclear fuel directly overhead, Prianichnikov imagined that radiation levels could be thousands of roentgen an hour, permitting him to safely spend only five or six seconds at his task; a full minute would be suicide. He dressed lightly, in cotton overalls and cloth respirator, planning on using speed alone to protect him against the gamma emissions from the remains of the reactor. But when he crawled through the hole with his radiation and temperature probes, his dosimeter failed, and the work took much longer than he expected. Then, just as he was completing the equipment installation, he felt something powdery shower on his head from above. Terrified, Prianichnikov scrambled from the blackened basement as fast as he could, tearing off his clothes as he went. He sprinted naked along the kilometer-long route to the administrative block, scattering astonished soldiers to the left and right along his path. Only after he arrived did he discover that the debris sprinkling on him had not been nuclear fuel but sand: his eventual dose from the few minutes he spent beneath the reactor was less than 20 rem.

By then, some four hundred miners from the Donbas and Moscow coal basins were at work on the massive heat exchanger project the scientists had planned to install in the earth beneath Reactor Number Four. Once again, the deadline set by the government commission seemed all but impossible: the entire undertaking would have to be designed and built, tested and ready, in a little more than a month. The miners began tunneling from a pit excavated more than 130 meters away from their objective, near the wall of the third reactor building, and worked around the clock in three-hour shifts. The shaft was just 1.8 meters in diameter, and swelteringly hot, but the earth over their heads shielded them from the worst of the radiation on the surface. They were forbidden to smoke underground but lingered at the tunnel entrance to snatch occasional cigarettes or drink water, where they were unwittingly bombarded with gamma rays from the dust and debris surrounding them. Digging with hand tools and pneumatic drills, pushing the spoil out along a miniature railway line, the miners soon arrived beneath the foundations of the reactor, where they began work on the thirty-square-meter chamber that would house the heat exchanger. Here, they could reach up and touch the foundation plate of the reactor building. The concrete felt warm beneath their palms. The designers from Sredmash warned them repeatedly that making even the slightest departure from their specifications for the chamber risked bringing the entire reactor vessel and its contents down on their heads, in a collapse that would inter them all instantly in a mass grave.

When the chamber was complete, Sredmash engineers moved in to install the heat exchanger, built in Moscow in sections limited in size by the narrow diameter of the tunnel, and conditions underground became even more infernal. Welding the pieces together filled the narrow, poorly ventilated space with toxic gas, and men began to choke and pass out. The forty-kilogram graphite blocks integral to the design were relayed hand to hand down the length of the tunnel by teenaged military conscripts. Temperatures in the tunnel reached 60 degrees centigrade, and the young men worked half naked until, at the end of each shift, they had to be pulled from the tunnel, spent with exertion.

Final assembly began in June, under the command of Vyacheslav Garanikhin, a towering Sredmash welding foreman with shaggy hair and a disheveled beard who at one point lumbered through the tunnel threatening construction workers with an axe. But long before the project was completed on June 24, temperature readings from Prianichnikov's sensors had declined yet further, and fear of the China Syndrome abated at last. The heat exchanger, with its intricate network of stainless steel tubing, ten kilometers of control cabling, its two hundred thermocouples and temperature sensors layered in concrete and sandwiched beneath a layer of graphite blocks—the result of weeks of frenzied work by hundreds of miners, soldiers, construction workers, electricians, and engineers—was never even turned on.

By the end of May, General of the Army Valentin Varennikov had been summoned back from Kabul, where he was in command of all ongoing armed operations in Afghanistan, to take charge of the military cleanup in Chernobyl. When the general arrived, he found more than ten thousand men of the Chemical Warfare Divisions alone laboring inside the zone and that hundreds of Ministry of Energy construction workers living nearby had been abruptly drafted. But it was clear that the cleanup operation would need yet more manpower. The Politburo now recognized that if the Union's young draftees—already plagued by alcoholism and drug abuse—continued to be sent into the high-radiation zone, the health of an entire generation of Soviet youth could be ruined, rendering the country incapable of defending itself in the event of an attack from the West. On May 29 the Politburo and the Council of Ministers of the USSR issued a decree unprecedented in peacetime: calling up hundreds of thousands more military reservists—men aged twenty-four to fifty—for a mobilization of up to six months. They were told they were required for special military exercises; many discovered the truth only when they were already in uniform. By the beginning of July, more than forty thousand of these troops were encamped around the perimeter of the Exclusion Zone, sleeping in rows of tents and driven toward the Special Zone each morning in convoys of open trucks. The journey was long and hot, along newly laid asphalt roads that glistened in the sunlight, sprayed with water by tankers to keep down the dust. The trees and fields that flashed past looked lush and green but were barricaded behind plywood fences marked with the warning "Do Not Pull Over—Contaminated."

Whether rising from the roadsides in the slipstream of endless columns of speeding trucks and concrete mixers or whirling in the downdraft of the heavy transport helicopters, the dust carried radiation throughout the zone. Lifted into the sky by the breeze, microscopic radioactive particles a few microns across traveled with insidious ease, settling nearby or brought down as heavy fallout a hundred kilometers away by rain. Physicists from the Ukrainian Academy of Sciences ventured into the zone to take air samples using gauze screens and ordinary vacuum cleaners and found that, in the wake of a single helicopter flight, radiation levels increased as much as a thousand times. The dust blanketed equipment, furniture, and documents in offices and found its way into the hair, lungs, and stomachs of those who worked in them. From inside the body, the damage inflicted by "hot" particles—the almost invisible fragments of nuclear fuel blown out of the reactor core—was exponentially greater than when kept outside: 1 microgram of plutonium could bombard the soft tissues of the esophagus or lungs with 1,000 rads of energetic alpha radiation, with lethal results. The liquidators wore berets and caps, and their electrostatic petal respirators hung everywhere; they tried to drink only from sealed bottles of mineral water. Those who understood the threat eventually developed an unconscious habit of picking the tiniest specks of dust off clothes and tabletops—flicking them away in a constant, reflexive process of personal decontamination. But others remained unaware of the invisible dangers around them: soldiers lounged in the sun close to the reactor, smoking cigarettes and stripped to the waist in the summer heat; a group of KGB officers arrived in the zone incognito, clad in tank crew overalls and carrying expensive Japanese-made dosimeters—but approached the ruins of Unit Four without knowing enough to turn the devices on. Only the fate of the crows that had come to scavenge from the debris but stayed too long—and whose irradiated carcasses now littered the area around the plant—provided any visible warning of the costs of ignorance.

Daily radiation surveys conducted across the length and breadth of the zone and beyond—by helicopter, plane, by armored vehicle, and on foot by troops clad in rubber suits and respirators—revealed that the contamination had spread far across Ukraine, Belarus, and Russia. The plume from Unit Four had cast a shadow not only across the cities of Pripyat and Chernobyl but also upon collective farms and industrial enterprises, small towns, isolated villages, forests, and great tracts of agricultural land. Dense traces of radioactivity reached north and west across the thirty-kilometer zone, but fallout composed of twenty-one different radionuclides forged inside Reactor Number Four—including strontium 89, strontium 90, neptunium 239, cesium 134, cesium 137, and plutonium 239—had also formed a leopard-spot pattern of intense contamination up to three hundred kilometers from the plant. The threat to the population from the radiation was twofold: external, from the fine irradiated dust and debris on the ground around them; and internal, from radioisotopes poisoning the food chain through the soil, crops, and farm animals. By the end of May, more than five thousand square kilometers of land—an area bigger than Delaware—had become dangerously contaminated. Wind and weather would only make matters worse, and dust from radioactive areas was being constantly redistributed into areas the troops had already cleaned, making early decontamination work almost entirely pointless.

The work of decontaminating the huge area beyond the boundaries of the power plant was complicated not only by meteorology and the Herculean proportions of the undertaking but by the varying topography and materials involved. Radioactive aerosols had seeped into concrete, asphalt, metal, and wood. Buildings, workshops, gardens, shrubs, trees, and lakes had all been in the path of the cloud that had drifted across the landscape for days and weeks. Roofs, walls, farmland, machinery, and forest would have to be washed, scraped, picked clean, or cut down and buried. The word "liquidation" was nothing more than a martial euphemism. The reality was that radionuclides could be neither broken down nor destroyed—only relocated, entombed, or interred, ideally in a place where the long process of radioactive decay might pose a less immediate threat to the environment.

This was a task on a scale unprecedented in human history, and one for which no one in the USSR—or, indeed, anywhere else on earth—had ever bothered to prepare. Yet now it was also subject to the routinely absurd expectations of the Soviet administrative-command system. When General Pikalov, commander of the chemical warfare troops, gave his initial situation report on the thirty-kilometer zone to the visiting leaders of the Politburo Operations Group, he forecast that decontamination work would take up to seven years to complete. Upon hearing this, the hardline Politburo member Yegor Ligachev exploded in fury. He told Pikalov he could have seven months.

"And if you haven't done it by then, we'll relieve you of your Party card!"

"Esteemed Yegor Kuzmich," the general replied, "if that is the situation, you needn't wait seven months to take my Party card. You can have it now."

On their return to Moscow, the Politburo Operations Group had confronted a new problem: finding a way to permanently isolate what remained of Reactor Number Four from the environment. Now that the graphite fire was finally going out, and the specter of the China Syndrome had receded, it was imperative to prevent further radioactive releases into the atmosphere around the plant—and also to restart Chernobyl's three remaining reactors as soon as possible. The electricity they generated may not have been crucial to the Soviet economy, but restoring them to operation would once more demonstrate the might of the Socialist state and reassure the people of its commitment to nuclear energy. And they could be safely brought back online only once the ruins of Unit Four had been sealed. Responsibility for this fell first to the Ministry of Energy, but its construction teams were soon overwhelmed by the magnitude of the tasks before them. On May 12 they simply gave up in despair.

But General Secretary Gorbachev was determined to smother the fiasco as quickly as possible. The USSR had been the first in the world to construct a nuclear power station, he told the Politburo. Now it must be the first to build a coffin for one. It was time to call in the experts who had written the rules of Soviet nuclear construction in blood: the specialists of the Ministry of Medium Machine Building. Sredmash chief Efim Slavsky arrived the next day with a team of ten men, flying into Kiev aboard his personal Tupolev Tu-104 jet and then circling the ruins of the plant by helicopter.

"What a mess," he said as he surveyed the wreckage from the air. His staff was aghast at the scene, which was far worse than the official reports had suggested. Smoke still drifted from the crater of the reactor, which now resembled a dormant volcano, threatening to roar to life at any moment. It was clear that whoever undertook the job of walling in the remains of Unit Four would be working in one of the most hostile environments mankind had ever encountered. The task promised levels of radiation almost beyond imagining, a construction site too dangerous to survey, and an impossible deadline: Gorbachev told Slavsky he wanted the reactor sealed up by the end of the year. Deaths were almost inevitable. The octogenarian nuclear minister turned to his men.

"Lads, you'll have to take the risk," he said.

The following afternoon, in an interview broadcast on Soviet Central Television, the head of the government commission, Ivan Silayev, outlined plans for the reactor's final resting place: a tomb in which the ruins of Unit Four would be interred forever. It would be "a huge container," he explained, "which will enable us to secure the burial of everything that remains . . . of this entire accident." The resulting structure would be monumental, built on a scale to last a hundred years or more, and, before the cameras, Silayev gave it a name resonant with history and ritual: *sarkofag*.

Sarcophagus.

In public, the Soviet government continued to assure its people that the catastrophe was under control and that the radiation already released posed no long-term threat. But in its secret sessions within the Kremlin, the Politburo Operations Group heard that the direct effect of the disaster on the population of the USSR was already reaching alarming heights. On Saturday, May 10, Ryzhkov learned that a total of almost 9,500 people had already been hospitalized in connection with the accident, at least 4,000 of them in the previous forty-eight hours alone. More than half of that number were children, 26 of whom had been diagnosed with radiation sickness. The levels of contamination in four of Russia's western regions had begun a rise that had yet to be explained, and the Soviet Hydrometeorology Department had decided to fly weather control missions over Kiev, using aircraft to release more cloud-seeding materials into the atmosphere in the hope of preventing radioactive rain from falling on the city.

The Soviet prime minister issued new instructions to insulate Moscow from the spreading threat. Civil defense troops threw roadblocks across all major routes leading into Moscow and checked every vehicle for radiation. Traffic backed up for hours, as furious drivers boiled in the unseasonal heat. Travelers arriving in the capital from both Belarus and Ukraine were hospitalized and decontaminated. Now Ryzhkov also instructed the state's agro-industrial enterprises to halt all deliveries of meat, dairy, fruit, and vegetable products from the affected areas until further notice.

Meanwhile, in Kiev, the Ukrainian government had formed its own task force to oversee the cleanup of the towns and villages inside the thirty-kilometer zone and introduced measures to protect people in neighboring areas from contamination. On May 12 it banned fishing and swimming as well as washing clothes, animals, or cars in the rivers and ponds of five districts up to 120 kilometers south of the station.

On every approach to Kiev, traffic police established washing and decontamination checkpoints to ensure that not a single car could enter the city without being checked for radiation. Municipal tanker trucks patrolled the streets, sluicing thousands of liters of water onto the roads and sidewalks, while troops sprayed down walls and trees to remove radioactive dust. But the Ukrainian leaders, wary of panic and of antagonizing their bosses in Moscow, had still made no decision about whether to evacuate the children of the city.

The Kremlin's chief scientists on radiation medicine and meteorology—Leonid Ilyin and Yuri Izrael—refused to provide a definitive answer about the long-term effect of the spreading contamination. Summoned from Chernobyl to an urgent meeting with the Ukrainian government task force, the experts said that the reactor had been covered and radioactive emissions steeply reduced; soon they would cease altogether. They insisted that current radiation levels didn't warrant an evacuation and recommended merely that the republic take further steps to keep the population informed of the ongoing measures to manage the crisis. But the Ukrainian leaders suspected that—whatever their true beliefs—Ilyin and Izrael also didn't want to take responsibility for endorsing an evacuation. So, in a special late-night session of the Ukrainian Politburo, Vladimir Scherbitsky ordered the two scientists to draw up a written and signed statement of their opinion. He then locked the document in his office safe and disregarded their advice.

That night, Scherbitsky unilaterally ordered that every child in Kiev from preschool to seventh grade, and all those who had already been relocated from around Chernobyl and Pripyat, be evacuated from the city to safe areas in the East for at least two months. The following night, the Ukrainian health minister, Romanenko, appeared once more on TV to reassure viewers that radiation levels in the republic remained within international limits. But now he advised that children should be allowed to play outside only for short periods and forbidden from ball games that might kick up dust. Adults should shower and wash their hair every day. He added that the school year would be concluding two weeks early, "to improve the health of the children of Kiev city and oblast."

The evacuation began five days later, sweeping up 363,000 children, as well as tens of thousands of nursing and expectant mothers, in an exodus of a half million people—the equivalent of a fifth of Kiev's total population. It was a logistical task that dwarfed the initial effort to evacuate the thirty-kilometer zone and, from the outset, was overshadowed by the specter of panic. Thirty-three special trains ran on a shuttle schedule, departing every two hours from Kiev station; gaggles of grade school students clustered on the platforms, numbered paper labels pinned to their shirts, in case they got lost, and extra planes were provided for those who could not make the journey by rail. When the torrent of women and children overwhelmed the Pioneer camps and sanatoria of Ukraine, Soviet holidaymakers in resorts across the Caucasus were informed that their vacations were canceled, and evacuees found temporary homes from Odessa to Azerbaijan. Three days later, Kiev had become a city without children. No one could say when they might return.

On May 22 Scherbitsky put his signature on a Party report describing the republic's handling of the accident. Despite widespread failures and carelessness—particularly the belated provision of safe limits for the population's radiation exposure—90,000 people had been evacuated successfully from the Ukrainian areas of the thirty-kilometer zone. All had been rehoused, and more than 90 percent were already back at work. They had been provided with compensation of 200 rubles per person, amounting to 10.3 million rubles in total. Of the more than 9,000 men, women, and children who had been hospitalized or quarantined across Ukraine in the days after the accident, 161 had been diagnosed with radiation sickness, including 5 children and 49 Interior Ministry troops. A total of 26,900 children had been sent to Pioneer camps in other parts of the Soviet Union, and breast-feeding women had been relocated to sanitaria within the Kiev region.

But for all this apparent care for its citizenry, the dark undertow of Soviet history was already tugging at the first victims of the accident. Just the day before, the Ukrainian health minister had received a telegram from his superior in Moscow. The message provided instructions on how to record diagnoses on patients exposed to radiation as a result of the accident. While those with severe radiation sickness and burns were to be described accordingly—"acute radiation sickness from cumulative radiation exposure"—the records of those with lower exposure and without severe symptoms were not to mention radioactivity at all. Instead, Moscow dictated that the hospital files of these patients were to state that they had been diagnosed with "vegetative-vascular dystonia." This was a psychological complaint with physical manifestations—sweating, heart palpitations, nausea, and seizures—triggered by the nerves or "the environment," unique to Soviet medicine but similar to the Western notion of neurasthenia. The memo decreed the same misleadingly vague diagnosis for liquidators who came for examination having received the maximum allowable dose of radiation for emergency workers.

Back inside the zone, the dogs and cats left behind by the fleeing population had themselves become a health hazard—the Soviet Agriculture Ministry feared the spread of rabies and plague. More immediately, starving and desperate, with their hopelessly irradiated fur, the abandoned pets were now toxic to anyone they encountered.

The Ukrainian Ministry of Internal Affairs turned to the republic's Society of Hunters and Fishermen for help, calling for twenty teams of local men to spread out across the contaminated territory to begin liquidating all the abandoned pets they could find. Each group would be composed of ten to twelve hunters, accompanied by two hygiene inspectors, a policeman, and a dump truck; four mechanical excavators would dig pits for burying the dead animals. As the spring wore on, the silence of the deserted Polesian countryside was broken by the crack of rifle fire, as the men of the Society of Hunters and Fishermen stalked their domesticated quarry across the Exclusion Zone.

Although the industrious Ukrainian sportsmen would eventually manage to eliminate a total of twenty thousand agricultural and domestic animals inside the thirty-kilometer zone, it proved impossible to kill them all. Some dogs managed to escape beyond the perimeter and were fed and adopted by the liquidators they found camped there. The soldiers may have been heedless of the radiation the animals brought with them but nevertheless christened them with bitter new names, more appropriate to their changed environment: Doza or Rentgen, Gamma and Dozimetr.

The men of General Pikalov's command who spread out across the zone in the summer of 1986 were participants in a vast and singular experiment. Soviet contingency plans for a nuclear power plant accident had envisaged a single, brief radioactive release from a damaged reactor—not one that continued for so long and, even as decontamination work began, had still not ceased entirely. The houses and buildings within the thirty-kilometer zone had all been contaminated in different ways, depending on their distance from the plant and the atmospheric conditions at the time the plume had reached them. There was no established methodology to follow. Radiation experts were summoned from Chelyabinsk-40, where their work on the enduring legacy of the Mayak disaster had made them uniquely qualified to clean up a radioactive landscape. But even they had never encountered anything like this.

At first, the chemical troops simply tried washing everything clean. Using a water cannon and fire hoses, they blasted farm buildings and houses with water and the decontamination solution SF-2U. But as the solution soaked into the ground, the fallout became concentrated, and the radioactive contamination of the soil beside the buildings more than doubled, so that the top layer had to be removed with bulldozers. Some materials proved more recalcitrant than others: tiled walls, in particular, were hard to clean, and reinforced concrete remained just as contaminated after washing as it was before, and the troops had to scrub it with brushes to remove even some of the radionuclides. In yards and gardens, the men stripped away the upper levels of earth and piled it into mounds, which they sealed beneath a layer of clay and seeded with fresh grass. The most heavily contaminated soil was trucked away and buried in specially excavated pits as nuclear waste. Many settlements were decontaminated twice, or three times, but homes that resisted the process for too long were simply demolished. Eventually whole villages would be bulldozed flat and buried, their locations marked with triangular metal signs bearing the universally recognized trefoil symbol indicating a radioactive hazard.

Soviet technicians tried whatever they could to strip radionuclides from buildings and the land around them: the troops boiled up polyvinyl alcohol solution in field kitchens to make a liquid that could be painted on walls to trap contaminants and dry into a film that could be peeled away. They sprayed the shoulders of the roads with bitumen to contain dust and laid kilometer after kilometer of fresh asphalt where highway surfaces proved impossible to clean; they fitted Mi-8 helicopters with massive barrels to carry glue that could be dispersed from the air to trap radioactive particles on the ground. Specialists from NIKIMT, the technical services division of the Ministry of Medium Machine Building, scoured factories across the Union for anything that might be used to keep the dust at bay—so long as it was cheap and available in enormous quantities. As the summer wore on, everything from PVA glue to *barda*—a pulp made from beetroot and waste products from timber processing—was shipped into the rail yards on the perimeter of the zone and sprayed from beneath the helicopters in thick, dark showers.

Meanwhile, the threat posed by the radiation to the rivers, lakes, and reservoirs that covered Ukraine had tested the ingenuity of Soviet engineers and hydrologists to its limit. Summoned to the zone from Moscow and Kiev in the first few days after the explosion, they fought to keep fallout from being washed into the Pripyat and seeping into the groundwater, and to prevent contamination that had already swept into the river from penetrating downstream—toward Kiev and the huge reservoir that supplied the city with drinking water. Military construction brigades and men from the Soviet Water Ministry built 131 new dams and filters, sank 177 drainage wells, and began work on a subterranean wall of clay—five kilometers long, up to a meter thick, and thirty meters deep—intended to trap contaminated water before it could reach the river.

Close to Pripyat, the pine forest that filled the sanitary zone between the city and the plant also lay directly in the path of heavy fallout released from the reactor in the first few days after the explosion. Thickly dusted with beta-emitting radionuclides that exposed them to massive doses of radiation—in some places, up to 10,000 rads—almost forty square kilometers of woodland had been killed outright almost immediately. Within ten days, the dense stands of pine lining the main route between Pripyat and the station turned an unusual color, as their foliage changed gradually from deep green to coppery red. The soldiers and scientists who sped down this road had no need to peer from the observation ports of their armored personnel carriers to know they had entered the "Red Forest"; even shielded by armor plate and bulletproof glass, the needles of their radiometers began to swing wildly amid the extraordinary levels of contamination. The forest posed such a threat that it, too, would soon be mown down by combat engineers and buried in concrete-lined tombs.

In the fields of the collective farms, agricultural scientists employed deep plowing to turn the topsoil and safely inter radionuclides out of harm's way. They tested some two hundred different varieties of plants and crops to establish which absorbed the greatest levels of radiation, and sowed the fields with lime and other calcium powders to chemically bind strontium 90 in the soil and prevent it from traveling further into the food chain. The specialists' optimistic forecast was that farming could begin inside the zone within the year.

But in a place where the leaves on the trees and the earth beneath their feet had become sources of ionizing radiation, the work was Sisyphean. Even the most gentle summer breeze recirculated dust carrying alpha- and beta-emitting particles into the air. Every rain shower washed radiation from the clouds and flushed more long-lived nuclear isotopes into the ponds and streams. The arrival of autumn would send radioactive leaves skittering across the ground. The Pripyat marshes—the largest swamp in Europe—had become a massive sponge for strontium and cesium, and the vast stretches of agricultural land proved too large to be scraped clean even by squadrons of earth-moving machines. Only ten square kilometers of the zone would ever be truly decontaminated. A total cleanup would have required nearly six hundred million tonnes of topsoil to have been removed and buried as nuclear waste. And, even with the seemingly un-limited manpower at the disposal of the Soviet Union, this was regarded as simply impossible.

By the beginning of June, the thirty-kilometer zone had become a radio-active battlefield encircled by a besieging army. The detritus of combat—abandoned vehicles, wrecked equipment, zigzagging trenches, and massive earthworks—lay everywhere around the plant. But even as dosimetrists in protective suits roamed the open landscape and military helicopters crisscrossed the sky overhead, the banished citizens of Pripyat began try-ing to return to their homes. Looting was already a problem, and each person had something they needed urgently to retrieve from the city. Some had left behind identity cards and passports; others, large sums of money; some simply wanted their everyday belongings. On June 6 alone, Ukrainian MVD troops turned back twenty-six of the city's former resi-dents either trying to pass checkpoints or cross the perimeter of the for-bidden zone without the requisite documentation.

On June 3 the leader of the government commission issued orders that attempts to make Pripyat habitable cease, with immediate effect. The members of the Pripyat city council—the *ispolkom*—had found a new temporary home in an abandoned office building on Sovietskaya Street in Chernobyl town, and it was here, a few days later, that a KGB officer came to find Maria Protsenko. He had served in Afghanistan, and—unlike so many of his colleagues in the secret police—struck her as warm and polite. He told the architect that he needed help creating a new map of Pripyat. They were going to put up a fence around the city, he said, and he'd like her advice on where it should go. Dutifully unfolding her 1:2000 scale map once more, Protsenko sketched yet another copy, and together they examined the best and shortest route to take: enclosing the main buildings but cutting off the cemetery; avoiding places where digging might sever the sewer pipes and electricity cables vital to the urban infrastructure. She asked the important questions: how the soldiers would dig the footings, what kind of equipment they would use, how they would drive in the posts. She told herself that they were simply protecting the city from thieves and looters.

On June 10, engineering troops of the Twenty-Fifth Motorized Rifle Division arrived in Pripyat with supplies of barbed wire, wooden posts, and tractors equipped with giant augers. Driven on by the knowledge that they were at work in a high-radiation zone, they moved with amazing dispatch, and, within seventy-two hours, their task was complete: Protsenko's beloved *atomgrad* was enclosed behind a twenty-strand fence 2 meters high and 9.6 kilometers in circumference, patrolled by armed guards. Soon afterward, a centralized electronic alarm system, devised by the Special Technical Division of the Ministry of Medium Machine Building, was installed inside the perimeter to keep intruders out of the city.

Around the edge of the 30-kilometer zone, the engineers also cut a track between 10 and 20 meters wide through swamps, forests, and across rivers—out of Ukraine and through Belarus, building bridges and digging culverts. Wild dogs ran through the unharvested wheat in the fields as the men drove seventy thousand posts into the ground and strung 4 million meters of barbed wire between them. In some places, they found the radiation levels so high that they decided to expand the edge of the zone as they went, improvising the perimeter to include the new hot spots of contamination. By June 24, they had completed a 195-kilometer alarmed fence around the entire Exclusion Zone. Pripyat and the Chernobyl Atomic Energy Station now lay at the center of a vast depopulated area of 2,600 square kilometers, patrolled by soldiers from the Interior Ministry and accessible only to holders of a government-issued pass.

Still, Maria Protsenko continued to hold tight to her faith in what the Party leaders had told her: the evacuation was only temporary. One day— perhaps not soon, but at some time in the future—the stain of radiation would be scrubbed from the city, and she and her family would be allowed to return to their homes on the banks of the river.

But as the summer days began to shorten, and Protsenko continued to conduct the work of the city *ispolkom* from exile in Chernobyl, her responsibilities focused more and more on the developing bureaucracy of the nuclear no-man's-land; she learned to sense which of the specialists visiting her office had come directly from the Special Zone around the reactor—by the scent of ozone rising from their clothes. At the same time, she received official instructions to help arrange for the evacuated citizens to visit their apartments and collect their furniture and personal belongings. A twelve-person committee met to agree on what could be removed and how it might be done. They made plans to summon 150 furniture trucks from all over the oblast, to deploy a team of 50 dosimetrists to take radiation measurements in people's apartments and at checkpoints, to find buses to transport the visitors across the zone, and to requisition a half million polyethylene bags to contain their effects. After two weeks of planning, the operation was ready to begin when someone pointed out that completing it was impossible: the citizens of Pripyat remained homeless and therefore had nowhere to put any of the belongings they would remove from the abandoned city.

Eventually, Protsenko befriended a group of physicists from the Ukrainian Academy of Sciences who had come to monitor levels of radiation in the zone, and it was they who finally told her the truth. General Pikalov's chemical troops would go on to conduct a five-month campaign of decontaminating the streets and apartment blocks of the *atomgrad*, but it was intended only to contain the further spread of dangerous radioactivity. The government commission calculated that cleaning the city to make it habitable once more would require a dedicated force of 160,000 men. The price of such an operation would be unimaginable.

"Forget it," the physicists told her. "You will never return to Pripyat."

15

The Investigation

When Sergei Yankovsky arrived on the accident scene shortly before dawn on April 26, he wondered why he had bothered.

Just thirty, slight and bucktoothed, the chief investigator of the Prosecutor's Office of the Kiev Region had been a detective for almost six years. He worked "crimes against the person": rape, assault, armed robbery, suicide, and murder, as well as criminal negligence at work. While the KGB was busy locking up people for telling jokes about Brezhnev, and nonideological crime was—according to the precepts of Marxism-Leninism—supposed to be a strictly capitalist problem, Yankovsky found plenty to keep him busy.

Vodka, in particular, was a powerful engine of violent or sudden death. Weddings and funerals often ended in fights, stabbings—or, in winter, with men falling asleep outdoors, only to be discovered frozen stiff the next morning; deadly workplace accidents were common. On one *kolkhoz* in Yankovsky's district, five combine harvester drivers passed out in a wheat field after a vodka-heavy lunch, unaware that a sixth would prove more conscientious; by the time he realized what had happened, his five comrades had all been cut to pieces beneath the blades of his combine. In 1981 alone, Yankovsky sent 230 bodies to the morgue.

It had been two in the morning when he was woken by a phone call from his boss, Valery Danilenko, the deputy regional prosecutor in charge of investigations. Twenty minutes later, the chief was waiting outside Yankovsky's Kiev apartment in the department's mobile crime laboratory: a minibus full of equipment, painted in *militsia* colors, with red and blue lights and a siren. There was a fire at the Chernobyl station, he said, and they were going to investigate.

The road to the plant was practically empty, and they made good time through the silent countryside, the inky outlines of trees and electricity pylons silhouetted sharply against the flat horizon. If he saw another car, the driver flicked on the siren. As they approached the plant, they sped past a column of fire engines heading in the same direction.

Yet when they arrived at the power station, drawing up two hundred meters from the fourth reactor, the scene seemed oddly quiet. It was not yet fully light, and Yankovsky could see some mist or fog hanging above the building. But there were no flames. There were the fire trucks, but he couldn't make out any signs of great catastrophe. The investigator spotted someone standing in the gloaming who was idly smoking a cigarette, watching the water cascading through the wreckage.

"Hey! What happened here?" Yankovsky asked.

"Oh, something blew up," the man replied. Casually—as if it happened all the time.

The locals could have handled this, thought Yankovsky.

"Why did they call us out?" he said to Danilenko. "Why did they get us up so early?" It all seemed like a waste of time.

"Wait—wait a minute," Danilenko said. "Something isn't right here."

Together they headed to the main administration building of the plant. The leading regional officials were already there. Malomuzh, the Party chief from Kiev, was in the middle of a briefing.

"What are you doing here?" Malomuzh asked the detectives. "We can deal with this ourselves. The fire is already out. And the unit will be running again in no time."

But when they drove over to Pripyat, they found the police station full of Ukrainian Interior Ministry bigwigs. More information was starting to come in: men had been admitted to Hospital Number 126, burned and vomiting; the KGB was out on the perimeter of the plant, looking for saboteurs. It was clear that something serious had happened. Danilenko went for a meeting with his supervisor, the top regional prosecutor. Meanwhile, the local policemen gave Yankovsky a car and an office to use.

It was around 6:00 a.m. when Danilenko returned. The regional prosecutor had made a decision.

"We're opening a case," he told Yankovsky. "We're pressing charges."

The detective sat down at a typewriter, drew a single piece of paper around the cylinder, and began to type.

The investigation into the causes of the accident in Unit Four of the Chernobyl nuclear power plant that began in the early hours of April 26 developed along two parallel paths. The first, the criminal inquiry, escalated in scope and importance over the course of the day, as the impact of the disaster slowly became apparent. By lunchtime, as Sergei Yankovsky and a handful of colleagues spread out across Pripyat and the plant site, interrogating the operators in the hospital and seizing documents from the control rooms of the station, it was no longer a regional investigation but a republican one. Then, just before nightfall, the deputy prosecutor general of the USSR arrived from Moscow with new instructions. He ordered the creation of a special investigative group within the Second Department of the prosecutor's office of the Soviet Union, the division dedicated to crimes committed within the state's closed military and nuclear installations. The entire investigation was henceforth classified as top secret.

That same evening, the government commission in Pripyat also launched a technical and scientific inquiry, entrusted to Academician Valery Legasov—but overseen by Alexander Meshkov, the deputy head of the all-powerful Ministry of Medium Machine Building, which had designed the reactor in the first place. Meshkov concluded quickly that the cause of the accident had surely been operator error. The water pumps had been overloaded, the backup cooling system had been switched off, the reactor had run dry, and some kind of explosion had resulted. This was the much-feared, but predictable, maximum design-basis accident that every member of the operational staff was trained to guard against.

But the following morning, a pair of experts on RBMK reactors from the Kurchatov Institute flew into Kiev from Moscow to begin a forensic analysis of the data from the reactor. On their way by road from Zhuliany Airport to Pripyat, the scientists were held up by an endless stream of buses coming in the opposite direction and didn't reach their destination until evening. The next day, they went to the bunker beneath the station, where they gathered the logbooks from Unit Four, the computer printouts from the reactor's diagnostic and registration system, and tapes recording the conversations of the operators in the minutes before the explosion. As they examined the data, the physicists discerned the broad sweep of events that led up to the accident: the reactor running at low power; the withdrawal of almost every one of the control rods from the core; muffled voices, a shout of "Press the button!", and the activation of the AZ-5 emergency system. Finally, they saw the pen trace lines showing reactor power beginning a steep ascent until suddenly they rose vertically and ran off the top of the page.

To one of the two specialists, Alexander Kalugin, who had dedicated his career to the RBMK project, it all seemed chillingly familiar. Two years earlier, he had attended a meeting of the reactor design bureau, NIKIET, at which someone had suggested that—under certain circumstances—the descending control rods might displace water from the bottom of the core and cause a sudden spike in reactivity. At that time, the institute's scientists had dismissed this concern as too improbable to worry about. Now, as Kalugin gazed in dismay at the fearsome geometry of the computer printouts from Reactor Number Four, it seemed all too possible.

But until the data could be subjected to detailed analysis, Kalugin's idea remained merely a discomfiting theory. In the meantime, the experts phoned Legasov with their initial analysis. On the afternoon of Monday, April 28, a telegram arrived at the Politburo in Moscow: CAUSE OF ACCIDENT UNRULY AND UNCONTROLLABLE POWER SURGE IN THE REACTOR.

Yet the question of how this power surge had been triggered remained unresolved. The search for appropriate scapegoats began immediately.

By the end of the first week in May, the team of reactor specialists from the Kurchatov Institute had returned to their campus in Moscow and started decrypting the information contained in sacks full of documents, punch card printouts, manuals, and reels of magnetic tape recovered from the recording and diagnostic systems of Unit Four. Every one of the computers at the institute was turned over to the task and began working twenty-four hours a day to decode the data and reconstruct the final hours of the reactor. Meanwhile, the investigators from the prosecutor's office and the

KGB continued to walk the wards of Hospital Number Six, interrogating the engineers and operators of the plant even as they began to lapse into comas and die.

Back at the station, Director Viktor Brukhanov remained at his post, outwardly as impassive as ever, but exhausted and stricken by the deaths of his men, crushed by the burden of responsibility he felt for the catastrophe all around him. Each day, he fulfilled the instructions of the government commission as best he could but was absorbed in replacing specialist staff who had already been hospitalized or become too irradiated to continue work at the plant. At the end of each day, he returned to the Fairy Tale Pioneer camp, where, together with his senior colleagues, he had bunked in the camp library. Lying between the bookshelves at night, they talked for hours about what could have caused the catastrophe, but barely slept.

When Sergei Yankovsky came to interrogate the director about his role in the accident, he found him in the infirmary. "Damn it," Brukhanov told him, "I trusted Fomin. I thought it was an electrical test. I didn't think it would turn out like this." The detective mocked him with a quotation from the Russian poet Sergei Yesenin, an infamous suicide: "Perhaps tomorrow, the hospital bed will bring me eternal peace."

Soon afterward, the nuclear engineer and author Grigori Medvedev visited the accident scene and stumbled upon Brukhanov loitering in the hallway at the headquarters of the government commission in Chernobyl town. The academicians, Velikhov and Legasov, were sharing an office down the corridor with the Soviet minister of nuclear energy, where they were still struggling to contain fears of the China Syndrome. Brukhanov wore the white overalls of the plant operator; his eyes were red, his skin chalky, and a dejected expression had settled into the deep creases of his face.

"You don't look well," Medvedev said.

"Nobody needs me," Brukhanov said. "I'm just bobbing around like a piece of shit in an ice hole. I'm no use to anybody here."

"And where is Fomin?"

"He went crazy. They sent him away for a rest."

Two weeks later, on May 22, Brukhanov submitted a request to the nuclear energy minister, Anatoly Mayorets, seeking permission to take time off to visit his wife, Valentina, and their son, Oleg, who had been evacuated to the Crimea. Mayorets gave his approval, and Brukhanov flew south for a week's holiday.

In his absence, the minister made arrangements to have Brukhanov permanently removed from his position as director of the Chernobyl nuclear power plant.

While the investigation continued, Soviet leaders suggested in public that the accident was the result of an all-but-impossible confluence of events, triggered by the operators. "The cause lies apparently in the subjective realm, in human error," Politburo member—and future Russian president—Boris Yeltsin told a correspondent from West German TV. "We are undertaking measures to make sure that this doesn't happen again."

"The accident was caused by a combination of highly improbable technical factors," Andranik Petrosyants, the chairman of the Soviet State Committee for the Utilization of Atomic Energy, wrote in a statement published by the Los Angeles Times. "We are inclined to believe that the personnel made mistakes that complicated the situation." Petrosyants pledged that as soon as the inquiry was complete, a full report into the causes of the disaster would be presented at an international conference of the IAEA at the organization's headquarters in Vienna.

The task of leading the Soviet delegation and collating the report for the conference—which promised an unprecedented glimpse into one of the most clandestine redoubts of Soviet science—and preparing it for public consumption was assigned to Valery Legasov. Hardliners inside the Ministry of Medium Machine Building opposed his appointment, fearing that he would be hard to control. The academician had returned home from Chernobyl for the second time, on May 13, a changed man, his hands and face darkened by a radioactive tan, his ideological confidence shaken. With tears in his eyes, he described to his wife how overwhelmed they had been by the accident, how unprepared they were to protect the Soviet people from its consequences: the lack of clean water, uncontaminated food, and stable iodine. An examination at Hospital Number Six revealed the toxic fingerprint of the reactor deep within Legasov's body: doctors found fission products, including iodine 131, cesium 134 and 137, tellurium 132, and ruthenium 103, in his hair, airways, and lungs. His health shattered, he suffered from headaches, nausea, digestive problems, and chronic insomnia. Nevertheless, Legasov threw himself into collating material for the report, which was compiled from the work of dozens of specialists and hundreds of documents. He worked day and night in his office at the Kurchatov Institute and continued at home, as he and his colleagues compared their statistics until he was certain everything was accurate. He covered the floor of his living room in the villa at Pekhotnaya 26 with piles of papers, which crept down the corridor and climbed the stairs.

Meanwhile, behind closed doors in Moscow, a bureaucratic battle had begun over the joint *Report on the Causes of the Accident in Unit Four of the Chernobyl AES*, the confidential version of events being prepared for the Politburo. In memos, meetings, and multiple interim documents, the barons of the Soviet nuclear industry—the scientists and the heads of the competing ministries that controlled it—competed to divert blame from themselves, ideally before the final report reached General Secretary Gorbachev.

The conflict was hardly an even match: on one side was the Ministry of Medium Machine Building, the nuclear design bureau NIKIET, and the Kurchatov Institute, each headed by its respective octogenarian Titan of Socialist science, all veteran apparatchiks of the old guard: former revolutionary cavalryman Efim Slavsky; Nikolai Dollezhal, designer of the first-ever Soviet reactor; and Anatoly Aleksandrov, the massive, bald-headed Buddha of the Atom himself. These were the men who had created the RBMK, but had also ignored more than ten years of warnings about its shortcomings. On the other side was the Ministry of Energy, represented by the fifty-six-year-old nuclear neophyte Anatoly Mayorets. His ministry had built the plant, operated the reactor—and was responsible for the training and discipline of the staff who had blown it up.

The disputes began almost at once, with the completion of the commission's preliminary report on the causes of the disaster, just ten days after the explosion, on May 5. Overseen by Meshkov, Slavsky's deputy at the Ministry of Medium Machine Building, it unsurprisingly laid the blame for the accident on the operators: they had disabled key safety systems, flouted the regulations, and conducted the test without consulting with the reactor designers; Senior Reactor Control Operator Leonid Toptunov had pressed the AZ-5 button in a desperate and futile bid to stop the accident after it had begun, triggered as a result of his and his colleagues' simple incompetence. Toptunov and Shift Foreman Alexander Akimov were unlikely to contest this version of events: both would be conveniently dead within ten days.

But the specialists from the Ministry of Energy refused to put their signatures on the joint investigation report. Instead, they produced a separate appendix, based upon their own independent investigation. This opinion suggested that—whatever the operators' mistakes—Reactor Number Four could never have exploded were it not for the profound defects in its design, including the positive void coefficient and the faulty control rods that made reactivity increase rather than decrease. Their detailed technical analysis raised the possibility that pressing the AZ-5 button, instead of safely shutting down the reactor, as it was supposed to do, may have *caused* the explosion.

In response, Aleksandrov convened two special sessions of the nuclear industry's Interagency Scientific and Technical Council to consider the causes of the accident. But, despite its name, the council was packed with Ministry of Medium Machine Building staff and former boosters of the RBMK and chaired by Aleksandrov, who held the patent on its design. The meetings went on for hours, yet Aleksandrov used all his considerable skill to squash talk of the reactor's failings and returned again and again to general discussion of the operators' mistakes. When that failed, Slavsky—"the Ayatollah"—simply shouted down those whose opinions he didn't want to hear. The representative from the state nuclear regulator was never even permitted to report on his proposed design revisions, intended to improve the safety of the reactor.

But Gennadi Shasharin, Mayorets's deputy responsible for nuclear matters at the Ministry of Energy, refused to concede defeat. Following the second Interagency Council meeting, he drafted a letter to Gorbachev outlining the actual reasons for the accident and describing the attempts by Aleksandrov and Slavsky to bury the truth about the reactor design faults. Shasharin acknowledged the failures of the plant staff but argued that concentrating on these faults merely revealed the lack of organization and discipline at the plant: "they do not bring us closer to identification of the real causes of the disaster." And the deputy minister explained that however hard they tried, they wouldn't be able to hide the truth forever. The global scale of the disaster ensured that the international scientific community would demand to learn the technical details of the accident sequence. "Sooner or later," Shasharin warned the general secretary, "they will become known to a broad circle of reactor specialists in our country and abroad."

Viktor Brukhanov returned from the visit to his family in Crimea at the end of May. On arriving in Kiev, he phoned the power plant and asked for a car to collect him from the airport. There was an awkward pause on the line, and he knew something was wrong. When he reached the plant, Brukhanov went up to his office on the third floor of the administrative building. There, he found the windows covered with sheets of lead and another man sitting behind his desk. In the first of what would be many public humiliations for the beleaguered manager, no one had bothered to inform him that he was no longer in charge.

"What are we going to do with Brukhanov?" the new director asked his chief engineer. The two men decided to invent a position for him: deputy head of the Industrial-Technical Department, a sinecure in a back office where he could be kept busy while he awaited his fate. They both knew it was only a matter of time before he would be called to answer for his crimes.

Inside the headquarters of the Second Department of the Soviet Prosecutor General's Office, in a secret high-security building on Granovskogo Street in Moscow, the interrogations continued. Sergei Yankovsky's inquiries had now broadened to include the designers and scientists who had created and overseen the RBMK, and the academicians were summoned for questioning just like everyone else. Yankovsky brought in the reactor designer Nikolai Dollezhal, and the aging nuclear baron assured the detective that the blame for the explosion lay solely with the operators; there was nothing at all wrong with his design.

By the end of the summer, the investigation of the reactor designers would be broken off into a separate criminal case, while that of the plant operators gained momentum. Yankovsky crisscrossed the Union in pursuit of information. He flew to Sverdlovsk to confiscate documents and interrogate staff at the factory where the giant main circulating water pumps used in Unit Four had been manufactured. He spent ten days in Gorky, where nuclear expert Andrei Sakharov was being kept in internal exile as punishment for his human rights campaigning, taking with him punch card printouts from the reactor's computer system in the hope that Sakharov could assist in their analysis. And, back in Ukraine, Yankovsky visited other nuclear power plants to gather evidence about previous accidents. Everywhere he went, he was shadowed by officers of the KGB, sent to ensure the continued secrecy of everything his investigation uncovered.

On Wednesday, July 2, Viktor Brukhanov was called back to Kiev and handed a plane ticket to Moscow, where his presence was required the next day at a meeting of the Politburo. Before leaving, he went to bid farewell to Malomuzh, the deputy Party secretary for the region. The secretary had never before treated Brukhanov with anything but icy formality yet now seized him in a sudden embrace. It wasn't a good sign, but by now, the deposed director had resigned himself to his fate.

At precisely eleven o'clock the following morning, the Politburo gathered in a gloomy third-floor conference hall in the Kremlin. The room was filled with small desks, and Brukhanov found himself among the venerable leaders of the Soviet nuclear industry—including Aleksandrov, Slavsky, and Valery Legasov—all seated like so many errant schoolboys. The obligatory portrait of Lenin glared from the wall above them. General Secretary Gorbachev opened the session and asked Boris Scherbina to present his government commission's final report on the causes of the disaster.

"The accident was the result of severe violations of the maintenance schedule by the operating staff and also of serious design flaws in the reactor," the chairman began. "But these causes are not on the same scale.

The commission believes that the thing that triggered the accident was mistakes by the operating personnel."

This was the preferred narrative of the Ministry of Medium Machine Building—and yet Scherbina went on to admit that the failings of the reactor were extensive and insoluble. The RBMK was not up to modern safety standards and even before the accident would never have been permitted to operate beyond the borders of the USSR. In fact, he said, the reactor was so potentially hazardous that his specialists recommended existing plans to build any more should be scrapped.

By the time Scherbina had finished, Gorbachev was furious. His anger and frustration had been building for weeks as the catastrophe bloomed. He had struggled to find accurate information about what was happening, and his personal reputation in the West—as a reformer, a man one could do business with—had been tarnished by the fumbled attempts at a cover-up. He now accused Slavsky and Aleksandrov of presiding over a secret state and deliberately concealing from him the truth about why the accident had happened in the first place. "For thirty years, you told us that everything was perfectly safe. You assumed we would look up to you as gods. That's the reason why all this happened, why it ended in disaster. There was nobody controlling the ministries and scientific centers," he said. "And for the moment, I can see no signs that you have drawn the necessary conclusions. In fact, it seems that you are attempting to cover everything up."

The meeting blazed on for hours. Lunchtime came and went. Gorbachev asked Brukhanov if he knew about Three Mile Island and about the history of accidents at the Chernobyl station; the director was surprised by how polite the general secretary seemed. Slavsky continued to blame the operators, while Gorbachev's hardline deputy, Ligachev, clung to the flotsam of Soviet pride. "We showed the world that we're able to cope," he said. "No one was allowed to panic." The representatives of the Ministry of Energy admitted they had known that there were problems with the reactor, but Aleksandrov and Slavsky had nevertheless insisted on constantly expanding the nuclear power program.

At one point, Meshkov unwisely insisted that the reactor was still perfectly safe if the regulations were followed precisely.

"You astonish me," Gorbachev replied.

Then Valery Legasov admitted that the scientists had failed the Soviet people. "It is our fault, of course," he said. "We should have been keeping an eye on the reactor."

"The accident was inevitable. . . . If it hadn't happened here and now, it would have happened somewhere else," said Prime Minister Ryzhkov, who argued that the intoxicating power handed to Aleksandrov and Slavsky had proved their undoing. "We have been heading toward this for a long time."

As it approached seven o'clock in the evening—after almost eight hours of uninterrupted debate—Gorbachev presented his conclusions and proposed punishments for all those he held culpable. These were drafted into a resolution, which included a twenty-five-point action plan, and put to the vote before the Politburo eleven days later. In it, the Party leaders faulted Brukhanov and the chief engineer, Fomin, for tolerating rule breaking and "criminal negligence" inside the plant and for failing to safely prepare for the test during which the accident took place. They criticized the Ministry of Energy for its slipshod management, neglect of staff training, and for becoming complacent about the number of equipment accidents inside the nuclear plants under its jurisdiction. Finally, they attacked the state nuclear regulatory authority for its lack of effective oversight.

But the Politburo resolution also plainly recognized the true origins of the accident that destroyed Reactor Number Four. The catastrophe occurred "due to deficiencies in the construction of the RBMK reactor, which does not fully meet safety demands," it stated. Furthermore, although Efim Slavsky was well aware of these shortcomings and received numerous warnings, he had done nothing to address the failings of the reactor design.

The Politburo reserved the harshest discipline for the middle ranks of the *apparat*. Meshkov, deputy head of the Ministry of Medium Machine Building, and Shasharin, the deputy minister responsible for nuclear power at the Ministry of Energy, together with the deputy head of the nuclear design bureau, NIKIET, were all dismissed from their posts. Viktor Brukhanov was stripped of his Party membership and put on a plane back to Kiev in disgrace.

But they also proposed sweeping changes throughout the industries and organizations whose failings had been exposed by the accident. The resolution instructed the Ministry of Internal Affairs and the Ministry of Defense to equip and retrain troops and firefighters to deal with radiological emergencies and decontamination work. Gosplan and the Ministry of Energy should reexamine their long-term expectations of nuclear electricity. Standards of training and safety should be overhauled and oversight of nuclear energy unified under a new Ministry of Atomic Energy. Finally—in an implicit acknowledgment of all that was wrong with the reactor itself—the Party leaders decreed that all existing RBMK plants should be modified to bring them into line with existing safety standards. Plans to build further RBMK reactors were terminated immediately.

Yet those at the top of the nuclear industry, who had supervised the project from the start, escaped overt censure almost entirely. Slavsky—who, by now, was overseeing the construction of the Sarcophagus, intended to entomb the failed reactor forever—and Aleksandrov were merely reminded of their commitment to ensuring the safety of the peaceful atom. Nikolai Dollezhal's name wasn't mentioned at all.

At the end of the marathon meeting, Gorbachev underscored the profound international impact of the catastrophe. It had sullied the reputation of Soviet technology, and everything they did was now under global scrutiny. It was now imperative, he said, to be completely frank about what had happened, not only with fellow Socialist countries but also with the IAEA and the international community. "Openness is of huge benefit to us," he said. "If we don't reveal everything the way we should, we'll suffer."

Not everyone agreed. The following day, officers of the Sixth Directorate of the KGB circulated a list of subjects relating to the Chernobyl accident regarded as classified to varying degrees. Covering two sheets of typescript, the document listed twenty-six numbered items. At the top, marked *Sekretno*—"Secret"—was item one: "Information revealing the real reasons for the accident in Unit Four."

On his arrival in Kiev, Viktor Brukhanov was taken to the Leningrad Hotel and summoned the next morning to give a statement in the offices of the public prosecutor. The investigator gave him a list of questions, and Brukhanov wrote out his answers by hand. The statement eventually filled ninety pages, and when it was finished, he was driven back to the Fairy Tale Pioneer camp.

On the evening of Saturday, July 19, the official version of the Politburo's verdict was announced on *Vremya*. It was unequivocal and damning. Through the findings of the government commission, the anchor said, "it was established that the accident had been caused by a series of gross breaches of the operational regulations of the reactor by workers at the atomic power station. . . . Irresponsibility, negligence, and indiscipline led to grave consequences." The statement included a list of the ministers who had been dismissed and concluded with the news that Brukhanov had been expelled from the Party. The Prosecutor General's Office of the USSR had launched an investigation, and court proceedings would follow. There was no mention of any design faults in the reactor.

The next morning, the news was emblazoned on the front pages of *Pravda*, *Izvestia*, and all other Soviet newspapers; the Politburo statement would also be printed in full by the *New York Times*. That day in Moscow, a reporter for the Canadian *Globe and Mail* found a woman cleaning a statue of Lenin and asked her view of the guilty men. "They should all be thrown in jail," she said.

At home in Tashkent, where she lived with one of her three younger children, Brukhanov's elderly mother was watching TV when the news broke. When she learned of her eldest son's downfall, she staggered out of her apartment and into the street, where she suffered a heart attack and died. A few days later in Kiev, the Central Committee of the Communist Party of Ukraine handed down its own verdicts, ejecting Chief Engineer Nikolai Fomin for ordering the test that had led to the explosion, for "flagrant errors and omissions in work," and removing plant Party secretary Serhiy Parashyn from his position.

By the second week of August, Viktor Brukhanov had returned from his mother's funeral in Uzbekistan and, along with hundreds of other plant staff and liquidators, was billeted aboard one of eleven river cruise ships moored on a picturesque bend in the Dnieper some forty kilometers from the Chernobyl station. On August 12 the deputy chief engineer of the plant came back from a trip to Kiev bearing a summons with Brukhanov's name on it, instructing him to report, at 10:00 a.m. the next day, to room 205 of the prosecutor's office, on Reznitskaya Street in Kiev. There, after three further hours of interrogation and an hour's break for lunch, Brukhanov was formally charged, under Article 220, paragraph 2, of the Ukrainian Criminal Code, with "a breach of safety regulations in explosion-prone plants or facilities" and arrested. Led out through the back door by two men in civilian clothes, he was driven to a KGB holding cell, where he would spend much of the next year.

Two weeks later, on August 25, Valery Legasov, wearing a gray suit and striped tie, his face puffy and haggard behind thick glasses, took the floor on the opening day of the special technical conference at IAEA headquarters in Vienna. The mood was somber and tense, and the wood-paneled conference hall was packed. Six hundred nuclear experts from sixty-two different countries, accompanied by more than two hundred journalists, had come to discover the truth about the accident that, by now, transfixed the world. The burden upon Legasov was enormous: not only the reputation of all of Soviet science but also the future of the global nuclear industry was at stake. The disaster suggested that the USSR's technicians could not be trusted to build or operate their own reactors and that the technology itself was so intrinsically hazardous that, even in the West, nuclear power stations should be shut down or phased out.

Legasov had spent much of the summer collating the material he intended to present, drawn from the contributions of a team of twenty-three experts, half of them from the Kurchatov Institute, but also including the designers of the reactor, the chief of the USSR's environmental and meteorology agency, and the radiation medicine and decontamination specialists, Dr. Angelina Guskova and General Vladimir Pikalov.

And yet, glasnost or not, the organs of the Soviet state were no more ready to disclose the truth about the myriad failures of Socialist technology than they had ever been. When a draft copy of the report eventually reached the Central Committee, the head of the Energy Department was horrified by what he read. He forwarded it to the KGB with a note attached: "This report contains information that blackens the name of Soviet science. . . . We think it expedient that its authors face punishment by the Party and the criminal court."

While his notions of retribution may have been harsh, the fears of the Energy Department apparatchik were not unjustified. Revealing to the world the true roots of the disaster—the design of the reactor itself; the systematic, long-term failures and the culture of secrecy and denial of the Soviet nuclear program; and the arrogance of the senior scientists overseeing its implementation—was unthinkable. If the report acknowledged the design flaws of the RBMK reactor, responsibility for the accident could be traced all the way up to the chief designer and the chairman of the Academy of Sciences. In a society where the cult of science had supplanted religion, the nuclear chiefs were among its most sanctified icons—pillars of the Soviet state. To permit them to be pulled down would undermine the integrity of the entire system on which the USSR was built. They could not be found guilty.

Legasov's delivery was masterful. Speaking uninterrupted for five hours through translators, the academician held his audience spellbound. He outlined the design of the reactor in detail—acknowledging certain "drawbacks" but glossing over inconvenient facts—and described a minute-by-minute reconstruction of the accident sequence, far more terrifying than any of the Western experts had imagined. When he finished, he sat for several hours of questions, and Legasov and his team answered nearly every one. Pressed by reporters on the disadvantages in the reactor design he had mentioned, Legasov replied, "The defect of the system was that the designers did not foresee the awkward and silly actions by the operators." Nonetheless, he acknowledged that "about half" of the USSR's fourteen remaining RBMK reactors had already been shut down for technical modification, "to increase their safety."

Impressed by such apparently unprecedented candor from Soviet scientists, and reassured by the idea that the disaster was an extraordinary event that had little relevance for nuclear safety outside the USSR and that its health and environmental consequences fell within acceptable limits, the delegates left the hall confident in the future of Soviet atomic energy—and of their industry around the world. By the time they departed from Vienna at the end of the week, the mood was cheerful, almost jubilant. For the Soviet Union—and for Valery Legasov personally—the conference was, a prominent British physicist remarked in the *Bulletin of the Atomic Scientists*, "a public relations triumph."

On his return to Moscow, Legasov went directly to the Kurchatov Institute and ran up the steps to the third floor. "Victory!" he shouted to a friend.

Yet nagging questions remained.

Halfway through the conference, during a coffee break amid three days of sessions closed to the media, the MIT physicist Richard Wilson had buttonholed two members of the Soviet delegation with a question that puzzled him. In the copy of the report that Wilson had obtained, rush-translated by the US Department of Energy and dense with tables and statistics, the simple arithmetic in some places seemed wrong: the sum of figures plotting radioactive fallout in specific regions of the USSR didn't appear to match the total shown at the end. The two Soviet delegates had to admit that the numbers might not be quite right. Years afterward, Wilson learned that six pages of data on the contamination of Belarus and Russia had been excised from the report, on Legasov's instructions. He had doctored the report on the direct orders of Prime Minister Ryzhkov.

"I did not lie in Vienna," Legasov said to his colleagues in a report he delivered two months later at the Soviet Academy of Sciences. "But I did not tell the whole truth."

16

The Sarcophagus

In the darkened room just below the rooftop, the line of soldiers waited as their comrades checked their equipment. Over their olive-green uniforms, they strapped on knee-length lead aprons and bound pieces of the soft gray metal, cut from sheets three millimeters thick, to their chests, the backs of their heads, and along their spines; more was cupped at their groins and stuffed inside their boots. They wore green canvas hoods, cinched tight around their faces. The men pulled on heavy respirators and goggles to protect their eyes. Some put on plastic construction helmets.

"Are you ready?" asked General Tarakanov. His voice echoed from the concrete walls. The eyes of the first five men flashed with anxiety, and they headed toward the staircase. At the top, they turned right and followed their guide down a blackened passageway, on toward a ragged patch of blinding sky: a hole, blasted through the shell of the building with explosives, and just large enough to allow a man to pass through. This was the way out into Zone M, on the roof above Unit Three, where, months earlier, the firefighters had struggled to douse the burning debris thrown from Reactor Number Four.

General Tarakanov had divided the rooftops according to their height and level of contamination. He named each area after a woman in his life: Area K (Katya), where the gamma fields reached 1,000 roentgen; Area N (Natasha), up to 2,000 roentgen; and, finally, Area M (for Masha, the general's older sister). Here, the men spoke of the levels of radiation only in reverent whispers. Directly overlooking the gaping shell of Unit Four and the remains of the shattered reactor within, Area M was a shambles of scorched rubble and chunks of masonry thrown into the air by the force of the explosion. It was strewn with twisted concrete reinforcement rods and pieces of equipment hurled from the reactor hall, some weighing nearly a half tonne. Graphite blocks that had once formed the core of the reactor lay everywhere—some turned white, perhaps by the heat of the explosion, but otherwise intact. Around them, levels of radiation reached as much as 10,000 roentgen an hour: enough for a fatal dose in less than three minutes.

General Nikolai Tarakanov, deputy commander of the Civil Defense Forces of the USSR, was fifty-two years old, a balding and diminutive Cossack, the fifth of seven children, who as a boy had watched his village burned to the ground by the Nazis. He had lied about his age to enlist in the army and, fifteen years later, won a technical doctorate in military science. A specialist in post-atomic combat engineering, Tarakanov had written two Soviet armed forces textbooks on how to begin rebuilding in the aftermath of a nuclear strike. He studied detailed scenarios modeling the expected destruction of US missile attacks on the major cities in the USSR: gruesome visualizations projecting hundreds of thousands of deaths, a population struggling to survive in a poisoned landscape, and key industries rebuilt underground in the untracked hinterland of the empire. In 1970 he began practical experiments on a military proving ground in Noginsk, outside Moscow, where a small city had been laid out to simulate the postapocalyptic urban environment, complete with piles of debris and ruined buildings. There he helped develop the techniques and the protocols and the massive pieces of engineering equipment— the armored excavators and bulldozers, the IMR-2 vehicles, with their telescopic arms and mechanical pincers—that had been deployed in the most radioactive areas of Chernobyl's Special Zone at the beginning of May. But now it was September. In Area M, the plans and the technology had failed, and Tarakanov was sending his men into combat, armed only with shovels.

At the end of the passageway, the soldiers gathered themselves at the threshold. Their breath rasped in their rubber respirators. An officer started his stopwatch, and another five men stepped out into the light.

In the four months since Ivan Silayev had appeared on Soviet television and announced plans to build a sarcophagus to forever entomb the remains of Reactor Number Four, a fresh army of architects, engineers, and construction troops had been mustered in the zone and begun toiling around the clock to make the idea a reality. While their rivals in the Ministry of Energy were made responsible for restarting the remaining three reactors of the plant, the Sarcophagus project was entrusted to a construction unit formed specially for the purpose by the Ministry of Medium Machine Building and designated US-605. Designs for the new structure were solicited throughout the ministry's myriad acronym-bedecked agencies and subdivisions: the All-Union Scientific Research and Design Institute of Energy Technology, or VNIPIET; the ministry's main construction department, known as SMT-1; and NIKIMT, the experimental laboratory dedicated to research and development in nuclear building projects.

The final concept would eventually be chosen from a short list of eighteen designs. Engineers from the reactor design bureau, NIKIET, had suggested filling the ruined reactor with hollow lead balls. Others had proposed burying it beneath an enormous mound of crushed stone, or excavating a cavern beneath Unit Four large enough for the reactor to collapse into, so that the earth might swallow it whole. During the first meetings on the subject, Efim Slavsky, the pugnacious head of Sredmash, had presented his own, typically emphatic solution: drown the whole mess in concrete and forget about it. Big Efim's suggestion was met with an awkward silence, broken eventually by Anatoly Aleksandrov. The head of the Kurchatov Institute pointed out the inconvenient physics of Slavsky's approach: the continuing decay heat of the nuclear fuel remaining inside the reactor building made sealing it up impractical, if not impossible.

Attractive as it might have seemed to completely seal off the remains of Reactor Number Four from the atmosphere around it, the fuel mass would instead require both widespread ventilation, to allow it to continue cooling safely, and constant monitoring, to provide a warning in case a new chain reaction began. The ruins would have to be covered by a protective shell, although no one could yet say how this might be done. Unit Four sprawled across an area so large—nearly the size of a soccer field—that any roof would require support from vertical pillars built inside its perimeter. Yet this space was still a no-man's-land of collapsed walls, mangled equipment, and shattered concrete, much of it buried beneath the thousands of tonnes of sand and other materials dropped from General Antoshkin's helicopters. The engineers couldn't know for sure if anything within the ruins retained enough structural integrity to support the weight of even the flimsiest roof. And the radiation made it all but impossible to find out.

Among the proposed architectural solutions, there were some of soaring ambition, including a single arch with a span of 230 meters, and the suggestion to roll a series of prefabricated vaults across the entire width of the reactor hall; another was a massive single-span roof suspended from a row of inclined steel arms, raised into the air at six-meter intervals, a design the engineers referred to sardonically as "Heil Hitler." But these fantastical concepts could take years to complete, at astronomical cost—or lay beyond the known limits of Soviet engineering. The final design was instead dictated by a characteristically unrealistic Politburo timetable and the atrocious conditions at the construction site. Whatever was constructed around Unit Four had to be completed as quickly as possible—not in years, but in months—both to halt the spread of radioactivity and also to restart Units One, Two, and Three in relative safety, and thus salvage something of the USSR's tarnished technological prestige.

But the technical challenges were formidable, not least because the building would have to be put together remotely. Even after the ruins had been bombarded with sand and submerged in molten lead, radiation levels around Unit Four remained too high for anyone to work there for more than three minutes at a time. So the engineers planned to build the new structure from prefabricated sections, assembled using cranes and robots. And time was short. On June 5 Gorbachev gave Slavsky and his men until September to complete the new building: less than four months to accomplish one of the most dangerous and ambitious civil engineering feats in history. Work began at the site even before the engineers and architects in Moscow had agreed on a viable design.

To limit their overall radiation exposure, the teams of Sredmash US-605 rotated into the Chernobyl zone on two-month tours of duty. The first shift, beginning on May 20, had to clean up the mess left by the Ministry of Energy's abortive remediation efforts—a tangle of blocked roads, wrecked equipment, and half-finished concreting projects—and create the infrastructure necessary for the enormous undertaking to come. They had to prepare housing, food, and sanitation for a force of twenty thousand men, the majority of them military reservists conscripted into the service of Sredmash, who became known as *partizani*. The Ministry of Medium Machine Building regarded its technical experts—architects, engineers, scientists, electricians, dosimetrists—as irreplaceable. They had to be protected from overexposure so that they could work in the zone for as long as possible. But the often middle-aged *partizans* were perceived as ignorant, unskilled, and expendable. They were thrown into the front line wherever necessary to perform manual tasks in high-radiation zones, one platoon after another. These men caught their maximum dose in a matter of hours—or minutes—before being sent home and replaced with more cannon fodder.

The most important task for the first shift was to guarantee an uninterrupted supply of the Ministry of Medium Machine Building's chief protection against radiation: reinforced concrete. The rail lines and cement factories used during the construction of the first four Chernobyl reactors had been directly in the path of the first plumes of heavy fallout to rise from Unit Four and were so contaminated that they had to be abandoned. Before building work began, the Sredmash engineers laid thirty-five kilometers of new road and constructed decontamination stations, a rail interchange, a dock to receive shipments of a half million tonnes of gravel by river, and three new concrete plants.

The engineers then began to lay siege to the reactor, advancing on it slowly from behind a series of "pioneer walls" to shield building workers from the invisible fusillade of gamma rays streaming from within the ruins. At a safe distance, the engineers welded hollow steel forms, 2.3 meters square and almost 7 meters long, which they stacked like huge bricks on flatbed railcars. They pushed these into position around the reactor using armored combat engineering vehicles, and concreted them in place—railcars and all—using pumps stationed at least 300 meters away. More than 6 meters high and 7 meters thick, the resulting walls threw a "gamma shadow" in which workers could remain safely for up to five minutes at a time. The surface of the ground around them was also decontaminated: sprayed with dust-suppressant solution and gradually covered with another half-meter-thick layer of concrete.

The work was relentless, continuing twenty-four hours a day, seven days a week, in four shifts of six hours each. At night, the site was illuminated with searchlights and lit from overhead by a tethered blimp. The government commission measured the construction teams' progress on the Soviet scale, by the volume of reinforced concrete they poured each day, and kept them under incessant pressure. By the middle of the summer, a stupendous one thousand cubic meters of concrete—twelve thousand tonnes—was being churned out every twenty-four hours by the Sredmash plants. It was rushed to the remains of Unit Four in a relay of mixers and pump trucks, driven down the new roads at one hundred kilometers per hour by drivers who had to move fast to prevent their cargo setting in the summer heat—and who feared the radiation in the air around them. The roadsides were soon littered with the wrecks of overturned trucks.

In July and August, the second shift of Sredmash engineers filled in the space between the first pioneer wall and the walls of Unit Four itself with more concrete, rubble, and pieces of debris and contaminated equipment. From these foundations, they began building upward. Three high-capacity Demag cranes, including two mechanical monsters on caterpillar tracks bought from West Germany at a cost of 4.5 million rubles, arrived in the zone by rail. Capable of lifting almost twenty times the load of an ordinary crane, they were used to install huge prefabricated steel forms, which were backfilled with still more concrete, entombing the escarpment of highly radioactive debris that had tumbled from the northern side of the reactor building. This became the "Cascade Wall," which rose in a series of terraces—four colossal steps, each fifty meters long and twelve meters high—like the temple of a vengeful prehistoric god. The scale of the structure dwarfed the men and machines working in its shadow, and neither could remain near it for long. If brought too close, the engines of the concrete pumps guttered and died, and the dials of the dosimetrists' equipment went haywire, like compass needles in a magnetic field. It was a phenomenon the experts could never satisfactorily explain.

The steel forms of the Cascade Wall were preassembled into massive sections designed to be held in position by the cranes until buried in concrete. The work took weeks. Holes and voids in the walls of Unit Four meant thousands of cubic meters of liquid concrete poured uselessly into the ruins, filling the basement, corridors, and stairwells, until the gaps could be plugged. When the mortar set, radio-controlled explosive bolts released the crane cables, and lifting the next section could begin. But when the corner section of the Cascade Wall—a tower rising sixteen stories above the ground, in a seething gamma field—was finally fixed into position, the explosive bolts failed. The Sredmash specialists sought a volunteer from among the *partizans,* who agreed to be carried up on a separate crane and release the cables by hand. Before he embarked, they issued him three different dosimeters to record his exposure during the mission. It was an hour before he returned to the ground, where he was rewarded with 3,000 rubles, a case of vodka, and immediate demobilization. But he threw the dosimeters away, for fear of what they might tell him.

While the Sredmash engineers were at work on the Sarcophagus, a scientific task force from the Kurchatov Institute began trying to unravel the mystery of what had happened to the 180 tonnes of nuclear fuel they believed still lay somewhere within its rising walls. At first, the scientists had believed that most of the uranium had been blown out of the reactor vessel by the explosion and must be scattered inside what remained of the machine hall. But radiation detectors lowered into the ruins by helicopter revealed no evidence of it there. Academician Legasov now worried that if even a small amount of uranium fuel and graphite moderator remained intact and in the correct configuration within the reactor vault, it could lead to a further criticality: the beginning of a new nuclear chain reaction that no one could control, resulting in the release of further radionuclides into the atmosphere around the plant. His colleague, Velikhov, feared that the Sredmash construction crews, blindly pumping concrete over the scattered clusters of nuclear fuel, might inadvertently be building a colossal atomic time bomb.

Yet all their initial efforts to find the uranium fuel inside the reactor hall itself had failed. The members of the Kurchatov task force measured radiation exposure of thousands of roentgen an hour on all available routes through the debris toward the reactor vessel—from below, above, and both sides; they searched for molten lead and the melted residue of the sand, the boron carbide, or the dolomite thrown from the helicopters. But they found no evidence of any of it, and certainly no sign of the fuel.

Eventually the Kurchatov scientists reached one of the rooms in the basement of the reactor hall, three floors beneath—and far to the east of—the reactor vessel itself. Carrying a device that could measure dose rates up to 3,000 roentgen an hour, the team found relatively tolerable levels of radiation along their route. But then they pushed the sensor of the radiometer upstairs, into the space directly above where they stood. There, in compartment 217/2 on mark +6, it encountered a gamma field so hot that the instrument reached its maximum reading and then—its mechanism overwhelmed—burned out. Whatever lay inside was stupendously radioactive and represented a possible clue to the location of the hundreds of tonnes of lost fuel. Yet anyone entering the blackness of corridor 217/2 to find out what it was risked absorbing a lethal dose of gamma radiation in minutes, or seconds.

Alexander Borovoi, a heavy-set forty-nine-year-old neutrino physicist who had worked at the Kurchatov Institute for more than twenty years, arrived from Moscow to join the task force in late August. It was warm when he disembarked from the *Raketa* hydrofoil in Chernobyl, where he was issued a set of khaki overalls and two envelopes containing petal respirators—but no instructions on how to use them. That night, an old colleague from the institute just finishing his own shift stopped by to impart the "commandments" on how to survive in the high-radiation zones of the ruined station, accumulated through months of practical experience. To avoid getting lost, he told Borovoi, never enter any room not illuminated by electric light and always carry both a flashlight and a box of matches in case it failed; he warned him to beware of water falling from above, which could carry heavy contamination into the nose, eyes, or mouth; and, most important of all—the First Commandment—be alert for the smell of ozone. The lecturers back in Moscow might tell you that radiation has no odor or taste, he explained, but they've never been to Chernobyl. Intense gamma fields of 100 roentgen an hour and above—on the threshold for inducing acute radiation syndrome—caused such extensive ionization of the air that it left a distinctive aroma, like that after a lightning storm; if you smell ozone, his colleague said, run.

The following morning, under orders from Academician Legasov, Borovoi was sent on his first scouting mission into Unit Four.

As the Kurchatov task force continued its search for the fuel, and the Sredmash teams toiled to complete the Sarcophagus, the technicians of the Ministry of Energy raced to meet their own deadline: the Politburo had publicly promised that the first two of the three remaining Chernobyl reactors would be restored to life before winter set in. But now that the truth of the design faults in the RBMK had finally begun to emerge, the specialists first had to modify the reactors to make them safe to operate, improving their performance by altering the steam void coefficient and the functioning of the control rods. At the same time, they had to decontaminate the entire station from top to bottom, until the fabric of the building itself no longer endangered the operators who would work inside it. The basement cable tunnels that ran beneath the four reactors, flooded with radioactive water during the accident, were pumped dry and the concrete flooring and fireproof coatings chiseled out, ground away, and replaced. The plant's walls and floors were scoured with acid, stripped with quick-drying polymer solutions, or covered with thick sheets of plastic. The entire ventilation system was flushed of radioactive dust and hot particles or rebuilt, and every piece of electrical equipment in the giant complex was scrubbed clean with a solution of alcohol and

Freon, in a process that began in June and would continue for another three years.

But the most dangerous problem was above their heads. Four months after the explosion, the ziggurat roofs of Unit Three and the platforms of the red-and-white-striped ventilation stack looming over the carcass of Unit Four were still littered with fragments of graphite and reactor components, large and small. Fuel assemblies and ceramic pellets of uranium oxide, control rods, and zirconium channels lay where they had fallen, tangled among hoses abandoned by the firefighters who had died weeks before in Hospital Number Six. In some places, the debris was piled into treacherous mounds: a five-tonne concrete panel from the central hall, tossed into the air by the blast, had come to rest amid slick piles of reactor graphite. In others, where the bitumen had melted during the blaze, pieces of wreckage remained welded to the rooftop. All of it was intensely radioactive and would have to be removed before any operators could safely return to the rooms below to run the reactor and turbines of Unit Three.

The government commission turned once again to NIKIMT, the same Moscow laboratory that had suggested using the beetroot pulp now being sprayed over the zone to combat dust. The scientists responded with another ingenious yet thrifty solution, using rags produced as waste in the textile industry to make large mats, soaked in a cheap water-soluble glue and lowered onto the rooftops, where they stuck to the pieces of wreckage. When the glue dried, these "blotters" could be lifted away, bringing the radioactive debris with them, and then removed to be buried. The scientists' early tests proved successful: with a single square meter of blotter, they could retrieve two hundred kilos of wreckage from a height of seventy meters. But when they asked for permission to use the giant Demag cranes to carry the blotters up to the roof of Unit Three, the commission refused. The cranes were needed twenty-four hours a day for the construction of the Sarcophagus and could not be spared. The NIKIMT team conducted a second successful experiment, deploying its invention from helicopters, but was then denied permission to fly—because the rotor downdraft recirculated too much toxic dust.

In the meantime, Ministry of Energy technicians planned to clear the debris using robots: one, specifically designed to handle radioactive material, purchased from West Germany and nicknamed "Joker," and a pair of remote-controlled vehicles developed for use in the Soviet lunar exploration program, modified with small bulldozer blades. To save time and avoid having to move the wreckage to separate disposal sites, the technicians decided to simply shunt it off the edge of the roof, back into the bowels of Unit Four. But the sensitive electronics of Joker quickly failed in the gamma fields of Area M. And even the machines intended for use on the surface of the moon were no match for the inhospitable new landscape they encountered on the roof of the ruined plant. Their artificial brains scrambled, their wheels stuck in the bitumen, hung up on blocks of masonry or snarled in their own cables, one by one the robots all stuttered to a halt.

On September 16, General Tarakanov received a coded cipher summoning him to a meeting of the government commission in Chernobyl town. By now, Boris Scherbina—after presenting his report on the causes of the accident to Gorbachev in Moscow—had returned as full-time chairman of the commission. The meeting convened shortly after 4:00 p.m. in Scherbina's lead-lined office, in the district Party committee building on Lenin Street. The chief of the radiation scouts overseeing the cleanup operation above Unit Three spoke first. Yuri Samoilenko, a thickset Ukrainian with a shaggy mop of dark hair and a brooding gaze, looked haggard. His eyes were dark and pouchy. He chain-smoked constantly.

Using a sketch plan of the rooftops, annotated densely with numbered radiation readings and marked with red flags and stars to indicate the most acute hazards, he explained the situation they faced. All technical and automated means to clear the debris field had failed. Radiation levels were enormous. But the roofs had to be cleared before the Sarcophagus was sealed and the sole repository chosen for the most contaminated pieces of reactor debris was thus closed forever. Every other option had been exhausted. It was time, he said, to send in men to do the job by hand.

There was a heavy silence.

The campaign of the *bio-roboty*—the bio-robots—had begun.

Tarakanov's soldiers launched their operation three days later, on the afternoon of September 19. Their preparations were rushed, their equipment improvised. An initial test had been conducted in Area M by an Army Medical Corps radiologist who had stepped out onto the rooftop wearing experimental protective clothing and ten separate dosimeters to monitor his radiation exposure. Sheathed in his hood, lead apron, respirator, and pieces of lead sheet torn from the walls of government offices in Chernobyl, the radiologist sprinted across the roof, glanced around quickly, and then tossed five shovelfuls of graphite over the precipice into the ruin of Unit Four. In one minute and thirteen seconds, he absorbed a dose of 15 rem and won the Order of the Red Star. His outfit had reduced his exposure by slightly more than a third, but the gamma field was so powerful that the lead made little practical difference. To the soldiers who followed him, speed remained the best protection.

To prepare his troops for the battlefield, Tarakanov built a full-scale mockup of the rooftops: a new postapocalyptic training ground, this time drawn from life, modeled on aerial photographs of the plant, and scattered with dummy graphite blocks, fuel assemblies, and pieces of zirconium tubing. He issued them crude and hastily manufactured equipment, including shovels, rakes, and wooden stretchers to carry off large fragments of wreckage. They were told to use long-handled tongs to pick up pieces of nuclear fuel and given sledgehammers to break loose the blocks of debris stuck fast in melted bitumen. Tarakanov gathered the men in the room near the roof and used closed-circuit TV images fed from cameras above to brief them on their tasks. To every new detachment, he gave the same speech: "I'm asking any one of you who doesn't feel up to it or feels sick to leave the team!" Many were young, and reluctant. But if they didn't do it, who else would?

Years afterward, the general would insist that no man ever broke ranks.

Out on the roof, the men stumbled and ran, weighed down by their clumsy armor, their lead-lined boots slipping and skidding over the slick graphite rubble. They jogged along ramps and scrambled up ladders and paused to catch their breath in the gamma shadow of the vent stack. They scooped up a few fragments of radioactive waste, picked their way to the edge, and flung it out into the sky above what remained of Reactor Number Four. Each man's task was timed with a stopwatch to keep his estimated dose under the regulation 25 rem. Three minutes, two minutes, forty seconds—it was over quickly, marked by the wail of an electric siren or the clanging of a bell. They were supposed to go out only once, but some men returned to the rooftops again and again. Their eyes hurt, and their mouths filled with the taste of metal; they couldn't feel their teeth. In Area M, former combat photographer Igor Kostin was overcome by a mystical sensation, as if exploring another world. The radiation was so intense that afterward it became visible on film, seeping into Kostin's cameras, rising through the sprockets, leaving ghostly traces at the foot of his pictures, like high-water marks after a flood.

When the men came down, they felt as if their blood had been sucked dry by vampires. They curled up and couldn't move. Every soldier's work was recorded in a ledger by specialists from Obninsk, the attrition itemized with a grocer's precision:

Dudin N. S.—Threw down seven zirconium pipes weighing up to 30 kilos.

Barsov I. M.—Removed two pipes of diameter 80 mm, length 30–40 cm . . . 10 zirconium pipes . . . weighing 25 kilos.

Bychkov V. S.—With a sledgehammer, smashed a block of graphite baked into the bitumen.

Kazmin N. D.—Threw down pieces of graphite, up to 200 kilos.

For twelve days, Tarakanov's army of bio-robots relayed onto the roofs from eight in the morning until eight at night: 3,828 men in all, each of them eventually given a printed certificate and a small cash bonus, admitted for decontamination, and sent home. On October 1 the general declared the operation complete. At a quarter to five that afternoon, following months of repairs, modification, and safety tests, the reactor of Unit One came back online at last. For the first time in five months, the Chernobyl nuclear power plant was once again generating electricity.

On the roof of Unit Three, Tarakanov and the scientists overseeing the clearance staged a small ceremony to celebrate their success. They watched as a trio of radiation scouts wearing blue sneakers and canvas overalls jogged away across the empty expanse of Area M and climbed the ladder rising up the side of the giant vent stack. When they reached the top, 150 meters above the ground, the men lashed a flag to the guardrail and unfurled it in the breeze. From the open door of a helicopter hovering overhead, Igor Kostin captured the image: the red banner stiffening in the wind, a stirring symbol of man's triumph over radiation.

Eight days later, Tarakanov was climbing into his car outside the plant when he collapsed. After almost two weeks inside his command post watching his troops' progress on closed-circuit TV, and in repeated visits to the rooftops, the general himself had collected a dose of 200 rem.

On September 30, news of the completion of the Cascade Wall of the Sarcophagus was splashed on the front page of *Izvestia*. By that time, the third shift of Sredmash construction unit 605 had arrived in the zone, a force of eleven thousand men with orders to storm onward and finish the project. The chief engineer of the shift, Lev Bocharov, had worked for the Ministry of Medium Machine Building for almost thirty years. A cheerful fifty-one-year-old who strode across the Special Zone in a padded jacket and a black beret, Bocharov had won three state prizes and begun his career with one of the most monumental projects in Sredmash history: raising Shevchenko, a city of 150,000 people, beside a uranium mine on a remote desert peninsula in Kazakhstan. With a labor force of ten thousand Gulag prisoners who lived and worked behind barbed wire, Bocharov oversaw the construction of the city's uranium processing facilities, the world's first commercial "breeder" reactor, the planet's largest nuclear-powered desalination plant, and everything necessary to support the men and women who ran them, from cinemas to a toothpaste factory.

Bocharov's task in Chernobyl was the most demanding yet faced by the Sredmash engineers at the site. It was his responsibility to close the steel coffin around Unit Four—by roofing over the wrecked central hall and completing a thick concrete wall between Units Three and Four. This would isolate the ruined section of the building from the rest of the plant and allow the remaining reactors to resume normal operation. But the project had already fallen behind schedule, and the revised completion dates were as absurd as ever.

By now, Unit Four was no longer a recognizable part of a nuclear power plant, its shattered facade enveloped by sheer walls of burgundy-painted, mortar-streaked steel, approached by a muddy rampart and ministered by insect-like concrete pumps and Demag cranes. Above the central hall and the exposed reactor core, radiation levels remained so high that sending in riveters or welders was impossible. So the steel parts of the Sarcophagus were preassembled into the heaviest sections the cranes could lift and designed to be held in place by gravity alone—a massive steel house of cards. Colossal and ungainly, each piece was given a nickname by the engineers, according to its shape or size: the Cap, the Skirt, the Octopus, the Dog House, the Airplane, the Hockey Sticks, and, finally, the Mammoth, a single beam seventy meters long and weighing almost 180 tonnes, so large it had to be delivered to the site on specially manufactured trailers crawling along at four kilometers per hour.

Bocharov and his engineers set up their headquarters directly in front of Unit Four, in a building with meter-thick concrete walls, which before the accident had been intended as a storage facility for liquid radioactive waste. In the inside-out world of the zone, it was now one of the least contaminated places in the station complex, and it was here that Commission Chairman Boris Scherbina attended daily briefings on the project, forwarding updates to Gorbachev every twenty-four hours. Efim Slavsky, the octogenarian head of Sredmash, was a constant visitor. From inside the makeshift bunker, the engineers oversaw the assembly of the Sarcophagus using a network of remote-controlled television cameras. Keeping their eyes on a bank of TV screens displaying feeds from the deadliest parts of the construction site, they shouted instructions—"Up!" "Down!" "Left!" "Right!"—to the crane drivers through walkie-talkies. The Demag operators themselves drove blind, cocooned in their cabins by sheets of lead fifteen centimeters thick, seeing nothing but close-up black-and-white images of their own crane hooks flickering on a small monitor.

Bocharov, too, worked in the dark. Even as final assembly began, he had no blueprints of the Sarcophagus and could not take reliable measurements within the ruins. He worked instead from aerial photographs of Unit Four shot from helicopters or by satellite and looked through binoculars from the lead-lined observation post up on mark +67 in Unit Three. When it finally became impossible to proceed without surveying the scene in person, the technicians of NIKIMT produced yet another creative solution. The *batiskaf*—bathyscaphe—was a twenty-tonne lead cabin, with a single porthole of leaded glass thirty centimeters thick, which dangled on a five-meter cable from the hook of a Demag crane. With enough space inside to accommodate four men, the bathyscaphe, lifted a hundred meters in the air, could be "flown" by crane over Unit Four and allowed the engineers to descend into even the most radioactive areas of the site in relative safety.

The chief designer's plan for covering the reactor was simple but risky. He had proposed a roof built from twenty-seven massive steel pipes, laid side by side over beams supported by what remained of the walls of the reactor building and covered with concrete. But radiation made it impossible to assess how badly damaged these supporting walls were or estimate their ability to bear the weight of the new roof. If they collapsed, the physicists feared it could cause a new explosion.

When they hoisted the Airplane roof beam into position, it was so heavy that one of the main cables on the Demag crane snapped, parting abruptly with the crack of a cannon shot. According to Bocharov, the crane operator, fearing a fatal collapse, leapt from his lead-lined cab and fled in terror. It was another twenty-four hours before the cable could be replaced and a new driver found to winch the piece into position.

And when Bocharov took Boris Scherbina up to the observation post on mark +67 to show him the intended foundation site of the largest and most important beam—the 180-tonne Mammoth, designed to support the roof covering the entire southern side of the Sarcophagus—the chairman was horrified. All he could see to rest the beam upon was a tangled mess of wreckage, including not just broken concrete and tangled pipework but also the remains of office furniture jutting from the rubble. "Are you crazy?" he asked Bocharov. "It's impossible! Find another way."

But there was no other way. The completion of the entire structure now rested on the successful installation of this single, massive piece of

steel. If he couldn't make it work, they would have to start the Sarcophagus again from the beginning. Bocharov decided to scout the foundation
site in person, on foot.

By late autumn, tens of thousands of middle-aged *partizans* had been
drafted from across the Soviet Union and put to work in the high-
radiation areas of the zone until they reached their 25 rem limit. Afterward, they were decontaminated and demobilized and told to sign a
pledge of secrecy before being sent back to where they had come from,
clutching a small cardboard booklet: the official record of their total accumulated dose. Few regarded this document as accurate. Before leaving,
some were presented with awards for distinguished service and given a
choice of reward: cassette player or watch? Many returned to their homes
and sought to purge the radiation from their bodies with vodka. Regardless of the triumphant headlines in *Pravda* and *Izvestia*, the bitter truth
about the conditions they faced spread gradually through towns and cities across the USSR. As a result, when reservists received the draft notice
summoning them for "special training," they increasingly knew what it
meant. Some bribed the draft officer to stay at home: while a deferment
from the war in Afghanistan could reportedly be bought for 1,000 rubles,
escaping from duty in Chernobyl cost only half as much. And inside some
tented encampments on the perimeter of the zone, commanders faced
mutiny from their troops. One group of two hundred Estonian *partizans*,
told that their tour was being extended from two to six months, gathered
in a furious mob and refused to go back to work. Military police patrols
in Kiev picked up senior officers who had deserted their men, drunkenly
attempting to flee the city by train.

But there were still many who volunteered to work in Chernobyl,
attracted by word of the high wages, paid as a bonus for service in the
high-radiation zone, or who were motivated by scientific curiosity—or
the chance to sacrifice themselves for the motherland, as their fathers and
grandfathers had done in the Great Patriotic War.

Vladimir Usatenko was thirty-six years old when he was drafted on
October 17, one of eighty *partizans* flown from Kharkov to Kiev aboard

an Ilyushin-76 transport plane and driven in a fleet of trucks to a tent encampment near the power plant. An engineer who had performed his national service as a radio operator in the Soviet missile defense forces, he could have bribed his way out but chose not to do so. Inside the zone, he found total chaos: there were uniformed soldiers everywhere, scurrying to their tasks like green ants, but the senior officers seemed to have little idea what was going on. Gangs of troops lollygagged in high-radiation areas, awaiting instructions or watching while others worked, apparently ignorant of their mounting doses.

Usatenko assumed command of a platoon of men, and the noncommissioned officers who had been there a while warned him to look after himself: pay no attention to the commanders, and save your lads from the worst of the radiation. Almost immediately, they were assigned to work for Sredmash US-605 inside the machine hall, beneath the rising walls of the Sarcophagus. Usatenko took eight men up to level +24.5, where the concrete barrier between Units Three and Four was still under construction, and spent an hour nailing wooden boards along the wall. Everything they did was supposed to be secret, and the soldiers never learned the purpose of their work: here, a stack of boards; there, hammers; here, nails. Get to it. The jobs varied, but in the most important ways, they were all the same: backbreaking, manual, inexplicable. They hauled forty-liter sacks filled with water into the basement of the building as part of a chain of men mixing concrete by hand, threw abandoned fire hoses from the roof, and collected debris from beneath the bubbler pool—told to simply pick up whatever they could find and bring it out quickly.

It was dark and humid inside the Sarcophagus, and Usatenko's biggest fear was of losing his men somewhere inside the benighted labyrinth. But there was heavy radiation everywhere, and, in certain rooms, they could feel it popping against their eyeballs, like an invisible spray; in others, they found that Sredmash had installed speakers relaying a constant, low-frequency roar—an aural warning not to linger. Elsewhere, the construction specialists from US-605 slung garlands of 36-volt lamps along the walls and watched the *partizans'* progress through TV cameras from their lead-lined observation booths. At last, when given the command to work inside a room right beside the reactor—where, after a single minute, their dosimeters reached their maximum readings—Usatenko and his men rebelled. They went toward the room, but then pushed over the camera monitoring the entrance and hid in safety until the time allotted for the task was up. It took the US-605 technicians ten days to install a new video camera. By then, Usatenko and his men were gone.

Vladimir Usatenko would eventually complete twenty-eight missions inside Units Three and Four and spend a total of forty-four days inside the zone. But he found no great patriots there. Everyone he talked to wanted only to catch his regulation 25 rem and get home as soon as possible.

Led by a physicist from the Kurchatov Institute who knew the route and accompanied by a cameraman weighed down by cumbersome video equipment, Lev Bocharov, the chief engineer of the final shift of US-605, made his way into the ruins of Unit Four, toward the foundation site of the Mammoth beam. The men mounted a staircase, wrenched away from the wall by the explosion and now hanging in space at a fun-house angle. At mark +24, they turned down a darkened corridor and started to run. But the farther they went, the lower the ceiling became: they realized slowly that the hallway had been filled with errant Sredmash concrete. By the time they reached the end of the black passageway, Bocharov and his team had to crouch and then wriggled through a space forty centimeters high, each holding the legs of the man in front. On level +39, they finally saw daylight: an exit hole close to the place where the Mammoth would rest. Leaving the others behind, Bocharov sprinted out across the rubble. Three minutes later, he returned, with a large dose of radiation—and a plan.

Using a Demag crane, the bathyscaphe, a team of sixty *partizans* hand-picked for their speed and fitness, and a supply of fishing net flown in overnight from the Arctic port of Murmansk, Bocharov improvised a concrete platform, poured on top of the debris at level +51. A series of hasty loading tests assured the engineers that this foundation would be strong enough to bear the weight of the Mammoth. At ten in the evening on the first of November, the massive beam was finally lowered into place. For the first time since the liquidation began, Efim Slavsky was seen to smile.

After that, work moved quickly: with the poisonous maw of the reactor finally covered, the Sredmash teams installed a ventilation system to stabilize the atmosphere inside the Sarcophagus and connected a network of radiation- and temperature-monitoring devices to a freshly decontaminated room nearby filled with computer equipment. There was still no sign of the missing 180 tonnes of uranium from the reactor core, and Academician Legasov and the other scientists remained concerned about the possibility of a new chain reaction. So inside the new structure the Sredmash engineers also installed a sprinkler system, supervised by the Kurchatov Institute, designed to spray the ruins with neutron-absorbing boron carbide solution and coat everything with a film to suppress any new criticality as soon as it started. Finally, the roof and windows of the Unit Four machine hall were layered with steel plates, and the western end of the reactor hall was shored up with a row of ten massive steel buttresses, each forty-five meters high.

When Slavsky arrived to survey the project once more, on November 13, the Sarcophagus was all but complete—a terrible edifice of black angles, still and ominous, which perfectly expressed its purpose, like a medieval fantasy of a prison to hold Satan himself. It was an extraordinary achievement, a technical triumph in the face of horrifying conditions, and a new pinnacle of Soviet gigantomania: the engineers boasted that the structure contained 440,000 cubic meters of concrete, 600,000 cubic meters of gravel, and 7,700 tonnes of metal. The costs had risen to more than 1 million rubles—or $1.5 million—a day. As he gazed up at his masterpiece, a cathedral of brutalism in concrete and steel, it was said that tears welled in the old man's eyes.

It would be Slavsky's final achievement as the leader of the sprawling Sredmash empire. A week later, Prime Minister Ryzhkov summoned him to his office in the Kremlin and asked for his resignation. Slavsky scribbled a single sentence in his distinctive blue pencil: "I've become deaf in my left ear, so please dismiss me," a truculent parting shot that made clear his feelings about being forced to step down when he felt he had so much more to give. Slavsky was eighty-eight years old and six months short of celebrating thirty years at the head of Sredmash. When news of his departure reached the headquarters of the Ministry of Medium Machine Building on Bolshaya Ordynka Street in Moscow, his staff wept with grief.

The document formally commissioning the Sarcophagus received its final signature on November 30, 1986, just seven months and four days after the first explosions tore through Reactor Number Four. On December 3, Lev Bocharov completed his tour in the Special Zone. Winter had come to Ukraine, and the first snow would soon fall on the Sarcophagus. He arrived at the railway station in Kiev muffled in the cold-weather jacket and striped undershirt issued to the troops in Afghanistan. With a handful of colleagues, Bocharov boarded the overnight train to Moscow carrying a large cardboard box filled with bottles of vodka. On the way home, they all had a drink.

When the train pulled into Moscow early the following morning, Bocharov thought they would be greeted at the station like conquering heroes, but there were no crowds on the platform to meet them. He saw only his wife, a friend who had driven her in his car to collect him—and a soldier back from the Afghan quagmire, who recognized the engineer's fur-collared camouflage.

"Kandahar?" the soldier asked.

"Chernobyl," Bocharov said.

The soldier put an arm around his shoulder. "Brother, you had a tougher job."

17

The Forbidden Zone

By the beginning of August 1986, the number of graves in the special section of the clean new cemetery near the village of Mitino in the suburbs of Moscow had risen to twenty-five. They stood in two rows, fifty meters from the yellow-tiled crematorium at the entrance, with space for more. Some had white marble headstones, with inscriptions lettered in gold and adorned with a Soviet star; others, so fresh that they were little more than dirt mounds, were scattered with flowers and marked with pieces of cardboard. Crows circled overhead. When inquisitive Western reporters made the trip out to the graveyard and tried to record the names of the dead, police officers confiscated their notebooks and silently led them away.

In September, Dr. Angelina Guskova announced that a total of thirty-one men and women were now dead as a direct result of the explosion and fire in Unit Four of the Chernobyl nuclear power plant. This number would henceforth be regarded as the official death toll of the accident. Anything higher was treated as evidence of bourgeois Western propaganda. The body of pump operator Valery Khodemchuk, killed immediately by the blast or by falling debris, remained buried beneath the wreckage of the reactor hall; his colleague Vladimir Shashenok, who had died as a result of physical trauma and thermal burns a few hours later in the Pripyat hospital, had been laid to rest in the graveyard of a small village near the power station. Since then, twenty-nine more victims—operators, firemen, and security staff—had succumbed to the effects of acute radiation syndrome in the radiology wards of Kiev and the specialized clinic in Moscow. Of the thirteen patients who had been treated with bone marrow transplants by Robert Gale and the Soviet specialists, all but one had died—so many that Guskova would eventually dismiss the technique as useless for managing ARS. Yet many of those who had sustained terrible injuries in the first hours of the disaster, after months of agonizing treatment in Hospital Number Six, had at last begun to recover.

Deputy Chief Engineer Anatoly Dyatlov, who had insisted on proceeding with the fateful turbine experiment over the objections of his subordinates and then spent hours wandering in disbelief through the wreckage of Unit Four, had sustained ghastly beta radiation burns on his lower legs and absorbed a total dose of 550 rem but was released from the hospital at the start of November. He returned to Kiev and soon after was arrested and placed in pretrial detention. Major Leonid Telyatnikov, who had commanded the Chernobyl plant's fire brigade on the night of the accident, had not been told about the deaths of his men until July, when he was released from isolation to walk the hallways of the hospital unaided, wearing a gauze mask to protect him from infection. In August, he was discharged and sent to recuperate at a resort on the Latvian coast with his wife and two children, but told to avoid too much sunshine—and fatty foods, because of radiation damage to his liver. The following month, he had recovered sufficiently to visit his parents in Kazakhstan.

The doctors regarded the survival of some of the most badly exposed operators as almost miraculous. One electrical engineer, Andrei Tormozin, had been only 120 meters from the reactor when it exploded, and then spent hours in highly radioactive areas of the machine hall, working to stop feed pumps and extinguish oil fires. He had absorbed what Guskova and the other specialists had always understood to be a mortal dose of gamma and beta radiation: almost 1,000 rem. His body rejected a bone marrow transplant; he contracted blood poisoning and radiation-induced hepatitis and was not expected to live. But at the end of May, his blood counts began to rally, and—for reasons the doctors could never fully explain—he eventually made a complete recovery.

Alexander Yuvchenko, who had listened as the machines sustaining his friends in adjacent rooms fell silent one by one, had himself lingered close to death throughout the month of May. For weeks, his wife, Natalia, woke each morning in a nearby hostel, fearing what might have happened overnight, and asked her mother to phone the hospital. Superstitious, Yuvchenko hoped that if she didn't call the doctors herself, the news about her husband's condition would be better. When his bone marrow function collapsed, the physicians kept him going with blood transfusions, and Natalia scoured the city for scarce and expensive ingredients to keep up his strength. She brought black caviar sandwiches to his bedside; his friend Sasha Korol came to visit and insisted he try ketchup instead. But Yuvchenko proved unable to eat anything, and he was placed on an IV.

It was June before Yuvchenko's bone marrow began functioning again, the first white cells reappeared in his bloodstream, and it seemed certain he would live. But it also seemed possible that the radiation burns—especially the ones on his arm and shoulder—would never heal fully, and the surgeons had to cut away repeatedly at both skin and muscle to remove the rotting black tissue from his shoulder blade. The agonizing open wounds left where beta particles had eaten into the flesh of his elbow made it unlikely he would ever be able to live a normal life again.

But in the second half of September, the doctors allowed Yuvchenko to go home for a short time to the new apartment his family had been granted by the government, in a well-heeled neighborhood near Moscow State University. He looked gaunt and skinny and had become addicted to the narcotics the doctors had used to stifle the terrible pain of his burns. While the doctors wanted to wean Yuvchenko off the painkillers, they also had to encourage him to learn how to live for himself after weeks of around-the-clock care. But the radiation was far from finished with him. New burns continued to reveal themselves on his legs and arms even months after the explosion, and he was admitted once more to Hospital Number Six for further treatment.

While the surviving victims of ARS lay in their beds in Moscow, the evacuees from Pripyat remained in limbo, not knowing if or when they might return to their homes in the abandoned *atomgrad*. Just outside the Exclusion Zone, in the town of Polesskoye, thousands of the displaced citizens, without clean clothes or money, outfitted themselves in whatever they could, including bathrobes and the white overalls of nuclear plant workers. Their belief in the power of vodka to protect the body against radiation led them to break down the doors of the settlement's liquor store, and *samogon* was soon changing hands for up to 35 rubles a liter—the price of a good cognac in Kiev. In the meantime, the state struggled to provide them with new jobs, and schools for their children. In May the Soviet Red Cross Society contributed a one-off payment of 50 rubles per person to every refugee from the catastrophe. Later that month, the Soviet government provided a further lump sum of 200 rubles for every member of each family of displaced people. Fifteen cashiers distributed the millions of rubles this required, the cash brought in bags from the bank to a municipal office in Polesskoye each morning, under the eyes of *militsia* officers armed with machine guns. And still, throughout June and July, the people returned to the offices of their city-council-in-exile on Sovietskaya Street in the town of Chernobyl to ask: "When can I go home?"

On July 25 they received an answer: that morning, the first busloads of Pripyat evacuees set out to return to their city—but only as part of an official program to reclaim what they could from their apartments and seek compensation for what they could not. Arriving at a checkpoint on the perimeter of the thirty-kilometer Exclusion Zone, they were issued cotton overalls, shoe covers, petal respirators, and thick polyethylene bags. After a document check at the entrance to Pripyat, they were permitted to spend three or four hours in their abandoned apartments and walking the streets of the city, where yellow sand banked against the curbs and grass was already sprouting through cracks in the chalky asphalt. Sixty-nine men and women stepped off the buses that first morning, and hundreds more returned every day for months after that, to salvage what they could from their former homes.

The refugees were allowed to reclaim only property within strictly defined categories. Large pieces of furniture and any object that gathered a great deal of dust—including rugs and TV sets—were prohibited, as were all children's belongings and toys and anything at all that registered radiation readings higher than 0.1 milliroentgen per hour. Both electricity and water supplies to the apartment buildings had been disconnected, and the tart smell of cigarette smoke and human sweat that had once lingered in the stairwells and hallways was already gone. Despite *militsia* patrols and the alarm system fitted at the entrance of each building, many found that their apartments had been looted. Their fridges were filled with the putrefying remains of the food bought in anticipation of the May Day feast at the beginning of the long, hot summer. Some found it hard to hold back tears as they considered their abandoned belongings, standing in musty rooms they now realized they might never see again.

It was September when Natalia Yuvchenko returned to the two-room flat she and Alexander had shared with their son on Stroiteley Prospekt. She found Kirill's stroller, broken and lying outside the entrance to their stairwell, and went upstairs afraid of what she might find. But when she reached the apartment, everything was just as she had left it: the first thing she saw was the forgotten carton of milk that Sasha Korol had brought for the boy on the morning of the evacuation, still resting on the saddle of Alexander's bicycle. She didn't take much but gathered a handful of slides and photographs, including the one of her and Alexander posing in hats on his birthday the year before, and the comic verses that her neighbor had written that night to mark the occasion. Other residents retrieved seemingly random possessions—a plastic bag of science-fiction novels, a handful of flatware—in a hurried tussle of utility and sentiment. Each visitor was allowed a maximum of four hours inside for deciding what to rescue from their previous life, before climbing back aboard the bus to leave. Valentina Brukhanov, by now living at the riverbank Zeleny Mys settlement and working double shifts in her job at the plant while her husband sat in a KGB holding cell in Kiev, recovered her most treasured possessions: a pair of crystal glasses they had been given for their twenty-fifth wedding anniversary; a family portrait taken when their son was small; a coveted sheepskin coat, which she eventually gave to a neighbor; and a few books, which she wiped down with vinegar in the belief that it would help neutralize radiation.

It was often late at night by the time each group of visitors returned to the dosimetric checkpoint on the perimeter of the zone, where their belongings were checked by teenaged nuclear engineering students from the MEPhI campus in Obninsk. They manned the barriers in all weathers, waving the wands of their radiometers over boxes of china, tape recorders, books, cameras, clothing, and bric-a-brac. When their belongings proved too contaminated to pass, some people tried to bribe their way through with cash or the other currency in common use throughout the forbidden zone—vodka. The young students were astonished to discover that even former workers from the Chernobyl plant were ignorant of the dangers of radioactive dust, and surprised by furtive strangers who materialized from the shadows to offer cases of alcohol in exchange for a few moments riffling through the piles of confiscated belongings, which they planned to sell in secondhand markets outside the zone.

The visits to the deserted city continued for exactly four months and ceased on October 25, 1986. By that time, 29,496 people had returned to their apartments in Pripyat. Some had been back more than once, but others hadn't been at all, and their belongings remained unclaimed. The town council planned a program of further trips for the fall of the following year, but the government commission refused permission for these to proceed. A state decree provided for compensation for lost property to be paid to the displaced: a flat payment of 4,000 rubles for a single person and 7,000 for a family of two. At the time, a new car—for anyone lucky enough to find one—cost 5,000 rubles. The *ispolkom* received hundreds of applications for compensation every day throughout the summer, and by the end of the year, the claims for the domestic property lost by the residents of Pripyat to the ravages of the peaceful atom—and excluding cars, garages, dachas, and motorboats—had reached a total of 130 million rubles. That autumn, the furniture stores of Kiev experienced a prolonged boom in trade, as evacuees attempted to rebuild their lives, beginning with a hollow quest to replace almost every major possession they had ever owned.

At first, the plight of those who had been banished from their homes by the fallout from Unit Four had elicited widespread sympathy across the USSR. At the end of April, the government established a relief fund at the state bank—named, with traditional Soviet austerity, Account No. 904—into which well-wishers could deposit donations to help support the victims. In May a charity rock concert—the first ever in the Soviet Union—was held in the Olympic Stadium in Moscow before an audience of thirty thousand people, with a live TV link to a studio in Kiev, where miners, plant operators, and other liquidators gathered, and firefighters recited the names of their comrades who had died in the wards of Hospital Number Six. In early August, the chairman of the state bank reported that Account No. 904 had already received donations of almost 500 million rubles, sent in by individuals and collectives and drawn from wages, pensions, and bonuses, as well as foreign-currency transfers from abroad.

But the permanent resettlement of 116,000 people—the specialists and their families evacuated from Pripyat, the residents of Chernobyl, and the farmers from the dozens of small settlements that now fell within the thirty-kilometer Exclusion Zone, all of whom needed new jobs, schools, and homes—was more complicated. In June the Politburo passed a resolution making the fate of the evacuees a political priority and called on the republican governments of Ukraine and Belarus to build tens of thousands of new apartments before winter arrived. In Ukraine, fifty thousand men and women arrived from all over the republic, and shock-work construction began at once. The first settlement, 150 brick houses near the enormous Gorky collective farm, a little more than a hundred kilometers south of Chernobyl, was unveiled at an elaborate ceremony in August. Each home was reportedly provided with furniture, gas cylinders, electric light, towels and linens, and a concrete cellar stocked with potatoes. In total, the Ukrainian republic alone undertook the building of 11,500 new single-family houses, with the intention of completing all of them by October 1.

But the Politburo task force in Moscow also requisitioned an additional 13,000 newly completed apartments in Kiev and other cities across Ukraine—snatching them from under the noses of families who had spent years on waiting lists—and handed the keys to evacuees from Pripyat. Specialists from the Chernobyl plant and their families were transferred to the remaining three Ukrainian nuclear stations in Konstantinovka, Zaporizhia and Rovno, where they were given new jobs and moved into brand-new flats. When they arrived, they were not all welcomed warmly by their colleagues, who saw no justice in having to give up their hard-won places in line to fellow nuclear specialists apparently chased out of their homes by the consequences of their own incompetence. In Kiev, several large apartment construction projects that had been due for completion by winter—and would have been ripe for requisition for the evacuees—mysteriously came to a halt. In the end, many of the former residents of Pripyat were found homes in the same sprawling complex of high-rise buildings in Troiesh-chyna, a remote and isolated suburb on the northeastern edge of the city.

There, they were shunned by their new neighbors, who both resented the refugees and feared the invisible contagion of radioactivity. At school, other children were forbidden by their parents from sharing desks with pupils evacuated from Pripyat—and not without good reason. The radiation readings in the stairwells and hallways of the new apartment blocks in Troieshchyna were soon found to be hundreds of times higher than elsewhere in Kiev.

Back in Chernobyl, the government commission remained zealously committed to overcoming the handicaps of operating a power station in the heart of the radioactive zone. With the first reactor back online at the beginning of October, the new station director announced plans to have the second generating electricity imminently. Unit Three remained so contaminated that the plant's chief engineer and specialists from the Kurchatov Institute all advised that it would be too expensive—and cost too many operators their health—to recommission it. But their objections were overruled, and the third Chernobyl reactor was scheduled for reconnection to the grid in the second quarter of 1987. The commission even issued orders to resume the construction work on Reactors Five and Six, which—although it was close to completion—had been stopped dead on the night of the accident.

In the meantime, *Pravda* reported ambitious plans to build another *atomgrad* to accommodate the workers who would operate the resurrected Chernobyl plant and their families. This was to be a new city of the future, fit for the twenty-first century, located forty-five kilometers northeast of Pripyat in the middle of the forest on a remote bank of the Dnieper. Named Slavutych, the city would be filled with modern conveniences, and particular care would be taken to integrate it into the natural environment. Planned around a central market square, it would have a statue of Lenin and, nearby, a museum dedicated to the heroes of Chernobyl.

In Moscow, the propaganda narrative of the disaster had now crystallized around the gallant sacrifice of the firefighters from the Chernobyl station and Pripyat brigades and their commander, Major Telyatnikov. In September, a photograph of Telyatnikov—still bald from the effects of radiation sickness—appeared on the front page of *Izvestia* under the headline "Thank You, Heroes of Chernobyl," and the state media announced that he and General Antoshkin, commander of the helicopter troops, had been granted the state's highest award, Hero of the Soviet Union. The two young lieutenants who had led their men to the roofs of the reactor buildings to pour water on the fragments of fuel assemblies and chunks of blazing graphite, Vladimir Pravik and Viktor Kibenok, were given the award posthumously. The leaders of Sredmash US-605, who had built the Sarcophagus, were made Heroes of Socialist Labor. Once his auburn hair grew back, Telyatnikov was sent abroad, where he was greeted as a celebrity. He was presented with awards by fellow firefighters in the United States and Britain, interviewed by *People* magazine, and, in London, granted an audience with Prime Minister Margaret Thatcher.

At a televised award ceremony in January the next year, the grizzled apparatchik Andrei Gromyko, the ceremonial leader of the USSR, delivered a speech in which he lionized the firemen, the liquidators from the armed forces, and the Sredmash construction chiefs who had entombed the smoldering reactor in sand and concrete. "Tens upon tens of millions of people around the world, near and far, followed your shock-work with hope," he said. "This feat is a mass feat, a feat of the whole people. . . . Yes, Chernobyl was pain we shared together. But it has become a symbol of the victory of Soviet man over the elements. . . . At the same time, our Party renders due tribute to each individual. There are no nameless heroes. Each of them has his own face, his own character, his own specific exploit."

Yet some heroes would prove more equal than others. There was still no public acknowledgment for the engineers and operators of the Chernobyl station who had put out the fires and prevented further explosions inside the turbine hall, or for those who had toiled in vain amid lethal fields of gamma radiation to cool the doomed reactor. The few awards granted to plant workers were processed in total secrecy. At one point, Anatoly Dobrynin, the Central Committee secretary in charge of foreign affairs, came to visit the injured operators in the wards of Hospital Number Six, but the trip went unreported in the press. Instead of recognition of their loved ones' heroism, the families of Alexander Akimov and Leonid Toptunov were notified that, under Article 6, paragraph 8, of the Ukrainian Criminal Code, the accused would escape criminal prosecution for their crimes only on account of their recent deaths.

Throughout the winter of 1986, disgraced Chernobyl plant director Viktor Brukhanov remained in his KGB jail in Kiev, awaiting his impending trial. He was permitted no visitors, but once a month his wife, Valentina, could bring him a five-kilogram parcel of food, which she packed with sausages, cheese, and butter. Occasionally Brukhanov had a cellmate—a counterfeiter or a burglar—but mostly he spent endless weeks in solitude and passed the time reading books from the prison library and learning English. For a while, Valentina was allowed to bring him English-language newspapers, until their son wrote a furtive four-word message to his father inside one of them: "I love you, Daddy." Then that privilege, too, was revoked.

At first, Brukhanov refused to retain a lawyer to conduct his defense in court, because he understood that the final verdict of the trial had been decided long ago. But his wife persuaded him otherwise, and in December Valentina traveled to Moscow, where she found an attorney willing to represent Brukhanov—a specialist permitted to act in cases involving the closed facilities of the Soviet nuclear complex, with the top-secret clearance necessary to view evidence gathered behind the Ministry of Medium Machine Building's wall of silence. That same month, as part of the discovery procedures dictated by Soviet law, the investigators brought Brukhanov the materials they had uncovered during the course of their inquiries, which would be used in the case against him. Buried among the files, the director found a letter written by one of the Kurchatov Institute experts, which revealed to him the existence of the secret history of the RBMK reactor—the trail of perilous design faults that the scientists had known about all along but kept hidden from Brukhanov and his staff for sixteen years.

On January 20, 1987, after Brukhanov had sat in isolation for six weeks poring over the details of his case, the investigators from the prosecutor's office filed their closing indictment with the Supreme Court of the USSR. They sent a total of forty-eight files of evidence to Moscow, all of which were classified top secret. Fifteen, containing documents taken directly from the plant, were so contaminated with radioactive dust that the lawyers had to wear protective clothing to read them.

Brukhanov—along with four other senior members of the plant staff, including Dyatlov and Fomin—was charged formally under Article 220, paragraph 2, of the Ukrainian Criminal Code, under which he was accused of "a breach of safety regulations" resulting in loss of life or other serious consequences at "explosion-prone plants or facilities." This was an inventive legal gambit—never before had Soviet jurists considered a nuclear power station an installation likely to explode—and the first of many logical contortions necessary to confine responsibility for the accident to the handful of selected scapegoats. To bolster the case, Brukhanov and Fomin were also charged under Article 165 of the Criminal Code with abuse of power. Brukhanov was accused of deliberately underreporting the radiation readings at the station on the morning of the accident, thereby delaying the evacuation of the station and of Pripyat, and knowingly sending men into the most dangerously contaminated areas of the reactor building. If found guilty, the three most senior members of the plant staff each faced a maximum of ten years in prison.

Scheduled to begin on March 18, 1987, the trial was delayed when it became obvious that Deputy Chief Engineer Nikolai Fomin remained too mentally unstable to take the stand. Placed under arrest at the same time as Brukhanov, he had attempted suicide in prison, shattering his spectacles and using the fragments to slash his wrists. While the wretched technician was sent to the hospital to recuperate, the trial was postponed until later in the year.

Maria Protsenko returned to Pripyat for almost the last time in January 1987. Bundled against the cold in a padded cotton jacket and pants, thick felt *valenki* on her feet, she led a small squad of soldiers up the snow-covered steps of the White House. They went from room to room in the deserted city council building, cleaning out every cupboard and safe, filling sacks with documents too contaminated to be consigned to the archives of the bottomless Soviet bureaucracy but too sensitive to leave behind. When they had finished, Protsenko collected the key to every office door in the building, while the soldiers threw the sacks into the back of a truck. They then drove them away to be buried in one of the growing patchwork of eight hundred radioactive waste dumps spreading across the Exclusion Zone.

On April 18 that year, local council elections were held for the new *atomgrad* in Slavutych, the Pripyat administration was formally dissolved, and the city bureaucratically ceased to exist. After working for almost a full year inside the Exclusion Zone with barely a day off, Protsenko was now transferred to a new job in Kiev. In recognition of everything she had endured in the long months since the accident, she was finally permitted to apply for membership in the Communist Party. At the end of the year, Protsenko checked in to a hospital in Kiev and stayed there for more than a month, suffering the symptoms of what the doctors described as "nervous tension" related to overwork. They marked her medical records with the diagnosis "Ordinary illness: not related to ionizing radiation." Back in Pripyat, the empty *ispolkom* building became the headquarters of the Kombinat Industrial Association, a new state enterprise established to manage the long-term research and liquidation efforts within the thirty-kilometer zone. Now sole masters of the empty town, the new authority reopened Pripyat's main swimming pool to give the liquidators somewhere to relax and established an experimental farm in the city's hothouses, where green-thumbed technicians grew strawberries and cucumbers in irradiated soil.

As the cleanup inside the thirty-kilometer zone ground on, morale among the tens of thousands of liquidators who had been shipped in to perform the dangerous and apparently inexorable task sank ever lower. Dust from highly contaminated areas continued to blow into those that had been cleaned, rendering weeks of work pointless; Kombinat appeared to have been making progress in Pripyat until the KGB learned that their specialists had been reporting readings from only the cleanest areas, and thus underestimating the true levels of radiation in the city by more than ten times. The secret policemen also noted that the liquidators' food was bad, radiation safety was lacking, workers weren't even being paid on time, and one radioactive waste dump was being flooded regularly with river water. The Kombinat leaders would eventually be reprimanded by the Party for tolerating nepotism, theft, and drunkenness.

In the meantime, looting from inside the zone had begun on an industrial scale, often initiated by the liquidators themselves and sometimes with the collusion of their commanders. One night, radiation reconnaissance officer Alexander Logachev watched in amazement as a renegade group of soldiers loaded one truck after another with gas stoves and building supplies taken from a heavily radioactive construction warehouse, in the shadow of the plant. "Guys, are you fucking crazy?" he asked, but they carried on regardless, and, by dawn, two massive Antonov 22 heavy lift transport aircraft were on their way to the Siberian military district filled with toxic contraband. Lieutenant Logachev himself joined the pilfering soon enough—although he remained sufficiently professional to decontaminate his stolen goods before taking them beyond the perimeter of the zone.

In Pripyat, the cars and motorcycles left behind by the fleeing population—more than a thousand of them in all, corralled in the center of town—fell victim to scavengers, who stole their windshields and destroyed their bodywork. Some cars had been requisitioned to provide transport for the scientists and technicians within the zone, organized into an improvised motor pool of brightly colored Ladas, Zhigulis, and Moskvitches. The hood and door of each one was painted with a number in a white circle, and details on who was using which one, and where, were noted in a log that Maria Protsenko had carefully kept until her last day on the job. The hundreds of vehicles that remained, too contaminated ever to be returned to their owners, were removed to a radioactive waste zone, where they were crushed, bulldozed into a trench, and buried.

As the first anniversary of the disaster approached in April 1987, in Moscow the members of the Politburo considered a series of propaganda measures designed to show the USSR's handling of the disaster in the best possible light. The proposals included story ideas for use on TV and in the scientific press and publications for foreign distribution. The official Soviet report to the IAEA had included seventy pages of detailed radio-medical information provided by Dr. Angelina Guskova and her colleagues—including the collective radiation dose they expected the seventy-five million people in the western USSR to receive as a result of the accident. But the Soviet report had not forecast the total number of additional deaths or sicknesses the contamination might cause, and Western specialists had filled this gap with estimates that infuriated the Soviet doctors. Robert Gale told the press they could expect to see another seventy-five thousand people die of cancers directly attributable to the effects of the disaster, forty thousand of them within the Soviet Union, and the remainder abroad.

So, in spite of the increasing freedoms afforded to the editors and producers of the Party-controlled media by Gorbachev's glasnost, in this case, the truth would not be allowed to impede a directive to "smash the hostile, prejudiced statements in the Western press." The deputy chief of the USSR State Committee for Radio and Television proposed a list of twenty-six stories to be carried by the TASS agency, including "Birthplace: Chernobyl," which followed three hundred children born to evacuees from the Exclusion Zone, showing that they were under constant medical observation, with no evidence of illness; "What Was the Scent of the Winds of April?," in which the chairman of the Soviet meteorological agency presented data to refute the idea that dangerous radioactive particles fell on Western Europe; and "Palettes of the Spring Marketplace," a report on the spring fruit and vegetables arriving in Kiev, with dosimetry results showing a reassuring absence of radionuclides.

The final plan, approved on April 10, included counterpropaganda measures to be sent to Soviet embassies abroad and a proposal to allow a delegation of foreign journalists to report directly from the Exclusion Zone. Newspaper reporters from the *New York Times* and *Chicago Tribune* finally entered the area in late June to bear witness to the sterile moonscape of concrete and asphalt surrounding the Sarcophagus, the desiccated pines of the Red Forest, and the empty streets of Pripyat.

Here, more than a year after the accident, the streetlights still came on at night, and operatic music sometimes crackled from the speakers mounted along Kurchatov Street. But the bright pennants that twitched in the breeze above the central square were sun-bleached and tattered, and the laundry on the apartment balconies had begun to rot. Yet still the authorities maintained the illusion that the city was not dead but only sleeping and one morning would be awoken by the footsteps of its returning population.

18

The Trial

The trial of Viktor Brukhanov and the other five men who stood accused of causing the disaster at the Chernobyl Atomic Energy Station began on July 7, 1987. Soviet law dictated that a criminal hearing take place within the same district where the alleged crime occurred. But since Pripyat had become a radioactive ghost town, the proceedings were held in the nearest alternative, fourteen kilometers from the plant, in Chernobyl itself. Although subjected to months of decontamination, the city remained at the heart of the thirty-kilometer Exclusion Zone and was accessible only to those issued a government pass. While the trial was nominally open to the public, attendance was therefore restricted to those who worked inside the zone or whom the authorities saw fit to grant entry. Both within the borders of the Soviet Union and beyond, those transfixed by the world's worst nuclear accident awaited justice, but the Party did not want its legal pantomime disrupted by an intrusive audience. A handful of representatives of the international press, including correspondents from BBC radio and Japanese TV, were invited to attend but would be brought in by bus to witness only the opening and closing of the proceedings, in which nothing but scripted statements would be read. The formerly dilapidated Palace of Culture on the corner of Sovietskaya and Karl Marx streets had been refurbished and repainted to provide the venue for the trial: the theater auditorium filled with new chairs, the walls hung with lustrous gray curtains, and a radiation checkpoint installed at the entrance.

At 1:00 p.m., Judge Raimond Brize of the Supreme Court of the USSR took his place on the stage, and the six defendants, escorted by uniformed Internal Ministry troops, stepped into the dock. For two hours, they sat and listened as Brize read the indictment aloud. Collectively, the six men stood accused of negligence in conducting a dangerous and unsanctioned experiment on Reactor Number Four of the Chernobyl Atomic Energy Station, resulting in the total destruction of the unit, the release of radioactive fallout, the evacuation of 116,000 individuals from two separate cities and dozens of villages, and the hospitalization of more than two hundred victims of radiation sickness, of whom at least thirty were already dead. The court also heard that the Chernobyl plant had suffered a long history of accidents that had not been addressed or even reported and that what had been regarded as one of the best and most advanced nuclear installations in the USSR had, in fact, operated constantly on the edge of catastrophe as a result of its lax and incompetent management. There was no mention of any faults in the design of the RBMK-1000 reactor.

All five men accused of breaching the safety regulations of an "explosion-prone facility"—including Boris Rogozhkin, the chief of the night shift at the time of the accident, and Alexander Kovalenko, the head of the workshop who had signed off on the test—entered pleas of not guilty. But Brukhanov and Fomin admitted to criminal failures in their performance of official duties, under Article 165—the lesser offense, which carried a sentence of five years in prison. "I think I'm not guilty of the charges brought against me," Brukhanov told the court. "But as a manager, I was negligent in some ways."

The hearings began at eleven each morning and continued until seven in the evening, with an hour's break for lunch. The brassy summer sun beat on the low roof of the small brick courtroom, thickening the atmosphere inside with heat and tension. Yet Brukhanov remained as composed and impassive as ever. Wearing a suit jacket but no tie, he sat with his head slightly raised, listening attentively as the witnesses and experts took the stand. He gave an account of his actions on the night of the accident yet did little to defend himself. He stood by the safety record of the plant and described the impossible nature of his job: the difficulty of recruiting trained staff and the overwhelming burden of being responsible for every detail of both the plant and the city. And yet he told the court it had been beyond his power to order an evacuation of Pripyat; he had not intended to conceal the true levels of radioactivity. Brukhanov claimed that he had not closely read the critical statement on radiation levels around the plant and town before he signed it. When asked by the prosecutor how he could have failed in such an important duty, he remained silent.

During cross-examination, one of the director's fellow defendants asked him if there was any documentary evidence that the plant had ever been categorized as "explosion prone." Brukhanov carefully demurred. "The answer to this question is provided in the investigation materials," he said.

Despite all the humiliations and hardships that had befallen him, and the apparent inevitability of his fate, Brukhanov remained a creature of the system that had molded him. He understood the role the Party expected him to play on the stand and stuck almost unswervingly to the script.

"Who do you think is guilty?" a people's assessor asked.

"The court will decide," Brukhanov replied.

"Do you consider yourself the principal offender?" said the prosecutor.

"I think that the shift personnel are—as well as Rogozhkin, Fomin, and Dyatlov."

"But what about you, as the senior administrator?"

"Me, too."

Chief Engineer Nikolai Fomin, the once-imperious Party apparatchik, the electrical engineer who had learned nuclear physics by correspondence course, sat before the court a broken man, frowning to himself or staring owlishly away into space. His face pale and glistening with flop sweat, he stood to read aloud from a sheet of prepared remarks. He explained how he had been incapacitated by his car crash months before the accident, staggered under his workload, and appealed in vain to the Ministry of Energy to allow him to restructure the management of the plant. He admitted that he had approved the fateful test program for Reactor Number Four without notifying the nuclear safety authorities or the reactor designers in Moscow and had not even told Brukhanov it was taking place. He described how he had arrived in the bunker at around four on the morning of the accident and yet remained unaware of the scale of destruction or the grievous injuries of his men. The prosecutor said he found the chief engineer's level of ignorance at this moment "incomprehensible."

When asked who had caused the accident, Fomin replied, "Dyatlov and Akimov, who deviated from the program."

Of all the defendants who came before the court, Deputy Chief Engineer Anatoly Dyatlov proved the most confrontational. He sat forward in his seat, rigid and alert, waiting to pounce with questions, corrections, demands, and requests for clarifying references to specific documents and directives. He had an acute command of the technical aspects of the testimony, learned more every day from the information disclosed in evidence, and proved combative and truculent. At one point, when pressed by one of the assembled experts about the reactivity margin of the reactor, he replied, "Is this a physics exam? I'll ask *you* to answer the question!"

From the beginning, Dyatlov contended that the Chernobyl operators bore no blame for what had happened to Reactor Number Four and addressed in detail every one of the charges brought against him. He said responsibility for the accident rested with those who had failed to warn the plant staff that they were operating a potentially explosive reactor and that he personally had given no instructions that violated any regulations. Despite being contradicted by several witnesses, Dyatlov also insisted that he had not been present in the control room of Unit Four at the crucial moment when Leonid Toptunov let the power of the reactor fall almost to zero before the test, that he had not given orders to raise it, and that he did not send the two now deceased trainees to the reactor hall to lower the control rods by hand.

But it soon became clear that neither the design of the reactor nor the long trail of accidents and institutional cover-ups that preceded the disaster would be considered by the court. Although none of the accused was tortured to confess or brought to the stand to denounce counterrevolutionary activity, no one doubted the outcome of the proceedings: it effectively became one of the final show trials in the history of the Soviet Union. Although the chief prosecutor relied upon the official government commission's accident report to make his case against the operators, he ignored what it had to say about problems with the reactor design. Reporters were told that a separate case against the designers would be prosecuted at a later date.

Yet many of the expert witnesses called to the stand were drawn from the very state agencies—including NIKIET and the Kurchatov Institute—responsible for the original design of the RBMK-1000. Unsurprisingly, the physicists absolved themselves of blame, arguing that the quirks of their reactor became dangerous only in the hands of incompetent operators. The court stifled any dissent from this view. When one nuclear specialist began to explain that Toptunov, Akimov, and Dyatlov could not have known about the positive void coefficient that helped precipitate the explosion of the reactor, the prosecutor dismissed him abruptly from the stand. Dyatlov submitted twenty-four written questions to be put to the expert witnesses about the specifications of the reactor and whether it conformed to the regulations of the USSR's State Committee for Nuclear Safety. But the judge simply ruled them out of order, without further explanation.

On July 23 the prosecutor delivered his closing remarks. He was merciless. Senior Reactor Control Engineer Leonid Toptunov, who had died of radiation poisoning three months short of his twenty-sixth birthday, had been "a weak specialist." His boss, Shift Foreman Alexander Akimov, was "soft and indecisive" and afraid of Dyatlov—who was described as intelligent but disorganized and cruel. The prosecutor regarded the deputy chief engineer as a "nuclear hooligan" who "thoughtlessly broke the canons and commandments of nuclear safety" and whose criminal acts were directly responsible for the catastrophe. Chief Engineer Fomin had been in a position to halt the accident before it began, yet failed to do so.

But the prosecutor reserved his most caustic assessment for the director of the Chernobyl nuclear power plant, who he implied had lied to his superiors in the hope of concealing the magnitude of the accident and clinging to his position, and in so doing endangered the lives of not only his staff but also every citizen of Pripyat. "There is no reason to believe that Brukhanov did not know the true radiation situation," he said. The director's behavior revealed "the moral collapse of Brukhanov, as a leader and as a man."

In response, the defendants' lawyers made their own arguments, and the accused spoke for themselves. Brukhanov's lawyer said that his client was a decent man who knew he must assume the blame. They both recognized that, according to the operational regulations of the power station, the director was formally responsible for everything that happened within the plant. But privately they hoped Brukhanov might escape with only a conviction for administrative negligence and not the more serious charges of abuse of power. Fomin accepted his guilt and pleaded for the mercy of the court. Dyatlov expressed sorrow for the dead and sympathy for the injured but remained defiant. The three other members of the staff on trial—Rogozhkin, Kovalenko, and Yuri Laushkin, the plant's nuclear safety inspector—asked to be acquitted on all charges.

But the Soviet people had been well prepared to expect harsh justice for the men whose corruption and incompetence had despoiled the land of three republics and poisoned thousands of innocent victims. *Pravda's* science editor, Vladimir Gubarev, had already published *Sarcophagus: A Tragedy*, a play about an accident at a fictional nuclear plant, in which he laid blame on a broken system but also upon the station officials: the unnamed director has approved the building of a dangerously flammable roof so that construction can finish on time, and when a radioactive explosion takes place, he evacuates his own grandchildren, leaving the population of his city to their fate.

When the decorated firefighter Major Leonid Telyatnikov was asked his opinion of the defendants, he was unequivocal: "Of course, they should be punished," he said. "According to the government commission, it was human error. It was their fault. The consequences were very severe." Others went further still. During an adjournment in the trial, Valentina Brukhanov waited on a park bench in Kiev beside an old man who had fought in the Great Patriotic War. When talk turned to the Chernobyl proceedings, the veteran explained that some people thought that the defendants should be sent to prison, but in his opinion that wouldn't be just. He said they should all be shot.

On Tuesday, July 29, another fiercely hot day, Judge Brize delivered his verdict. All six men were found guilty: Yuri Laushkin was given two years in prison; Alexander Kovalenko, three; and Boris Rogozhkin, five. All three were taken into custody in the courtroom. Brukhanov, Fomin, and

Dyatlov each received the maximum sentence: ten years' confinement in a penal colony. Every one of them remained stoic except Fomin, who wept in the dock. Valentina Brukhanov fainted. Afterward, one of the investigators told her, "Now you can terminate your marriage at any time."

Driven from the Palace of Culture in a black van with bars on the windows, the former director of the Chernobyl Atomic Energy Station was sent to serve out his sentence in a penitentiary in Donetsk, in the far east of Ukraine. He was shipped there by rail, aboard one of the Soviet prison system's notoriously barbaric *stolypin* cars, and was lucky to survive the journey: during the two weeks it took to travel the seven hundred kilometers, he was sustained mostly by rations of pickled herring. When he finally arrived at the prison, every one of his fellow convicts turned out into the yard to see the face of the infamous culprit behind the world's worst nuclear disaster: a small, frail figure swallowed in drab blue overalls.

As the end of 1987 approached, the new *atomgrad* for the Chernobyl workers and their families in Slavutych was almost ready to begin receiving its first residents, due to arrive from both the workers' shift camp on the Dnieper and from apartments where they had been living in Kiev. Built at frantic speed and garlanded with publicity, Slavutych had been intended as a showcase for Soviet unity, with five districts built in different regional styles by architects from the republics of the Caucasus, Ukraine, Russia, and the Baltic states. But even this prestige project had been dogged by the usual bureaucratic obstructions, construction delays, labor disputes, and slipshod work. At the last minute, the model city's central heating system broke down, rendering it uninhabitable before spring.

In preparation for the arrival of the new citizens, in September a radiation survey of Slavutych had been conducted by scientists from the USSR's hydrometeorological monitoring service, the Ministry of Health, and the Ministry of Defense. They found that the town was being built on land contaminated with cesium 134, cesium 137, ruthenium 106, and cerium 144; the woods nearby contained isotopes of cesium, strontium, and plutonium. They reported that the resulting annual radiation exposure fell within official limits permitted for populations living near nuclear power stations but recommended asphalting paths, regularly washing streets and yards, and—especially in the surrounding forest, where people might be expected to take walks and gather mushrooms—cutting down trees and bagging fallen leaves.

On December 4, 1987, after more than eighteen months of decontamination, repairs, and modifications, the last of the three surviving reactors of the Chernobyl Atomic Energy Station once again began providing electricity to the Soviet grid. Unit Three, although now separated from its entombed twin by a wall of concrete and lead, remained so radioactive that reluctant engineers were rotated in from other reactors—to prevent them being overexposed during the course of their shifts. Despite the sacrifices of General Tarakanov and his bio-robots, uranium fuel pellets were still scattered on the roof of the building, and the turbine operators who worked in the machine hall below did so from protective concrete cabins fitted with portholes of lead glass.

The three Chernobyl reactors—along with the twelve other RBMK-1000 units operating elsewhere in the USSR—had all been subjected to the extensive technical refit proposed in the secret Politburo resolution the previous July. In what amounted to a tacit admission of the designers' culpability in the accident, each RBMK was now fueled with more highly enriched uranium; modified with scores of extra control rods, which reduced the positive void coefficient; and featured a faster and more effective emergency shutdown system. The authorities revised instruction programs for reactor operators and made money available to build computer simulators to prepare them for accident scenarios. Yet little had really changed: more than a year after the disaster, the Politburo received a report showing that Soviet atomic power stations continued to be bedeviled by bad construction, poor staff discipline—and hundreds of minor accidents.

At the Chernobyl plant, the operators manning the three remaining reactors were also demoralized by the way their dead colleagues had been blamed for the accident. Although they went to work dutifully every day, many believed the true causes of the disaster had not been properly considered; some remained convinced that the same thing could easily happen to them. Almost none of them wanted to live in Slavutych.

In public, Valery Legasov continued to hold the Party line about the safety of the USSR's nuclear industry. He said that he did not fault Soviet reactors, which had been designed to take into account all but the most unforeseeable circumstances. The academician insisted that nuclear power represented the zenith of atomic science and was essential for the future of civilization. But privately, Legasov had been struck by what he had heard Prime Minister Ryzhkov tell Gorbachev and rest of the Politburo more than a year earlier: that the explosion in Chernobyl had been inevitable, and that if it hadn't happened there, it would have happened at another Soviet station sooner or later. It was only then that Legasov had finally recognized the true scope of the decay at the heart of the nuclear state: the culture of secrecy and complacency, the arrogance and negligence, and the shoddy standards of design and construction. He saw that both the RBMK reactor and its pressurized water counterpart, the VVER, were inherently dangerous. He began to investigate the problems in more detail and advocated at Sredmash for a new generation of reactor, cooled with molten salt. But his suggestions were met with fury and indignation: Efim Slavsky, at that time still head of the Ministry of Medium Machine Building, told Legasov he was technically illiterate and should keep his nose out of matters that didn't concern him.

By then, Legasov's health problems had intensified, and over the following year, he made repeated visits to Hospital Number Six, where he was treated for neurosis, erratic white blood cell counts, and problems with his heart and bone marrow. Although the doctors made no formal diagnosis of acute radiation syndrome, the scientist's wife had no doubt what was happening to him. Nonetheless, inspired by the rising winds of perestroika, Legasov now went to work on a series of proposals to modernize the monolithic structures of Soviet science. The report he delivered to his colleagues at the Academy of Sciences challenged the hegemony of some of the most powerful forces of the state, and to anyone else would have appeared an obvious political risk. He proposed that the Ministry of Medium Machine Building should be broken up into smaller units that would compete with one another in an internal market; that the research of the Kurchatov Institute should be funded with a new rigor, focused on practical results; and that the old men who now controlled the budgets and made all the appointments to its many jobs-for-life should be replaced with younger and more vital scientists. Legasov had good reason to believe the report would be well received. Not only had he distinguished himself in the liquidation of the Chernobyl disaster and in defending the reputation of the nuclear industry in Vienna. He was, after all, also Aleksandrov's anointed successor to lead the Kurchatov Institute and had powerful backing within the Politburo.

Yet Legasov's proposals were ignored. He failed to realize that he and his ideas would alienate not just the old guard, whose comfortable positions in the status quo he threatened, but also his more reform-minded colleagues—who saw Legasov as a creature of the Era of Stagnation, with a privileged background that had eased a frictionless rise to the top. Even his role at Chernobyl became controversial, as his fellow scientists began to question the wisdom of the operation to cover the burning reactor with sand and lead. In the spring of 1987, the Central Committee of the Communist Party ordered that the staff of the Kurchatov Institute undertake a perestroika of its own and elect a supervisory Science and Technology Council. Legasov, citing his poor health—and also recognizing that votes against him could jeopardize his path to succeed Aleksandrov as director—did not wish to stand. But Aleksandrov insisted, and when the results were declared, Legasov discovered how little he understood his colleagues' feelings about him. Of the 229 votes, only 100 had been cast in his favor; 129 were against. Legasov was thunderstruck. At fifty, it was the first reversal of his gilded career.

At a meeting of his Party cell on June 10, old man Aleksandrov had better news to report: he told the room that they should congratulate Legasov. The director explained that he had seen the final list of those who would be honored by the Politburo for their heroism in Chernobyl, and the name of his first deputy was near the top: Legasov would be granted the one award that had so far eluded him and made a Hero of Socialist Labor. But when the final list was published, Legasov's name was no longer on it. Word went around that Gorbachev had decided at the last minute that no one from Kurchatov should receive a state award for actions in containing a disaster that the institute had helped precipitate. The following day, Legasov phoned his secretary from home. Before he ended the call, he asked her to look out for his two children, and she became concerned. Colleagues hurried to the house on Pekhotnaya 26, where Legasov was found unconscious, a bottle of sleeping pills at his side.

Although rescued from his attempted suicide, Legasov returned to work profoundly changed. The playful light in his eyes had dimmed, and he shuffled up the stairs like an old man. Attending a scientific conference in England that summer, he met his old friend and *Pravda* science editor Vladimir Gubarev, whose play *Sarcophagus* was being performed at the National Theatre in London. Gubarev tried to encourage the academician to make the most of the foreign trip, to find some girls or perhaps see the West End production of *Cats*. But Legasov wanted only to get back to his hotel. That autumn, for the first time, he began reading the Bible. Using a new Japanese Dictaphone he received as a gift from an old friend, he made a series of tape recordings about his experience in Chernobyl, gathering material for a memoir. But he told those close to him that his career was finished. He tried, unsuccessfully, to commit suicide again.

Afterward, Gubarev attempted to raise his friend's spirits by suggesting he articulate his ideas about nuclear safety in an article for *Pravda*. Legasov finished the piece in a matter of days and, once it was published, called Gubarev every day for news on how it had been received. When this, too, was ignored, Legasov took a more drastic step. He gave an interview to the liberal literary journal *Novy Mir* in which he warned that—contrary to everything he had said before—another Chernobyl catastrophe could occur at any one of the other RBMK stations in the USSR, at any time; he told the reporter that many scientists were aware of the danger, but no one would act to stop it. In a separate interview with *Yunost*, another Soviet journal where the shackles of censorship were being loosened by glasnost, Legasov went further still.

Turning his back on every political orthodoxy he had believed in since he was a teenager, the academician said that Soviet science had lost its way. The men and women behind the great triumphs of Soviet technology—who had created the first nuclear power plant and launched Yuri Gagarin into space—had been striving for a new and better society and acted with a morality and strength of purpose inherited from Pushkin and Tolstoy. But the thread of virtuous purpose had run through their fingers, leaving behind a generation of young people who were technologically sophisticated but morally untethered. It was this profound failure of the Soviet social experiment, and not merely a handful of reckless reactor operators, that Legasov believed was to blame for the catastrophe that had bloomed from Reactor Number Four.

By the beginning of 1988, Legasov had given up hope of ever succeeding Aleksandrov as director of the Kurchatov Institute of Atomic Energy. Instead, as Gorbachev's reforms gathered pace and public criticism of the state increased, the academician organized a Council on Ecology and proposed to set up his own institute of nuclear safety—an autonomous body that would bring truly independent regulation to the Soviet atomic industry. He submitted his plans to the Academy of Sciences, optimistic that they would be approved in recognition, if nothing else, of his role in liquidating the consequences of the greatest nuclear accident in history.

But when the hearings were finally held toward the end of April, his mentor Aleksandrov gave the idea only tepid support, and Legasov's proposal was declined. The academician received the news on April 26, 1988, exactly two years to the day after the accident. That afternoon, Legasov's daughter, Inga, picked up her son from kindergarten as usual. When they reached home, she was delighted to find her father waiting beside his car outside the entrance to her apartment building. Inga invited him upstairs for something to eat, but he said he had to go. "I'm coming from the Academy of Sciences," he said. "I just dropped by for a moment to look at you." It was the last time she saw him alive.

At lunchtime the next day, Legasov's son, Alexey, returned from work to the family home at Pekhotnaya 26 and discovered his father's body hanging in the stairwell, a noose around his neck. He had left no note. When a colleague from the Kurchatov Institute checked Legasov's office for radioactivity, he found every one of his belongings too contaminated to be returned to his family. Gathered into a series of large plastic bags, they were buried instead. Soon afterward, when an official visited Anatoly Aleksandrov in his office to discuss candidates to assume some of Legasov's duties, the eighty-five-year-old director broke down and cried. "Why did he abandon me?" he said. "Oh, why did he abandon me?"

Two weeks after Legasov's death, the Soviet minister of health delivered the opening address of an international conference on the medical consequences of the accident, convened in Kiev and attended by representatives of the International Atomic Energy Agency and the World Health Organization (WHO). For the first time, Soviet scientists admitted that 17.5 million people, including 2.5 million children under seven, had lived in the most seriously contaminated areas of Ukraine, Belarus, and Russia at the time of the disaster. Of these, 696,000 had been examined by Soviet medical authorities by the end of 1986. Yet the official tally of deaths ascribed to the disaster to date remained the same as that announced the previous year: 31. The health minister said that they had not discovered a single case of injury in the general population due to radiation. "One must say definitely," he told the assembled delegates, "that we can today be certain there are no effects of the Chernobyl accident on human health."

But the citizens of the Soviet Union no longer trusted their scientists. In Kiev, even two years after the accident, young couples were afraid to have children, and people ascribed every kind of minor illness to the effects of radiation. Ukrainian *Pravda* began publishing what it promised would be weekly reports on radioactivity in the three major cities nearest to the station, updated regularly, like the weather forecast. But the mandarins of the atomic industry still failed to appreciate the degree to which they had lost the public's confidence. Accustomed to their place as revered icons of the Socialist utopia, they found themselves regarded with suspicion and enmity yet clung to their convictions with righteous disdain.

Speaking at a press conference on the closing day of the medical convention in Kiev, the head of the Soviet Institute of Biophysics chastised those scientists who had publicly predicted that thousands of cancers would develop as a result of the accident. "They inflict great damages, because they forget there are many variables," he said. "We never speak of any number of cases. That is immoral."

He dismissed reports of illnesses caused by the long-term consequences of the explosion as the result of a new psychological syndrome, which he called "radiophobia."

For the final rulers of the USSR, the most destructive forces unleashed by the explosion of Reactor Number Four were not radiological but political and economic. The cloud of radiation that spread out across Europe, making the catastrophe impossible to conceal, had forced the touted openness of Gorbachev's glasnost on even the most reluctant conservatives in the Politburo. And the general secretary's own realization that even the nuclear bureaucracy had been undermined by secrecy, incompetence, and stagnation convinced him that the entire state was rotten. After the accident, frustrated and angry, he confronted the need for truly drastic change and plunged deeply into perestroika in a desperate bid to rescue the Socialist experiment before it was too late.

But once the Party relaxed its rigid grip on information, it proved impossible to fully regain its former levels of control. What began with more open reporting from Chernobyl—the news stories in *Pravda* and *Izvestia* were followed by TV documentaries and personal testimonies in popular magazines—widened to include open discussion of long-censored social issues, including drug addiction, the abortion epidemic, the Afghan war, and the horrors of Stalinism. Slowly at first, but then with gathering momentum, the Soviet public began to discover how deeply it had been misled—not only about the accident and its consequences but also about the ideology and identity upon which their society was founded. The accident and the government's inability to protect the population from its consequences finally shattered the illusion that the USSR was a global superpower armed with technology that led the world. And, as the state's attempts to conceal the truth of what had happened came to light, even the most faithful citizens of the Soviet Union faced the realization that their leaders were corrupt and that the Communist dream was a sham.

Soon after Valery Legasov's suicide, *Pravda* published an edited extract of the Chernobyl memoir the academician had recorded on tape, in which he described the hopeless lack of preparation for the catastrophe and the long history of safety failures that had led up to it. "After I had visited

Chernobyl NPP, I came to the conclusion that the accident was the inevitable apotheosis of the economic system which had been developed in the USSR over many decades," he wrote, in a testament that appeared under the headline "It's My Duty to Say This." By September 1988, in a sign of how quickly the system was changing, the Politburo had bowed to public concern and abandoned work on two new nuclear plants, even though one of them—on the outskirts of Minsk—was almost complete.

Ten months later, the Soviet nuclear engineer Grigori Medvedev published a sensational exposé of the accident in *Novy Mir*. In spite of glasnost, it had taken Medvedev two years to get his account into print, fighting a clandestine battle against the KGB and a Chernobyl censorship commission organized specifically to keep the most sensitive information about the accident from the public. Behind that stood Boris Scherbina, the chairman of the government commission, who rightly feared what Medvedev would reveal about his actions in Pripyat. A minute-by-minute reconstruction of the events of April 26 based on his own visits to the scene and dozens of interviews with witnesses, Medvedev's *Chernobyl Notebook* was explosive. It depicted Viktor Brukhanov as a spineless fool, the mandarins of the Soviet nuclear industry as callous and incompetent, and showed Scherbina unnecessarily delaying the evacuation of the doomed *atomgrad*. An introduction to the story was provided by the USSR's most famous dissident, Andrei Sakharov, recently released from internal exile by Gorbachev. In a letter he personally addressed to the general secretary, Sakharov had threatened that if the Central Committee did not allow Medvedev's story to be published, he would personally see to it that the information it contained was disseminated as widely as possible. "Everything that pertains to the Chernobyl disaster, its causes and consequences, must become the property of glasnost," Sakharov wrote in his introduction. "The complete and naked truth is necessary."

In February 1989, almost three years after the accident, a prime-time report on *Vremya* revealed to the Soviet people that the true extent of radioactive contamination beyond the perimeter of the thirty-kilometer Exclusion Zone had been covered up—and that the total area of contamination outside the zone was, in fact, even larger than that within it. "'Glasnost wins after all' is the way we might begin this story," the correspondent said, standing in front of multicolored maps showing that the most heavily radioactive hot spots lay as far as three hundred kilometers from the station, across the border in Belarus, in the districts of Gomel and Mogilev, where witnesses had watched black rain fall in April and May 1986. The land was so poisoned that the Belarusian government estimated another hundred thousand people would have to be evacuated, and planned to request the equivalent of $16 billion in further aid from Moscow.

A few weeks later, just as the last Soviet troops slunk home in defeat from Afghanistan and amid worsening news about the domestic economy, General Secretary Gorbachev traveled to the scene of the accident for the first time. He pulled on a white coat and protective cap to tour Reactor Number Two of the Chernobyl plant, his wife, Raisa, at his side, and visited Slavutych. In Kiev, he spoke to Party officials, announcing a program to protect the environment, and promised to hold a public referendum on any controversial new projects. He pleaded for patience with the growing shortages and the faltering economy and warned that any Soviet republics thinking of leaving the Union were "playing with fire." But environmental issues were already becoming a focal point for nascent independence movements in Latvia and Estonia and would soon provide a platform for Zelenyi Svit, the Green World opposition party in Ukraine. When Gorbachev climbed from his limousine in Kiev for one of his choreographed walkabouts and began to talk about the need to support perestroika, the crowd veered off script. "People are afraid," one woman told him. When he tried to respond, another woman interrupted the general secretary to ask his view of two new nuclear reactors under construction in Crimea.

As the third anniversary of the disaster approached, *Moscow News* reported from a collective farm in the Zhitomir region of Ukraine, forty kilometers west of the Exclusion Zone, where radioactive hot spots of strontium 90 and cesium 137 had been discovered. Farmers in the area had observed a steep rise in the number of birth defects in their livestock since the accident, describing piglets with froglike eyes and malformed skulls, and calves born without legs, eyes, or heads. One member of a team visiting from the Ukrainian Academy of Sciences in Kiev told the press their findings were "terrifying" and insisted the region be evacuated immediately. A representative from the Kurchatov Institute dismissed any connection between such deformities and the accident, blaming excessive use of fertilizers and improper farming methods instead. In October 1989 the newspaper *Sovietskaya Rossiya* reported that hundreds of tonnes of pork and beef contaminated with radioactive cesium had been secretly mixed into sausages and sold to unsuspecting shoppers throughout the Soviet Union in the years since 1986. Although the workers at the meat factory responsible had been paid a bonus to compensate them for their exposure to radiation, a follow-up report to the Politburo insisted that the Chernobyl sausage was perfectly safe to eat and had been processed "in strict accordance with the recommendations of the Ministry of Health of the USSR."

Inside the zone, as thousands of soldiers continued to scour the landscape of radionuclides, bulldoze ancient settlements into the ground, and toss contaminated furniture from the windows of apartments in Pripyat, scientists began to notice strange new phenomena in the wildlife they found there. Hedgehogs, voles, and shrews had become radioactive, and mallards had developed genetic abnormalities; in the cooling reservoir of the plant, silver carp grew to monstrous sizes; the leaves of the trees around the Red Forest had swelled to supernatural proportions, including giant conifers with pine needles ten times their usual size and acacias with "blades as large as a child's palm." The authorities announced their intention to set up a wildlife preserve in Belarus and an international research center inside the zone to study the long-term effects of radiation on the environment.

But money was short. The Soviet economy, after decades of spending on the Cold War arms race, was now staggering under the burden of the botched market reforms of Gorbachev's perestroika, the high price of withdrawing and demobilizing troops from Afghanistan, and the collapse in the international oil market. And the financial cost of Chernobyl—the irradiation and destruction of equipment, the evacuations, the medical care, and the loss of factories, farmland, and millions of kilowatts of electricity—continued to rise. The price for the construction and operation of the Sarcophagus alone was 4 billion rubles, or almost $5.5 billion.

One estimate put the eventual bill for all aspects of the disaster at more than $128 billion—equivalent to the total Soviet defense budget for 1989. The bleeding was slow but proved impossible to stanch—one more open wound that the state could no longer shrug off as the Soviet colossus sank slowly to its knees.

In July 1989 Gorbachev gave a speech signaling to the people of the Soviet satellite countries of Eastern Europe—East Germany, Czechoslovakia, Romania, and the rest—that he would not intervene if they chose to unseat their leaders or even break with the brotherhood of Socialism altogether. Four months later, the Berlin Wall fell, and the Soviet empire began to unravel.

Within the borders of the USSR, ethnic division and opposition to Moscow rule gathered pace amid chronic shortages and an imploding economy. Riots and civil disobedience roiled through the fifteen Soviet republics. In Lithuania, six thousand people encircled the Ignalina nuclear power plant, where the two new RBMK-1500 reactors had become a target of nationalist anger, triggering the start of protests that soon led the three Baltic states to declare independence from the Union. In Minsk, a reported eighty thousand people marched on the headquarters of the Belarusian government, demanding to be relocated from contaminated territory. "Our leaders have been lying to us for three years," one participant told a Soviet reporter. "And now they have deserted this land damned by God and Chernobyl."

In the West, public confidence in nuclear energy—which had never fully recovered after Three Mile Island—was finally shattered by the explosion in Reactor Number Four. The disaster unleashed a wave of popular mistrust, and opposition to the industry spread across the globe. In the twelve months following the accident, the governments of Sweden, Denmark, Austria, New Zealand, and the Philippines all pledged to permanently abandon their nuclear programs, and nine other nations either canceled or delayed plans for further reactor construction. Opinion polls suggested that, since Chernobyl, two-thirds of people around the world opposed any further development of nuclear energy. The United States faced a complete collapse in reactor building, and the very name of the Ukrainian plant became embedded in the international lexicon as shorthand for the failings of technology and a well-justified suspicion of official information.

In Ukraine, the Soviet Ministry of Energy's continuing construction of new nuclear power stations had become a focal point of regional opposition to Moscow. When Kiev called for an end to work on the controversial plant in Crimea, building went on regardless—until the local authorities sanctioned strikes and cut off funding for the project at the state bank. On March 1, 1990, the Supreme Soviet of Ukraine passed a series of sweeping environmental protections for the republic, among them an agreement to close down all three of the remaining reactors at the Chernobyl plant within five years. On August 2, republican legislators placed a moratorium on the construction of any new nuclear plants in Ukraine. In Moscow, the Ministry of Energy was forced to consider who would control the Soviet Union's network of nuclear plants if its Unionwide decision-making power was suddenly devolved to the republics.

In the Exclusion Zone, hundreds of thousands of tonnes of debris from the reactor, radioactive soil, vegetation, furniture, cars, and equipment had now been interred in roughly eight hundred waste disposal sites, known as *mogilniki*, or "burial grounds"—concrete-lined trenches, pits, and mounds sprayed with polymer solutions and then seeded with grass. But the warren of nuclear dumps had been hurriedly excavated and poorly maintained. Nobody had bothered to keep track of what had been buried where, and by the beginning of 1990, the liquidation was being starved of manpower. Even offered twice the average Soviet salary and bonus payments paid directly into their bank accounts, when military reservists were told they were being sent to Chernobyl, many refused to go. The continued mobilizations created a public outcry, and, at last, the Soviet military authorities decided to stop sending troops to the zone. In December 1990 the liquidation effectively came to a halt.

By the end, it was almost impossible to calculate the total number of liquidators who served in the forbidden zone—in part because the figures were falsified by the Soviet government. By the beginning of 1991, as many as six hundred thousand men and women from across the Soviet Union had taken part in cleanup work in the radioactive netherworld surrounding the site of Reactor Number Four and would be officially recognized

Captain Sergei Volodin, the first helicopter pilot on the scene of the accident on April 26, in the cockpit of his aircraft.

Major General Nikolai Antoshkin, chief of staff of the Soviet Air Defense Forces of the Kiev region, in the cabin of one of the helicopters under his command.

Captain Piotr Zborovsky in 1986. Nicknamed Los—or "Moose"— on account of his strength, Zborovsky commanded the operation to pump water from the basement of Unit Four in the hope of heading off a second explosion at the plant—one scientists feared could be orders of magnitude larger than the first.

A photograph of the Polisye Hotel (*left*) and the Pripyat *ispolkom* building—the "White House" (*right*)—annotated by Antoshkin to show the sighting position from which his officers guided helicopters into their bombing runs over Unit Four.

Maria Protsenko, chief architect of Pripyat, in her office in Chernobyl after the accident; visible behind her desk is a map of the evacuated city.

6

Two of the US hematology specialists who flew to Moscow in early May to help treat the victims of the accident, photographed with their Soviet counterparts at Hospital Number Six: (*left to right*) Dr. Richard Champlin, Dr. Robert Gale, Dr. Alexander Baranov, and Dr. Angelina Guskova.

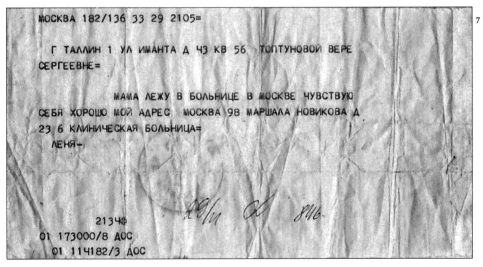

The telegram sent by Leonid Toptunov from his bed in Hospital Number Six to his parents in Estonia on April 29. "MAMA I AM IN HOSPITAL I FEEL OK," it reads, and provides the address in Moscow where they could find him.

The STR-1, one of a handful of Soviet-designed robots deployed to remove radioactive debris from the roof of Unit Three in the late summer of 1986, shunting a block of graphite from the reactor core off the edge of the building. When the radiation proved too much for the robots, more than 3,000 men were sent to do the job instead.

Deputy Prime Minister Ivan Silayev (*left*), the second leader of the government commission sent from Moscow to take charge of the crisis, examines aerial photographs of the ruined plant, with the scientists Yuri Izrael (center) and Evgeny Velikhov (*right*) in May 1986.

Deputy Prime Minister Boris Scherbina (*second from left*), the original leader of the government commission, and Academician Valery Legasov (*fourth from left*) on their return to Chernobyl to deal with the ongoing liquidation in September 1986.

Efim Slavsky—also known as "Big Efim" or "the Ayatollah"—(*left*), head of the Ministry of Medium Machine Building, with design and engineering specialists Vladimir Kurnosov (*center*) and Ilya Dudorov (*right*) at the scene in September 1986. Slavsky had established a dedicated construction unit, Sredmash US-605, to encase the ruins of the reactor in a 'Sarcophagus' of steel and concrete.

The Sarcophagus under construction in mid-October 1986. The upper levels of the Cascade Wall are being backfilled through the pipes of a chain of pumping trucks; on the right, a caterpillar-tracked Demag crane is at work on the tower which will anchor the western end of the building.

The completed Sarcophagus in November 1986.

The trial of those accused of causing the disaster began in the Palace of Culture in Chernobyl in July 1987. Seated between two MVD guards are (*left to right*): plant director Viktor Brukhanov, Deputy Chief Engineer for Operations Anatoly Dyatlov, and Chief Engineer Nikolai Fomin.

One of the equipment graveyards to which buses, fire trucks, helicopters, armored vehicles, and construction equipment—too contaminated to be removed from the exclusion zone surrounding the plant—were consigned after the accident. By 1997, according to one estimate, the total losses resulting from the accident had reached $128 billion.

The "Elephant's Foot," the huge congealed mass of once-molten sand, uranium fuel, steel, and concrete, found in the autumn of 1986 by the scientists of the Chernobyl Complex Expedition in the basement beneath the ruins of Reactor Number Four. It remained so radioactive that less than five minutes in its presence was enough to guarantee an agonizing death.

17

The city of Pripyat in April 2016, with the Chernobyl station and the arch of the New Safe Confinement building visible on the horizon. Thirty years after the accident, the *atomgrad* had been almost completely reclaimed by nature.

as Chernobyl liquidators. In acknowledgment of their service, many were issued special identity cards and an enameled medal depicting the Greek letters for alpha, beta, and gamma surrounding a scarlet drop of blood. All understood that, as with the veterans of the Great Patriotic War, their sacrifices had earned them a lifetime of care from their motherland. In Kiev, the Soviet Ministry of Health set up a dedicated clinic—the All-Union Radiation Medicine Research Center—to provide treatment for everyone who had been exposed to radiation. But as the first of the demobilized liquidators began to fall ill, arriving in clinics with complaints that seemed inexplicable, unpredictable, or premature, they found the state's doctors reluctant to connect their symptoms to the conditions they had endured inside the thirty-kilometer zone. The bankrupt state could ill afford to provide the specialist care it had promised to more than a half million new potential invalids, and so doctors wrote up their notes in code; their medical records were classified as secret. All but the most extreme cases were dismissed with the same diagnosis given to Maria Protsenko: "Ordinary illness: not related to ionizing radiation."

Early in December 1991, in a national referendum called by the parliament in Kiev four months earlier, the Ukrainian people voted to declare independence from the USSR, and Mikhail Gorbachev lost the battle to hold together the union of the twelve remaining Soviet republics. Returned briefly to his position as head of state after a failed reactionary coup in August, he had been forced to watch as Russian president Boris Yeltsin stripped him of his powers and announced that he was suspending the activities of the Communist Party. On Christmas Day, Gorbachev appeared on television to deliver an emotional resignation address, and the red banner of the Soviet Union was lowered for the last time from the flagpole above the Kremlin. Amid the chaos of the collapsing empire, most of the men and women who had fought the Battle of Chernobyl were forgotten—the final defenders of a nation that had seemingly vanished overnight.

In the years that followed, many of those who had lived through the catastrophe became invalids in middle age—struck down by mysterious clusters of symptoms, including high blood pressure, cataracts, kidney trouble, and chronic fatigue. Captain Sergei Volodin, the first helicopter pilot on the scene, who had inadvertently flown his machine through the plume of radioactive vapor rising from the reactor, developed a fear of heights and took an air force desk job. Those unable to work at all subsisted on diminishing state pensions and struggled to find medical care. Some died of heart disease and blood disorders, including leukemia, in hospitals in Kiev and Moscow. Major Telyatnikov, who commanded the firefighters on the night of the accident, was killed by cancer of the jaw in December 2004 at the age of fifty-three. For others, the psychological burden of enduring the disaster only to be set helplessly adrift in a world so abruptly transformed proved too much to bear. The electrical engineer Andrei Tormozin, who had miraculously survived exposure to apparently lethal levels of radiation, a failed bone marrow transplant, and blood poisoning, emerged alive from Hospital Number Six but afterward sank into depression and drank himself to death.

Almost two decades after the accident, in February 2006, in a desolate cafe near the apartment block where he lived in Kiev, I met the physicist Veniamin Prianichnikov. A big man wearing a three-piece suit and a polka-dot tie, sixty-two years old, he was animated and emphatic, his speech rich with metaphor and arid humor. He recalled with exacting clarity the flecks of graphite on the leaves of his wife's strawberry plants and the fight against the China Syndrome in the basement of Unit Four. Of the five men who had gathered temperature and radiation readings from the bowels of the reactor in the most terrifying days of May 1986, he told me that four were already dead. "So twenty percent survived," he said with a dark smile. "If you include me."

The liquidators who lived on did so with the fear that they'd returned from the battlefield with fatal wounds no one could ever see. "We know that the invisible enemy is eating away inside us like a worm," said General Nikolai Antoshkin, whose helicopter crews fought to extinguish the nuclear inferno. "For us, the war continues, and, little by little, we are slipping away from this world."

When I visited Alexander and Natalia Yuvchenko in their apartment not far from Moscow State University in 2006, the engineer's arms and back were scarred violet-red with the results of skin-grafting operations so numerous he had stopped counting at fifteen. He had gone back to work as quickly as he could after being discharged from Hospital Number Six but later spent weeks in a hospital in Germany being treated by military doctors and still had to undergo two weeks of medical testing every year. Yuvchenko had recently started a new job that returned him to his chosen field, nuclear engineering, for the first time since 1986. He was delighted to be able to make business trips to Ukraine, where he visited the country's remaining atomic power stations and worked once more with the same colleagues he had known since his days at university in Odessa.

Yet when he began to talk about the accident, rivulets of sweat ran through his close-cropped hair, and the handkerchief he kneaded in his fist was soon soaked through. Yuvchenko didn't know if the radiation made him infertile, although the doctors assured the couple that they could safely have had more children. But Natalia didn't trust them and questioned their motives: she didn't want to become the unwitting subject of some callous experimental study.

So their son, Kirill, then studying to become a doctor, remained an only child, and they had adopted a Siamese cat named Charlie—born on April 26, which they agreed was a good omen. But Alexander said that the effects the radiation had on his health weren't as bad as people might have imagined. "The doctors keep telling me I've survived—so I can carry on now without worrying," he said. "But when I went back to Ukraine, they started telling me about people who had died. Was it due to radiation? I don't know. I don't understand anything about statistics. But when my friends ask me about it, I tell them: the less you think about it, the longer you'll live."

19

The Elephant's Foot

The afternoon of Monday, April 25, 2016, was beautiful and warm in Pripyat—more like summer than spring. The city was still and empty: poplar seeds floated and danced on the breeze, the heavy green silence broken only by birdsong. Almost exactly thirty years after the explosion of Unit Four, the *atomgrad* that Viktor Brukhanov conjured from an empty field was finally being reconquered by nature. The fine Baltic roses planted on the director's orders had long ago run wild, and their blackened hips rotted in a ragged tangle of bushes in the middle of the city square; a forest of willow, pine, and wild pear filled the soccer stadium; a single silver birch sapling burst from the shattered front steps of the White House; and clusters of oak and acacia had transformed a broad stretch of Kurchatov Street into a dappled woodland path. On the lampposts, the hammer and sickle symbols endured, but the Soviet stars were rusted and bent, pushed aside by thrusting branches, and the images on the crosswalk signs had been erased by decades of sun and rain. A blanket of thick moss stretched beneath the disintegrating Ferris wheel.

Open to the elements, attacked by water, frost, and corrosive lichens, many of the buildings of the city were in danger of collapse. On Sportivnaya Street, the entrance to one apartment house was cut off by massive panels of fallen concrete. Looters and scavengers had stripped the city of almost every piece of metal they could find, leaving behind rooms furnished only with the dark silhouettes of vanished radiators and holes in the streets where they had torn steel pipes and cables from deep underground. In the blocks on Lenina Prospekt, the stairwells were carpeted with broken glass, and paper curled from the bedroom walls in limp sheets, powdery and leached of color. On the fourth floor of Building 13/34 overlooking

Kurchatov Street, the front door of Brukhanov's family home had been torn from its hinges and lay in the hallway beneath a heavy coat of pale gray dust. The big corner apartment was almost bare, and there was little evidence of who had once lived there: a child's picture of a vintage car stuck to the tiles in the darkened bathroom, a single high-heeled shoe on the kitchen floor. But from the balcony of a room overlooking the square, the words on the roof of the ten-story apartment block opposite were still just visible above the trees: *Hai bude atom robitnikom, a ne soldatom!*

"Let the atom be a worker, not a soldier!"

Three kilometers away, the construction cranes above Reactors Five and Six of the Chernobyl plant remained frozen where they had stopped when work there was abandoned on the night of the accident. But inside the main part of the power station, a skeleton staff labored on. When the newly independent republic of Ukraine began receiving its first bills for electricity generated in Russia, the government reversed its decision to close the plant's remaining three reactors, and the last of them was not finally shut down until 2000. Since then, a dwindling workforce had tended to the cooling, decommissioning, and dismantling of Units One, Two, and Three, commuting to work every day on a dedicated electric train from an hour away in Slavutych.

When I first visited the Chernobyl station, one morning in the depths of winter, the heating had stopped working, and the inside of the complex was bitterly cold. Up on mark +10, on the "dirty" side of the sanitary lock, a light snow blew past the windows, and two men in white overalls and thick coats walked briskly down the deaerator corridor, their breath condensing in clouds. In Control Room Number Two, three engineers stood around their desks, smoking cigarettes and making murmured telephone calls. Many of the dials and annunciator alarms on the desk were plastered with paper labels, printed with the legend "Taken out of service." In the center of the mothballed control panel that had once twitched constantly with readouts from the reactor, a small color television screen displayed closed-circuit images from the machine hall, where the giant turbines were being slowly dismantled. Farther west down the long corridor that once connected all four of the plant's reactor units, the skeleton staff melted away. Inside the dormant hulk of Unit Three, an oppressive quiet gathered in the dim machine rooms. The floors were still covered with the yellowing plastic mats laid during decontamination in 1986. Drab, gray light seeped through the filthy glazing, heavy pipework hung from the ceiling, and a tang of machine oil and ozone lingered in the damp air.

Down one more staircase and along a windowless corridor, it became impossible to go farther. The presence of something close by but unseen, something monstrous, became almost tangible. The cluster of thick pipes running overhead came to a sudden halt, their severed ends hanging in the air just short of what had once been the entrance to a passageway, now sealed with solid concrete. Against this wall, surrounded by a puddle of milky liquid, stood a red marble monument bearing a bronze bas-relief: the dark shape of a man wearing the cylindrical cap of a power plant worker, one arm outstretched in alarm, as if reaching for help that would never come. This was the tombstone of Valery Khodemchuk, the first man to die as a result of the explosion in Unit Four. His surviving colleagues had built the memorial as close as they dared to the place where they believed his body lay buried. Whatever remained of the vanished machinist lay immediately beyond this point, on the other side of three meters of concrete and a layer of lead, beneath thousands of tonnes of rubble, sand, and twisted debris. Somewhere in there with him, too, was the melted heart of Reactor Number Four, a protean mass of uranium, zirconium, and other core elements that remained almost as enigmatic and deadly as it was on the day the catastrophe began almost thirty years before.

From the very start, the small band of Kurchatov Institute specialists who began exploring inside the rising Sarcophagus in the summer of 1986 had faced terrible obstacles. Sent from Moscow to locate the hundreds of tonnes of nuclear fuel that had once fired Reactor Four, they were hampered by fields of gamma radiation, collapsed wreckage, cascades of freshly pumped Sredmash concrete, and malfunctioning equipment. Attempts to use robots initially ended in the same way as so many others had in the Special Zone. At its first test, one device, created at enormous expense to explore the ruins, proved incapable of navigating even minor obstructions; it had to be rescued repeatedly by its operators and ultimately stopped dead in a high-radiation zone. In a scene captured on video and screened that night before the assembled task force, the robot then unexpectedly came back to life and—in a ridiculous pantomime of flashing lights and waving appendages—fled down the corridor, before it screeched and fell on its side and had to be retrieved by its handlers in a blizzard of invective.

Eventually, rudimentary reconnaissance began with the help of a miniature plastic tank bought by one of the scientists for 12 rubles (the equivalent, then, of $5) from the toy store Detsky Mir—Children's World—in Kiev. The toy was controlled by a battery-operated box on the end of a long cable and modified to carry a dosimeter, a thermometer, and a powerful flashlight. The scientists used it like a radiosensitive hunting dog which ran ten meters ahead of them and warned of imminent danger. Although fully aware of the perils surrounding them, the members of the Kurchatov task force were urged on by the gravity of their search for the missing fuel—necessary to guarantee that a new chain reaction could not begin—but also by scientific curiosity. Inside the Sarcophagus, they were explorers on the frontier of an alien world, where they found gamma radiation fields scaling heights no one had witnessed before and strange new materials forged at temperatures of more than 10,000 degrees centigrade in the crucible of a disintegrating nuclear reactor.

In the autumn of 1986, the members of the Kurchatov team made one of their first and most memorable discoveries, when they finally entered the mysterious corridor 217/2, where months before their remote radiation probe had run off the scale and burned out. To reach it, the scientists now wriggled through a narrow tunnel formed in the ruins, armed with flashlights and clad in thin plastic suits to protect them from radioactive dust. What they found there was a massive, globular, stalagmite-like formation of some mysterious substance. It appeared to have flowed down from somewhere above their heads before solidifying into an anthracite-black glassy mass. The formation, which they named the Elephant's Foot, stood half as tall as a man and weighed as much as two tonnes. Its surface was emitting an astonishing 8,000 roentgen per hour, or 2 roentgen a second: five minutes in its presence was enough to guarantee an agonizing death. Nonetheless, orders came down from the government commission for photographs and a full analysis.

Unable to find the fuel from the reactor, the scientists had also found
no sign so far of the more than sixteen thousand tonnes of materials
dropped into the reactor by General Antoshkin's heroic helicopter crews
and therefore hoped that the Elephant's Foot might contain some of the
lead intended to cool the core. But it did not readily surrender samples for
testing. The substance proved too hard for a drill mounted on a motorized
trolley, and a soldier who volunteered to attack it with an axe left the room
empty-handed—and so overexposed he had to be evacuated immediately
from Chernobyl. Finally, a police marksman arrived and shot a fragment
of the surface away with a rifle. The sample revealed that the Elephant's
Foot was a solidified mass of silicon dioxide, titanium, zirconium, magnesium, and uranium—a once-molten radioactive lava containing all the radionuclides found in irradiated nuclear fuel that had somehow flowed into
the corridor from the rooms nearby. But it contained no trace of the lead
dropped into Unit Four from the air during the frantic early days of May.

By measuring the temperature of the air inside the subreactor spaces,
the Kurchatov experts found evidence that there might be more lava, still
hot from the energy of radioactive decay, in the room at the end of the
corridor, which had housed the gigantic stainless steel cross—Structure
S—supporting the reactor vessel and everything in it. But, once again,
they found their path blocked by rubble and radioactivity. At almost every
meeting of the government commission, Boris Scherbina and his officials
reproached them for their failure to find the fuel and questioned them
about the continuing danger of a new self-sustaining chain reaction.

At the beginning of 1988, they formed a new group, an interdisci-
plinary team of 30 scientists dedicated to exploring the Sarcophagus and
mapping what it contained, backed by a force of up to 3,500 construction
workers: the Chernobyl Complex Expedition. Using horizontal drilling
rigs, extending up to 26 meters in length, operated by technicians from
Sredmash and the Soviet mining industry, the Complex Expedition began
boring deep within Unit Four, extracting core samples from the debris to
examine the very fabric of the building. By the late spring of 1988, almost
exactly two years after the explosion, the drilling had reached the reactor
vessel itself. On May 3 a drill bit burst through the outer concrete wall of
the shaft, passed through the layer of sand ballast, the steel walls of the

inner protection tank, and, at last, into the reactor vault. The scientists pushed a probe through the hole and toward the center of the vault in an attempt to gauge the parameters of the destruction in the graphite stacks and fuel assemblies of the reactor core—the origin point of the accident. But the probe encountered no resistance, and traveled smoothly on across the full 11.8-meter diameter of the former active zone, without pause or interruption.

The scientists were bewildered. Where was the fuel? The following day, they inserted a periscope and a powerful lamp to illuminate the scene within and were astonished by what they saw: the giant vault of Reactor Number Four, which had once contained 190 tonnes of uranium fuel and 1,700 tonnes of graphite blocks, and was since supposed to have been filled with load after load of sand, lead, and dolomite, was almost completely empty.

In the course of an operation that would continue for years, as they penetrated farther and farther into the warren of ruins within Unit Four, taking samples and photographs and shooting video footage of what they found, the men of the Complex Expedition unraveled the mystery of what had happened inside the building during the most feverish period of the Battle of Chernobyl. They discovered that only the smallest trace of the 17,000 tonnes of material eventually targeted on the building from above had found its mark inside the reactor vault. Instead, the majority of the loads were discovered, in mounds up to fifteen meters high, scattered elsewhere among the debris filling the central hall. Some lead ingots had struck the red-and-white-striped vent stack, and the roof of Unit Three, almost a hundred meters away from the target, had been smashed in by cargo dropped from the air. Moreover, it seemed that almost all of the 1,300 tonnes of graphite that had not been thrown from the reactor by the explosion had been consumed by fire: in the end, the blazing reactor had simply burned itself out. The Soviet fliers' courageous efforts to smother it with sand from two hundred meters up had been almost entirely pointless.

But the Complex Expedition also revealed that the theoretical physicists' terror of the China Syndrome, initially dismissed as nearly impossible by the practically minded nuclear wildcatters of Sredmash, had not been so far-fetched after all. They established that the two tonnes of deadly compounds in the Elephant's Foot represented just a fraction of an incandescent river of radioactive lava that had formed inside the reactor in the minutes after the accident began and had flowed slowly downward into the basement of the building until the reactor vault itself was practically empty.

The zirconium cladding of the fuel assemblies had melted first, reaching a temperature of more than 1,850 degrees Celsius within a half hour of the explosion, and dissolved the uranium dioxide pellets it contained, pooling into a hot metal soup that then absorbed parts of the reactor vessel itself—including stainless steel, serpentinite, graphite, and melted concrete. The radioactive lava, now containing some 135 tonnes of molten uranium, then began eating its way through the lower biological shield of the reactor, a massive steel disk filled with serpentinite gravel, weighing 1,200 tonnes. It burned clean through the shield and its contents, absorbing around a quarter of its total mass, and ran smoothly into the rooms below. The steel cross supporting the entire reactor vessel—Structure S—reached its yield point and buckled; the shield fell out of the bottom of the reactor, and the searing corium began burning a hole in the floor of the subreactor space. The pipes of the steam suppression system, which had proved unable to save Reactor Number Four in its death throes, now provided a convenient path for the lava as it attempted to leave the building entirely, spreading out to the south and east along four separate routes, melting metalwork, flowing through open doorways, slowly filling corridors and rooms, and creeping down through the plumbing, one floor after another, toward the foundations of Unit Four.

By the time the corium reached the steam suppression pools, emptied at such great cost by Captain Zborovsky and the specialists from the plant, it had burned and seethed through three floors of the subreactor space, engorging itself with more parts of the building and structural debris until it reached a combined weight of at least a thousand tonnes. In some rooms, molten metal gathered in puddles fifteen centimeters deep and solidified where it lay. And yet—despite the best efforts of Zborovsky and his men—several hundred cubic meters of water remained gathered in the suppression compartments, and it was not until the flow of corium reached this that the China Syndrome finally came to an end. When the lava dropped into the dregs of the water in the suppression pool, it cooled harmlessly, and the molten heart of Reactor Number Four at last ended its journey—a gray ceramic pumice floating on the surface of a radioactive pond, but just centimeters away from the foundations separating the building from the earth below.

It would take the men of the Complex Expedition until 1990 to find most of the melted fuel and deliver a report assuring the government commission that the ghost of the reactor was unlikely—"for the time being"—to rise again. Even four years after the accident, the temperature inside some of the fuel clusters remained as high as 100 degrees Celsius. But unless it became saturated with water, the scientists calculated that a new criticality was almost impossible, and they installed a new monitoring system to give them an early warning if one did begin. By that time, the final implosion of the Soviet Union had started, and financial and political attention shifted far from Chernobyl.

Increasingly forgotten and starved of resources, the Complex Expedition slogged on regardless. Its leader, the portly neutrino physicist Alexander Borovoi, would eventually make more than one thousand trips into the Sarcophagus. Borovoi said that he had always lacked physical courage and simply understood the dangers of radiation well enough to manage the risks. But each time he entered the great, black building, it evoked in him the same feelings as the first movement of Tchaikovsky's Sixth Symphony: an ominous prelude to a struggle between life and oblivion. The physicists lacked computers and protective equipment, and whenever he went inside, Borovoi carried a roll of medical tape with which to patch his plastic overalls in case of tears.

Eventually the men ran short even of underwear, and, following the screening of a documentary about the expedition by the BBC, the scientists began receiving care packages of hand-knitted socks from the West. Borovoi handed these out as rewards to members of the group who distinguished themselves inside the Sarcophagus. And yet the scientists found their work so fascinating and important that some refused to leave at the end of their assignments; many deliberately left their dosimeters in the office to avoid officially registering their maximum dosages and being sent home. After the Elephant's Foot, they came upon other formations of solidified lava inside the ruins of Unit Four, frozen into uncanny shapes, which they granted individual nicknames: the Drop, the Icicle, the Stalagmite, and the Heap. Among their unique discoveries was a new substance they christened Chernobylite—a beautiful, but deadly, blue crystal silicate composed of zirconium and uranium, which they chipped from the ruins. It could be examined safely only for short periods of time, and samples had to be removed from Unit Four in lead-lined containers. And, as money ran out and fears of a new self-sustaining chain reaction subsided, the Complex Expedition became aware of a new problem. The Sarcophagus itself, built under conditions of great secrecy with all the haste and ingenuity the USSR could muster, was not proving quite the triumph of engineering the Soviet propaganda machine had claimed.

Although the limitations of the building had been carefully concealed during its construction, Borovoi and his team now found gaps in the walls large enough for a man to pass through and cracks through which water could penetrate and radioactive dust could escape. They also began to fear that, inside the Sarcophagus, what remained of the concrete skeleton of Unit Four could soon collapse. By the time Borovoi was recalled to Moscow, it had become clear that a new means of protecting the world from the remains of the still-hot reactor would have to be found.

Former plant director Viktor Brukhanov walked free from prison on September 11, 1991, having served five years of his ten-year sentence, most of it in the penal colony in Donetsk. He was released early for good behavior, under the rules of the Soviet judicial system, and was permitted to spend the closing months of his punishment in compulsory labor—known as khimiya, or "chemistry"—in Uman, a town closer to his wife, Valentina, in Kiev. At fifty-five, he emerged from incarceration shattered and emaciated. The good Czech overcoat Valentina had bought him after her move to Kiev hung off him like a sack. When he came home, she showed him around their new apartment—in the same complex in the city where so many of his invalided colleagues and employees had been rehoused, by then becoming known as Little Pripyat—and introduced him to the five-year-old granddaughter he had never met.

Brukhanov emerged into a profoundly changed world. The rigid certainties of the Party he had served so unfailingly were dissolving, and even his popular infamy as the man responsible for the Chernobyl disaster was being swept away amid the revelations of higher and more unspeakable crimes. On December 26, 1991, the president of the Collegium on Criminal Cases of the USSR Supreme Court sent a two-line letter to Brukhanov's lawyer in Moscow, notifying him that the appeal against his client's sentence had been returned without review, since the state responsible for imposing it no longer existed.

At first, Brukhanov longed to return to Pripyat and—in spite of everything that had happened—may have hoped to find a new job at the nuclear plant he had built. Eventually Vitali Sklyarov, the sardonic former apparatchik still overseeing the Ukrainian Ministry of Energy, now part of the independent national government, found Brukhanov a position with the ministry's body for international trade in Kiev. In early 1992 he quietly went back to work, a forgotten man.

The fallen director was the last of the Chernobyl plant staff punished for their part in the explosion of Reactor Number Four to be set free. Kovalenko and Rogozhkin had successfully petitioned for early release and already returned to jobs at the plant. The former nuclear safety inspector, Laushkin, had also been released but died of stomach cancer soon afterward. Chief Engineer Nikolai Fomin had never fully recovered from the shock of the accident. Two years after his arrest, in 1988, he was diagnosed with a "reactive psychosis" and transferred to a psychiatric hospital. Granted early release for health reasons in 1990, Fomin found work again at the Kalinin nuclear power plant north of Moscow—although even then his mental state reportedly remained fragile.

Anatoly Dyatlov, the dictatorial deputy chief engineer, had spent his years of incarceration contesting the verdict of the Soviet court, writing letters and giving interviews from prison in an attempt to publicize what he'd learned about the failings of the RBMK reactor and clear his name and those of his staff. He wrote directly to Hans Blix at the International Atomic Energy Agency in Vienna to point out the failings of their technical analysis, but also to the parents of Leonid Toptunov, describing how their son had stayed at his post to try to save the crippled reactor and how he had been unjustly blamed for causing the accident. He explained that the reactor should never have been put into operation and that Toptunov and his dead colleagues were the victims of a judicial cover-up. "I fully sympathize with you, and grieve with you," Dyatlov wrote. "There is nothing more unbearable than losing one's child."

In prison, Dyatlov continued to suffer from the terrible radiation burns he had sustained as he wandered through the wreckage of Unit Four on the night of the accident and in October 1990 was also granted early release due to his declining health. From his flat in Troieshchyna, the increasingly frail engineer continued a campaign to reveal the truth about the design faults of the reactor and the way the accident had been whitewashed by Academician Legasov and the Soviet delegation to the IAEA.

Despite the publication of popular accounts, including Grigori Medvedev's *Chernobyl Notebook*, which called into question the official Soviet version of the story, the government commission's reports into the causes of the accident remained classified, and the public perception of events remained unchanged—that of a perfectly reliable reactor blown up by incompetent operators. But as the bonds of the secret state began to loosen further, the truth about the origins of the explosion in Unit Four at last started to emerge. In the face of opposition from NIKIET and the Ministry of Atomic Energy, the state's independent nuclear safety board belatedly launched its own inquiry into the causes of the accident. The board sought advice from both midlevel members of the RBMK design team and former specialists from the Chernobyl plant. The leader of the board of inquiry began an intensive correspondence with Dyatlov about the events immediately leading up to the explosion. Incidental documents submitted by the Soviet authorities to the IAEA had already begun to undermine the official version, and in July 1990 one senior member of the original Soviet delegation to Vienna admitted publicly that the designers had been chiefly to blame for the catastrophe, and the operators' actions played only a minor role.

The report that the nuclear safety board of inquiry delivered to the Soviet Council of Ministers in January 1991 completely contradicted the story Legasov had told the IAEA in 1986, of a delicate piece of equipment blown up by feckless operators who breached one vital safety regulation after another. Their findings couldn't entirely vindicate Dyatlov, who maintained not only that the plant personnel had nothing to do with the accident but also that at the fateful moment in its development—when Leonid Toptunov raised the reactor power from zero so that the rundown test could take place—he hadn't even been in the room. But they made clear that, although the actions of the operators contributed to what happened, they should not be held responsible for a disaster that was decades in the making.

In May 1991, while the report was still under review by the Fuel and Energy Department of the Council of Ministers, one of its principal authors, former Chernobyl chief engineer Nikolai Steinberg, presented his findings to the inaugural International Congress on Human Rights held by Moscow's Sakharov Center. He told the delegates that the origins of the Chernobyl disaster lay in a combination of "scientific, technological, socioeconomic, and human factors" unique to the USSR. The Soviet nuclear industry, lacking even rudimentary safety practices, had relied upon its operators to behave with robotic precision night after night, despite constant pressure to beat deadlines and "exceed the plan" that made disregard for the letter of the regulations almost inevitable. He reported that Dyatlov and the now deceased operators in Control Room Number Four had brought the reactor into an unstable condition, but only on account of the acute pressure they felt to complete the test on the turbine.

"Under those circumstances," Steinberg said, "the unit operators and managers made a decision that, in all probability, predetermined the subsequent accident." But nobody could know for sure, because they had still not determined whether, once the test began, the reactor might have been shut down without risking disaster. Although Dyatlov, Shift Foreman Akimov, and Senior Reactor Control Engineer Toptunov had violated some operating regulations, they were ignorant of the deadly failing of the RBMK-1000 that meant that insertion of the control rods, instead of shutting down the reactor at the end of the test, could initiate a runaway chain reaction.

Every one of the investigators behind the report now agreed that the fatal power surge that destroyed the reactor had begun with the entry of the rods into its core. "Thus the Chernobyl accident comes within the standard pattern of most severe accidents in the world. It begins with an accumulation of small breaches of the regulations. . . . These produce a set of undesirable properties and occurrences that, when taken separately, do not seem to be particularly dangerous, but finally an initiating event occurs that, in this particular case, was the subjective actions of the personnel that allowed the potentially destructive and dangerous qualities of the reactor to be released."

Steinberg recognized that the origins of the accident lay with those who had designed the reactor and the secret, incestuous bureaucracy that had allowed it to go into operation. But he concluded that it was no longer constructive to allocate blame—whether with "those who hang a rifle on the wall, aware that it is loaded, or those who inadvertently pull the trigger."

But the barons of the atomic industry had no more appetite for the truth than they had five years earlier. The Kremlin's nuclear minister did not immediately accept the findings in Steinberg's report but instead ordered up a second investigation from a new commission, this time packed with the same men behind the Vienna report of 1986. Under pressure from the reactor designers in NIKIET, the compass needle of their accusations began to swing once more toward the operators. Two members of the commission chosen from the state nuclear safety board resigned in protest, and their boss refused to put his signature on the new document. The issue was still unresolved when, in August 1991, Yeltsin and Gorbachev faced down the plotters of the abortive coup and the USSR began its final slide toward extinction.

It wasn't until the following year, after the Soviet nuclear safety team's organization had been dissolved, that its findings were published as an appendix to an updated version of the original IAEA report on the Chernobyl accident. Seeking to redress the inaccuracies of their 1986 account based on what they described as "new information," the IAEA experts revealed at last the true magnitude of the technical cover-up surrounding the causes of the disaster: the long history of previous RBMK accidents, the dangerous design of the reactor, its instability, and the way its operators had been misled about its behavior. In dense scientific detail, it described the inherent problems of the positive void coefficient and the fatal consequences of the control rod "tip" effect.

Although the IAEA panel continued to find the behavior of the Chernobyl operators "in many respects . . . unsatisfactory," it acknowledged that the primary causes of the worst nuclear disaster in history rested not with the men in the control room of Unit Four but with the design of the RBMK reactor. Six years after their burial in Mitino cemetery, the report at last went some way to rehabilitating the reputations of Alexander Akimov, Leonid Toptunov, and the other operators who had died in Hospital Number Six. But by then, the turgid technicalities of the revised report attracted little attention outside specialist circles. Back in Kiev, former deputy chief engineer Dyatlov, still not satisfied, kept up his lonely struggle for exoneration in the press, until his own death—from cancer of the bone marrow, at the age of sixty-four—in December 1995. In 2008, more than two decades after the accident, Akimov and Toptunov—along with twelve other engineers, electricians, and machinists from the plant staff—were finally recognized for their heroism on the night of April 26, when President Viktor Yanukovych posthumously awarded each of them the Ukrainian Order for Courage, Third Class.

20

A Tomb for Valery Khodemchuk

E arly one evening in October 2015, I returned to the brick apartment building off Vernadsky Prospekt in Moscow where almost ten years earlier I had met Alexander and Natalia Yuvchenko. The sun had already set, and it was cold outside, but the first snow of the year had not yet fallen. The Yuvchenkos' ninth-floor flat had been expensively renovated, with a glossy modern kitchen and a new bathroom, but seemed spartan and cold. Charlie the cat was gone. Natalia said that she had spent a lot of time in Germany recently, where she worked as a cosmetologist, and visited her Moscow home only occasionally. At fifty-four, pale and birdlike, Yuvchenko wore a short-sleeved green sweater decorated with pink piping, her hair dyed a dark auburn and teased into a bouffant. She made herb tea, offered a plate of sweet pastries, and, after a while, told the story of what had happened to her husband.

Alexander had seemed fine when I visited in 2006, and it wasn't until the end of that year that Natalia first noticed him looking thinner. Even then, she thought it was good for her husband to lose some weight; it made him look younger. His new nuclear engineering work was going well, and he seemed fit and happy. In early October they had taken a vacation in Crete, and one day he came down to the beach holding a canoe paddle: "Natasha," he said, "I want you to keep me company!" Although he had left behind competitive rowing more than twenty years before, his love for the sport hadn't left him, and he never missed a race on TV.

Now he had found a dinghy but needed a partner. Natalia had never rowed in her life, but her husband was insistent: just across the bay; it would take only a few minutes. She climbed into the bow, and Alexander got in behind. She dipped her paddle into the blue water. It wasn't easy.

She was so slight and inexperienced, and her husband, still only forty-four, remained broad backed and powerful. She had to paddle frantically to keep up with his long arms and practiced stroke, but she found a rhythm and threw herself into it. And when they reached the beach at last, Natalia turned to find her husband behind her, delighted and panting—exhausted by having to keep up with his wife's extraordinary pace. "You're a champion!" he said. In place of a trophy, he bought her a pair of fine aquamarine earrings from a nearby jewelry store.

At the end of the vacation, the couple were on the plane back to Moscow when Alexander suddenly felt faint, and the blood drained from his face. He put it down to the change in air pressure in the cabin and thought little more about it. By the time they were settled in at home again, he seemed well enough, although he continued to look very pale. But the results of his regular blood tests were clear, and Natalia thought he had been traveling too much for work. He probably just needed to slow down.

After the New Year holiday, in early January 2007, Alexander came down with a high fever. They thought it was probably a virus, and he took medication to bring it down. But his temperature continued to seesaw—low in the mornings, spiking at night—and they realized that there was some underlying problem. Their son, Kirill, called the doctor.

They discovered that Alexander's spleen had expanded to several times its normal size—a common symptom of leukemia. His blood counts had proved misleading, and by the time he was admitted to a hospital, Yuvchenko's bone marrow had started to fail. He returned to Hospital Number Six, now renamed the Burnasyan Medical Center, where the two doctors who had overseen his treatment in 1986—Angelina Guskova, by then eighty-two, and Anzhelika Barabanova, seventy-two—continued to work as consultants. Initially, Natalia Yuvchenko hoped that with the right medication, her husband's disease could be managed and he would have a few more years of normal life. But over the next eighteen months, he developed a tumor that grew to a size and malignancy that thwarted all therapy, including new and experimental drugs imported from Switzerland.

At the end of the summer of 2008, Natalia prepared herself for the worst. She believed that, by then, he had no more than five days to live, and—together with Kirill and Alexander's brother, Vladimir—sat with him in shifts around the clock. Natalia made food for him at home, took it to the hospital, and fed him herself. They were stunned when, at the end of August, he pulled through and was suddenly back on his feet. The doctors allowed him to return home on weekends, where he was able to take walks, drive, and shop for fresh vegetables in the market. And even when Alexander returned to the hospital, he kept working from his bed. He insisted Natalia take a business trip to Paris she had planned for November. But, although he remained active and determined to carry on just as before, his condition continued to worsen. His face and body swelled so much that he no longer looked like himself.

Early in October 2008, when they were quite certain of the prognosis, the doctors allowed Alexander to return home to spend two weeks with his family. He came to Natalia's office every afternoon in his car to pick her up from work and drive her home. On October 25 he turned forty-seven. He drank champagne, and friends and colleagues, including many he hadn't heard from in years, called from all over the world to congratulate him and wish him well.

The following week, more than twenty-two years after he had first crossed the threshold of the brown brick building on Marshal Novikov Street, Alexander Yuvchenko returned to the hallways of the former Hospital Number Six for the last time. That same day, he called Natalia to say that he was being taken into intensive care for surgery and wouldn't be able to phone again. By now, Kirill—twenty-five years old and a trainee surgeon—was working at the hospital himself and continued to see his father every day. But when a quarantine was declared on the ward, visitors were prohibited, and Kirill warned Natalia not to even try coming to the clinic. On Saturday the two men laughed and joked together, but on Monday, November 10, Alexander fell into a coma. Eight hours later, shortly before midnight, Kirill telephoned his mother.

"Father is dead," he said.

Almost twenty-five years after the explosion of Reactor Number Four, in February 2011, the thirty-kilometer Exclusion Zone surrounding the power station remained deeply contaminated. Levels of radiation everywhere varied wildly and unpredictably: an invisible patchwork of fallout leached deep into the landscape. The site of the Red Forest—where the poisoned trees had been cut down by Soviet engineering troops, buried in concrete-lined pits, covered with fresh soil and sand, and replanted with grass and new pines—had proved so radioactive that the road through it was abandoned. Instead, traffic to the plant was directed down a new route, laid several hundred meters to the east. On a sandy path that cut past stands of healthy-looking conifers toward the patches of spindly and deformed pines beyond, the electronic clicks of a Geiger counter rose from a light patter to a steady torrent, until a guide told me we should go no farther. Beyond this point lay a barren clearing carpeted with dead needles and broken branches, where almost nothing would grow and the Geiger counter would make a sound no one wanted to hear: the continuous squall of white noise indicating levels of radiation thousands of times higher than normal.

In the decades since the pioneer troops of the Twenty-Fifth Motorized Rifle Division first staked out its boundaries with fence posts and barbed wire in the summer of 1986, the territory of the Exclusion Zone had expanded repeatedly, as the newly independent governments overseeing it revised the Soviet norms of what constituted dangerous levels of radiation downward to meet Western standards. In 1993 the main section of contaminated land in Belarus—designated as the Polesia State Radiological and Ecological Reserve—was enlarged to include an extra 850 square kilometers of territory. In 1989 Ukraine added another 500-square-kilometer stretch of polluted countryside to the western edge of its part of the zone, encompassing the newly evacuated areas of the Polesia and Narodichi regions, to create a single administrative entity named the Zone of Exclusion and the Zone of Unconditional (Mandatory) Resettlement. Together, by 2005, the contiguous parts of the Belarusian and Ukrainian zones made up a total area of more than 4,700 square kilometers of northwestern Ukraine and southern Belarus, all of it rendered officially uninhabitable by radiation.

Beyond the borders of the evacuated land, the contamination of Europe with radionuclides from the explosion had proved widespread and long lasting: for years after the accident, meat, dairy products, and produce raised on farms from Minsk to Aberdeen and from France to Finland were found laced with strontium and cesium and had to be confiscated and destroyed. In Britain, restrictions on the sale of sheep grazed on the hill farms of North Wales would not be lifted until 2012. Subsequent studies found that three decades after the accident, half of the wild boar shot by hunters in the forests of the Czech Republic were still too radioactive for human consumption.

At the same time, a remarkable story had emerged from within the Exclusion Zone—a fairytale narrative of ecological rebirth and renewal. Far from enduring decades of inevitable sickness and death in an atomic wasteland, the plants and animals left behind in the evacuated area after the accident had apparently made an amazing recovery. The first evidence for the phenomenon came from three cows and a bull found wandering near the reactor after the explosion. Taken to an experimental farm near Pripyat, all four animals—named Alpha, Beta, Gamma, and Uranium by researchers—had initially been rendered infertile by the acute doses of radiation received after the accident. But they slowly recovered, and the radioactive farm's first calf was born in 1989. And when the experimental herd was expanded, to thirty or more cattle, including some raised on uncontaminated land outside the zone, the research team examined the blood work of the two groups of animals.

They expected to find at least some evidence in the analysis of the two groups' differing levels of radiation exposure; they found none.

After the breakup of the USSR, as the economies of Ukraine and Belarus pitched into steep decline, the appetite for funding further Chernobyl research dried up. But one scientist—Sergei Gaschak, a former liquidator who in the summer of 1986 had spent twelve hours every day for six weeks washing the radioactive dust from cars and trucks close to the plant—stayed on in the zone. Venturing deep into the forests and swamps of the abandoned landscape, Gaschak began to spot creatures long since eliminated from the rest of Ukraine and Belarus by hunting and collective farming: wolves, elk, brown bears, and rare birds of prey. His observations helped foster a new notion of the zone, as appealing as it was counterintuitive: demonstrating that nature was apparently capable of healing itself in new and unpredictable ways. In the absence of man, plants and animals were thriving in a radioactive Eden.

The idea of the miracle of the zone took hold through TV documentaries and books that told a story of how chronic exposure to the relatively low levels of radioactivity left in many areas was proving apparently harmless—or, in some cases, even beneficial—to animal populations. Yet scientific evidence for the thesis was thin, or contradictory. Gaschak himself lacked the funding to conduct large-scale studies of the wildlife population of the zone and based his theories on estimates. And a team of independent researchers, led by Timothy Mousseau from the United States and Anders Pape-Moller from Denmark, published dozens of papers that contradicted his results and instead showed patterns of early death and malformation among the plants and animals in the zone.

What seemed clear from much of the research into low-level radiation conducted since 1986 was that different species and populations reacted to chronic exposure in varying ways. Pine trees coped with it less well than birch. Moller and Mousseau found that migrant barn swallows were apparently very radiosensitive; resident birds less so. Winter wheat seeds taken from the Exclusion Zone in the days after the disaster and then germinated in uncontaminated soil had produced thousands of different mutant strains, and every new generation remained genetically unstable, even twenty-five years after the accident. Yet a 2009 study of soybeans grown near the reactor seemed to show that the plants changed at a molecular level to protect themselves against radiation.

Meanwhile, the World Health Organization asserted confidently that no inherited or reproductive effects on nearby populations would result from the accident. This bore out decades of earlier research showing that, although mammalian fetuses exposed to radiation while in the womb might suffer from birth defects, the risk of it causing inheritable mutations in human beings was almost certainly too small to detect. But some researchers insisted that no one could be truly certain where mankind might fall on the continuum of DNA damage and long-term adaptation found in lower organisms, and finding the answer could take decades or even centuries. They argued that the genetic effects of chronic radiation exposure on each species studied had often been subtle, varied, and demonstrated conclusively only after many generations; the potential genetic changes in human beings—who, by 2011, had only recently produced their third generation, as the children of the liquidators themselves began to raise families—might take hundreds of years to fully unravel. "That's what we want to know," Moller explained. "Are we more like barn swallows or soybeans in terms of radiation-induced mutation?"

As the twenty-fifth anniversary of the Chernobyl disaster approached in 2011, the government of Ukraine pressed ahead with plans to open the Exclusion Zone as a tourist attraction. "The Chernobyl zone is not as scary as the whole world thinks," a spokeswoman told a British reporter. "We want to work with big tour operators and attract Western tourists, from whom there is great demand." The authorities had already tolerated the surreptitious return of more than a thousand peasants to their ancestral homes inside the zone, where they chose to live out their old age in remote seclusion, "aborigines of the nuclear reservation," subsisting on fruit and vegetables they grew themselves. Now the researchers working in the area feared that this new initiative was a prelude to reopening the area for full repopulation and were appalled—Sergei Gaschak, because he hoped the zone could become a permanent wildlife reserve where elk and lynx could live beyond the reach of hunters; Moller and Mousseau, because they feared for the long-term health of a human population exposed to the mutagens remaining in the environment.

But after a quarter century, the collective memory of the world's most devastating atomic accident had dimmed and softened. In the harsh light of climbing oil costs and global warming, governments were reconsidering the viability of nuclear power. The first contract to build a new nuclear power plant in the United States in more than thirty years was already under way. At the beginning of March 2011, Ukraine announced plans to begin construction of two new reactors not far from Chernobyl. The government in Kiev was still plotting the future of the forbidden zone when, on March 11, 2011, the news came in from the Tokyo Electric Power Company's nuclear plant in Fukushima, Japan.

The disaster involving the three General Electric–built reactors on the northeastern coast of Honshu followed a now familiar course, this time played out live on television: a loss of coolant led to reactor meltdown, a dangerous buildup of hydrogen gas, and several catastrophic explosions. No one was killed or injured by the immediate release of radiation, but three hundred thousand people were evacuated from the surrounding area, which will remain contaminated for decades to come. During the early stages of the emergency cleanup, it became clear that robots were incapable of operating in the highly radioactive environment inside the containment buildings of the plant. Japanese soldiers were sent in to do the work, in another Pyrrhic victory of bio-robots over technology.

Sweeping away the convenient fallacy that what had happened in Chernobyl had been a once-in-a-million-years fluke, the Fukushima accident stifled the nuclear renaissance in the cradle: the Japanese government immediately took all of its remaining forty-eight nuclear reactors off-line, and Germany shut down eight of its seventeen reactors, with the announced intention of closing the rest by 2022 as part of a move to renewable energy. Existing plans for all new reactors in the United States were suspended or canceled.

Yet nuclear power endured. More than seven years after the Japanese disaster, the United States still had a hundred licensed and operational power reactors—including one at Three Mile Island. France continued to generate 75 percent of its electricity from nuclear plants, and China had recently embarked on a reactor building spree, with twenty new units under construction and thirty-nine already in operation. Some environmentalists argued that humanity could not afford to turn its back on the promise and terrors of the peaceful atom. The global need for electricity was increasing exponentially: humanity was predicted to double the amount of energy it used by 2050. Despite the growing certainty that burning fossil fuels was the cause of devastating climate change—making the stabilization of carbon emissions imperative—coal remained the most widely used source of energy in the world. The fine particulates from fossil fuel plants in the United States killed more than thirteen thousand people a year; worldwide, three million people died annually as a result of air pollution released by coal- and oil-fueled power stations. Even to begin to head off climate change, all the extra power-generating capacity that the world would need to create over the coming thirty-five years would have to be clean, yet neither wind, solar, hydroelectric, nor geothermal power—nor any combination of them—had the potential to bridge the gap.

Nuclear power plants emit no carbon dioxide and have been statistically safer than every competing energy industry, including wind turbines. And at last, more than seventy years after the technology's inception, engineers were finally developing reactors with design priorities that lay not in making bombs but in generating electricity. In principle, these fourth-generation reactors would be cheaper, safer, smaller, more efficient, and less poisonous than their predecessors and could yet prove to be the technology that saves the world.

Less than a month before the explosion of Reactor Number Four in 1986, a team of nuclear engineers at Argonne National Laboratory–West in Idaho had quietly succeeded in demonstrating that the first of these, the integral fast reactor, was safe even under the circumstances that destroyed Three Mile Island 2 and would prove disastrous at Chernobyl and Fukushima. The liquid fluoride thorium reactor (LFTR), an even more advanced concept developed at Tennessee's Oak Ridge National Laboratory, is fueled by thorium. More plentiful and far harder to process into bomb-making material than uranium, thorium also burns more efficiently in a reactor and could produce less hazardous radioactive waste with half-lives of hundreds, not tens of thousands, of years. Running at atmospheric pressure, and without ever reaching a criticality, the LFTR doesn't require a massive containment building to guard against loss-of-coolant accidents or explosions and can be constructed on such a compact scale that every steel mill or small town could have its own microreactor tucked away underground.

In 2015 Microsoft founder Bill Gates had begun funding research projects similar to these fourth-generation reactors in a quest to create a carbon-neutral power source for the future. By then, the Chinese government had already set seven hundred scientists on a crash program to build the world's first industrial thorium reactor as part of a war on pollution. "The problem of coal has become clear," the engineering director of the project said. "Nuclear power provides the only solution."

———

By the time the thirtieth anniversary of the accident approached, the Exclusion Zone had opened to regular sightseeing tours from Kiev, and it seemed that the international scientific community had reached a reassuring consensus on the long-term health effects of the Chernobyl catastrophe. With the Soviet medical record fragmented and compromised by secrecy and cover-up, the mantle of scientific authority on the disaster had been assumed by the numerous nongovernmental organizations operating under the umbrella of the United Nations. And with each successive five-year milestone after the catastrophe, the World Health Organization, the United Nations Scientific Committee on the Effects of Atomic Radiation (UNSCEAR), and the IAEA all marched together toward the same conclusion: the public health effects of the Chernobyl accident "were not nearly as substantial as had at first been feared."

The Chernobyl Forum, a United Nations study group cooperating with the governments of Ukraine, Belarus, and Russia, estimated that by 2005, about four thousand people who were children at the time of the accident had developed thyroid cancer caused by iodine 131 from the reactor, leading to nine deaths. Their estimates suggested that up to five thousand fatal cancers might occur in the most heavily contaminated regions of the former USSR as a result of radiation released by the accident, making up part of a projected twenty-five thousand additional cancers attributable to the disaster in Europe as a whole. In a population of more than five million people living in the affected areas, the UN scientists viewed these numbers as barely statistically significant. They instead attributed the majority of disease in the fallout zones to psychological factors—a "paralyzing fatalism"—the latest incarnation of Soviet "radiophobia." In a follow-up report ten years later, the WHO noted that the recent discovery of a pattern of cataracts in liquidators had led to a downward revision of the safe-dose limits imposed on nuclear workers by the International Commission on Radiological Protection. They also noted an increase in cardiovascular disease among liquidators exposed to chronic low-dose radiation—but cautioned that this might also be the result of other factors, including poor diet, lack of physical activity, and stress.

Dr. Robert Gale, whose work in Hospital Number Six had made him both a minor celebrity and a big name in radiation medicine, had already announced that, in medical terms, it was time to move on. "Basically nothing happened here." he said, "Nothing happened here . . . and nothing is going to happen here."

Yet these conclusions were drawn almost entirely from studies conducted on groups of liquidators, often exposed to large doses of radiation, and thyroid cancer sufferers, or from broad risk-projection models. Little effort had been made to establish an internationally recognized body of data on the long-term consequences of the accident on the population at large, to replicate the seventy-year study of the Japanese survivors of the atomic bomb attacks in 1945. The UN agencies used the unreliable nature of dosimetry conducted among civilians as reason enough to abandon any further effort at lifespan studies, and, as a result, the opportunity to understand the long-term impact of low-dose radiation on human beings was lost. In the absence of large-scale epidemiological work, independent scientists from countries around the world continued to log "endocrinological, musculoskeletal, respiratory, and circulatory problems and a rise in malignant tumors, especially of the breast and prostate," among residents of the affected areas.

And in the space that remained, anxiety and misunderstanding about the real threats of radioactivity and nuclear power continued to multiply.

In Moscow, Kiev, and Minsk, and in towns and villages throughout the former Soviet Union, the surviving witnesses to the events of April 1986 carried on, in old age and declining health.

In the eastern Ukrainian city of Dnipro, I talked to Colonel Boris Nesterov, who had led the first helicopter pilots in their bombing raids over the reactor. He said that surgeons had already removed a fifth of his intestines, but he was still flying at the age of seventy-nine.

In the garden of his dacha in the countryside outside Kiev, a former KGB major explained that he had taken sick the night before and had intended to cancel our meeting, but his wife had persuaded him otherwise: it might be his last chance to share what he knew. Sitting in his snowbound cottage on the edge of a national park, Alexander Petrovsky, who had helped fight the fires on the roof of Unit Three, credited a life of fresh air and daily swims in the nearby river with saving him from the depression and alco-holism that plagued his former comrades. Yet Piotr Khmel, the firefighter who had sped to the scene of the explosion while drinking from a bottle of Soviet champagne, was still at work and insisted on serving celebratory cognac from a pistol-shaped decanter he kept on the desk in his office.

When I first met Maria Protsenko, the former chief architect of the city of Pripyat was approaching seventy. She lived alone with six cats in an apartment in the suburbs of Kiev and moved with difficulty on a pair of battered aluminum canes. Following a fall from the fourth floor of her building—she had locked herself out and was climbing into her apart-ment from a neighbor's balcony, a feat she had often pulled off success-fully in the past—the doctors had told her she was unlikely to walk again. But she had proved them wrong and continued to commute to the Salva-dor Dalí Art Institute in the city, where she taught interior design. Wear-ing a smart dark-gray suit and cream blouse, with the pin of the Union of Soviet Architects at her lapel, she said that immediately after the catastro-phe, she had been afraid to talk about everything she had seen—"because I knew how I could end up . . . the example of my grandfather was enough for me." But now she described everything in vivid detail, with the fond nostalgia of a military veteran coloring all but the darkest of episodes. She was still mourning the death of her husband and her son—both lost to cancer. The daughter who had whiled away her final afternoon in Pri-pyat watching a movie with her father simply didn't like to discuss what had happened. When we met again the following year, Protsenko brought along homemade Easter gifts, her original liquidator's pass, and the note-book she had used during her long months in the zone. "It still stinks of radiation . . . like rain—ozone," she said. When I couldn't identify the scent, she bent over the table and, to my obvious alarm, blew dust from the pages directly into my nostrils.

"*Tcha!*" she spat, her eyes twinkling with mischief. "If you needed to be afraid, I wouldn't have brought it!"

I found Viktor Brukhanov, one autumn morning shortly before his eight-ieth birthday, in the fourth-floor apartment he and his wife, Valentina,

had occupied together since his release from prison. Brukhanov had retired from his job at the Ukrainian Ministry of Energy when his eyesight
began to fail, and had become increasingly reclusive. Two strokes had
left him nearly blind and rendered his face all but immobile; his mind
remained sharp. He recalled the optimism and high expectations of his
first weeks in Chernobyl and his struggles with the Party bosses; he talked
about the demands of raising a city from scratch in the Pripyat marshes,
and the plans that grew ever larger, for more and bigger reactors, and a
second phase of the plant on the other side of the river. But when conversation turned to the night of the accident that destroyed Unit Four, he
rose slowly from his chair and withdrew to another room, leaving his wife
to take up the story.

When I returned to visit them again a few months later, Brukhanov
had suffered a third stroke. He had fallen badly and broken his left arm,
which doctors had strapped across his abdomen in a complicated gray
foam sling. Lying on a green velour couch in a small back room of the
apartment, his head propped on a pile of pillows, Brukhanov was wearing a
light-blue T-shirt, navy track pants, and thick socks. His hair was white and
cropped short; his skin dry, papery, the color of parchment; his dark-blue
eyes stared, unfocused, into the middle distance, and his free hand trembled. But when he spoke, although his words were mangled by numbed
lips and a slack tongue, they tumbled out as quickly as before. He defended
his actions on the night of the explosion and maintained that he had first
learned of the total destruction of Reactor Number Four only the next day,
when he circled it in a helicopter. At the trial, he had admitted his culpability as a manager for what had happened simply because that was his job.
"The director has primary responsibility for everything that's happening at
the plant and with the staff. So I had to."

He hadn't bothered to defend himself in court, he insisted, because he
knew that was how the Party had decided it would be. And once the USSR
collapsed, he'd had no interest in petitioning the Ukrainian authorities
to clear his name. "It was just pointless," he said. "Nobody will ever do
anything about it."

But if he acknowledged that he was still troubled by his own responsibility for the accident, he also talked almost as if the issue were an ad-

ministrative technicality. "I still feel responsible for the people and for the installation," he said.

And when I asked him to name his greatest regret, the ghosts of long-dormant ambition seemed to stir in him. He struggled to sit up.

"What I regret most is that I didn't live to see my office built at the top of the ten-story building, so I could watch over the first and second stages of the Chernobyl nuclear power plant," he said.

Valentina, apparently bewildered by this flash of technocratic hubris from the Soviet past, dabbed spittle from the corners of her husband's mouth with a polka-dot handkerchief. "I don't understand, Vitya," she told him. "I don't understand."

"According to the plan, a ten-story building was to be erected . . . ," he began, but trailed off. "I'm joking, of course."

Then the old man's sightless eyes found mine, his gaze hard and sapphire blue. For a moment, I felt plant director Viktor Brukhanov—winner of the Order of the Red Banner for Socialist Labor, holder of the Order of the October Revolution—staring straight back at me, and it seemed quite possible that there was no humor in this joke at all.

By the morning of April 26, 2016, the beautiful weather in Pripyat had turned suddenly cold; a frigid wind scythed down the river toward the plant, and rain whipped from a leaden sky. Beneath a massive arch erected a few hundred meters from the ailing Sarcophagus, the Ukrainian president, former chocolate magnate Petro Poroshenko, stood before a microphone. His amplified voice boomed off the stainless steel roof overhead, echoing like Zeus in a cheap film production of the Greek myths:

"Satan sleeps beside the Pripyat.

"He lies, damn him, in the guise of a dry willow on the bank of the Pripyat, on the bank of a river that was once blue and clear."

Directly in front of the president's podium, a group of construction workers in blue-gray jackets, corralled behind a fence of fluorescent orange tape, stamped their feet for warmth.

"And a black candle flickers for him in the atomic block.

"And villages lie in ruin and woe for him.

"His clawed roots snag in the sand.

"And the wind whistles into his hollow ear."

Behind Poroshenko, heavy trucks and excavators toiled up a muddy slope, and men wearing rubber boots and face masks prowled in the shadow of a new structure rising around the ruins of Unit Four. With each new gust of icy air blowing toward the president, the level of gamma radiation spiked sharply. The alarm on a pocket dosimeter beeped insistently; the site was still so contaminated that eating and drinking outside were forbidden.

"He scrawled obscenities upon the houses. Stole the icons, lost his respirator. And now he wants to rest.

"This is his kingdom. He is an emperor here."

His invocation complete, Poroshenko launched into a speech, carried live on national television, to mark exactly thirty years since the disaster. He spoke of the accident's catalytic role in Ukrainian independence and the breakup of the USSR and placed it on the continuum of events that threatened the state's very existence, halfway between the Great Patriotic War and the Russian invasion of Crimea in 2014. He described the enduring costs of the accident, the 115,000 people he said would never return to their homes in the Exclusion Zone, the 2.5 million more living on land contaminated by radionuclides, and the hundreds of thousands of disabled *Chernobyltsi* who continued to require support from the state and society. "The issue of the aftermath of the catastrophe is still open," he said. "Its heavy burden lay on the shoulders of the Ukrainian people, and we are, unfortunately, still very far from overcoming it for good."

The president turned, then, to the glinting arch that rose on the construction site above and around him, known by its designers as the New Safe Confinement—a new structure that, he announced, "will cover the Sarcophagus like a huge dome." The still-incomplete project had its roots in the fears raised first by the men of the Kurchatov Institute's Complex Expedition in 1990 and plans formulated by the G7 countries in 1997 but delayed for more than a decade by wrangling over who would foot the bill. The official cost of its completion would ultimately triple, to at least 1.5 billion euros—donated by a group of forty-three countries around the world—even though the funds were meticulously managed to prevent them from being siphoned away by the corruption submerging the government of Ukraine. Designed to stabilize and hermetically seal the decaying Sarcophagus, the building was one of the most ambitious civil engineering tasks ever undertaken: a giant steel arc 108 meters high—tall enough to contain the Statue of Liberty—packed with ventilation and dehumidifying equipment and three times as large as St. Peter's Basilica in Rome.

Its architects faced problems not encountered on any other construction project since the specialists of Sredmash US-605 had set down their tools in the winter of 1986. Reactor Number Four still remained too radioactive for work to take place around it, so the arch was being constructed on a separate site four hundred meters away, and the French contractors would then slide it into place using rails and dozens of hydraulic pistons. At thirty-six thousand tonnes, it would be the biggest land-based movable structure ever built. Even protected by a specially constructed concrete radiation shield, every one of the workers at the site had to be monitored for exposure. Working time was limited to periods ranging from hours to seconds.

Yet Poroshenko expressed confidence that with international help—including a fresh infusion of 87.5 million euros from the European Bank of Reconstruction and Development—his country would see the project to completion and finally banish the disaster to history. "Ukrainians are a strong people," he said, "that can overcome even the nuclear demon."

Six months later, mist and snow once again blanketed the fields beside the Pripyat as Poroshenko stood beside executives from the European Bank, the French ambassador to Ukraine, and the eighty-eight-year-old former International Atomic Energy Agency chief Hans Blix, for a grand dedication ceremony. Inside a heated canvas marquee, close to the same spot where Viktor Brukhanov and the *nomenklatura* from Moscow had once embarked on their grand project with a ceremonial stake and cups of cognac, a crowd of well-fed men in dark suits celebrated with champagne, hors d'oeuvres, and platters of profiteroles. At the entrance, young women in matching navy uniforms accessorized with scarlet neckerchiefs handed out lanyards hung with dosimetry buttons to monitor guests' radiation exposure. Others stepped into the snow to take self-portraits against the backdrop of the marvel of civil engineering outside. Slid into place at last, the massive span of the New Safe Confinement had entirely swallowed the black silhouette of the Sarcophagus. When sun broke through the heavy cloud, the steel glinted in the autumn light.

A fresh testament to the power of gigantomania, the finished structure made up in bulk what it lacked in elegance, with the aesthetics of an aircraft hangar and the presence of a suburban shopping mall. Back in Moscow, the original architects of the Sarcophagus regarded it with derision and insisted it was an absurd boondoggle. But if it worked as intended, the design would seal up the ruins of Unit Four in complete safety for another hundred years. "We have closed a wound—a nuclear wound—that belongs to all of us," Hans Blix told the crowd. The new building would also serve as a final monument to the last resting place of Valery Khodemchuk—a radioactive mausoleum to memorialize for generations to come the first victim of the accident.

Around his remains, the engineers hoped, the New Safe Confinement would provide a secure space within which the wreckage of the molten core of Reactor Number Four might finally be dismantled. Yet, even as the final deadline for the structure's completion approached, no one seemed certain how that might be accomplished. At least one veteran nuclear expert feared that even now, more than thirty years after the catastrophe began, neither man nor machine could work in such a hostile environment.

Epilogue

Anatoly Aleksandrov retired as the head of the Soviet Academy of Sciences in October 1986 and as director of the Kurchatov Institute in early 1988, but kept working there until his death in February 1994, at the age of ninety. He never accepted responsibility for the explosion of Reactor Number Four, and in an interview given shortly before he died continued to fault the plant operators for what happened: "You are driving a car, you turn the steering wheel in the wrong direction—an accident!" he said. "Is the engine to blame? Or the designer of the car? Everyone will reply, 'The unskilled driver is to blame.'"

Major General **Nikolai Antoshkin** was transferred to Moscow in 1989, promoted to colonel general, and eventually founded the first Russian aerobatic display team. As commander of the frontline aviation units of the Russian Federation, he oversaw airborne operations during the war in Chechnya. He retired from the air force in 1998, and in 2002 became chairman of the Society for Heroes of the Soviet Union. In 2014, Antoshkin was elected to the State Duma as a member of the ruling United Russia party.

Hans Blix retired as the director general of the IAEA in 1997. Three years later he was recalled to the United Nations to serve as the leader of the UN Monitoring, Verification, and Inspection Commission, charged with overseeing Iraqi compliance with its obligations to rid itself of weapons of mass destruction. In February 2003, his commission concluded that there were no such weapons in Iraq; a US-led force of more than 125,000 troops nevertheless invaded the country the following month. Shortly afterward, Blix left the UN for good.

Lev Bocharov, chief engineer of the third shift of Sredmash US-605, reported on the integrity of the Sarcophagus until the fall of the Soviet Union. In 1996, he was part of a Russian group that submitted their own plans for a replacement structure to the president of Ukraine, but which were rejected in favor of the European proposals. Twenty years later, he remained sprightly at eighty-one, living with his wife in a large house he designed himself in Zvenigorod, to the west of Moscow.

Alexander Borovoi continued to oversee the exploration and monitoring inside the Sarcophagus for almost twenty years after the accident and eventually located all but 5 percent of the missing fuel inside the building. He helped devise the original concept for the New Safe Confinement project and has since dedicated himself to cataloging and preserving the documentary record and practical lessons of Chernobyl.

After the dissolution of the Pripyat city council, **Alexander Esaulov** was resettled in a new home in the Kiev suburb of Irpin and eventually found a job in the bureaucracy of the Ukrainian energy industry. He embarked on a career as a writer and has since published twenty-seven books—many of them children's adventure stories. He still keeps the official seal of the Pripyat mayor's office on his desk and in his spare time gives occasional tours of the abandoned city for foreign tourists.

Dr. Robert Gale returned repeatedly to Moscow and Kiev in the years after the accident and became a well-known face throughout the Soviet Union. In 1988, he published a memoir about his experiences, *Final Warning*, which was adapted into a TV movie starring Jon Voight as Gale and Jason Robards as Armand Hammer. With an international reputation as an expert in the medical response to nuclear disaster, he attended the scenes of several major radiation accidents, including those in Goiânia, Brazil, in 1987, and Fukushima in 2011.

Following his fall from power, **Mikhail Gorbachev** set up a charitable foundation and think tank based in Moscow and struggled to remain influential in Russian politics. In 1996, he ran for president of the Russian Federation but won less than 1 percent of ballots counted. He would later insist that it was the explosion of Reactor Number Four—and not his own bungled reforms—that proved the catalyst in the destruction of the Union he had so desperately wished to preserve. In April 2006, he wrote: "The nuclear meltdown at Chernobyl twenty years ago this month, even more than my launch of *perestroika*, was perhaps the real cause of the collapse of the Soviet Union five years later. Indeed, the Chernobyl catastrophe was an historic turning point: there was the era before the disaster, and there is the very different era that has followed."

Dr. Angelina Guskova published extensively on what she discovered from her treatment of the patients in Hospital Number Six and lectured the staff of nuclear power plants throughout Russia on the lessons they could learn from the accident. She remained an advocate of nuclear power generation for the rest of her life and continued working at the Burnasyan Medical Center until shortly before her death in 2015 at the age of ninety-one.

After completing his final tour of duty in the Exclusion Zone, **Alexander Logachev** began lobbying to have the leading role of the 427th Mechanized Regiment of Civil Defense on the front line of the accident acknowledged by Moscow. In 1987, he was granted a personal audience with Mikhail and Raisa Gorbachev to press his case. Eventually sixty-four members of the regiment were granted medals and awards, but after his meeting with the general secretary Logachev was given orders for immediate transfer to Siberia. He was demobilized from the Soviet Armed Forces in 1989 and now practices alternative medicine.

Veniamin Prianichnikov died of complications from stomach cancer in Kiev in May 2014, aged seventy.

Maria Protsenko still teaches art, design, and architecture in Kiev. Every year on April 26, she puts on the medal she was awarded as a liquidator and places flowers at the memorial to those who died in the accident. Afterward, she lectures and takes questions from her students on what she recalls of the disaster and its consequences. She has not returned to Pripyat in more than thirty years.

Cliff Robinson left his job in the laboratory at Forsmark nuclear power station in the autumn of 1986. He then embarked on a year of research into the radioactive rain that fell over Sweden that spring, in pursuit of a PhD, but eventually became a high school physics teacher in Uppsala, where he lives today.

Prime Minister **Nikolai Ryzhkov** gradually split with Gorbachev over the course of Soviet economic reform and suffered a heart attack at the end of 1990. The following year, he lost the race to become the first president of the Russian Federation to Boris Yeltsin and began advocating a resurrection of the USSR and the planned economy. In 2014, aged eighty-four, he was sanctioned by the US government for his role in the Russian annexation of Crimea.

Boris Scherbina continued overseeing the liquidation of the consequences of the Chernobyl accident until December 1988, when Gorbachev sent him to Armenia as head of a new commission to marshal the Soviet emergency response to the devastating earthquake that killed twenty-five thousand people. By the time Scherbina arrived, the radiation he absorbed in Chernobyl had profoundly damaged his health, and the strain of handling a new disaster proved too much for him. Six months later, he was relieved of his duties on the Council of Ministers and died in August 1990 at the age of seventy.

Vladimir Scherbitsky remained a fierce opponent of both glasnost and the rise of Ukrainian nationalism and clung to power for years after the accident. In September 1989 he was finally dismissed from the Politburo by Gorbachev, surrendered control of the Ukrainian Communist Party to his deputy, and announced his retirement. A broken man, his health failing, he died less than a year later, at the age of seventy-one, on February 16, 1990. In April 1993, a year-long inquiry by the Prosecutor General's Office of the newly independent Ukraine concluded that Scherbitsky— along with his senior ministers—had deliberately concealed the truth about the Chernobyl accident and resulting radiation levels in Ukraine and had failed in his duty to protect the population of the republic. Due to Scherbitsky's death and the expiration of the Ukrainian statute of limitations on the charges, the case—little more than political theater—was dropped before a trial could begin.

Ukrainian energy minister **Vitali Sklyarov** adapted quickly to the post-Soviet world and became an enthusiastic advocate of privatizing state energy utilities but opposed the construction of new nuclear stations on economic and ecological grounds. In 1993, after more than thirty years as a power man, he resigned his ministerial position to become an advisor to then-prime minister Vitaly Masol. His few days' work in the Exclusion Zone left him with a dose of 80 rem, but he remains in good health in his eighties and continues to spend time at his former government dacha in Koncha-Zaspa.

Following his enforced retirement, **Efim Slavsky** lived out his remaining years in a grand apartment in Moscow, increasingly deaf and surrounded by mementos of his decades in power. A furious witness to the disintegration of the political system to which he had devoted his life, he died at the age of ninety-three in November 1991.

Boris Stolyarchuk survived his radiation exposure in Control Room Number Four on the night of the accident and afterward returned to work in the nuclear industry. In 2017, he was named acting head of the Ukrainian State Nuclear Regulatory Inspectorate.

Following treatment for radiation-induced leukemia in Hospital Number Six, General **Nikolai Tarakanov** returned to work in Soviet disaster relief and attended the scene of the Armenian earthquake in 1988. He began writing poetry in the hospital, published thirty books, and toured the United States, giving lectures about his work in the Exclusion Zone and the continuing risk of nuclear accidents. In 2016, he turned eighty-two and announced that his latest book would be a biography of Vladimir Putin, entitled *Supreme Commander in Chief*.

Vladimir Usatenko used the 1,400 rubles he earned during his six weeks' work as a liquidator in the Special Zone to buy his first color television set. In 1990 he was elected to the Ukrainian parliament and appointed chairman of a subcommittee dedicated to the scientific, social, and legal issues created by the disaster and nuclear power in Ukraine.

Following the suicide of Valery Legasov and the retirement of Anatoly Aleksandrov, Academician **Evgeny Velikhov** was appointed director of the Kurchatov Institute in 1988. He became president of the organization in 1992 and led Russian participation in the ITER multinational project to develop an experimental plasma fusion reactor. In 2001 he was appointed by President Vladimir Putin to lead the Russian contribution to an international initiative to develop new nuclear power technologies proposed by Putin at the UN Millennium Summit.

Detective **Sergei Yankovsky** did not attend the trial of the six men he helped indict for their part in the disaster but returned to his regular caseload of murder and corruption. In 1995, he was transferred to work at the Ukrainian Rada and began campaigning to have the fifty-seven volumes of investigative materials on the Chernobyl accident returned to Kiev from Moscow. When he left the job eight years later, the boxes of documents and tape recordings remained in the basement of the Russian Supreme Court, still classified top secret. In the spring of 2017, he was sixty-one years old and recuperating from a recent illness on the grounds of a state sanatorium in Kiev. "There are so many things in those files that no one will ever know," he said.

Natalia Yuvchenko lives and works in Moscow, close to her son, Kirill, and his wife, and her three grandchildren.

After pumping out the basement of Unit Four, Captain **Piotr "Moose" Zborovsky** was promoted to major. He was awarded the Order of the Red Star, with a citation commending him "for mastering new equipment and weaponry." In 1993, he was transferred to the civil defense reserve and took work first as a caretaker, then as a security guard. But he had lost his vaunted strength, suffered repeated blackouts, and found his bones began to fracture easily. He died in 2007 at the age of fifty-five.

Acknowledgments

The origins of this project go back many years, with roots in the story that I first followed in the news as a teenager and returned to decades later as a magazine writer. During that time I've received invaluable help from friends and colleagues all over the world. It was inspired, and made possible, by the men and women whose lives were changed by the explosion of Reactor Number Four and who agreed to share with me the stories of their time in Pripyat, the Chernobyl plant, and the other atomic cities and installations of the USSR. From my first meeting with Alexander and Natalia Yuvchenko one gray afternoon in Moscow, I have been welcomed into the homes of people who showed me great kindness, hospitality, and patience, even when discussing the most traumatic of events; I'm grateful to all of them for agreeing to be cross-examined by a foreign stranger in the interests of bringing their experience wider attention. I'd also like to thank Anna Korolevska for helping me make contact with many of those witnesses to the disaster, and to Elena Kozlova, Tom Lasica, Maria Protsenko, and Nikolai Steinberg, who provided me with vital leads and introductions that proved essential to recreating an accurate account of what happened.

My first reporting from Chernobyl was made possible by my editors at the *Observer Magazine*, Allan Jenkins and Ian Tucker, who—after some spirited debate—agreed to trust me with a thick envelope of hard currency and dispatch me on that initial trip to Russia and Ukraine. Later, my editors at *Wired* in London and San Francisco helped me make subsequent forays into the Exclusion Zone, each of which I may well have sworn would be absolutely my last. As I embarked upon the reporting for the book itself, I was lucky to benefit from the advice of Piers Paul Read, who was generous with both his time and encouragement, and practical guidance in navigating the nations of the former USSR from Natalia Lentsi, Andrey Slivka, Micky Lachmann, Fiona Cushley, and Matt McAllester.

Katia Bachko, Peter Canby, David Kortava, Tali Woodward, Joshua Yaffa, and Polina Sinovets all helped guide me toward the outstanding Russian-speaking researchers and fact-checkers I've mentioned elsewhere; Eugenia Butska, Anton Povar, and Gennadi Milinevsky allowed me to explore otherwise inaccessible parts of the city of Pripyat, the Chernobyl nuclear power plant, and the Exclusion Zone that proved central to the narrative of the book. I'd also like to thank Andrea Gallo, Inna Lobanova-Heasley, Michael Wilson, Michael D. Cooper, and Gunnar Bergdahl for their help along the way.

Among the many other memorable experiences I've enjoyed while reporting in Ukraine over the years, I'd particularly like to thank Roman Shumeyko for his rightly legendary *plov* and a great evening of lakeside drinking. Closer to home, I'm grateful to Rose George, Greg Williams, James and Ana Freedman, Yudhijit Bhattarcharjee, Brendan Koerner, Julie Satow, Ted Conover, Evan Ratliff, Nick Thompson, and Keith Gessen for their valuable advice and support.

Ihor Kulyk and the staff of the Ukrainian Institute of National Remembrance, and Christian Ostermann and the researchers of the Wilson Center Cold War History Project, all helped me find and translate key documents in Ukraine; Melanie Locay and the staff of the New York Public Library's research division gave me access to materials I could not otherwise have hoped to get my hands on, and to the privileges of the Allen Room, where I enjoyed the time and silence necessary to use them properly.

Throughout the long process of making the initial idea into a book, I've been able to rely on the urbane, unflappable guidance of my agent, Edward Orloff; I'd also like to thank Millicent Bennett, Henry Vines, Michelle Kroes, and Scott Rudin, who were all early believers in the project. At Simon & Schuster, Jon Karp and Ben Loehnen shared my enthusiasm for the story from the beginning, provided exacting feedback on the manuscript—and, finally, a wonderful title; Amar Deol shepherded me expertly through the often bewildering logistics of publication, and

Kayley Hoffman, Phillip Bashe, and Josh Cohen saved me from numerous typographical and literal mistakes.

I'm greatly indebted to Katie Mummah, Rob Goldston, Frank Von Hippel, and Alexander Sich for talking me through the complexities of atomic physics and nuclear engineering and reviewing those parts of the manuscript dealing with the design and drawbacks of the RBMK-1000 reactor; Timothy Jorgensen displayed similar generosity and patience in helping me understand the science and terminology of radiobiology and radiation medicine; any errors or oversimplifications that remain are entirely my own.

For keeping me sane when I finally decamped to Jay Street to wrangle my mountains of material into a book, I'd like to thank Chris Heath, Lauren Hilgers, Nathan Thornburgh, and the good people of Roads and Kingdoms. And to all my friends who have endured the long, itinerant progress of my reporting this and many other projects over the years and given up their couches, mattresses, and spare rooms so I had somewhere to stay— Toby Amies, Andrew Marshall, Peg Rawes, and Tom Corby; David Keeps, Ian Tucker, Michael Odell, Dan Crane, Kate and John; Micky and Lisa; Rupert, Julie, Stella, Soren, and Nancy; Matt, Pernilla and Harry—thank you. And belated apologies for not doing more washing up.

Most of all, I'm grateful to my frankest critic and most tireless cheerleader, Vanessa Mobley, and our daughter, Isla, for their love and forbearance through years of long absences, endless revisions, and of having to enter our apartment through a hallway booby-trapped with potentially radioactive footwear and piles of books about the Soviet economy; I could not have got here without you.

New York, September 2018

Author's Note

This is a book of history, but also a work of reporting. In recreating the immediate experience of living through the disaster, I relied on interviews I began conducting with eyewitnesses in 2006, as well as upon published first-person accounts and declassified Soviet documents. This would not have been possible without the dedicated assistance of some superb translators, fixers, and researchers. These included Olga Ticush, Misha Smetnick, Anna Sorokin, and Artemis Davleyev in Russia; and Alex Livotka, Ostap Zdorovylo, Natalia Mackessy, Tetiana Vodianytska, and Dmytro Chumak in Ukraine. In New York, James Appell spent many months chasing elusive sources and translating documents and correspondence; Anna Kordunsky took on the work of fact-checking the manuscript and proved a tireless and unsparing collaborator on everything from the technical details of reactor design to the finer points of Russian etymology.

But the book would not exist at all without the expertise and support of Taras Shumeyko. My guide and reporting companion over the course of a dozen years, in visits to survivors of the catastrophe in provincial cities and snowbound villages, at international conferences and inside the Exclusion Zone, Taras helped track down many of the men and women whose testimony proved central to the narrative. He charmed and cajoled reluctant veterans and archivists and unearthed documents and conducted interviews that became fundamental to the story.

Several Russian-language books reported in the months after the accident provided an excellent source of first-person testimony that informs the narrative. Both Iurii Shcherbak's *Chernobyl* and Vasily Voznyak and Stanislav Troitsky's *Chernobyl: It Was Like This—The View from the Inside* furnished valuable eyewitness perspective on what happened, and the collection of personal essays written by participants in the liquidation gath-

ered by A. N. Semenov in *Chernobyl: Ten Years On: Inevitability or Accident?*
provided an essential guide to the experience of the ministers, nuclear engineers, and other specialists caught up in the catastrophe. Elena Kozlova's
The Battle With Uncertainty is the definitive account of the work surrounding the Sarcophagus, told almost entirely in the words of those who built it.

In English, the transcription taken contemporaneously from testimony at the trial of Viktor Brukhanov by Nikolay Karpan and translated
in his book *From Chernobyl to Fukushima* was indispensable in building
a picture of events and attitudes leading to the explosion. The unedited
transcripts of interviews conducted for the BBC TV series *The Second
Russian Revolution* in 1990 and 1991, held in the library of the London
School of Economics, revealed candid accounts of the accident by aging
members of the Politburo and other senior Soviet figures whose views
would otherwise have remained out of reach. My road map to the personal histories of those who worked at the Chernobyl plant was *Ablaze:
The Story of the Heroes and Victims of Chernobyl* by Piers Paul Read, who
interviewed almost every one of the principal actors in the story to create
one of the first comprehensive English-language accounts of what happened.

The collapse of the Soviet Union, the revolutions in Ukraine, and
the three decades since the disaster have helped open many archives of
once-classified official records. The minutes and transcripts of meetings
of the Politburo and of Prime Minister Nikolai Ryzhkov's Chernobyl Operations Group proved especially useful in separating facts from the Soviet
myth of the accident—although I have treated these with some caution.
The minutes of the Politburo were taken in different forms, varying in
length, detail, and potential veracity, and have emerged from the Russian
archives both as photographs of the original documents and, in some
cases, as transcripts copied down by researchers given privileged access to
archives that have opened briefly, only to be closed again. I'm grateful to
both Mark Kramer, program director of the Project on Cold War Studies
at the Davis Center for Russian and Eurasian Studies at Harvard, and the
Russian Chernobyl historian Vladimir Maleyev for their guidance in this
area. The paper trail of many important decisions and all the early deliberations of the government commission in Pripyat and Chernobyl remains

scant; some sources suggest that documents were destroyed either at the time or soon afterward to limit the spread of both information and contamination. I'm therefore greatly indebted to Anna Korolevska, deputy director for science of the Chernobyl Museum in Kiev, for allowing me access to the original Ukrainian Interior Ministry working record of the accident—maintained from the early hours of the morning on April 26 until May 6, 1986, and apparently preserved by the officer entrusted with its destruction—which provided an invaluable insight into the evolving response to events as they happened.

The wider story of the accident, stretching back to the initial decision to build a nuclear power station near Kiev in the late 1960s, through the catastrophe itself and up to the present day, is well represented in several collections of documents: notably "Fond 89: The Soviet Communist Party on Trial," part of the remarkable collection of more than nine million previously secret Party records photographed in the Russian State Archive of Contemporary History (RGANI) and preserved by the Hoover Institution, Stanford, a copy of which I was able to read in Harvard's Lamont Library. An even broader selection of primary sources has been collated by the Ukrainian historian Natalia Baranovska in her book *The Chernobyl Tragedy: Documents and Materials*; the Ukrainian Institute of National Remembrance has built a growing collection of documents pertaining to the disaster in an online archive; and the involvement of the Ukrainian KGB with the Chernobyl plant is related in detail by the 121 documents published by Yuri Danilyuk as "The Chernobyl Tragedy in Documents and Materials." Perhaps most important, I also had access to the primary materials—logbooks, personal letters, maps, photographs, and government records—curated over the course of more than twenty years by the Chernobyl Museum.

The volume of technical literature so far published on the disaster is overwhelming, but for an understanding of the ways in which the attitudes and imperatives of the Soviet nuclear industry opened the path to disaster inside Reactor Number Four, I relied upon the research of Sonja D. Schmid and Paul Josephson, in particular their respective books *Producing Power* and *Red Atom*. For an appreciation of the complexities of radiation science, I found useful guides in Robert Gale and Thomas

Hauser's *Final Warning* and Timothy Jorgensen's excellent *Strange Glow*. To accurately recreate the sequence of the accident itself, I drew on the abundant detail contained in the IAEA report known as INSAG-7 and the extraordinary work done by Dr. Alexander Sich for his PhD dissertation, "The Chornobyl Accident Revisited: Source Term Analysis and Reconstruction of Events During the Active Phase."

Glossary

Academy of Sciences: The most distinguished academic institute in the USSR, dedicated to basic research in mathematics and the natural, physical, and social sciences, with equivalents in each of the Soviet republics.

Aktiv: A group of Communist Party activists entrusted with enacting the decisions of the Party at local level, especially in the workplace.

Apparat: The bureaucracy of the Soviet state, particularly of the Communist Party itself.

Apparatchik: A functionary of the apparat.

Atomshchiki: Nuclear engineers; the elite of atomic power specialists, often trained in the military facilities of the Ministry of Medium Machine Building.

Central Committee: In theory, the highest decision-making body of the Communist Party, directing the activities of Party committees, ministries, and enterprises throughout the USSR. In practice, subordinate to the much smaller Politburo.

Council of Ministers: The Soviet cabinet, with counterparts at the top of the government in each republic, which enacted the decisions of the Politburo.

Energetiki: Power engineers, including those operating nuclear stations, directly employed by the Ministry of Energy and Electrification.

Era of Stagnation: The period after 1970, given its name by Mikhail Gorbachev, in which the Soviet economy and culture began to ossify.

General Secretary: Leader of the Communist Party of the Soviet Union and de facto head of state.

Gorkom: The district committee of the Communist Party, responsible for enacting Party decisions at town or city level.

Gosplan: The state committee for economic planning, the brain of the centralized economy.

Ispolkom: A local government committee appointed to manage the affairs of a town, city, or district; the equivalent of a Western municipal council.

Kolkhoz: A collective farm.

Komsomol: The All-Union Leninist Young Communist League, the youth wing of the Communist Party, with membership limited to ages fourteen through twenty-eight.

Kurchatov Institute of Atomic Energy: The USSR's principal agency for research and development in nuclear energy, originating as the secret Laboratory Number Two of the Academy of Sciences, which had been dedicated to building the Soviet atom bomb.

Militsia: The regular police force maintained by the MVD; despite its name, not to be confused with a militia.

Minenergo: The Ministry of Energy and Electrification, the Soviet civilian energy ministry.

MVD: *Ministerstvo vnutrennykh del*—the Ministry of Internal Affairs, the paramilitary agency responsible for the police, firefighting, and internal security.

NIKIET: *Nauchno-issledovatelskiy i konstruktorskiy institut energotekhniki*—the Scientific Research and Design Institute of Energy Technology, the agency where the designs of Soviet nuclear reactors, including the RBMK, were created.

NIKIMT: *Nauchno-issledovatelskiy i konstruktorskiy institut montazhnoy tekhnologii*—the Research and Design Institute of Assembly Technology, the Sredmash department devoted to developing technical solutions for nuclear construction.

Nomenklatura: The elite of the Communist Party, appointed to key positions with access to special privileges and high salaries, and the name given to the system of lists that assigned them.

Oblast: An administrative district or province. In 1986 Ukraine was divided into twenty-four oblasts, of which the Kiev oblast was the largest. Each oblast was further subdivided into *rayons*, or regions, similar to the bureaucratic hierarchy of US counties or city boroughs.

OPAS: *Gruppa okazaniya pomoschi atomnym stantsiyam pri avariyakh*—the Ministry of Energy's emergency response team dedicated to accidents at nuclear power plants.

Politburo: The "political bureau" of the Central Committee of the Communist Party of the USSR, intended to assume decision-making between meetings of the Central Committee. In fact, the seat of ultimate power in the USSR.

RBMK: *Reaktor bolshoy moschnosti kanalnyy*—a graphite-moderated boiling water reactor, a design unique to the Soviet Union.

Soyuzatomenergo: The All-Union Industrial Association for Nuclear Power, the civilian agency within the Ministry of Energy overseeing the operation of atomic power plants.

Sredmash: The Ministry of Medium Machine Building, responsible for the Soviet nuclear weapons program and all reactor technology.

Supreme Soviet: Nominally, the parliament of the USSR, responsible for formulating legislation for the entire Union, with equivalent subsidiary bodies in each of the member republics.

VNIIAES: *Vsesoyuznyy nauchno-issledovatelskiy institut po ekspluatatsii atomnykh elektrostantsiy*—All-Union Research Institute for the Operation of Nuclear Power Plants, the civilian body providing research support for atomic power stations.

VVER: *Vodo-vodyanoy energetichesky reaktor*—a water-cooled and water-moderated reactor, similar to the Western pressurized water reactor.

Young Pioneers: The Communist Party's politically indoctrinating answer to the Scouts, for children aged ten through fourteen.

ZhEK: *Zhilishchno-ekspluatatsionnaya kontora*—the housing and maintenance office of city apartment buildings.

Units of Radiation

There are numerous ways of measuring radiation and its effects, and the terminology has evolved repeatedly since the science emerged more than a century ago. Although scientists now use standardized international (SI) units, the ones used in this book are those prevailing in the USSR at the time of the Chernobyl accident, principally the roentgen and the rem. To aid the reader in understanding the meaning of these older units, and their relation to their replacements, each is described here.

curie (Ci): A quantity of *radioactivity*, originally defined by the number of radioactive disintegrations in one gram of radium (approximately 37,000,000,000 disintegrations per second). The Ci has been replaced by a newer standardized unit, the **becquerel.**

rad: Radiation absorbed dose, a measurement of the *dose* of ionizing radiation deposited in a given mass of matter—whether a brick, a pine tree, or a human organ. The SI unit of absorbed dose is the **gray (Gy)**, which represents one joule of ionizing radiation energy deposited per kilogram of matter. One hundred rad is equivalent to one gray.

roentgen (R): A measurement of *exposure* to X-rays and gamma radiation, based on the amount of energy deposited by ionizing radiation in a mass of air. A thousandth of a roentgen is a **milliroentgen (mR)**; a millionth of a roentgen is a **microroentgen (μR)**. Exposure over time can be expressed in **roentgen per hour (R/h)**. Normal background radiation in the USSR in 1986 was stipulated at between 4 and 20 microroentgen per hour.

rem: Roentgen equivalent man, quantifies the *health effects* of exposure to ionizing radiation. The rem measures *dose equivalency* and is calculated using different factors, including the absorbed dose and the type of radiation involved. It can be used to predict the biological effect of a dose, including cancer, regardless of whether it is delivered by alpha, beta, neutron, X-ray, or gamma radiation. One rem is a little less than the citizens of Denver, Colorado, absorb from natural background radiation in the course of a year; 5 rem is the annual exposure limit for US nuclear workers; 100 rem is the threshold of acute radiation syndrome; and an instantaneous dose of 500 rem to the whole body would be lethal to most people. The rem has been replaced by the standard international unit, the **sievert** (**Sv**), and its smaller subunits; these are the **millisievert** (**mSv**)—a thousandth of a sievert, and the **microsievert**—one millionth of a sievert (**μSv**), which is used on the displays of most modern dosimeters. One sievert is equivalent to 100 rem.

Notes

PROLOGUE

1 *Saturday, April 26, 1986:* Precise time given on Alexander Logachev's dosimetry map of Chernobyl station from April 26, 1986, archive of the Chernobyl Museum, Kiev, Ukraine.

1 *Senior Lieutenant Alexander Logachev loved radiation:* Alexander Logachev, Commander of Chemical and Radiation Reconnaissance, 427th Red Banner Mechanized Regiment of the Kiev District Civil Defense, author interview, Kiev, June 1, 2017; Yuli Khariton, Yuri Smirnov, Linda Rothstein, and Sergei Leskov, "The Khariton Version," *Bulletin of the Atomic Scientists* 49, no. 4 (1993), p. 30.

1 *Logachev knew how to protect himself:* Logachev, author interview, 2017.

1 *As he sped through the suburbs:* Alexander Logachev, *The Truth* [Истина], memoir, 2005, later published in another form in *Obozreniye krymskih del*, 2007; Colonel Vladimir Grebeniuk, commander of 427th Red Banner Mechanized Regiment of the Kiev District Civil Defense, author interview, Kiev, February 9, 2016.

2 *But as they finally approached the plant:* Logachev, *The Truth.*

2 *Their armored car crawled counterclockwise:* Logachev dosimetry map of Chernobyl station, the Chernobyl Museum.

3 *2,080 roentgen an hour:* Logachev, *The Truth.*

Part 1. Birth of a City

1. THE SOVIET PROMETHEUS

7 *At the slow beat:* Viktor and Valentina Brukhanov (husband and wife; director and heat treatment specialist at Chernobyl nuclear power plant in April 1986), author interviews, Kiev, September 2015 and February 2016. Author visit to Kopachi, Ukraine, February 17, 2006. Cognac and the driving of the stake are mentioned in the documentary film *The Construction of the Chernobyl Nuclear Power Plant* [Будівництво Чорнобильської АЕС], Ukrainian Studio of Documentary Chronicle Films, 1974. A still photograph of the ceremony is included in the documentary film *Chernobyl: Two Colors of Time* [Чернобыль: Два цвета времени], directed by I. Kobrin (Kiev: Ukrtelefim, 1989), pt. 3 mark 40:05, www.youtube.com/watch?v=keEcEHQipAY.

7 *If the Soviet central planners had their way:* Zhores A. Medvedev, *The Legacy of Chernobyl* (New York: Norton, 1990), 239; "Controversy Around the Third Phase of the Chernobyl NPP," *Literaturnaya Gazeta*, May 27, 1987, translated in "Aftermath of Chernobyl Nuclear Power Plant Accident, Part IV," Joint Publication Research Service, Soviet Union: Political Affairs (hereafter, JPRS, Soviet Union Political Affairs), 111.

7 *They had considered a few options:* Vitali Sklyarov, *Chernobyl Was . . . Tomorrow*, trans. Victor Batachov (Montreal: Presses d'Amérique, 1993), 22.

8 *A small but ancient town:* Alexander Sich, "The Chornobyl Accident Revisited: Source Term Analysis and Reconstruction of Events During the Active Phase" (PhD diss., Massachusetts Institute of Technology, 1994), 203.

8 *The town of Chernobyl:* Richard F. Mould, *Chernobyl Record: The Definitive History of the Chernobyl Catastrophe* (Boca Raton, FL: CRC Press, 2000), 312.

8 *Two and a half million citizens:* Between 1979 and 1989, Kiev's population grew from about 2.2 million to 2.6 million: V. A. Boldyrev, *Results of USSR Population Census [Итоги переписи населения СССР]* (Moscow: USSR State Committee on Statistics, 1990), 15: http://istmat.info/files/uploads/17594/naselenie_sssr ._po_dannym_vsesoyuznoy_perepisi_nas eleniya_1989g.pdf. Also see Sich, "The Chornobyl Accident Revisited," 196.

8 *Viktor Brukhanov had arrived:* Viktor and Valentina Brukhanov, author interview, 2015; author visit to the city of Chernobyl, April 25, 2016.

8 *The oldest of four children:* Viktor Brukhanov, interview by Oleg Nikolaevich, "Stories about Tashkent Natives: True and Sometimes Unknown. Part 1" [Истории о ташкентцах правдивые и не всем известные. Часть 1], *Letters about Tashkent*, April 29, 2016: http://mytashkent.uz/2016/04/29/istorii-o-tashkenttsah-pravdivye -i-ne-vsem-izvestnye-chast-1.

8 *He had an exotic look:* Major Vasily Lisovenko (head of the Third Division of the Sixth Department, Ukrainian KGB), author interview, September 2016.

9 *At the Ministry of Energy:* Lisovenko, author interview, 2016. Since the revolution, it had been common Soviet practice to appoint loyal Party members to the top jobs in technical enterprises and have them be advised by specialists. Grigori Medvedev, transcript of interview made during the production of *The Second Russian Revolution*, a 1991 BBC documentary film series: 2RR archive file no. 1/3/3, 16 (hereafter 2RR).

9 *In July 1969:* Neporozhny argued for this step in a letter to Soviet Premier Alexei Kosygin on July 4, 1969. Sonja D. Schmid, *Producing Power: The Pre-Chernobyl History of the Soviet Nuclear Industry* (Cambridge, MA: MIT Press, 2015), 34n97.

9 *He set ambitious targets:* Charles Dodd, *Industrial Decision-Making and High-Risk Technology: Siting Nuclear Power Facilities in the USSR* (Lanham, MD: Rowman & Littlefield, 1994), 73–74.

9 *The first atomic power plant in Ukraine:* V. A. Sidorenko, "Managing Atomic Energy," in V. A. Sidorenko, ed., *The History of Atomic Energy in the Soviet Union and Russia [История атомной энергетики Советского Союза и России]* (Moscow: Izdat, 2001), 219.

9 *400 million rubles:* The total cost estimate for building Chernobyl was 389.68 million rubles in 1967. See Document No. 1 in N. Baranovska, ed., *The Chernobyl Tragedy: Documents and Materials* [Чорнобильська Трагедія: Документи і матеріали] (Kiev: Naukova Dumka, 1996): "Appeal from the Council of Ministers of USSR to the Central Committee of Communist Party of Ukraine to approve the project of building the Central Ukrainian nuclear power station near the village of Kopachi, Chernobyl district, Kiev region," February 2, 1967.

9 *He drew up lists of the materials:* Viktor and Valentina Brukhanov, author interview, 2015. Brukhanov also described his responsibilities during these early days in an interview with Maria Vasyl, *Fakty i kommentarii:* "Former ChNPP director Brukhanov: 'Had they found legal grounds to have me shot, they would have done so.'" [Бывший директор ЧАЭС Виктор Брюханов: «Если бы нашли для меня расстрельную статью, то, думаю, расстреляли бы.»], October 18, 2000, http://fakty.ua/104690-byvshij-direktor-chaes-viktor-bryuhanov-quot-esli-by-nashlidlya-menya-rasstrelnuyu-statyu-to-dumayu-rasstrelyali-by-quot.

10 *Before Brukhanov could start:* Baranovska, ed., *The Chernobyl Tragedy*, Document No. 7: "The joint decision of subdivisions of the USSR Ministry of Energy and Electrification on constructing a temporary cargo berth for the Chernobyl NPP," April 29, 1970.

10 *A cluster of wooden huts:* Brukhanov, interview by Vasyl, *Fakty i kommentarii,* 2000.

11 *As shock-work brigades excavated:* Vasily Kizima (construction supervisor at Chernobyl), author interview, Kiev, Ukraine, February 2016. Gennadi Milinevsky (Kiev University student sent to Chernobyl construction site to assist with building work in the summer of 1971), author interview, Kiev, April 2016. "Shock workers"—*udarniki*—was the name given to those members of the Soviet workforce who regularly exceeded plan targets and participated in Communist labor competitions. By 1971 there were 17.9 million shock workers in the USSR: Lewis Siegelbaum, "Shock Workers;" Seventeen Moments in Soviet History, http://soviethistory.msu.edu/1929-2/shock-workers/.

11 *The first group of nuclear specialists to arrive:* Nikolai Steinberg, author interview, Kiev, Ukraine, September 2015.

11 *According to Soviet planning regulations:* Schmid, *Producing Power,* 19.

11 *Its residents began to build summer houses:* Alexander Esaulov (deputy chairman of the Pripyat *ispolkom*), author interview, Irpin, Ukraine, July 2015.

11 *Viktor Brukhanov's initial instructions:* Brukhanov, interview by Vasyl, *Fakty i kommentarii,* 2000; Steinberg, author interview, 2015.

12 *Enough to serve at least a million modern homes:* Electricity consumption varies according to many factors, including geographical location, but this conservative estimate is based upon figures provided by the Nuclear Regulatory Commission for twenty-first-century homes in the northeastern United States: "What Is a Megawatt?", February 4, 2012, www.nrc.gov/docs/ML1209/ML120960701.pdf.

12 *The deadlines set by his bosses:* Baranovska, ed., *The Chernobyl Tragedy*, document

No. 10: "Resolution of the USSR Ministry of Energy and Electrification on the organization and implementation of operations to oversee the physical and energy launch of the NPPs under construction on USSR territory," July 29, 1971. Steinberg, personal communication with author, August 6, 2018.

12 *The USSR was buckling under the strain:* Some Soviet historians estimate that, in real terms, the USSR's annual spending on troops and armaments before 1972 amounted to between 236 and 300 billion rubles a year—and by 1989 represented almost half of the state budget. Yevgenia Albats, *The State Within a State: The KGB and Its Hold on Russia—Past, Present, and Future*, trans. Catherine Fitzpatrick (New York: Farrar, Straus and Giroux, 1999), 189.

12 *From the beginning:* Baranovska, ed., *The Chernobyl Tragedy*, document No. 13: "Resolution of the Communist Party of Ukraine and the Council of Ministers of the USSR on the Construction Progress of the Chernobyl Nuclear Power Plant," April 14, 1972.

12 *Key mechanical parts:* Schmid, *Producing Power*, 19; George Stein, "Pipes Called 'Puff Pastry Made of Steel,'" *Los Angeles Times*, May 16, 1986; Piers Paul Read, *Ablaze: The Story of the Heroes and Victims of Chernobyl* (New York: Random House, 1993), 30 and 46–47.

12 *The quality of workmanship:* Sklyarov, *Chernobyl Was . . . Tomorrow*, 163. (The Russian term *pred-montazhnaya reviziya oborudovaniya* is translated here as "pre-erection overhaul," but "installation" is closer to the sense of "montage" in the original. See the original Russian edition of the book, *Завтра был . . . Чернобыль*. Moscow: Rodina, 1993, 165.)

12 *Throughout late 1971:* Baranovska, ed., *The Chernobyl Tragedy*, document no. 13; Vladimir Voloshko, "The Town That Died at the Age of Sixteen" [Город, погибший в 16 лет], undated, Pripyat.com, http://pripyat.com/people-and-fates/gorod-pogibshii-v-16-let.html.

13 *And still there was more:* These construction targets were mandated by Ukrainian Party leaders for the period 1972 to 1974. See Baranovska, ed., *The Chernobyl Tragedy*, document no. 13.

13 *For an appointment with his boss:* Brukhanov's supervisor Artem Grigoriants headed the Ministry of Energy's directorate for nuclear power (Glavatomenergo), tasked with overseeing Chernobyl construction and enforcing deadlines.

13 *The Party had originated:* The genesis of the Communist Party is described in detail in Robert Service, *A History of Modern Russia* (Cambridge, MA: Harvard University Press, 2010), 47–99.

13 *"a classless society":* Raymond E. Zickel, ed., *Soviet Union: A Country Study* (Washington, DC: US Government Printing Office, 1991), 281.

13 *Known collectively as the* nomenklatura: Theodore R. Weeks, *Across the Revolutionary Divide: Russia and the USSR, 1861–1945* (Chichester, UK: Wiley-Blackwell, 2010), 77.

14 *This institutionalized meddling:* The confusion and infighting of the early years of Party bureaucracy was revealed in the captured archives described by Merle

Fainsod in *Smolensk Under Soviet Rule* (Cambridge, MA: Harvard University Press, 1958).

14 *By 1970, fewer than one in fifteen:* Communist Party membership in 1970 was approximately 13.4 million. A. M. Prokhorov, ed., *Great Soviet Encyclopedia* [Большая Советская Энциклопедия], vol. 24 (Moscow, 1997), 176.

14 *Viktor Brukhanov joined in 1966:* Viktor Brukhanov, interview by Sergei Babakov, "I don't accept the charges against me . . ." [«С предъявленными мне обвинениями не согласен . . .»], *Zerkalo nedeli,* August 27, 1999, https://zn.ua /society/c_predyavlennymi_mne_obvineniyami_ne_soglasen.html.

14 *"Had the face of a truck driver":* Read, *Ablaze,* 31.

14 *The humiliation:* Sklyarov, *Chernobyl Was . . . Tomorrow,* 172.

15 *Seated at the top:* Vladimir Shlapentokh, *A Normal Totalitarian Society: How the Soviet Union Functioned and How It Collapsed* (Armonk, NY: M.E. Sharpe, 2001), 56; Stephen Kotkin, *Armageddon Averted: The Soviet Collapse, 1970–2000,* 2nd ed. (New York: Oxford University Press, 2003), 67.

15 *A game of chance:* Angus Roxburgh, *Moscow Calling: Memoirs of a Foreign Correspondent* (Berlin: Birlinn, 2017), 28–30.

16 *Crops rotted in the fields:* David Remnick, *Lenin's Tomb: The Last Days of the Soviet Empire* (New York: Vintage Books, 1994), 249.

16 *Soft spoken but sure of himself:* See Sklyarov, *Chernobyl Was . . . Tomorrow,* 119 and 122. Vitali Sklyarov, Ukrainian Energy Minister at the time of the accident, encountered Brukhanov frequently both in the years leading up to April 26, 1986, and immediately afterward.

16 *Yet when Brukhanov arrived:* Viktor and Valentina Brukhanov, author interviews, 2015 and 2016.

16 *Thirteen years later:* Viktor and Valentina Brukhanov, author interview, 2016. A photograph of an earlier November 7 celebration in Pripyat, in 1984, is available at "Pripyat Before the Accident. Part XIX," Chernobyl and Pripyat electronic archive titled *Chernobyl—A Little About Everything* [Чернобыль: Обо Всём Понемногу], November 14, 2012, http://pripyat-city.ru/uploads/posts/2012-11 /1352908300_slides-04.jpg.

16 *Hailed for his illustrious achievements:* Viktor and Valentina Brukhanov, author interview, 2016.

16 *Construction was also well under way:* Zhores Medvedev, *Legacy of Chernobyl,* 239; Lyubov Kovalevska, "Not a Private Matter" [Не приватна справа], *Literaturna Ukraina,* March 27, 1986, online at www.myslenedrevo.com.ua/uk/Sci/Hist Sources/Chornobyl/Prolog/NePryvatnaSprava.html.

17 *The Soviet counterpart to MIT:* Paul R. Josephson, *Red Atom: Russia's Nuclear Power Program from Stalin to Today* (Pittsburgh, PA: University of Pittsburgh Press, 2005), 55.

17 *A glossy book:* Yuri Yevsyukov, *Pripyat* [Припять] (Kyiv: Mystetstvo, 1986), available online at http://pripyat-city.ru/books/57-fotoalbom.html.

17 *The average age:* Vasily Voznyak and Stanislav Troitsky, *Chernobyl: It Was Like*

This—The View from the Inside [Чернобыль: Так это было—взгляд изнутри] (Moscow: Libris, 1993), 223.

17 *Watched as a pair of elk:* Viktor and Valentina Brukhanov, author interview, 2015.

17 *Financed directly from Moscow:* Esaulov, author interview.

17 *In the Raduga—or Rainbow—department store:* Raduga sold everything from furniture to toys. Natalia Yuvchenko (teacher at School Number Four, Pripyat; wife of Chernobyl senior mechanical engineer Alexander Yuvchenko), author interview, Moscow, Russia, October 2015.

18 *An office on the fifth floor of the* ispolkom: Svetlana Kirichenko (chief economist of the Pripyat *ispolkom*), author interview, Kiev, April 2016.

18 *Trouble was confined:* Alexander Esaulov, *The City That Doesn't Exist* [Город, которого нет] (Vinnytsia: Teza, 2013), 14; Viktor Klochko (head of Pripyat department of the KGB), interview by Taras Shumeyko, Kiev, September 2015.

18 *Each spring, the river:* Anatoly Zakharov (fire engine driver and lifeguard in Pripyat), author interview, Kiev, Ukraine, February 2016.

18 *Could look back on a year of triumphs:* Viktor and Valentina Brukhanov, author interview, 2016.

19 *Speaking unselfconsciously:* Remnick, *Lenin's Tomb*, 144–47.

19 *Hammered on the anvil:* Sklyarov, *Chernobyl Was . . . Tomorrow*, 123.

19 *Orders of merit and pay bonuses:* For example, the Party granted medals to seven Chernobyl engineers involved in the launch of Unit Four in December 1983. "Resolution 144/2C of the Central Committee of the Communist Party of the Soviet Union" [Постановление Секретариата ЦК Коммунистической Партии Советского Союза № СТ 144/2C], March 6, 1984, Microfilm, Hoover Institution, Russian State Archive of Contemporary History (RGANI), Opis 53, Reel 1.1007, File 33.

20 *Not even being manufactured in the USSR:* Kizima, author interview, 2016.

20 *"Go build it!":* Viktor Brukhanov, interview by Vladimir Shunevich, "Former director of the Chernobyl Atomic Power Station Viktor Brukhanov: 'At night, driving by Unit Four, I saw that the structure above the reactor is . . . gone!'" [Бывший директор Чернобыльской Атомной Электростанции Виктор Брюханов: «Ночью, проезжая мимо четвертого блока, увидел, что верхнего строения над реактором . . . Нету!»], *Fakty i kommentarii*, April 28, 2006, http://fakty.ua/104690-byvshij-direktor-chaes-viktor-bryuhanov-quot-esli-by-nashlidlya-menya-rasstrelnuyu-statyu-to-dumayu-rasstrelyali-by-quot.

20 *He found the extra funds:* Viktor and Valentina Brukhanov, author interview, 2015.

20 *The last day of December:* The commissioning date of Unit Four is specified in Nikolai Karpan, *From Chernobyl to Fukushima*, trans. Andrey Arkhipets (Kiev: S. Podgornov, 2012), 143.

20 *At the beginning of the 1980s:* Sich, "The Chornobyl Accident Revisited," 148.

20 *A dense network:* Schmid, *Producing Power*, 34.

20 *Brought forward by a year:* David R. Marples, *Chernobyl and Nuclear Power in the USSR* (New York: St. Martin's Press, 1986), 120.

20 *Labor and supply problems:* Kovalevska, "Not a Private Matter." David Marples translates and discusses excerpts of Kovalevska's article in *Chernobyl and Nuclear Power in the USSR*, 122–24. Also see an interview with Kovalevska by the journalist Iurii Shcherbak in his book *Chernobyl: A Documentary Story*, trans. Ian Press (New York: St. Martin's Press, 1989), 15–21.

20 *A team of dedicated KGB agents:* By the time of the accident the station was monitored by 91 KGB agents, 8 residents, and 112 "authorized persons," according to Volodymyr Viatrovych, head of the Ukrainian Institute of National Remembrance (lecture in Kiev, April 28, 2016, www.youtube.com/watch?v=HJpQ4SWx HKU). For an example of KGB reports on Chernobyl supply and construction problems, see Document No. 15, "Special report of the 6th department of the UkSSR KGBM . . . concerning the facts of shipping of poor-quality equipment for the Chernobyl NPS from Yugoslavia," from January 9, 1984, in Yuri Danilyuk, ed., "The Chernobyl Tragedy in Documents and Materials" [Чорнобильська трагедія в документах та матеріалах], Special Issue, *Z arkhiviv VUChK–GPU–NKVD–KGB* 1, no. 16 (2001).

20 *Brukhanov received instructions:* Viktor Kovtutsky (chief accountant at Chernobyl construction department), author interview, Kiev, Ukraine, April 24, 2016.

21 *An office at the top:* Viktor and Valentina Brukhanov, author interview, 2016.

21 *Any time of the day or night:* Sklyarov, *Chernobyl Was . . . Tomorrow*, 123.

21 *If something went wrong:* Author interviews: Viktor and Valentina Brukhanov, 2015; Steinberg, 2015; Serhiy Parashyn (Communist Party secretary of ChNPP), Kiev, November 30, 2016.

21 *Despite his technical gifts:* Author interviews with Parashyn and Kizima.

21 *Derided him as a "marshmallow":* Vasily Kizima, account in Grigori Medvedev, *The Truth About Chernobyl*, trans. Evelyn Rossiter (New York: Basic Books, 1991), 141.

21 *A moral decay:* Steinberg, author interview, 2017; Valery Legasov, "My duty is to tell about this," a translated recollection of the Chernobyl accident provided in full in Mould, "Chapter 19: The Legasov Testament," *Chernobyl Record*, 298.

21 *The USSR's economic utopianism:* Alec Nove, *The Soviet Economy: An Introduction*, 2nd rev. ed. (New York: Praeger, 1969), 258.

21 *The inexorable construction work:* Brukhanov described twenty-five thousand construction workers in need of steady employment in his interview with *Fakty i kommentarii* in 2000. Details on Jupiter employees, many of them women, come from Esaulov, *The City That Doesn't Exist*, 13, as well as the author's interview with Esaulov in 2015.

21 *The highly qualified technical elite:* Schmid, *Producing Power*, 87.

22 *The "power men":* Ibid., 90.

22 *"you could spread it on bread":* Alexander Nazarkovsky (senior electromechanical engineer at Chernobyl), author interview, Kiev, February 2006.

22 *"like a samovar":* Legasov, "My duty is to tell about this," 300.

22 *Drank from glassware:* Anna Korolovska (deputy director for science of the Chernobyl Museum), author interview, Kiev, July 2015.

22 *Listlessly filled out their shifts:* Read, *Ablaze*, 45.

22 *Group of Effective Control:* Ibid.

22 *At the top:* Steinberg, author interview, 2017; Schmid, *Producing Power*, 153. For a discussion of how staff turnover amplified the problem, see Marples, *Chernobyl and Nuclear Power in the USSR*, 120.

22 *The chief engineer:* Grigori Medvedev, *The Truth About Chernobyl*, 44.

22 *Pushed through by the Party:* Gennadi Shasharin, who in 1986 served as Deputy Minister of Energy and Electrification of the USSR, discusses Fomin's appointment in "The Chernobyl Tragedy," in A. N. Semenov, ed., *Chernobyl. Ten Years On. Inevitability or Accident?* [Чернобыль. Десять лет спустя. Неизбежность или случайность?] (Moscow: Energoatomizdat, 1995), 98.

22 *Physics through a correspondence course:* Expert testimony in trial transcript, Karpan, *From Chernobyl to Fukushima*, 148.

22 *The decision had already been taken:* Shasharin, "The Chernobyl Tragedy," 98.

23 *The news would be announced:* Esaulov, author interview, 2015.

23 *A lush variety of trees and shrubs:* "Chernobyl NPP: Master Plan of the Settlement" [Чернобыльская АЭС: Генеральный план поселка], Ministry of Energy and Electrification of the USSR, 1971, 32.

23 *But Brukhanov was especially fond of flowers:* Author interviews, 2015: Esaulov, Kirichenko, and Viktor and Valentina Brukhanov; Viktor Brukhanov, interview by Anton Samarin, "Chernobyl hasn't taught anyone anything" [Чернобыль никого и ничему не научил], *Odnako*, April 26, 2010, www.odnako.org/magazine /material/chernobil-nikogo-i-nichemu-ne-nauchil-1/.

23 *On the elevated concrete plaza:* Maria Protsenko (chief architect for the city of Pripyat), author interview, Kiev, September 5, 2015. A photo of the placeholder monument can be seen at "Pripyat Before the Accident: Part XVI," Chernobyl and Pripyat electronic archive, December 2011, http://pripyat-city.ru/uploads/posts /2011-12/1325173857_dumbr-01-prc.jpg.

2. ALPHA, BETA, AND GAMMA

25 *Unimaginably dense:* Robert Peter Gale and Eric Lax, *Radiation: What It Is, What You Need to Know* (New York: Vintage Books, 2013), 12.

25 *"the strong force":* Robert Peter Gale and Thomas Hauser, *Final Warning: The Legacy of Chernobyl* (New York: Warner Books, 1988), 6.

25 *"neither mass nor energy":* Ibid.

25 *In 1905 Albert Einstein:* Ibid., 4–6.

26 *580 meters above the Japanese city:* Richard Rhodes, *The Making of the Atomic Bomb* (New York: Simon & Schuster, 1988), 711.

26 *The bomb itself was extremely inefficient:* Emily Strasser, "The Weight of a Butterfly," *Bulletin of the Atomic Scientists* website, February 25, 2015; Jeremy Jacquot, "Numbers: Nuclear Weapons, from Making a Bomb to Making a Stockpile to Making Peace," *Discover*, October 23, 2010.

26 *Some seventy-eight thousand people died instantly:* As a result of the chaos and de-

struction caused by the bombing, and uncertainty about the number of people present in the city at the time, figures for the total death toll directly attributable to the explosion vary significantly, and the true numbers will never be known. These figures are part of a "best estimate" for casualties from Paul Ham, *Hiroshima Nagasaki: The Real Story of the Atomic Bombings and Their Aftermath* (New York: Thomas Dunne Books/St. Martin's Press, 2014), 408.

26 *The atoms of different elements vary by weight:* Gale and Hauser, *Final Warning*, 6.

26 *Adding to or removing neutrons:* Fred A. Mettler Jr., and Charles A. Kelsey, "Fundamentals of Radiation Accidents," in Igor A. Gusev, Angelina K. Guskova, Fred A. Mettler Jr., eds., *Medical Management of Radiation Accidents* (Boca Raton, FL: CSC, 2001), 7; Gale and Hauser, *Final Warning*, 18.

26 *Radiation is all around us:* Craig Nelson, *The Age of Radiance: The Epic Rise and Dramatic Fall of the Atomic Era* (New York: Simon & Schuster, 2014), 3–4.

27 *Radon 222, which gathers as a gas:* Gale and Lax, *Radiation*, 13 and 17–18.

27 *Polonium 210, a powerful alpha emitter:* Ibid., 20.

27 *It was also the poison:* John Harrison et al., "The Polonium-210 Poisoning of Mr Alexander Litvinenko," *Journal of Radiological Protection* 37, no. 1 (February 28, 2017). The FSB was organized in 1995 as the principal Russian state security service and the successor to the KGB.

28 *Gamma rays—high frequency electromagnetic waves:* Gale and Hauser, *Final Warning*, 18–19.

28 *Severe exposure to all ionizing radiation:* Mettler and Kelsey, "Fundamentals of Radiation Accidents," 7–9; Dr. Anzhelika Barabanova, author interview, Moscow, October 14, 2016.

28 *To the atomic pioneers:* Gale and Lax, *Radiation*, 39.

28 *"I have seen my own death!":* Timothy Jorgensen, *Strange Glow: The Story of Radiation* (Princeton, NJ: Princeton University Press, 2016), 23–28.

28 *In 1896 Thomas Edison devised:* Ibid., 31–32; US Department of the Interior, "The Historic Furnishings Report of the National Park Service, Edison Laboratory," 1995, 73, online at www.nps.gov/parkhistory/online_books/edis/edis_lab_hfr .pdf. A photograph of the box can be found in Gilbert King, "Clarence Dally: The Man Who Gave Thomas Edison X-Ray Vision," Smithsonian.com, March 14, 2012.

28 *Even as the damage caused:* Jorgensen, *Strange Glow*, 93–95.

29 *In 1903 Marie and Pierre Curie:* Gale and Lax, *Radiation*, 43–45.

29 *Because radium can be mixed:* Jorgensen, *Strange Glow*, 88–89.

29 *A successful lawsuit:* Gale and Lax, *Radiation*, 44.

29 *Some radionuclides:* Timothy Jorgensen, associate professor in the Department of Radiation Medicine at Georgetown University, author interview, telephone, June 19, 2016.

29 *The survivors of the atom bomb:* National Research Council, *Health Risks from Exposure to Low Levels of Ionizing Radiation: BEIR VII Phase 2* (Washington, DC: National Academies Press, 2006), 141.

30 *Of those who lived through the initial explosion in Nagasaki:* Data provided by Masao Tomonaga (head of the Atomic Bomb Disease Institute at Nagasaki University), cited in Gale and Lax, *Radiation*, 52–57.

30 *The effect of ionizing radiation:* James Mahaffey, *Atomic Awakening: A New Look at the History and Future of Nuclear Power* (New York: Pegasus, 2009), 286–89 and 329–33. Also see Dwayne Keith Petty, "Inside Dawson Forest: A History of the Georgia Nuclear Aircraft Laboratory," *Pickens County Progress*, January 2, 2007, online at http://archive.li/GMnGk.

30 *On August 21, 1945:* Daghlian's estimated whole-body dose was 5,100 millisieverts, equal to 510 rem. Jorgensen, *Strange Glow*, 111; James Mahaffey, *Atomic Accidents: A History of Nuclear Meltdowns and Disasters: From the Ozark Mountains to Fukushima* (New York: Pegasus Books, 2014), 57–60.

31 *Admitted himself to medical care:* A colleague of Daghlian's at Los Alamos, Joan Hinton, recalls driving him to the hospital after he walked out of the building just as she happened to pull up in her car. See Ruth H. Howes and Caroline L. Herzenberg, *Their Day in the Sun: Women of the Manhattan Project* (Philadelphia: Temple University Press, 1999), 54–55.

31 *Attributed his death to burns:* "Atomic Bomb Worker Died 'From Burns,'" *New York Times*, September 21, 1945. See also Paul Baumann, "NL Man Was 1st Victim of Atomic Experiments," *The Day*, August 6, 1985.

31 *"the Article":* David Holloway, *Stalin and the Bomb: The Soviet Union and Atomic Energy, 1939–1956* (New Haven, CT: Yale University Press, 1996), 213. The Soviet bomb's American predecessor, the device detonated in New Mexico's Jornada del Muerto desert in 1945, was known by the scientists who constructed it as "the Gadget."

31 *Noted by his secret police minders:* Svetlana Kuzina, "Kurchatov wanted to know what stars were made of—and created bombs" [Курчатов хотел узнать, из чего состоят звезды. И создал бомбы], *Komsomolskaya Pravda*, January 10, 2013, www.kp.ru/daily/26012.4/2936276.

32 *Initially modeled on the ones in Hanford:* Although the production reactor was at first slated to follow a horizontal design like those in Hanford, it was ultimately based on a vertical design by the Soviet engineer Nikolai Dollezhal (Holloway, *Stalin and the Bomb*, 183; Schmid, *Producing Power*, 45).

32 *Kurchatov had succeeded:* Mahaffey, *Atomic Awakening*, 203. The full title of the book was *Atomic Energy for Military Purposes: The Official Report on the Development of the Atomic Bomb Under the Auspices of the United States Government, 1940–1945*. Fifty thousand copies of the Russian translation were printed and made available to Soviet scientists (Josephson, *Red Atom*, 24).

32 *Nuclear work was the responsibility:* In Russian, the First Main Directorate was known as PGU, short for *Pervoye Glavnoye Upravleniye*. Roy A. Medvedev and Zhores A. Medvedev, *The Unknown Stalin*, translated by Ellen Dahrendorf (New York: I. B. Tauris, 2003), 133; Simon Sebag Montefiore, *Stalin: The Court of the Red Tsar* (New York: Knopf, 2004), 501–2.

32 *By 1950, the First Main Directorate:* Medvedev and Medvedev, *Unknown Stalin*, 134 and 162.

32 *In direct proportion to the punishment:* Holloway, *Stalin and the Bomb*, 218–19.

32 *Only as a concession:* Ibid., 347.

32 *It was not until the end of 1952:* Josephson, *Red Atom*, 20–26.

33 *Theoretically capable of wiping out humanity:* Gale and Lax, *Radiation*, 48.

33 *Kurchatov was shaken:* Holloway, *Stalin and the Bomb*, 307 and 317.

33 *Part of an attempt to mollify:* Stephanie Cooke, *In Mortal Hands: A Cautionary History of the Nuclear Age* (New York: Bloomsbury, 2010), 106–11.

33 *No one was especially surprised:* Josephson, *Red Atom*, 173.

33 *It marked the first time in twenty years:* Cooke, *In Mortal Hands*, 113.

33 *Just enough to drive a locomotive:* Schmid, *Producing Power*, 97.

34 *Arrested, imprisoned, and shot:* Montefiore, *Stalin*, 652.

34 *The First Main Directorate was reconstituted:* Schmid, *Producing Power*, 45 and 230n29.

34 *Newly appointed Soviet premier:* Josephson, *Red Atom*, 11.

34 *With the success of Atom Mirny-1:* Ibid., 4–5.

34 *The physicists who worked on AM-1:* Paul Josephson, "Rockets, Reactors, and Soviet Culture," in Loren Graham, ed., *Science and the Soviet Social Order* (Cambridge, MA: Harvard University Press, 1990), 174.

34 *"near-mythic figures":* Josephson, *Red Atom*, 11. The Great Patriotic War was the name given to the Soviet fight against Nazi Germany, beginning with the German invasion of the USSR in June 1941.

34 *Not all that it seemed:* Ibid., 25; Schmid, *Producing Power*, 45.

35 *Atom Morskoy, or "Naval Atom":* Ibid., 46.

35 *Inherently unstable:* Josephson, *Red Atom*, 26–27.

35 *Has even aligned spontaneously in nature:* Evelyn Mervine, "Nature's Nuclear Reactors: The 2-Billion-Year-Old Natural Fission Reactors in Gabon, Western Africa," *Scientific American*, July 13, 2011.

36 *30 billion fissions every second:* Ray L. Lyerly and Walter Mitchell III, *Nuclear Power Plants*, rev. ed. (Washington, DC: Atomic Energy Commission, 1973), 3; Bertrand Barré, "Fundamentals of Nuclear Fission," in Gerard M. Crawley, ed., *Energy from the Nucleus: The Science and Engineering of Fission and Fusion* (Hackensack, NJ: World Scientific Publishing, 2016), 3.

36 *"shakes," for a "shake of a lamb's tail":* Chuck Hansen, *U.S. Nuclear Weapons: The Secret History* (Arlington, TX: Aerofax, 1988), 11.

36 *Fortunately, among the remaining 1 percent of neutrons:* World Nuclear Association, "Physics of Uranium and Nuclear Energy," updated February 2018, www .world-nuclear.org/information-library/nuclear-fuel-cycle/introduction/physics -of-nuclear-energy.aspx; Robert Goldston and Frank Von Hippel, author interview, Princeton, NJ, February 2018.

36 *By inserting electromechanical rods:* Goldston and Von Hippel, author interview, 2018.

37 *To generate electricity:* The first reactor in the United Kingdom was the GLEEP, or Graphite Low Energy Experimental Pile, which began operation at Harwell, Oxfordshire, in 1947. In the United States, the first experimental boiling water reactor was created by the Argonne National Laboratory in 1956. See "Nuclear Development in the United Kingdom," World Nuclear Association, October 2016; and "Boiling Water Reactor Technology: International Status and UK Experience," Position paper, National Nuclear Laboratory, 2013.

37 *The first Soviet reactors:* Frank N. Von Hippel and Matthew Bunn, "Saga of the Siberian Plutonium-Production Reactors," Federation of American Scientists Public Interest Report, 53 (November/December 2000), https://fas.org/faspir/v53n6 .htm; Von Hippel and Goldston, author interview, 2018.

37 *It was a risky combination:* Mahaffey, *Atomic Awakening,* 206–7.

37 *Three competing teams:* Josephson, *Red Atom,* 25; Schmid, *Producing Power,* 102.

37 *But the Soviet engineers' work:* Holloway, *Stalin and the Bomb,* 347.

37 *The more experimental:* Josephson, *Red Atom,* 56.

37 *The first major defect with their design:* Ibid., 27.

37 *When working normally:* "RBMK Reactors," World Nuclear Association, June 2016, www.world-nuclear.org/information-library/nuclear-fuel-cycle/nuclear-power -reactors/appendices/rbmk-reactors.aspx.

38 *Revealed a fantastical vision:* Igor Kurchatov, "Speech at the 20th Congress of the Communist Party of the Soviet Union," in Y. N. Smirnov, ed., *Igor Vasilyevich Kurchatov in Recollections and Documents* [Игорь Васильевич Курчатов в воспоминаниях и документах], 2nd ed. (Moscow: Kurchatov Institute/Izdat, 2004), 466–71.

39 *Four different reactor prototypes:* V. V. Goncharov, "The Early Period of USSR Atomic Energy Development" [Первый период развития атомной энергетики в СССР], in Sidorenko, ed., *The History of Atomic Energy,* p. 19; Schmid, *Producing Power,* 20.

39 *But before construction could begin:* Schmid, *Producing Power,* 22 and 26–27.

39 *Like their counterparts in the West:* Ibid., 18.

39 *"too cheap to meter":* This infamous phrase was coined in September 1954 by Lewis Strauss, chairman of the U.S.' Atomic Energy Commission, in a speech to the National Association of Science Writers, and has haunted the nuclear industry ever since. Thomas Wellock, " 'Too Cheap to Meter': A History of the Phrase," United States Nuclear Regulatory Commission Blog, June 3, 2016.

39 *Yet the sheer size of the Union:* Schmid, *Producing Power,* 22.

40 *Apparently convinced by Kurchatov's promises:* Ibid., 21. The first station, in its design stage in 1956, would become Novovoronezh nuclear power plant. The second, under construction since 1954, was the Beloyarsk Nuclear Power Station (103 and 275n125).

40 *Construction costs skyrocketed:* Ibid., 29.

40 *In the meantime, the Ministry of Medium Machine Building:* Ibid., 106 and 266n41; Holloway, *Stalin and the Bomb,* 348.

40   *The EI-2, or "Ivan the Second":* Holloway, *Stalin and the Bomb*, 348 and 443n16.

40   *A new emphasis:* Schmid, *Producing Power*, 34.

40   *A toothy, young Queen Elizabeth II: The Atom Joins the Grid*, London: British Pathé, October 1956, www.youtube.com/watch?v=DVBGk0R15gA.

41   *A costly fig leaf:* In the film *Windscale 1957: Britain's Biggest Nuclear Disaster* (Sarah Aspinall, BBC, 2007), British journalist Chapman Pincher says, "I believe there were times when it was taking electricity out of the grid rather than pumping it in." See also Lorna Arnold, *Windscale 1957: Anatomy of a Nuclear Accident* (New York: St. Martin's Press, 1992), 21; and Mahaffey, *Atomic Accidents*, 181.

41   *Two thousand tons of graphite:* For a diagram of the Windscale reactor, see Mahaffey, *Atomic Accidents*, 163.

41   *It burned for two days:* Rebecca Morelle, "Windscale Fallout Underestimated," October 6, 2007, BBC News; Arnold, *Windscale 1957*, 161.

41   *A board of inquiry:* Arnold, *Windscale 1957*, 78–87.

41   *Did not fully acknowledge the scale:* Mahaffey, *Atomic Accidents*, 181. The unexpurgated account of the Windscale fire, known as the Penney Report, was declassified and released to the public in January 1988. Mahaffey gives a detailed description of the fire in *Atomic Accidents*, 159–81.

42   *They began using gamma rays:* Josephson, *Red Atom*, 4, 142–43, 147, and 248. It is only fair to add that US scientists also pursued food irradiation with some enthusiasm, with the FDA approving bacon irradiated using a cobalt 60 source for human consumption, in 1963 (160). The physicist Edward Teller was also a keen— but frustrated—proponent of "peaceful nuclear explosions," or PNEs, and the US military developed several mobile reactors of their own.

42   *A dedicated Communist:* Ibid., 113–17.

42   *Ozymandian dreams:* Ibid., 117–18 and 246–49.

43   *"Big Efim" and "the Ayatollah":* Sklyarov, *Chernobyl Was . . . Tomorrow*, 10–11. Also see recollections about Slavsky collated by V. Y. Bushmelev, "For Efim Pavlovich Slavsky's 115th Birthday" [К 115-летию Ефима Павловича Славского], Interregional Non-Governmental Movement of Nuclear Power and Industry Veterans, October 26, 2013, www.veteranrosatom.ru/articles/articles_276.html.

43   *Although as young men:* Angelina Guskova, interview by Vladimir Gubarev, "On the Edge of the Atomic Sword" [На лезвии атомного меча], *Nauka i zhizn*, no. 4, 2007; Igor Osipchuk, "The legendary academician Aleksandrov fought with the White Guard in his youth" [Легендарный академик Александров в юности был белогвардейцем], February 4, 2014, *Fakty i kommentarii*, http://fakty.ua/176084 -legendarnyj-prezident-sovetskoj-akademii-nauk-v-yunosti-byl-belogvardejcem.

43   *A massive nuclear empire:* Schmid, *Producing Power*, 53; "The Industry's Evolution: Introduction" [Эволюция отрасли: Введение], Rosatom, www.biblioatom.ru /evolution/vvedeniye.

43   *"post office boxes":* Fedor Popov, *Arzamas–16: Seven Years with Andrei Sakharov* [Арзамас-16: семь лет с Андреем Сахаровым] (Moscow: Institut, 1998), 52; Schmid, *Producing Power*, 93.

43 *Led by Slavsky:* Schmid, *Producing Power,* 50 and 234n55.

43 *The clandestine impulse persisted:* Although most nuclear research would eventually be conducted by scientists answering to the seemingly transparent State Committee for the Utilization of Atomic Energy, this was also simply a public front for Sredmash. Nikolai Steinberg recalls that, long before the fall of the USSR, this false distinction between Sredmash and the State Committee was well known to foreign specialists—"as they say, 'everything is confidential, but nothing is secret.'" Georgi Kopchinsky and Nikolai Steinberg, *Chernobyl: On the Past, Present and Future* [Чернобыль: О прошлом, настоящем и будущем] (Kiev: Osnova, 2011), 123. Later, the Soviet government established a regulatory body—the State Committee on Safety in the Atomic Power Industry, which sent representatives to oversee operating conditions at every nuclear power plant in the Union. But the committee never published reports and operated under conditions of strict secrecy. Zhores Medvedev, *Legacy of Chernobyl,* 263–64; Schmid, *Producing Power,* 50–52, 60, and 235n58.

44 *As one of the twelve founding members:* David Fischer, *History of the Atomic Energy Agency: The First Forty Years* (Vienna: IAEA, 1997), 40 and 42–43.

44 *The safest nuclear industry in the world:* In contrast to the representatives of the apparently more troubled industries in the United States, Britain, and France, the Soviet delegation never reported a single incident at a reactor or reprocessing plant (Medvedev, *Legacy of Chernobyl,* 264–65).

44 *At 4:20 p.m on Sunday, September 29, 1957:* Kate Brown, *Plutopia: Nuclear Families, Atomic Cities, and the Great Soviet and American Plutonium Disasters* (Oxford: Oxford University Press, 2015), 232.

44 *An underground waste storage tank:* G. Sh. Batorshin and Y. G. Mokrov, "Experience in Eliminating the Consequences of the 1957 Accident at the Mayak Production Association," International Experts' Meeting on Decommissioning and Remediation After a Nuclear Accident, IAEA, Vienna, Austria, January 28 to February 1, 2013.

45 *Light rain and a thick, black snow:* Brown, *Plutopia,* 239.

45 *It took a year to clean up:* Ibid., 232–36.

45 *In the remote villages:* V. S. Tolstikov and V. N. Kuznetsov, "The 1957 Radiation Accident in Southern Urals: Truth and Fiction" [Южно-уральская радиационная авария 1957 года: Правда и домыслы], *Vremya* 32, no. 8 (August 2017): 13; Brown, *Plutopia,* 239–44.

45 *Up to a half million people:* Some scientists estimate that 475,000 people could have been exposed (Mahaffey, *Atomic Accidents,* 284), but others, in particular official Russian sources, cite much lower numbers of about 45,500 people. See Russian Ministry of Emergency Situations, "The Aftermath of the Man-Made Radiation Exposure and the Challenge of Rehabilitating the Ural Region" [Последствия техногенного радиационного воздействия и проблемы реабилитации Уральского региона], Moscow, 2002, http://chernobyl-mchs.ru/upload/program _rus/program_rus_1993-2010/Posledstviy_Ural.pdf.

45 *Shot down by a Soviet SA-2:* Oleg A. Bukharin, "The Cold War Atomic Intelligence Game, 1945–70," *Studies in Intelligence* 48, no. 2 (2004): 4.

3. FRIDAY, APRIL 25, 5:00 P.M., PRIPYAT

46 *Almost everyone was looking forward to the long weekend:* Kovtutsky, author interview, 2016.

46 *"Let the atom be a worker, not a soldier!":* Maria Protsenko, author interview, Kiev, September 2015. A photograph of the slogan can be seen at "Pripyat Before the Accident: Part IX," Chernobyl and Pripyat electronic archive, March 25, 2011, http://pripyat-city.ru/uploads/posts/2011-04/1303647106_50008255-pr-c.jpg.

47 *Yuvchenko had worked at the station for only three years:* Alexander Yuvchenko, author interview, 2006.

47 *It was only over the objections of his trainer:* Alexander's younger brother Vladimir chose rowing and represented the USSR in the 1988 Olympics in Seoul. Natalia Donets et al., *25 Years of the National Olympic and Sports Committee of the Republic of Moldova* [25 de ani ai Comitetului National Olimpic si Sportiv din Republica Moldova] (Chisinau: Elan Poligraph, 2016), 16.

47 *Something futuristic and spectacular:* Natalia Yuvchenko, author interview, Moscow, October 2015.

47 *In spite of the long hours he put in at work:* Natalia Yuvchenko, email correspondence with author, December 2015; Maria Protsenko, author interview, April 2016.

47 *Borrowed a neighbor's small motorboat:* Natalia Yuvchenko, author interview, 2015.

48 *Like all newly qualified Soviet specialists:* Alexander Yuvchenko, author interview, 2006.

48 *When their son was born:* Natalia Yuvchenko, author interview, 2015; author visit to the Yuvchenkos' flat in Pripyat, April 27, 2016.

49 *Still, with no family nearby to provide help:* Read, *Ablaze*, 61; Natalia Yuvchenko, author interview, 2015.

50 Rich Man, Poor Man: Natalia Yuvchenko, email correspondence with author, 2015.

50 *He seemed restless:* Natalia Yuvchenko, author interview, 2015.

50 *She left the TV:* Read, *Ablaze*, 61.

50 *A few hundred meters away:* Alexander (Sasha) Korol, author interview, Kiev, September 2015.

51 *Toptunov was born within sight:* Vera Toptunova, author interview, Kiev, September 2015.

51 *Toptunov's father liked to boast:* Korol, author interview, 2015.

51 *When Toptunov was thirteen:* Toptunova, author interview, 2015.

51 *Established with the patronage of Kurchatov:* Originally established as the Moscow Mechanical Institute of Munitions in 1942, the university shifted its focus almost exclusively to nuclear physics after the war, with Kurchatov's encouragement and support. "History," National Research Nuclear University MEPhI, https://mephi.ru/about/index2.php.

51 *The examination was notoriously difficult:* Andrei Glukhov, author interview, Chernobyl nuclear power plant, February 2016.

52 *The course work was tough:* Alexey Breus, author interview, Kiev, July 2015.

52 *Long banished from television:* Kristin Roth-Ey, *Moscow Prime Time: How the Soviet Union Built the Media Empire That Lost the Cultural Cold War* (Ithaca, NY: Cornell University Press, 2011), 258–59.

52 *Shy-looking, with glasses and a lingering hint of puppy fat:* Author interviews, 2015: Toptunova, Breus, and Glukhov.

52 *At MEPhI, Toptunov took up:* Korol, author interview, 2015.

52 *Against the advice of his tutors:* Breus, author interview, 2015.

52 *One night after class:* Korol, author interview, 2015.

52 *After four years of study:* Korol, interviews by the author (2015) and Taras Shumeyko (April 2018, Kiev).

53 *Like all other new engineers:* Breus, author interview, 2015.

53 *During the summer and autumn of 1983:* Ibid.; Korol, author interview, 2015. Date of first criticality is provided in Sich, "The Chornobyl Accident Revisited," 83.

54 *Toptunov organized a gym:* Korol, author interview, 2015; Toptunova, author interview, 2015; R. Veklicheva, "A Soviet way of life. The test" [Образ жизни— Советский. Испытание], *Vperiod* (official newspaper of Obninsk Communist Party Committee), June 17, 1986, Vera Toptunova's personal archive.

54 *He had a girlfriend:* Korol, author interview, 2015; Josephson, *Red Atom*, 6–7. Radionuclides accumulated in the water during its passage through the reactors were intended to become trapped in sediment that fell to the bottom of the cooling pond, filtering them out before the water reached the Pripyat River (Zhores Medvedev, *Legacy of Chernobyl*, 92). Because reactor outfall maintained the pond at a year-round temperature of 24 degrees centigrade, in 1978 local Soviet authorities decided to develop the radioactive lake for commercial fish farming. Subsequent tests established that fish caught there contained potentially dangerous levels of strontium 90, and three years later their sale was forbidden. Local fishermen continued regardless. Danilyuk, ed., *Z arkhiviv*, document no. 6: "Report of the UkSSR KGBM on Kiev and Kiev region to the UkSSR KGB concerning violations of the radiation safety requirements while studying the feasibility to use the Chernobyl NPS cooling pond for the purposes of industrial fishery," March 12, 1981.

54 *The operators spoke their own coded language:* Sergei Yankovsky, author interview, Kiev, February 2016.

54 *There were thick stacks of manuals:* The operators' manual is described by Anatoly Kryat (chief of the nuclear physics laboratory at the plant), in court testimony reproduced by Karpan, *From Chernobyl to Fukushima*, 190.

54 *After one of these safety exams:* Korol, author interview, 2015.

54 *Only after all of this study:* Anatoly Kryat, author interview, Kiev, February 2016.

55 *"Let's go together":* Korol, author interview, 2015.

55 *A warm, sultry night:* Svetlana Kirichenko (chief economist of the Pripyat *ispolkom*), author interview, April 24, 2016; recollections of Pripyat residents quoted in

Vasily Voznyak and Stanislav Troitsky, *Chernobyl: It Was Like This—The View from the Inside* [Чернобыль: Так это было—взгляд изнутри] (Moscow: Libris, 1993).

55 *On the bus:* Boris Stolyarchuk, author interview, Kiev, July 2015; Iurii Shcherbak, "Chernobyl: A Documentary Tale" [Чернобыль: Документальная повесть], *Yunost*, nos. 6–7 (1987), translated by JPRS Political Affairs as "Fictionalized Report on First Anniversary of Chernobyl Accident," Report no. JPRS-UPA-87-029, September 15, 1987 (hereafter, "Report on First Anniversary of Chernobyl Accident"), pt. 1, 24.

55 *Taking the polished marble stairs:* Read, *Ablaze*, 61; author visit to the Chernobyl nuclear power plant, February 10, 2016. The route and routine for entering the building had remained largely unchanged since 1986.

56 *The turbine hall housed:* Sich diagrams a cross-section of the turbine hall in "The Chornobyl Accident Revisited," 192.

56 *The deaerator corridor:* Author visit to the Chernobyl nuclear power plant, 2016; Steinberg, personal communication with author, August 6, 2018; Sich, "The Chornobyl Accident Revisited," 191. The deaerator corridor became more widely known after the accident as the "gold" corridor.

57 *A little more than five hundred meters away:* Anatoly Zakharov, author interviews, February 2006 and February 2016; Piotr Khmel, author interview, July 2015; author visit to Fire Station Number Two, Pripyat, April 25, 2016; Leonid Shavrey, testimony in Sergei Kiselyov, "Inside the Beast," trans. Viktoria Tripolskaya-Mitlyng, *Bulletin of Atomic Scientists* 52, no. 3 (1996): 47.

57 *They had already been called:* Kiev Region Fire Department Dispatch Log for April 25–26, 1986 (Chernobyl Museum archive, Kiev), 109–11.

58 *There was never much action:* Alexander Petrovsky, author interview, Bohdany, Ukraine, November 2016; Zakharov, author interview, 2016.

58 *Behind the doors:* Piotr Khmel, author interview, 2016.

58 *At the back of the building:* Zakharov, author interview, 2016.

58 *The third watch lacked discipline:* Piotr Khmel, author interview, 2015; Leonid P. Telyatnikov, account, in Iurii Shcherbak, *Chernobyl* [Чернобыль], trans. Ian Press (London: Macmillan, 1989), 26–27; Shcherbak, trans. JPRS, "Report on First Anniversary of Chernobyl Accident," 46–66.

58 *"Major Telyatnikov will get back":* Piotr Khmel, author interview, 2015.

59 *"people's champagne":* Graham Harding, "Sovetskoe Shampanskoye—Stalin's 'Plebeian Luxury,'" *Wine As Was*, August 26, 2014.

59 *At around eleven:* Khmel, author interview, 2015.

59 *Over at the power station:* Alexander Yuvchenko, author interview, 2006.

4. SECRETS OF THE PEACEFUL ATOM

60 *On September 29, 1966:* International Atomic Energy Agency, International Nuclear Safety Advisory Group, "The Chernobyl Accident: Updating of INSAG-1," Safety series no. 75-INSAG-7, 1992 (hereafter INSAG-7), 32; Schmid, *Producing Power*, 111.

60 *A direct descendant of the old Atom Mirny-1:* For a discussion of how economies of scale influenced this choice, see Marples, *Chernobyl and Nuclear Power in the USSR*, 111.

60 *Twelve meters across:* International Atomic Energy Agency, INSAG-7, 40; the weight of graphite in the core is provided in Zhores Medvedev, *Legacy of Chernobyl*, 5.

60 *These channels contained:* Zhores Medvedev, *Legacy of Chernobyl*, 236; Alexander Sich, telephone interview, May 2018.

60 *The power of the reactor:* Sich, "The Chornobyl Accident Revisited," 185.

61 *To help protect the plant:* Reactor pit dimensions (21.6 m × 21.6 m × 25.5 m) are given in Sich, "The Chornobyl Accident Revisited," 429. See also 179 for a cross-sectional view of the reactor vault. Also see USSR State Committee on the Utilization of Atomic Energy, "The Accident at the Chernobyl Nuclear Power Plant and Its Consequences," information compiled for the August 1986 IAEA Experts' Meeting in Vienna (hereafter "USSR State Committee Report on Chernobyl"), "Part 2: Annex 2," 7 and 9. Sich (244) describes serpentinite as a hydrous magnesium silicate.

61 *A biological shield:* Alexander Sich provides a detailed breakdown of the reactor shaft construction materials, showing that the combined mass of Structure E was at least two thousand tonnes ("Chornobyl Accident Revisited," 427). The same number is noted by the International Atomic Energy Agency in its report on Chernobyl (INSAG-7, 9). These calculations revise the lower estimate of a thousand tonnes cited in 1987 by the US Nuclear Regulatory Commission in *Report on the Accident at the Chernobyl Nuclear Power Station* (NUREG-1250), 2–12.

61 *Pierced by ducts:* USSR State Committee Report on Chernobyl, Part 2 Annex 2, 7, and 9; Sich, "Chornobyl Accident Revisited," 196.

61 *Known by the plant staff as the* pyatachok: Grigori Medvedev, *The Truth About Chernobyl*, 73–74.

61 *The RBMK was a triumph:* Alexander Sich notes that the typical core of a 1,300-megawatt PWR reactor used in the West was 3.4 meters in diameter and 4.3 meters high ("Chornobyl Accident Revisited," 158). See also Josephson, *Red Atom*, 299t6.

61 *Soviet scientists proclaimed it:* Sich, "Chornobyl Accident Revisited," 156–57; Schmid, *Producing Power*, 115 and 123.

61 *Personally taken credit for its design:* Schmid, *Producing Power*, 290n124.

61 *In contrast to its principal Soviet competitor:* Ibid., 123; Josephson, *Red Atom*, 36.

62 *Aleksandrov also saved money:* Schmid, *Producing Power*, 112.

62 *A less costly solution:* Zhores Medvedev, *Legacy of Chernobyl*, 236.

62 *A break in the pressure tubes:* Nikolai Steinberg, author interview, September 2015.

62 *Worse accidents were theoretically possible:* International Atomic Energy Agency, INSAG-7, 9.

63 *Yet the designers saw no need:* Charles K. Dodd, *Industrial Decision-Making and High-Risk Technology: Siting Nuclear Power Facilities in the USSR* (Lanham, MD: Rowman & Littlefield, 1994), 83–84.

63 *The Ministry of Medium Machine Building ordered:* Schmid, *Producing Power*, 110.

63 *One scientist from the Kurchatov:* The physicist was Vladimir Volkov (ibid., 145).

63 *Another recognized that the hazards:* This expert was Ivan Zhezherun, also of the Kurchatov Institute (Zhores Medvedev, *Legacy of Chernobyl*, 258–59).

63 *But by then, the government:* Schmid, *Producing Power*, 110 and 124; International Atomic Energy Agency, INSAG–7, 37.

63 *It was not until 1968:* Schmid, *Producing Power*, 110–11.

63 *So, to save time:* International Atomic Energy Agency, INSAG–7, 37; Anatoly Dyatlov, *Chernobyl: How It Was* [Чернобыль: Как это было] (Moscow: Nauchtekhizdat, 2003), online at http://pripyat-city.ru/books/25-chernobyl-kak-yeto-bylo .html, 27.

63 *Construction began:* Dodd, *Industrial Decision-Making and High-Risk Technology*, app. A.

64 *The republic's new 2,000-megawatt atomic energy station:* At this point, Sredmash officials had yet to decide which type of reactor they would build at this second site. They considered three options: a gas-cooled graphite model known as the RK-1000, the VVER, and the RBMK. At first, they dismissed the RBMK as technically and economically the worst of them all. They chose the more advanced, and safer, gas-cooled RK-1000 instead. But by mid-1969, Moscow's ambitious nuclear construction targets were already slipping out of reach, and time was precious. Sredmash recognized that—whatever its limitations—the graphite-water colossus could be manufactured more quickly than the more sophisticated gas-cooled model. It reversed its decision, and went with the RBMK after all. Six months later, at the dawn of a new decade, Viktor Brukhanov was summoned to the headquarters of the Ministry of Energy and Electrification in Moscow and given instructions to build the first two RBMK-1000 reactors of the Chernobyl nuclear power plant (International Atomic Energy Agency, INSAG–7, 32–33; Schmid, *Producing Power*, 120–25).

64 *The first RBMK unit:* Construction start dates for RBMK units across the USSR are given in Sich, "Chornobyl Accident Revisited," 148.

64 *But the initial Leningrad reactor:* The Leningrad station's Unit One reached full power eleven months after its start-up, on November 1, 1974 (Schmid, *Producing Power*, 114).

64 *The first problem arose:* International Atomic Energy Agency, INSAG–7, 35–37.

65 *Even as it went into full-scale:* Ibid., 37.

65 *The RBMK was so large:* Ibid., 6.

65 *One specialist compared it to a huge apartment building:* Veniamin Prianichnikov, author interview, Kiev, February 13, 2006.

65 *Isolated hot spots:* Kopchinsky and Steinberg, *Chernobyl*, 140; International Atomic Energy Agency, INSAG– 7, 39–40.

65 *"experience and intuition":* International Atomic Energy Agency, INSAG–7, 4–5.

65 *A third fault:* Ibid., 43; Sich, "Chornobyl Accident Revisited," 185. The original design documentation for the RBMK proposed seven-meter-long RCPS rods with a

seven-meter absorber and displacer, which would completely span the core top to bottom when lowered; 68 of these would be Emergency Protection System (EPS) rods. But in the final design, none of the rods would be long enough to completely span the core, and instead of 68 EPS rods, there would be only 21. For second-generation RBMK reactors, this was amended to 24 EPS rods, and the total number of rods increased to 211.

65 *Yet the AZ-5 mechanism was not designed:* International Atomic Energy Agency, INSAG-7, 45.

66 *Starting from their fully withdrawn position:* Ibid., 41.

66 *Fuel assemblies became stuck in their channels:* Schmid, *Producing Power*, 114.

66 *The valves and flow meters in other RBMKs:* Kopchinsky and Steinberg, *Chernobyl*, 140–41.

66 *On the night of November 30, 1975:* Vitali Abakumov, a former Leningrad station engineer, shared details and personal recollections of the accident in "Analyzing the Causes and Circumstances of the 1975 Accident on Unit One of Leningrad NPP (Perspective of an Engineer-Physicist, Participant and Witness to the Events)" [Анализ причин и обстоятельств аварии 1975 года на 1-м блоке ЛАЭС (комментарий инженера-физика, участника и очевидца событий)], April 10, 2013, http://accidont.ru/Accid75.html. Also see Valentin Fedulenko, "Versions of the Accident: A Participant's Memoir and an Expert's Opinion" [Версии аварии: мемуары участника и мнение эксперта], September 19, 2008, www.chernobyl .by/accident/28-versii-avarii-memuary-uchastnika-i-mnenie.html.

66 *But the commission knew otherwise:* Kopchinsky and Steinberg, *Chernobyl*, 161.

67 *Sredmash suppressed the commission's findings:* Ibid.; International Atomic Energy Agency, INSAG-7, 48–49.

67 *The day after the Leningrad meltdown:* The government decree no. 2638 R was issued on December 1, 1975 (International Atomic Energy Agency, INSAG-7, 33).

67 *On August 1, 1977:* Nikolai Steinberg, author interview, Kiev, May 28, 2017.

67 *Ukraine's first nuclear electricity:* Kopchinsky and Steinberg provide the date in *Chernobyl*, 116. For a history of Ukraine's electricity grid, see "Section 3: Ukraine's Unified Power Grid," in K. B. Denisevich et al., *Book 4: The Development of Atomic Power and Unified Electricity Systems* [Книга 4: Развитие атомной энергетики и объединенных энергосистем] (Kiev: Energetika, 2011), http://energetika.in.ua /ru/books/book-4/section-2/section-3.

67 *Together they sang:* Kopchinsky and Steinberg, *Chernobyl*, 139–40. (The preceding part of the chant was a sardonic recognition that the more sophisticated VVER reactors that had been planned to go into operation were mired in production problems and delays.)

67 *Making dozens of adjustments every minute:* Steinberg, author interview, 2015.

68 *Rumors reached them:* Ibid.; Kopchinsky and Steinberg, *Chernobyl*, 140.

68 *"How can you possibly control":* Georgi Reikhtman (at the time of this exchange, a trainee reactor operator at Chernobyl Unit One), author interview, Kiev, September 2015.

68 *The reactor was riddled with faults:* Kopchinsky and Steinberg, *Chernobyl*, 140–42.

68 *In 1980 NIKIET completed:* International Atomic Energy Agency, INSAG–7, 48–49.

68 *The report made it clear:* Ibid., 82.

68 *But the staff of Soviet nuclear power plants:* Schmid, *Producing Power*, 62–63; Read, *Ablaze*, 193.

69 *One of the new directives stipulated:* International Atomic Energy Agency, INSAG–7, 72.

69 *Deprived of information:* Ibid., 48–50.

69 *Early on the evening of September 9, 1982:* The KGB's report on the incident the following day is provided in Danilyuk, ed., *Z arkhiviv*, Document no. 9: "Report of the UkSSR KGBM on Kiev and Kiev Region to the 2nd KGB Head-office of the USSR and the 2nd KGB Managing the UkSSR Concerning the Emergency Stoppage of the Chernobyl' NPS Power Unit 1 on September 9, 1982," September 10, 1982.

69 *Nikolai Steinberg was sitting at the desk:* Steinberg, author interview, 2015.

70 *"to prevent the spread of panic-mongering":* The purported lack of radioactive releases was mentioned in a KGB report from September 13, 1982, available in Danilyuk, ed., *Z arkhiviv*, Document no. 10: "Report of the UkSSR KGBM on Kiev and Kiev Region to the USSR KGB and the UkSSR KGB Concerning the Results of Preliminary Investigation of the Cause of the Emergency Situation on the Chernobyl' NPS as of September 9, 1982," September 13, 1982. The fact that radiation was indeed released was noted by KGB on September 14. See KGB of the Ukrainian SSR, *Report of USSR KGB on the number of foreigners from capitalist and developing countries in the Ukrainian SSR, England-Based Combatants of the Organization of Ukrainian Nationalists, the Consequences of the Accident at the NPP* [Информационное сообщение КГБ УССР о количестве иностранцев из капиталистических и развивающихся стран в УССР, ОУНовских боевиках в Англии, последствиях аварии на АЭС на 14 сентября 1982 г.], September 14, 1982, declassified archive of the Ukrainian State Security Service, http://avr.org .ua/index.php/viewDoc/24447/.

70 *In fact, radioactive contamination:* Danilyuk, ed., *Z arkhiviv*, Document no. 12: *Report of the UkSSR KGBM on Kiev and Kiev Region to the USSR KGB and UkSSR KGB concerning the Radioactive Contamination of Chernobyl' NPS Industrial Site Due to the Accident on 9 September 1982*, September 14, 1982; and Document no. 13: *Report of the Chief of the UkSSR KGBM on Kiev and Kiev Region to the Chairman of the UkSSR KGB Concerning the Radiation Situation Which Occurred on the Chernobyl' NPS Industrial Site Due to the Accident on September 9, 1982*, October 30, 1982; Viktor Kovtutsky, chief accountant at Chernobyl construction department, author interview, Kiev, April 2016; Esaulov, *City That Doesn't Exist*, 19.

70 *When the reactor had been brought back online:* Read, *Ablaze*, 43–44.

70 *Workers carried blocks:* Andrei Glukhov, author interview, Slavutych, Ukraine, 2015.

70 *The incident was classified:* Author interviews with Steinberg and Glukhov.

70 *Nikolai Steinberg would wait years:* Kopchinsky and Steinberg, *Chernobyl*, 141; Steinberg, author interview, 2015.

70 *In October 1982 a generator exploded:* Grigori Medvedev, *Truth About Chernobyl*, 19 and 44–45.

71 *Both incidents were concealed:* Author interviews with Steinberg, Glukhov, and Kupny; Grigori Medvedev, *Truth About Chernobyl*, 19.

71 *The severity of this "positive scram" effect:* The ORM, or operational reactivity margin, measured the total number of control rods—or their equivalent in power-quenching capacity—inserted into the core at any one time. For example, an ORM of 30 might indicate 30 rods completely inserted, 60 rods inserted to half their length, or 120 inserted to a quarter.

71 *If more than 30 of these rods:* International Atomic Energy Agency, INSAG–7, 39–43.

72 *The source of the positive scram effect:* Steinberg, author interview, 2017; Sich, "Chornobyl Accident Revisited," 159.

72 *Like all the manual control rods:* Steinberg, author interview, 2017; Kopchinsky and Steinberg, *Chernobyl*, 144; International Atomic Energy Agency, INSAG–7, 42–44, 90n24. The letter (ref. no. 33-08/67) was dated December 23, 1983.

72 *But they weren't:* At a 1983 interdepartmental meeting in Moscow hosted by Aleksandrov, Efim Slavsky reacted with fury when discussion turned to the shortcomings of the RBMK. His stance helped "close the door to a serious conversation about this type of reactor," recalls Georgi Kopchinsky, then the head of the nuclear energy sector at the Central Committee, who attended the meeting. See Kopchinsky and Steinberg, *Chernobyl*, 145. There is more detail on the failures to address the known design faults of RBMK reactors in Nikolai Karpan, *Chernobyl: Revenge of the Peaceful Atom* [Чернобыль: Месть мирного атома] (Kiev: CHP Country Life, 2005), 399–404.

72 *Some partial modifications:* INSAG–7 (45) reports that modification of the control rods was proposed by NIKIET as early as 1977 but implemented in only a few RBMK reactors. Kopchinsky mentions that the idea came from the atomic power plant in Kursk and "was never incorporated into the reactor blueprint." Instead, alteration of each RBMK unit required individual confirmation from NIKIET, a process that "dragged on for months" (Kopchinsky and Steinberg, *Chernobyl*, 144).

73 *The news never reached the reactor operators:* Steinberg, author interview, 2015; Alexey Breus, author interview, Kiev, July 2015; Kopchinsky and Steinberg, *Chernobyl*, 144; Andrei Glukhov recalled that the staff of the nuclear safety department at Chernobyl received a document notifying them of the "tip effect" in 1983 but that it was marked for limited distribution only, and the operating instructions of the reactor were not revised to mention the phenomenon (Glukhov, telephone interview, July 2018).

73 *The last day of 1983:* The commissioning date of Unit Four is confirmed by Ni-

kolai Fomin, chief engineer of Chernobyl NPP, in Karpan, *From Chernobyl to Fukushima*, 143.

73 *Early in the morning of March 28, 1979:* There is a brisk account of the accident in Mahaffey, *Atomic Accidents*, 342–50, and Mahaffey, *Atomic Awakening*, 314–17. More detail can be found in Mitchell Rogovin and George T. Frampton Jr. (NRC Special Inquiry Group), *Three Mile Island: A Report to the Commissioners and to the Public* (Washington, DC: Government Printing Office, 1980).

74 *News of Three Mile Island was censored:* Grigori Medvedev, *Truth About Chernobyl*, 7.

74 *The failings of capitalism:* William C. Potter, *Soviet Decisionmaking for Chernobyl: An Analysis of System Performance and Policy Change* (report to the National Council for Soviet and East European Research, 1990), 6; Edward Geist, "Political Fallout: The Failure of Emergency Management at Chernobyl'," *Slavic Review* 74, no. 1 (Spring 2015): 107–8.

74 *Its operators were far better trained:* Leonid Bolshov, a senior physicist at the Kurchatov Institute at the time, recalls that the official line went as follows: the American operators were poorly educated former navy cadets, who never completed college, while the Russian operators were all nuclear science majors from top universities with superior training (author interview, 2017). Also see Zhores Medvedev, *Legacy of Chernobyl*, 272–73.

74 *Privately, Soviet physicists began to analyze:* Virtually the only public admission of these professional doubts that made it to print was a 1980 article coauthored by four physicists—Legasov, Sidorenko, Babayev, and Kuzmin—who wrote: "Under certain circumstances, despite existing safety measures, there might arise conditions for an accident at a [nuclear power plant] that would damage its active zone and release a small amount of radioactive substances to the atmosphere." It was promptly criticized by Sredmash as alarmist. "Safety Issues at Atomic Power Stations" [Проблемы безопасности на атомных электростанциях], *Priroda*, no. 6, 1980. For discussions of the USSR's evolving state oversight committees for nuclear power, see Schmid, *Producing Power*, 59–60 and 92.

74 *But neither Sredmash nor NIKIET:* International Atomic Energy Agency, INSAG-7, 34–35.

74 *A ten-page report on the wonders of nuclear energy:* "Nuclear Power Industry," *Soviet Life* 353, no. 2 (Washington, DC: Soviet Embassy, February 1986), 7–16.

74 *"In the thirty years":* Valery Legasov, Lev Feoktistov, and Igor Kuzmin, "Nuclear Power Engineering and International Security," *Soviet Life* 353, no. 2, 14.

74 *"one in 10,000 years":* Vitali Sklyarov, interview by Maxim Rylsky, "The Nuclear Power Industry in the Ukraine," *Soviet Life* 353, no. 2, 8. When I spoke to Sklyarov in 2017, he claimed to have no memory of this statement or the article it appeared in.

5. FRIDAY, APRIL 25, 11:55 P.M., UNIT CONTROL ROOM NUMBER FOUR

75 *A rancid haze of cigarette smoke:* Alexey Breus, author interview, Kiev, July 2015. Smoking was forbidden in most parts of the plant, and the control rooms were one

of the few places where it was tolerated. The reactor control engineers smoked at the control panels, and Leonid Toptunov (along with an overwhelming number of others in the USSR at the time) was a smoker.

75 *Now entering his second day without sleep:* Dyatlov's defender, cross-examining Fomin in court, in trial proceedings reproduced in Karpan, states that Dyatlov managed the operations of Unit Four alone for two days. Fomin responds that Dyatlov had gone home for a "break" on the afternoon of April 25, but remained available by telephone. *Chernobyl to Fukushima*, 148.

75 *Exhausted and unhappy:* Boris Stolyarchuk, author interview, Kiev, July 2015.

75 *It was one scenario:* A total loss of power of this kind had already occurred at the Kursk nuclear power plant in 1980: Zhores Medvedev, *Legacy of Chernobyl*, 269.

75 *The station had emergency diesel generators:* Sich, "Chornobyl Accident Revisited," 225.

75 *The rundown unit:* Nikolai Steinberg, former senior engineer at Chernobyl NPP, recalls that three similar tests took place before 1986—not one of them able to generate the amount of electricity that a total external power failure would require (author interview, 2015). The International Atomic Energy Agency report (INSAG-7) summarizes the history of the rundown test on p. 51. The fact that Reactor Number Four was commissioned without this test being completed is noted by the trial judge in the transcript of Chernobyl court proceedings in Karpan, *Chernobyl to Fukushima*, 143.

76 *Factories and enterprises throughout Ukraine:* This practice of *shturmovshchina* before important production quota deadlines—literally, taking them "by storm"—was a regular feature of Soviet working life: Zhores Medvedev, *Legacy of Chernobyl*, 25–26. See also the account of Igor Kazachkov, who supervised the first shift at Unit Four that day, in Shcherbak, *Chernobyl*, 34.

76 *The dispatcher said:* Gennadi Metlenko (chief engineer of Dontekhenergo), court testimony, in Karpan, *Chernobyl to Fukushima*, 178.

76 *By midnight on Friday:* Metlenko, questioned by Dyatlov in court testimony, in Karpan, *Chernobyl to Fukushima*, 180.

76 *He hadn't shown up at all:* Karpan, *Chernobyl to Fukushima*, 146 and 191; Glukhov, author interview, 2015.

76 *Twenty-five-year-old:* Leonid Toptunov's date of birth (August 16, 1960) is provided by his mother, Vera Toptunova, in author interview, 2015.

76 *If the test wasn't completed that night:* Zhores Medvedev, *Legacy of Chernobyl*, 28.

76 *At fifty-five:* Anatoly Dyatlov's date of birth (March 3, 1931) is given in the court verdict reproduced in Karpan, *Chernobyl to Fukushima*, 194.

76 *A veteran physicist:* Dyatlov, court testimony, in Karpan, *Chernobyl to Fukushima*, 151.

76 *The son of a peasant:* Read, *Ablaze*, 33–34 and 46.

77 *As the head of the classified Laboratory 23:* V. A. Orlov's and V. V. Grischenko's recollections, parts III and V in "Appendix 8: Memories about A. S. Dyatlov," in Dyatlov, *Chernobyl: How It Was*, 183 and 187. For a history of the Lenin Kom-

somol shipyard, see "Komsomol'sk-na-Amure," Russia: Industry: Shipbuilding, GlobalSecurity.org, November 2011.

77 *By the time he arrived in Chernobyl:* Dyatlov, court testimony, in Karpan, *Chernobyl to Fukushima,* 156.

77 *These small marine reactors:* Dyatlov, *Chernobyl: How It Was,* 25–32; Dyatlov, court testimony, in Karpan, *Chernobyl to Fukushima,* 152.

77 *But the manner Dyatlov had developed:* Anatoly Kryat's recollections, part IV in Appendix 8 of Dyatlov, *Chernobyl: How It Was,* 186; Steinberg, author interview, 2015; Glukhov, author interview, 2015.

77 *Even those colleagues:* Valentin Grischenko, who worked with Dyatlov at both the Lenin Komsomol shipyard and Chernobyl, notes that out of all Dyatlov's colleagues in Chernobyl, only one—another longtime collaborator, Anatoly Sitnikov—could be considered a close friend of his. Grischenko's recollections in Dyatlov, *Chernobyl: How It Was,* 187.

77 *He could be high-handed:* Korol, author interview, 2015.

77 *He demanded that any fault:* Dyatlov, court testimony, in Karpan, *Chernobyl to Fukushima,* 152; Breus, author interview, 2015.

78 *Even when overruled from above:* Grischenko's recollections, in Dyatlov, *Chernobyl: How It Was,* 187.

78 *And Dyatlov's long experience:* Dyatlov, *Chernobyl: How It Was,* 25–26.

78 *Dyatlov had fulfilled:* Steinberg, author interviews, 2015 and 2017; Read, *Ablaze,* 47; Grischenko's recollections in Dyatlov, *Chernobyl: How It Was,* 187.

78 *Only long afterward would his secret emerge:* Anatoly Dyatlov, personal letter (unpublished) to Toptunov's parents, Vera and Feodor, June 1, 1989, from the personal archive of Vera Toptunova; Sergei Yankovsky (investigator for the Kiev Region Prosecutor's Office), author interview, Kiev, February 7, 2016; Read, *Ablaze,* 47. Dyatlov describes the dose of radiation he received during his time at the shipyard—without attributing it directly to the accident—in an interview given to A. Budnitsky and V. Smaga, "The Reactor's Explosion Was Inevitable" [Реактор должен был взорваться], *Komsomolskoye Znamya,* April 20, 1991, reproduced in Dyatlov, *How It Was,* 168.

78 *Many admired him:* Description in Read, *Ablaze,* 47; V. V. Lomakin's recollections, part VI in Appendix 8 of Dyatlov, *Chernobyl: How It Was,* 188; Kopchinsky and Steinberg, *Chernobyl,* 151. Georgi Reikhtman, a submarine reactor room officer cashiered from the Soviet navy as part of a political purge, said that Dyatlov was a fair man, instrumental in having Reikhtman hired to work at the Chernobyl station when no one else would give him a job (Reikhtman, author interview, September 2015).

78 *Eager to learn, they believed:* Steinberg, author interview, 2015.

78 *For all his hours spent poring:* Nikolai Steinberg recalls that Dyatlov "would sometimes make offhand remarks about the RBMK being unknowable. We, the young, found this odd. We thought that Dyatlov, for one, knew everything." Kopchinsky and Steinberg, *Chernobyl,* 151; Steinberg, author interview, 2017.

79 *On the left sat:* The acronym SIUR stands for *starshiy inzhener upravleniya reaktorom.* Author visit to Chernobyl nuclear power plant Control Room Number Two, and interview with Alexander Sevastianov, February 10, 2016; Stolyarchuk, author interview, 2016.

79 *Forming a wall:* Author visit to Chernobyl nuclear power plant Control Room Number Two, and interview with Alexander Sevastianov, February 2016; Stolyarchuk, author interview, 2016.

80 *In the strict technical hierarchy:* Glukhov, author interview, 2015.

80 *Akimov, a gangling thirty-two-year-old:* Akimov's date of birth is given as May 6, 1953, in "List of Fatalities in the Accident at Chernobyl" [Список погибших при аварии на Чернобыльской АЭС], undated, Chernobyl Electronic Archive, available at http://pripyat-city.ru/documents/21-spiski-pogibshix-pri-avarii.html.

80 *He and his wife, Luba:* Read, *Ablaze,* 38–39.

80 *Akimov was clever:* Steinberg, author interview, 2015. When tasked with calculating the hypothetical probability of a serious accident occurring at Chernobyl, Akimov had estimated it at one in ten million a year (Read, *Ablaze,* 43).

80 *Control Room Number Four had now grown busy:* See the account of Yuri Tregub, who supervised the second shift at Unit Four that day, in Shcherbak, *Chernobyl,* 39; Grigori Medvedev, *Truth About Chernobyl,* 72.

80 *Perhaps assuming:* This explanation was suggested by Fomin in court testimony provided by Karpan, *Chernobyl to Fukushima,* 146.

80 *Akimov, a copy of the test protocol:* Stolyarchuk, author interview, 2016; Tregub, court testimony, in Karpan, *Chernobyl to Fukushima,* 180–81.

81 *When Toptunov had assumed responsibility:* International Atomic Energy Agency, INSAG-7, 4–5.

81 *So Toptunov began the process:* Sich, "Chornobyl Accident Revisited," 211; Alexander Sich, author telephone interview, December 2016. The subsequent accident report by a working group of Soviet experts likewise faults Toptunov for the power fall: INSAG-7, "Annex I: Report by a Commission to the USSR State Committee for the Supervision of Safety in Industry and Nuclear Power," 1991, 63. However, the authors of the INSAG-7 report mention that Dyatlov himself attributed the incident to equipment malfunction (International Atomic Energy Agency, INSAG-7, 11).

81 *Now Toptunov watched in dismay:* International Atomic Energy Agency, INSAG-7, p. 73; Tregub, testimony in Shcherbak, *Chernobyl,* 40.

81 *"Maintain power!":* Tregub, court testimony, in Karpan, *Chernobyl to Fukushima,* 181.

81 *At this point:* On two previous occasions when operators had attempted to raise power in ChNPP reactors without waiting out the poison override time, word had reached nuclear safety inspectors in Moscow about what was happening, and they called Brukhanov immediately to have the power increases halted. Yuri Laushkin, court testimony, in Karpan, *Chernobyl to Fukushima,* 175.

81 *Dyatlov himself would maintain:* In his memoir, Dyatlov describes leaving the con-

trol room before Toptunov transferred the system to Global Automatic, in order to "more thoroughly inspect areas with heightened radiation risk," which he thought had been made safer by the reduction of the reactor's power. He states that he did not return to the control room until 00:35 a.m. (*Chernobyl: How It Was*, 30).

82 *The recollections of others:* Tregub also testifies to Dyatlov's presence in the control room at this time (court testimony in Karpan, *Chernobyl to Fukushima*, 180–81). So does Metlenko (179), who says that Dyatlov stepped away from the console at about 00:28 a.m. "mopping his brow."

82 *"I'm not going to raise the power!":* Grigori Medvedev, *Truth About Chernobyl*, 55–56. Medvedev writes that Toptunov would recount his thoughts at this moment while in the medical center in Pripyat, less than twenty-four hours later. This view is expanded upon by Shcherbak's interviews with Igor Kazachkov and Arkady Uskov, quoted in *Chernobyl*, 366–69 and 370–74.

82 *The deputy chief engineer withdrew:* Metlenko, testimony in Karpan, *Chernobyl to Fukushima*, 179.

82 *"Why are you pulling unevenly?":* Tregub's account in Shcherbak, *Chernobyl*, 41. The elapsed time period is noted in Sich, "Chornobyl Accident Revisited," 212.

83 *By one in the morning:* International Atomic Energy Agency, INSAG-7, 71.

83 *Yet the engineers knew:* Ibid.

83 *Running the pump system:* Stolyarchuk, author interview, 2016.

83 *The rush of water:* Sich, "Chornobyl Accident Revisited," 212–14.

83 *A few moments later:* International Atomic Energy Agency, INSAG-7, 8.

83 *Some of the operators:* Razim Davletbayev (deputy chief of the Turbine Department), court testimony in Karpan, *Chernobyl to Fukushima*, 188; Dyatlov, *Chernobyl: How It Was*, 31.

83 *Ten men now stood:* Grigori Medvedev, *Truth About Chernobyl*, 71–72.

84 *"What are you waiting for?":* Davletbayev, court testimony, in Karpan, *Chernobyl to Fukushima*, 188.

84 *Simulating the effects:* Stolyarchuk, author interview, 2016. The test was initiated by the electrical shop, according to Fomin's court testimony provided in Karpan, *Chernobyl to Fukushima*, 142.

84 *The test program closely duplicated:* Metlenko states in his court testimony that the 1984 test was conducted on Turbo-generator no. 5, which was in Unit Three: Karpan, *Chernobyl to Fukushima*, 178.

84 *The chief engineer, had ordered that test himself:* Fomin, court testimony, in Karpan, *Chernobyl to Fukushima*, 142–44; International Atomic Energy Agency, INSAG-7, 51–52; Supreme Court of the USSR, court verdict for Brukhanov, Dyatlov, and Fomin, July 29, 1987, provided in Karpan, *Chernobyl to Fukushima*, 198.

84 *Fomin made two important changes:* Fomin, court testimony, in Karpan, *Chernobyl to Fukushima*, 145.

85 *Dyatlov, Akimov, and Metlenko:* Tregub's account in Shcherbak, *Chernobyl*, 41.

85 *Upstairs at mark +12.5:* Room description is from author visit to Main Circulating Pump room of Unit Three, Chernobyl nuclear power plant, February 10, 2016.

85 *164 of the 211 control rods:* Figure II-6, International Atomic Energy Agency, INSAG–7, 119.

85 *"Oscilloscope on!":* Dyatlov, *Chernobyl: How It Was,* 40.

85 *Inside the reactor, the cooling water:* International Atomic Energy Agency, INSAG–7, 8.

85 *Toptunov's control panel revealed nothing unusual:* International Atomic Energy Agency notes in INSAG–7 (p. 66): "Neither the reactor power nor the other parameters (pressure and water level in the steam separator drums, coolant and feedwater flow rates, etc.) required any intervention by the personnel or by the engineered safety features from the beginning of the tests until the EPS-5 button was pressed."

86 *"SIUR—shut down the reactor!":* Court testimonies of Yuri Tregub and Grigori Lysyuk (senior foreman of the electrical shop), in Karpan, *Chernobyl to Fukushima,* 182 and 184; Dyatlov, *Chernobyl: How It Was,* 40. Although Lysyuk maintains that the power surge was reported by Toptunov *before* the AZ-5 button was pressed, Dyatlov would say that this happened afterward, which also corresponds to further testimony and evidence from the computer data recovered after the accident.

86 *A transparent plastic cover:* Description of AZ-5 button provided by Nikolai Steinberg in author interview, 2017.

86 *"The reactor has been shut down!":* Tregub, court testimony, in Karpan, *Chernobyl to Fukushima,* 182.

86 *Oustripped the recording capacity:* International Atomic Energy Agency, INSAG–7, 66.

86 *For one scant second:* Ibid., 119.

86 *But then the graphite tips:* Dyatlov, *Chernobyl: How It Was,* 48; International Atomic Energy Agency, INSAG–7, 4 (section 2.2).

86 *The chain reaction began to increase:* International Atomic Energy Agency, INSAG–7, 67; Sich, "Chornobyl Accident Revisited," 220.

86 *A frightening succession of alarms:* International Atomic Energy Agency, INSAG–7, 55.

86 *Electric buzzers:* Author visit to Control Room Number Two and interview with Alexander Sevastianov, February 10, 2016.

86 *"Shut down the reactor!":* Dyatlov, *Chernobyl: How It Was,* 41.

86 *Standing at the turbine desk:* Tregub's account in Shcherbak, *Chernobyl,* 41–42.

86 *But the reactor was destroying itself:* Sich, "Chornobyl Accident Revisited," 231m; International Atomic Energy Agency, INSAG–7, 67.

87 *The channels themselves broke apart:* International Atomic Energy Agency, INSAG–7, 67–68.

87 *The AZ-5 rods jammed:* Dyatlov, *Chernobyl: How It Was,* 31.

87 *Out on a gantry at level +50:* Sich, "Chornobyl Accident Revisited," 219 and 230nl; Grigori Medvedev, *Truth About Chernobyl,* 73–74.

87 *At Toptunov's control panel, the alarm sounded:* International Atomic Energy Agency, INSAG–7, 55.

87 *The walls of the control room:* Tregub's account in Shcherbak, *Chernobyl*, 42.

87 *A rising moan:* Stolyarchuk, author interview, 2015.

87 *As the fuel channels failed:* Sich, "Chornobyl Accident Revisited," 221–22.

87 *The temperature inside the reactor rose:* Karpan, *Chernobyl to Fukushima*, 63.

87 *The lights of the Selsyn dials flared:* Grigori Medvedev, *Truth About Chernobyl*, 71.

87 *Releasing the AZ-5 rods from their clutches:* Dyatlov, *Chernobyl: How It Was*, 57.

88 *A mixture of hydrogen and oxygen:* This hypothesis is supported by the USSR Vienna report (USSR State Committee on the Utilization of Atomic Energy, "The Accident at the Chernobyl Nuclear Power Plant and Its Consequences"), 21; and Sich, "Chornobyl Accident Revisited," 223. For an alternative explanation, and discussion of the location of the second explosion, see Karpan, *Chernobyl to Fukushima*, 62–63.

88 *As much as sixty tonnes of TNT:* Estimates of the force of the explosion that destroyed the reactor vary enormously. An estimate of around twenty-four tonnes of TNT is cited in K. P. Checherov, "Evolving accounts of the causes and processes of the accident at Block 4." Valery Legasov's estimate would be just three to four tonnes of TNT (Legasov Tapes, Part One, p. 12). Karpan (*Chernobyl to Fukushima*, 62) says thirty tonnes, citing "Expert Conclusions" of state investigators reached on May 16, 1986. Finally, the KGB report of May 15, 1986, specified "no less than 50 to 60 tonnes" (Danilyuk, ed., *Z arkhiviv*, Document no. 34: "Report of the UkSSR OG KGBM and the USSR KGB in the town of Chernobyl' to the USSR KGB concerning the radioactive situation and progress in investigating the accident at the Chernobyl' NPS").

88 *The blast caromed off the walls of the reactor vessel:* Although this continues to be a matter of debate, a 1989 report by the USSR State Committee on Safety in the Atomic Power Industry explains that the shield was thrown far enough in the air to be flipped over: A. Yadrihinsky, "Atomic Accident at Unit Four of Chernobyl NPP and Nuclear Safety of RBMK Reactors" [Ядерная авария на 4 блоке Чернобыльской АЭС и ядерная безопасность реакторов РБМК], Gosatomenergonadzor Inspectorate at the Kursk Nuclear Power Station, 1989, 10–11. See also US Nuclear Regulatory Commission in "Report on the Accident (NUREG-1250)," 2–16 and 5–6. Further details of damage caused by the initial explosion are given in Sich, "Chornobyl Accident Revisited," 84–85.

88 *Almost seven tonnes of uranium fuel:* Sich, "Chornobyl Accident Revisited," 84. Radioactive fallout from the accident was detailed in a top-secret report on May 21 by Yuri Izrael, chairman of the State Committee for Hydrometeorology and Environmental Monitoring (Goskomgidromet) to Nikolai Ryzhkov, chairman of the USSR Council of Ministers: "Regarding the assessment of the radioactivity situation and radioactive contamination of the environment by the accident at the Chernobyl NPP" [Об оценке радиационной обстановки и радиоактивного загрязнения природной среды при аварии на Чернобыльской АЭС], May 21, 1986, Microfilm, Hoover Institution, *Russian State Archive of Contemporary History* (*RGANI*), Opis 51, Reel 1.1006, File 23.

88 *1,300 tonnes of incandescent graphite rubble:* Sich, "Chornobyl Accident Revisited," 405.

88 *Inside his workspace:* Alexander Yuvchenko, author interview, 2006.

88 *Over in the turbine hall:* Yuri Korneyev (turbine operator, fifth shift, Unit Four Chernobyl nuclear power plant), author interview, Kiev, September 2015.

89 *Former nuclear submariner Anatoly Kurguz:* Karpan, *Chernobyl to Fukushima,* 21.

89 *Tiles and masonry dust fell:* Dyatlov, *Chernobyl: How It Was,* 49.

89 *A gray fog bloomed:* Stolyarchuk, author interviews, 2015 and 2016.

89 *Outside the plant:* Karpan, *Chernobyl to Fukushima,* 11–12.

6. SATURDAY, APRIL 26, 1:28 A.M., PARAMILITARY FIRE STATION NUMBER TWO

91 *A purple cone of iridescent flame:* Eyewitness testimony in Karpan, *Chernobyl to Fukushima,* 12.

91 *In the telephone dispatcher's room:* Anatoly Zakharov, author interview, Kiev, February 2016.

91 *Many of the fourteen men:* Ibid. They arrived at the scene at one thirty in the morning, according to the Kiev Region Fire Department Dispatch Log, archive of the Chernobyl Museum.

91 *Pravik gave the order to go:* Alexander Petrovsky, author interview, Bohdany, Ukraine, November 2016. Times and details of alarms and equipment are provided in the Kiev Region Fire Department Dispatch Log, archive of the Chernobyl Museum.

92 *Two extra tanker trucks:* The Pripyat brigade was called out at 1:29 a.m. (Kiev Region Fire Department Dispatch Log). See also Leonid Telyatnikov's account in Shcherbak, "Report on First Anniversary of Chernobyl," trans. JPRS, pt. 1, 18.

92 *A number three alarm:* Pravik confirmed the number three alarm by telephone at 1:40 a.m., according to the Kiev Region Fire Department Dispatch Log, Chernobyl Museum. Additional details: V. Rubtsov and Y. Nazarov, "Men of the Assault Echelon," *Pozharnoye delo,* no. 6 (June 1986), translated in JPRS, *Chernobyl Nuclear Accident Documents,* 24–25.

92 *Now the giant superstructure of the plant:* Zakharov, author interview, 2016; Petrovsky, author interview, 2016.

92 *Everyone was talking at once:* Stolyarchuk, author interviews, 2015 and 2016.

92 *A constellation of warning lamps:* This description is based on author visit to the control room for Unit Number Two at the Chernobyl plant, February 10, 2016, and Dyatlov's court testimony in Karpan, *Chernobyl to Fukushima,* 157.

93 *In desperation, Dyatlov turned:* Dyatlov, *Chernobyl: How It Was,* 49; Anatoly Dyatlov interviewed by Michael Dobbs, "Chernobyl's 'Shameless Lies,'" *Washington Post,* April 27, 1992.

93 *"Lads," he said:* Read, *Ablaze,* 68; Dyatlov, *Chernobyl: How It Was,* 49. In his memoir, Dyatlov denied giving the order to supply water to the reactor, insisting it was

given after his departure from the control room by Chief Engineer Fomin. Dyatlov, *Chernobyl: How It Was*, 53.

93 *Engulfed in dust, steam, and darkness:* Yuvchenko, author interview, 2006.

93 *From beyond the shattered doorway:* Alexander Yuvchenko, testimony in *Zero Hour: Disaster at Chernobyl*, directed by Renny Bartlett, Discovery, 2004. Further description of the second operator, Alexander Novik, is from Alexander Yuvchenko, interview by Michael Bond, "Cheating Chernobyl," *New Scientist*, August 21, 2004.

94 *Then Yuvchenko saw:* Alexander Yuvchenko, author interview, 2006; Yuvchenko, interview by Bond, *New Scientist*, 2004. The moon was full at the time of the explosion, according to www.moonpage.com. Tregub's own memory of the sequence of events, recalled in an interview with Yuri Shcherbak, differs from Yuvchenko's (Shcherbak, *Chernobyl*, 42–43).

94 *The two men turned into the transport corridor:* Alexander Yuvchenko, author interview, 2006. Although Yuvchenko would later be convinced the ethereal glow he saw was the result of Cherenkov's Light, this phenomenon is visible only in mediums with a high refractive index, such as water—and unlikely to have been possible in the open air above Reactor Number Four (Alexander Sich, author interview, 2018).

95 *"Tolik, what is it?":* Zakharov, author interview, 2016.

95 *A scene of total chaos:* This description is drawn from the testimonies of Korneyev and Shavrey (in Kiselyov, "Inside the Beast," 43 and 47) and Razim Davletbayev (account in Kopchinsky and Steinberg, *Chernobyl*, 20).

96 *Pravik and Shavrey:* Shavrey, account in Kiselyov, "Inside the Beast," 47; Kiev Region Fire Department Dispatch Log, Chernobyl Museum.

96 *By two in the morning:* Kiev Region Fire Department Dispatch Log, Chernobyl Museum.

96 *Established a crisis center:* "Order No. 113: The measures concerning the emergency at the Chernobyl NPP" [О мерах в связи с ЧП на Чернобыльской АЭС], signed by Major General V. M. Korneychuk, April 26, 1986, "Operational Group of the Department of Internal Affairs of Kiev Oblast Lettered File on Special Measures in Pripyat Zone" [Оперативный Штаб УВД Киевского облисполкома, Литерное дело по спецмероприятиям в припятской зоне], April 26 to May 6, 1986, 5–6, Archive of Chernobyl Museum.

96 *"There's a fire in Unit Four":* Piotr Khmel, author interview, Kiev, 2016.

96 *In his flat on Lenina Prospekt:* Viktor and Valentina Brukhanov, author interview, 2015.

97 *I'm going to prison:* Viktor and Valentina Brukhanov, author interview, 2015.

97 *Designed as a refuge:* Details from author's visit to the bunker, February 2016.

97 *Brukhanov went upstairs:* Details of Brukhanov's movements here are drawn from a transcript of Brukhanov's trial testimony on July 8, 1987, taken contemporaneously in shorthand by Nikolai Karpan and published in *Chernobyl to Fukushima*

(126–34). The plant's civil defense chief, Serafim Vorobyev, states that Brukhanov instructed him to personally see to the opening of the bunker (Shcherbak, *Chernobyl*, 396). Brukhanov's order to announce a General Radiation Accident is confirmed by the telephone operator L. Popova, in Evgeny Ignatenko, ed., *Chernobyl: Events and Lessons* [Чернобыль: события и уроки] (Moscow: Politizdat, 1989), 95. When she tried to turn the automated system on, Popova discovered that the tape player was broken, so she began making the calls herself, one by one.

97 *The mayor of Pripyat arrived:* Brukhanov, court testimony in Karpan, *Chernobyl to Fukushima*, 128–29. The head of the Pripyat City Executive Committee—the Soviet equivalent of the mayor—was Vladimir Voloshko. Major V. A. Bogdan, whose formal title was the plant's head of security, is identified as a KGB officer in a KGB memo from May 4: Danilyuk, ed., *Z arkhiviv*, Document no. 26: "Report of the UkSSR KGB 6th Department to the USSR KGB Concerning the Radioactive Situation and Progress in Investigating the Accident at the Chernobyl' NPS."

98 *"There's been a collapse":* Parashyn, testimony in Shcherbak, *Chernobyl*, 76.

98 *Then he informed:* For a list of his calls, see Brukhanov, court testimony, in Karpan, *Chernobyl to Fukushima*, 129.

98 *Soon afterward, the director took the first damage reports:* Brukhanov, court testimony in Karpan, *Chernobyl to Fukushima*, 129; Parashyn, account in Shcherbak, *Chernobyl*, 76.

98 *There were soon thirty or forty men:* Parashyn, account in Shcherbak, *Chernobyl*, 76.

98 *After witnessing the horror:* Alexander Yuvchenko, author interview, 2006; Yuvchenko, interview by Bond, *New Scientist*, 2004; Vivienne Parry, "How I Survived Chernobyl," *Guardian*, August 24, 2004.

99 *Standing emergency regulations:* Karpan, *Chernobyl to Fukushima*, 18.

100 *Amid the radioactive steam:* Ibid., 18 and 20–22; Razim Davletbayev, "The Last Shift" [Последняя смена], in Semenov, ed., *Chernobyl: Ten Years On*, 371–77.

100 *Remained locked in a safe:* Karpan, *Chernobyl to Fukushima*, 25.

100 *Razim Davletbayev told himself:* Davletbayev, "The Last Shift," 377–78.

100 *Busy shutting down Turbine Number Eight:* Yuri Korneyev, testimony in Kiselyov, "Inside the Beast," 44; Korneyev, author interview, Kiev, 2015. Further details of Baranov's actions are provided at "Materials: Liquidation Heroes" [Материалы: Герои-ликвидаторы], the website of the Chernobyl NPP, http://chnpp.gov.ua/ru/component/content/article?id=82.

101 *The engineers had begun to comb the rubble:* Karpan, *Chernobyl to Fukushima*, 19; Nikolai Gorbachenko (radiation monitor at Chernobyl NPP), testimony in Grigori Medvedev, *Truth About Chernobyl*, 99.

101 *So three men picked their way:* Grigori Medvedev, *Truth About Chernobyl*, 101.

101 *Their path was strewn with wreckage:* Gorbachenko, testimony in Kiselyov, "Inside the Beast," 45.

101 *The fire escape that zigzagged:* Zakharov, author interview, 2015; Petrovsky, author interview, 2016.

101 *A handful of men:* Telyatnikov's account in David Grogan, "An Eyewitness to Di-

saster, Soviet Fireman Leonid Telyatnikov Recounts the Horror of Chernobyl,"
People, October 5, 1987; "Firefight at Chernobyl," transcript of Telyatnikov's appearance at the Fourth Great American Firehouse Exposition and Muster, Baltimore, MD, September 17, 1987, online at Fire Files Digital Library, https://fire
.omeka.net/items/show/625.

101 *Dozens of small fires:* A detailed description of the location of the fires is provided
in Karpan, *Chernobyl to Fukushima*, 12–15.

102 *Ignited by fragments of blazing debris:* Description of fires in Telyatnikov, "Firefight
at Chernobyl"; and Felicity Barringer, "One Year After Chernobyl, a Tense Tale of
Survival," *New York Times*, April 6, 1987.

102 *The air was filled with black smoke:* Telyatnikov in Barringer, "One Year After
Chernobyl."

102 *In the darkness around their feet:* Karpan, *Chernobyl to Fukushima*, 13.

102 *A more tangible threat:* Piotr Khmel, author interview, 2015.

102 *"Give me some pressure!":* Zakharov, author interview, 2016.

102 *Just as they had been trained to do:* Ibid. A report of laying foam hoses to the Unit
Three roof was made by Pravik to the dispatcher and noted in the Kiev Region Fire
Department Dispatch Log at 3:00 a.m.

102 *Kibenok had a separate line:* Zakharov, author interview, 2016.

102 *Even then, the handful of men:* Petrovsky, author interview, 2016; Rogozhkin, re-
calling a conversation with Telyatnikov in court testimony reproduced in Karpan,
Chernobyl to Fukushima, 170.

102 *Pellets of uranium dioxide:* Karpan, *Chernobyl to Fukushima*, 13. According to the
US National Institutes of Health, uranium fires cannot be effectively extinguished
with water, unless the burning material is submerged in liquid: "Even this will
not immediately extinguish the fire because the hot uranium metal dissociates
the water into H_2 and O_2, providing fuel and oxygen for the fire. If the quantity
of water is sufficient, eventually the water will provide enough cooling to extinguish the fire, but a significant amount of water can boil away in the process"
("Uranium, Radioactive: Fire Fighting," NIH, US National Library of Medicine,
webWISER online directory).

103 *Down on the ground:* Petrovsky, author interview, 2016.

104 *A fatal dose of radiation:* Estimates of what constitutes a fatal dose are based on
a "median lethal dose," or LD50, that which—if sustained instantaneously over
the whole body and left untreated—kills half of the individuals irradiated. Based
on data drawn from the bombings of Hiroshima and Nagasaki, these estimates
ranged from 3.5 to 4.0 Gy—or 350 to 400 rem. But experience with the victims
of Chernobyl led to an upward revision of these estimates, suggesting that, with
medical treatment, healthy men could survive whole-body doses of at least
5.0 Gy, or 500 rem. Gusev et al., eds. *Medical Management of Radiation Accidents*,
54–55.

104 *At a rate of 3,000 roentgen an hour:* Radiation levels on the rooftops are described
by Starodumov, commentary in *Chernobyl 1986.04.26 P. S.* [Чернобыль.1986.04.26

P. S.] (Kiev: Telecon, 2016); B. Y. Oskolkov, "Treatment of Radioactive Waste in the Initial Period of Liquidating the Consequences of the Chernobyl NPP Accident. Overview and Analysis" [Обращение с радиоактивными отходами первоначальный период ликвидации последствий аварии на ЧАЭС. Обзор и анализ], Chornobyl Center for Nuclear Safety, January 2014, 36.

104 *"Fuck this, Vanya!":* Petrovsky, author interview, December 2016.

104 *On the other side of the complex:* Leonid Shavrey, testimony in Kiselyov, "Inside the Beast," 47.

104 *The first men to arrive:* Vladimir Prischepa, recollections quoted in Karpan, *Chernobyl to Fukushima*, 15–16.

104 *Had arrived to take command:* Leonid Shavrey would later recall that Telyatnikov smelled of vodka and seemed completely intoxicated, although Petrovsky contested this. He insisted that Telyatnikov barely drank at all: "Maybe a shot at home—but at work? Never." Petrovsky, author interview, 2016.

104 *Still drunk on the People's Champagne:* Piotr Khmel, author interviews, 2006 and 2016.

105 *In the bunker:* Parashyn, account in Shcherbak, *Chernobyl*, 76; Brukhanov, court testimony in Karpan, *Chernobyl to Fukushima*, 140.

105 *Yet not everyone had succumbed:* Serafim Vorobyev, account in Shcherbak, *Chernobyl*, 397; Grigori Medvedev, *Truth About Chernobyl*, 152–54.

106 *Less than a hundred meters away:* Valentin Belokon, ambulance doctor, remembers seeing people coming from Unit Three toward the main administrative building a few minutes after 2:00 a.m. See Belokon's account in Shcherbak, "Report on First Anniversary of Chernobyl," trans. JPRS, pt. 1, 26–27.

106 *At 3:00 a.m., Brukhanov called:* Kiev Region Fire Department Dispatch Log, Chernobyl Museum.

106 *Vorobyev stood by and listened:* Vorobyev, account in Shcherbak, *Chernobyl*, 397.

106 *Yet Vorobyev knew:* Ibid., 398.

107 *"There's no mistake":* Ibid.; Grigori Medvedev, "Chernobyl Notebook" [Черно быльская тетрадь], *Novy Mir*, no. 6 (June 1989), trans. JPRS Economic Affairs, October 23, 1989, 35.

107 *Dyatlov assured him:* Read, *Ablaze*, 68–69; Grigori Medvedev, *Truth About Chernobyl*, 95.

108 *Outside in the corridor:* Dyatlov, *Chernobyl: How It Was*, 50.

108 *Running back upstairs:* Ibid., 53–54; Arkady Uskov, account in Shcherbak, *Chernobyl*, 71–72.

108 *By now, the levels of radiation:* Bagdasorov (shift foreman, Unit Three, Chernobyl NPP), account in Kopchinsky and Steinberg, *Chernobyl*, 17; Dyatlov, *How It Was*, 17.

108 *It was 5:15 a.m.:* Viktor and Valentina Brukhanov, author interview, 2016; Parashyn, recollections in Shcherbak, *Chernobyl*, 76.

109 *Ignoring instructions from above:* The shift foreman was Yuri Bagdasarov, who disobeyed the order of the plant's chief engineer, Boris Rogozhkin, to keep his reactor

going. See Bagdasarov's recollections in Kopchinsky and Steinberg (*Chernobyl*, 17) and the operating log of Unit Three in Dyatlov, *Chernobyl: How It Was*, 56–57.

109 *At the other end of the plant:* Uskov, account in Shcherbak, *Chernobyl*, 71–72.

109 *"Get that water in!":* Viktor Smagin (Unit Four foreman on the 8:00 a.m. shift, the "Second Shift" following Akimov's), recollections in Vladimir M. Chernousenko, *Chernobyl: Insight from the Inside* (New York: Springer, 1991), 62.

109 *Inside the narrow pipeline compartment:* Arkady Uskov's sketch of the scene, collection of the Chernobyl Museum, Kiev.

109 *Their white overalls:* Karpan, *Chernobyl to Fukushima*, 19.

110 *Akimov barely had enough strength:* Uskov, recollections in Kopchinsky and Steinberg, *Chernobyl*, 19.

110 *Helped from the compartment:* Uskov, account in Shcherbak, *Chernobyl*, 71–72.

110 *Gushed uselessly from shattered pipes:* Stolyarchuk, author interview, Kiev, December 2016; Dyatlov, *Chernobyl: How It Was*, 76; International Atomic Energy Agency, INSAG-7, 45.

110 *Thirty-seven fire crews:* Zhores Medvedev, *Legacy of Chernobyl*, 42.

110 *I'm so young:* Stolyarchuk, author interview, 2016.

7. SATURDAY, 1:30 A.M., KIEV

111 *For all the comforts:* Vitali Sklyarov, author interview, Kiev, February 2016; author visit to Koncha-Zaspa, February 6, 2016; Vitali Sklyarov, *Chernobyl Was . . . Tomorrow* (Montreal: Presses d'Amérique, 1993), 21–24.

111 *At fifty, Sklyarov had been a power man:* Sklyarov, *Chernobyl Was . . . Tomorrow*, 8 and 27; Vitali Sklyarov, *Sublimation of Time* [Сублимация времени] (Kiev: Kvic, 2015), 62–83.

112 *He had been kept informed of problems at the station:* In public, Sklyarov naturally stuck to the official line. See chapter 4.

112 *Even in conventional stations:* Sklyarov, author interview, 2016; Sklyarov, *Chernobyl Was . . . Tomorrow*, 27–28; Sklyarov, *Sublimation of Time*, 496–500.

112 *"There's been a series":* Sklyarov, author interview, 2016.

112 *Sklyarov immediately called:* Sklyarov, author interview, 2016; Vitali Cherkasov, "On the 15th anniversary of the atomic catastrophe: Chernobyl's sores" [К 15-летию атомной катастрофы: язвы Чернобыля], *Pravda*, April 25, 2011, www .pravda.ru/politics/25-04-2001/817996-0.

112 *From Kiev:* "Special Report" [Спецсообщение], handwritten document signed by Major General V. M. Korneychuk, April 26, 1986, document 1 in File on Special Measures in Pripyat Zone, maintained by the local *militsia* (Department of Internal Affairs of the Party Committee of Kiev Oblast), archive of the Chernobyl Museum.

113 *"One, two, three, four":* Boris Prushinsky, "This Can't Be—But It Happened (The First Days After the Catastrophe)" [Этого не может быть—но это случилось (первые дни после катастрофы)], in A. N. Semenov, ed., *Chernobyl: Ten Years On. Inevitability or Accident?* [Чернобыль. Десять лет спустя. Неизбежность

или случайность?] (Moscow: Energoatomizdat, 1995), 308–9. OPAS is the Russian acronym for *gruppa okazaniya pomoschi atomnym stantsiyam pri avariyakh*, or "group for rendering assistance to nuclear power plants in case of accidents."

113 *At 2:20 a.m., a call from the central command desk:* Read, *Ablaze*, 94; Sergei Akhromeyev and Georgi Korniyenko, *Through the Eyes of a Marshal and a Diplomat: A Critical Look at USSR Foreign Policy Before and After 1985* [*Глазами маршала и дипломата: Критический взгляд на внешнюю политику СССР до и после 1985 года*] (Moscow: Mezhdunarodnye otnosheniya, 1992), 98–99.

113 *The head of the USSR's civil defense:* Read, *Ablaze*, 93.

113 *By the time he left Moscow:* B. Ivanov, "Chernobyl. Part 1: The Accident" [Чернобыль. 1: Авария], *Voennye Znaniya* 40, no. 1 (1988), 32; Edward Geist, "Political Fallout: The Failure of Emergency Management at Chernobyl," *Slavic Review* 74, no. 1 (Spring 2015): 117.

114 *For the first time in its existence:* Leonid Drach, author interview, Moscow, April 2017.

114 *Knew the plant and staff well:* Kopchinsky's job titles at Chernobyl were deputy chief engineer for science (1976–1977) and deputy chief engineer for operations (1977–1979).

114 *Kopchinsky called for a car:* Kopchinsky, author interview, 2016.

114 *As the members:* Kopchinsky and Steinberg, *Chernobyl*, 8–9.

115 *Vladimir Marin was still at home:* Grigori Medvedev, *Truth About Chernobyl*, 152–54. In *The Legacy of Chernobyl*, Zhores Medvedev speculates that Brukhanov had orders to inform Party leaders before anyone else in the event of major industrial accidents (*The Legacy of Chernobyl*, 47). Piers Paul Read expands upon this in *Ablaze*, 77.

115 *As dawn broke:* Sklyarov, *Chernobyl Was . . . Tomorrow*, 32; Grigori Medvedev, *Truth About Chernobyl*, 117.

115 *Recalled from his holiday:* The deputy energy minister was Gennadi Shasharin.

115 *Ryzhkov told Mayorets:* Nikolai Ryzhkov, interview transcript, 2RR archive file no. 3/7/7, 16.

116 *"There's nothing left to be cooled!":* Kopchinsky and Steinberg, *Chernobyl*, 8–9; Kopchinsky, author interview, 2016. Kopchinsky believes that the telephone connection was deliberately cut by the KGB operative on the switchboard at the Chernobyl plant, as part of the effort to keep details of the accident secret.

116 *From his office in Kiev:* Sklyarov, *Chernobyl Was . . . Tomorrow*, 32.

117 *"The station might not be Ukrainian":* Sklyarov, author interview, 2016; Sklyarov, *Sublimation of Time*, 105.

117 *"What happened? What happened?":* Serhiy Parashyn, author interview, Kiev, November 2016. The scene inside the bunker is also described by Parashyn in Shcherbak, *Chernobyl*, 75–78.

117 *Samples taken by technicians:* Nikolay Karpan, "First Days of the Chernobyl Accident. Private Experience," www.rri.kyoto-u.ac.jp/NSRG/en/Karpan2008English .pdf, 8–9; Karpan, *Chernobyl to Fukushima*, 29–30.

117 *By 9:00 a.m.:* Alexander Logachev, interview by Taras Shumeyko, Kiev, June 2017; and Alexander Logachev, *The Truth [Истина]*. The time of Malomuzh's arrival is specified by Parashyn (in Shcherbak, *Chernobyl*, 76) as somewhere between seven and nine in the morning on April 26.

117 *Inside Brukhanov's office:* The meeting is detailed by Serafim Vorobyev, the head for civil defense at the plant, in Shcherbak, *Chernobyl*, 400.

118 *"Sit down":* Ibid.

118 *Malomuzh instructed Brukhanov:* Parashyn in Shcherbak, *Chernobyl*, 76–77; Karpan, *Chernobyl to Fukushima*, 26.

118 *The document was brief:* "On the Accident at the V. I. Lenin Nuclear Power Plant in Chernobyl" [Об аварии на Чернобыльской АЭС имени В. И. Ленина], signed by Viktor Brukhanov, April 26, 1986, classified, in the archive of the Chernobyl Museum. Brukhanov would later say that he knew radiation levels reached at least 200 roentgen per hour around the plant but signed the letter anyway because he "did not read [it] closely" (Brukhanov, court testimony, in Karpan, *Chernobyl to Fukushima*, 133).

118 *But it failed to explain:* Nikolai Gorbachenko and Viktor Smagin, testimonies in Grigori Medvedev, *Truth About Chernobyl*, 98–99 and 170; Dyatlov, *Chernobyl: How It Was*, 51–52.

118 *Just as a military transport plane was lifting off:* Prushinsky, "This Can't Be—But It Happened," 311–12. The time of the plane's departure is given as between eight thirty and nine in the morning by G. Shasharin, "The Chernobyl Tragedy" [Чернобыльская трагедия] in Semenov, ed., *Chernobyl: Ten Years On*, 80.

118 *He usually made it into his office:* Ryzhkov, interview transcript, 2RR, 17–18. According to Read, *Ablaze*, 95, Mayorets's team departed at ten o' clock. The text of the decree establishing the commission was provided to the author by Leonid Drach.

119 *In charge of all fuel and energy:* Drach, author interview, 2017.

119 *Ryzhkov located him:* V. Andriyanov and V. Chirskov, *Boris Scherbina [Борис Щербина]* (Moscow: Molodaya Gvardiya, 2009), 287.

119 *Academician Valery Legasov:* Margarita Legasova, *Academician Valery Alekseyevich Legasov [Академик Валерий Алексеевич Легасов]* (Moscow: Spektr, 2014), 111–13; Valery Legasov, "On the Accident at Chernobyl AES" [Об аварии на Чернобыльской АЭС], transcript of five cassette tapes dictated by Legasov in early 1988 (henceforth *Legasov Tapes*), http://lib.web-malina.com/getbook.php?bid=2755, Cassette One, 1–2.

120 *The head of the Communist Party Committee at the Kurchatov Institute:* Leonid Bolshov, author interview, Moscow, April 2017.

121 *Aleksandrov liked to drop round:* Inga Legasova, author interview, Moscow, April 2017.

121 *Only one man stood in his way:* Bolshov, author interview, 2017; Evgeny Velikhov, *Strawberries from Chernobyl: My Seventy-Five Years in the Heart of a Turbulent Russia*, trans. Andrei Chakhovskoi (CreateSpace Independent Publishing Platform, 2012), 5–12.

121 *He had traveled abroad:* Frank Von Hippel and Rob Goldston, author interview, Princeton, NJ, March 2018; Frank Von Hippel, "Gorbachev's Unofficial Arms-Control Advisers," *Physics Today* 66, no. 9 (September 2013), 41–47.

121 *"Tell him less about your successes":* Legasova, author interview, 2017.

121 *When he arrived:* Margarita Legasova, *Academician Valery A. Legasov*, 113.

121 *Despite the fine weather:* Read, *Ablaze*, 96–97, 197; Legasov, account in Shcherbak, *Chernobyl*, 414.

121 *The first of the planes from Moscow:* Grigori Medvedev, *Truth About Chernobyl*, 142; Shasharin, "Chernobyl Tragedy," 80; Drach, author interview, 2017; Anzhelika Barabanova, author interview, Moscow, October 2016.

121 *When he landed, Prushinsky learned:* Prushinsky, "This Can't Be—But It Happened," 312–13.

122 *The shell-shocked director:* For example, Vladimir Marin, who supervised the nuclear energy sector in the Central Committee apparatus in Moscow and arrived in Pripyat by early evening on April 26, writes that at five o'clock on Saturday afternoon, Brukhanov reported that the reactor was under control and being cooled (V. V. Marin, "On the Activities of the Task Force of the Politburo of the CPSU Central Committee at the Chernobyl NPP" [О деятельности оперативной группы Политбюро ЦК КПСС на Чернобыльской АЭС], in Semenov, ed., *Chernobyl: Ten Years On*, 267–68).

123 *It passed from the General Department:* Dmitri Volkogonov and Harold Shukman, *Autopsy for an Empire: The Seven Leaders Who Built the Soviet Regime* (New York: Free Press, 1999), 477.

123 *"An explosion occurred":* "Urgent Report, Accident at Chernobyl Atomic Power Station," April 26, 1986, History and Public Policy Program Digital Archive, Volkogonov Collection, Manuscript Division, Library of Congress. Translated for NPIHP by Gary Goldberg, http://digitalarchive.wilsoncenter.org/document/115341.

123 *The second—more senior—wave:* This party also included Alexander Meshkov (deputy minister of Sredmash) and Viktor Sidorenko (vice chairman of Gosatomenergonadzor, Sredmash's nuclear energy oversight committee) (Shasharin, "Chernobyl Tragedy," 80–81; Sklyarov, *Chernobyl Was . . . Tomorrow*, 33.

123 *"Don't try to frighten us":* Grigori Medvedev, *Truth About Chernobyl*, 154; Sklyarov, author interview, 2016.

123 *Bumping down on a dirt airstrip:* Sklyarov, *Chernobyl Was . . . Tomorrow*, 37–39; Shasharin, "Chernobyl Tragedy," 80–81; Colonel General B. Ivanov, "Chernobyl. Part 2: Bitter Truth Is Better" [2: Лучше горькая правда], *Voennye Znaniya* 40, no. 2 (1988): 22.

124 *Conducting a radiation reconnaissance:* This radiation surveillance work had been hampered by the continuing secrecy surrounding the station. When Logachev, the civil defense lieutenant responsible for scouting the power plant, was given his orders, he pointed out that Unit Four didn't appear to be shown anywhere on the schematic he had. Malomuzh himself took a pen and scribbled a rough outline of the reactor in an otherwise blank space in the middle of the map (Alexander Lo-

gachev, author interview, 2016). Logachev's dosimetry map of Chernobyl station from April 26, 1986, is in the archive of Chernobyl Museum.

124 *Circling the reactor at low altitude:* Prushinsky, "This Can't Be—But It Happened," 315.

125 *Fomin finally conceded:* Gennadi Berdov, Ukraine's deputy minister of internal affairs, testimony in Voznyak and Troitsky, *Chernobyl: It Was Like This*, 199.

125 *Worse news was to come:* Karpan, *Chernobyl to Fukushima*, 28.

125 *His armored personnel carrier:* The vehicle's top speed and weight are given in Logachev, *The Truth*.

125 *"You mean milliroentgen, son":* Logachev, interviews by author and Taras Shumeyko, 2016; Logachev, *The Truth*; Logachev, dosimetry map of Chernobyl station, Chernobyl Museum.

126 *7:20 p.m. on Saturday evening:* Alexander Lyashko, *The Weight of Memory: On the Rungs of Power* [Груз памяти: На ступенях власти], vol. 2 in a trilogy (Kiev: Delovaya Ukraina, 2001), 351.

126 *They were greeted by a delegation:* Legasov Tapes, Cassette One, 5.

126 *Boundless levels of marsh:* Author visit to Pripyat, April 25, 2016.

126 *But in Pripyat, Scherbina:* Drach, author interview, 2017; Sklyarov, *Chernobyl Was . . . Tomorrow*, 40.

126 *Intelligent, energetic, and hardworking:* Drach, author interview, 2017; Sklyarov, author interview, 2016; Scherbina's date of birth is given as October 5, 1919, in Andriyanov and Chirskov, *Boris Scherbina*, 387. This description also draws on Kopchinsky and Steinberg, *Chernobyl*, 53.

126 *Respect and admiration:* Drach, author interview, 2017.

126 *"So, are you shitting your pants?":* Sklyarov, *Chernobyl Was . . . Tomorrow*, 40; Sklyarov, author interview, 2016.

127 *"We have to evacuate the local population":* Prushinsky, "This Can't Be—But It Happened," 317. In spite of this exchange, and a weight of conflicting testimony from others, Prushinsky records that the issue of evacuation was resolved "without any delay" at the meeting that followed.

127 *The first meeting of the government commission:* Description of the crowd, tension: Vasily Kizima, author interview, Kiev, February 2016. Description of the room, smoking: Alexander Logachev, interview by Taras Shumeyko, Kiev, June 2017. Legasov gives the time of his arrival as around 8 p.m., and Prushinsky writes that the first meeting began two hours later (Prushinsky, "This Can't Be—But It Happened," 317).

127 *Academician Legasov listened:* Legasov Tapes, cassette One, 5.

127 *They said only:* Ibid., cassette One, 4.

127 *Scherbina divided the members:* Ibid., 5.

128 *Like the station's own physicists:* Shasharin, "Chernobyl Tragedy," 85–86; Karpan, *Chernobyl to Fukushima*, 78.

128 *But the Rovno station director was reluctant:* Sklyarov, author interview, 2017; Sklyarov, *Chernobyl Was . . . Tomorrow*, 41–42.

128 *At the same time, Legasov realized:* Read, *Ablaze*, 105–6.

129 *Everyone knew that* something *must be done:* Kizima, author interview, 2016.

129 *They found the figures alarming:* Logachev, interview by Taras Shumeyko, 2017.

129 *"They never evacuated people there!":* Drach, author interview, 2017. This account is confirmed by Legasov and General Berdov: Voznyak and Troitsky, *Chernobyl: It Was Like This*, 218.

129 *According to the state document:* The document (Russian name "Критерии для принятия решения по защите населения в случае аварии атомного реактора") is cited in Voznyak and Troitsky, *Chernobyl: It Was Like This*, 219.

129 *Even the regulations:* Geist, "Political Fallout," 115–16.

129 *He had little reason to believe:* Less than a year earlier, a radio station in Kiev conducting a training exercise had accidentally broadcast a recorded message announcing that the city's hydroelectric dam had burst, and urged citizens to gather their belongings and immediately leave their homes for higher ground. The announcement was met with inaction and indifference. Such was Kievans' distrust of official news sources that, instead of fleeing the supposed catastrophe, more than eight hundred people phoned the radio station to ask if the report was true. Nigel Raab, *All Shook Up: The Shifting Soviet Response to Catastrophes, 1917–1991* (Montreal: McGill-Queen's University Press, 2017), 143–44.

129 *By daybreak on Saturday:* Esaulov recalls hearing about the KGB order to cut off the phones from his boss, Pripyat *ispolkom* chairman Vladimir Voloshko, sometime in the early morning (Esaulov, *City That Doesn't Exist*, 16–17).

130 *The plume of vapor from the reactor had been drifting:* Read, *Ablaze*, 101–2; Akhromeyev and Korniyenko, *Through the Eyes of a Marshal and a Diplomat*, 100.

130 *It would bring down radioactive fallout:* Zhores Medvedev, *Legacy of Chernobyl*, 141.

130 *He resolved to wait until morning:* Ivanov, "Chernobyl. Part 3: Evacuation" [3: Эвакуация], *Voennye Znaniya* 40, no. 3 (1988): 38.

130 *A ruby glow:* Karpan, "First Days of the Chernobyl Accident," 2008.

130 *Taking samples from the coolant canal:* Kopchinsky and Steinberg, *Chernobyl*, 65; Armen Abagyan (head of VNIIAES), account in Voznyak and Troitsky, *Chernobyl: It Was Like This*, 213.

131 *"He's a panicker!":* Sklyarov, author interview, 2016; Sklyarov, *Sublimation of Time*, 105–6. In the author interview, Sklyarov recalled that Scherbina used the phrase *"skandal na ves' mir,"* which may be literally translated as "a scandal before the whole world," but *skandal* in Russian carries a combined meaning of a "humiliation" and a "mess." In *Sublimation of Time*, Sklyarov explains that he had first included an account of this episode in the manuscript of his 1991 memoir *Tomorrow . . . Was Chernobyl* but removed it before publication at the request of Vladimir Ivashko, who by then had succeeded Scherbitsky as Ukraine's Communist Party chief.

8. SATURDAY, 6:15 A.M., PRIPYAT

132 *It was sometime after 3:00 a.m.:* Alexander Esaulov, author interview, Irpin, July 2015; Shcherbak, "Report on First Anniversary of Chernobyl Accident," trans. JPRS, pt. 1, 30.

132 *There was always something going wrong:* Esaulov, *The City That Doesn't Exist*, 11–12.

132 *On the phone was Maria Boyarchuk:* Ibid., 16.

133 *But then another ambulance sped past:* Eventually a total of four ambulances were involved, according to Vitaly Leonenko, director of Medical-Sanitary Center No. 126 (author interview, Vepryk, Ukraine, December 2016). Arkady Uskov (account in Shcherbak, *Chernobyl*, 69) would remember passing two ambulances on his way to the plant at four thirty, and Piers Paul Read writes that by five o'clock, the vehicles were performing "a shuttle service" (*Ablaze*, 85).

133 *Esaulov began to suspect:* Esaulov, author interview, 2015; Esaulov, *City That Doesn't Exist*, 16.

134 *"Good morning, Boris":* Andrei Glukhov, author interview, Slavutych, Ukraine, 2015.

134 *Glukhov climbed to the top floor:* Glukhov, author interview, 2015; author visit to Toptunov's apartment in Pripyat, April 25, 2016.

135 *The Pripyat hospital:* Leonenko, author interview, 2016; author visit to Hospital No. 126, April 27, 2016.

135 *Formally diagnosed radiation sickness:* Ibid.; according to Angelina Guskova, the staff of the hospital initially reported to her that the injuries were the result of a chemical fire. Angelina Guskova, interview by Vladimir Gubarev, "On the Edge of the Atomic Sword" [На лезвии атомного меча], *Nauka i zhizn*, no. 4 (2007): www.nkj.ru/archive/articles/9759.

135 *The men and women arriving from the plant:* Tatyana Marchulaite (medical assistant at Hospital No. 126), account in Voznyak and Troitsky, *Chernobyl: It Was Like This*, 202–5.

135 *At first, Dyatlov refused treatment:* Read, *Ablaze*, 85–86.

136 *"You don't vomit at fifty":* This account was shared by Alexander Yuvchenko in Vivienne Parry, "How I Survived Chernobyl," *Guardian*, August 24, 2004, https://www.theguardian.com/world/2004/aug/24/russia.health.

136 *Burns and blisters covered his body:* Read, *Ablaze*, 85.

136 *"Get away from me":* Marchulaite, testimony in Voznyak and Troitsky, *Chernobyl: It Was Like This*, 205. Shashenok's time of death is specified by Nikolai Gorbachenko (radiation monitor at ChNPP) in Kiselyov, "Inside the Beast," 46.

136 *It was not yet 8:00 a.m.:* Natalia Yuvchenko, author interview, 2015; Read, *Ablaze*, 85 and 91.

136 *Around the corner, at home:* Maria Protsenko, author interview, Kiev, September 2015.

137 *When persuasion wouldn't work:* Anatoly Svetetsky, head of technological safety systems reactor and turbine departments of Units Three and Four, Chernobyl Nuclear Power Plant, interview by Taras Shumeyko, Kiev, May 28, 2017.

137 *"proletarian aesthetics"*: For an exploration of the role of proletarian aesthetics in Soviet energy-related construction, see Josephson, *Red Atom*, 96–97.

138 *As many as two hundred thousand people*: Sich, "The Chornobyl Accident Revisited," 204; Igor Kruchik, "Mother of the Atomgrad" [Мати Атомограда], *Tizhden*, September 5, 2008, http://tyzhden.ua/Publication/3758.

138 *"We're going to chase away the* shitiki*!"*: See the interview with Vasily Gorokhov (Chernobyl deputy director for decontamination from July 1986 to May 1987) for evidence that liquidators, too, believed in *shitiki*: Alexander Bolyasny, "The First 'Orderly' of the First Zone" [Первый «санитар» первой зоны], *Vestnik* 320, no. 9 (April 2003): www.vestnik.com/issues/2003/0430/koi/bolyasny.htm.

138 *"You've got a phone call"*: Protsenko, author interview, 2015.

138 *Hundreds of members of the* militsia: "Background information on the Town of Pripyat," April 26, 1986, Pripyat *militsia*, File on Special Measures in Pripyat Zone, 14, archive of Chernobyl Museum.

138 *An emergency meeting*: Description of the city administration meeting on Saturday morning: Protsenko and Esaulov, author interviews, 2015.

139 *Warming up for an afternoon match*: The fixture, part of the semifinal competition for the top soccer teams in the Kiev region, was canceled later that day ("Soccer in Pripyat: The History of the 'Builder' Soccer Club" [Футбол в Припяти. История футбольного клуба «Строитель»], *Sports.ru* blog, https://www.sports.ru/tribuna/blogs/golden_ball/605515.html., April 27, 2014.

139 *Malomouzh had arrived*: Parashyn, account in Shcherbak, *Chernobyl*, 76; Zhores Medvedev, *Legacy of Chernobyl*, 37.

139 *"There has been an accident"*: Protsenko, author interview, 2015.

139 *In the meantime, he explained*: Shcherbak, "Report on First Anniversary of Chernobyl Accident," trans. JPRS, pt. 1, 48.

139 *Naturally, there were questions*: Ibid., 37.

139 *"And please do not panic"*: Protsenko, author interview, 2015.

140 *A solitary armored car*: The time of the column's arrival is given in Vladimir Maleyev, *Chernobyl. Days and Years: The Chronicle of the Chernobyl Campaign* [Чернобыль. Дни и годы: летопись Чернобыльской кампании] (Moscow: Kuna, 2010), 21. Additional details: Colonel Grebeniuk (commander of the 427th Mechanized Regiment), interviews by author and Taras Shumeyko, Kiev, July 2016. By this time Senior Lieutenant Alexander Logachev's machine, which had suffered brake failure on the road from Kiev, had replaced at the head of the column by another vehicle; in an attempt to catch up with his comrades, Logachev drove directly toward the station instead. (Logachev, author interview, Kiev, 2017.)

140 *Controlled tightly by the KGB*: Kotkin, *Armageddon Averted*, 42.

140 *Protsenko sat down*: Protsenko, author interview, 2015.

140 *225th Composite Air Squadron*: Sergei Drozdov, "Aerial Battle over Chernobyl" [Воздушная битва при Чернобыле], *Aviatsiya i vremya* 2 (2011), www.xliby.ru/transport_i_aviacija/aviacija_i_vremja_2011_02/p6.php.

140 *In the pilot's seat*: Sergei Volodin, author interview, Kiev, July 2015.

140 *An airborne radiation survey:* Colonel Lubomir Mimka, author interview, Kiev, February 2016.

140 *On the way there:* Sergei Volodin, unpublished memoir, undated.

141 *And although he and his crew:* Volodin, author interviews, 2006 and 2015.

141 *Volodin knew Chernobyl well:* Ibid.; Volodin, unpublished memoir.

142 *At the construction headquarters:* Kovtutsky, author interview, 2016.

142 *From her desk in the White House:* Protsenko, author interview, 2016.

143 *A manager working on the Fifth and Sixth Units:* Grigori Medvedev, *Truth About Chernobyl,* 88–89 and 149–51.

143 *The technician's next-door neighbor:* Ibid., 150.

143 *Knowing that the KGB:* The engineer was Georgi Reikhtman (author interview, September 2015), who told his wife to pack all of the family's winter clothing. Because it was late spring, she thought he was talking nonsense and ignored him.

143 *Another persuaded Director Brukhanov:* This engineer was Nikolai Karpan: *Chernobyl to Fukushima,* 32–33.

143 *Arrving at Yanov station:* Veniamin Prianichnikov, author interview, Kiev, February 2006.

144 Militsia *officers everywhere:* An internal Ukrainian Interior Ministry memo specifies that by nine o'clock on Saturday morning, 600 *militsia* troops and 250 authorized "civilian persons" were deployed to the Pripyat area from local and regional bases. "Background Information on the Town of Pripyat," April 26, 1986, File on Special Measures in Pripyat Zone, 14, archive of Chernobyl Museum.

144 *Prianichnikov suspected:* Prianichnikov, author interview, 2006.

144 *When the civil defense major returned:* Volodin, author interview, 2006. The time of the first helicopter flight for air radiation reconnaissance on April 26 is specified by Major General M. Masharovsky in "Operation of Helicopters During the Chernobyl Accident," in Current Aeromedical Issues in Rotary Wing Operations, Papers Presented at the RTO Human Factors and Medicine Panel (HFM) Symposium, San Diego, October 19–21, 1998, RTO/NATO, 7-2.

145 *On his right, he could see the village:* Ibid.; Volodin, unpublished memoir.

146 *Natalia Yuvcehnko had spent all morning:* Natalia Yuvchenko, author interview, 2015.

147 *Vodka, cigarettes, and folk remedies:* Read, *Ablaze,* 87–88.

147 *Alexander said:* Natalia Yuvchenko, author interview, 2015.

147 *At 4:00 p.m., the members of the OPAS medical team:* Voznyak and Troitsky, *Chernobyl: It Was Like This,* 207.

147 *"Many are in grave condition":* Esaulov, *City That Doesn't Exist,* 23–24.

148 *After some debate:* Leonenko, author interview, 2016.

148 *Second Secretary Malomuzh summoned Esaulov:* Esaulov, *City That Doesn't Exist,* 25.

148 *By nightfall on Saturday:* Protsenko, author interview, 2016; David Remnick, "Echo in the Dark," *New Yorker,* September 22, 2008.

148 *When the boxes were silenced:* Protsenko, author interview, 2016; Kovtutsky, author interview, 2016.

149 *And then officials:* Author interviews: Natalia Yuvchenko, 2015; Natalia Khodemchuk, 2017; Alexander Sirota, 2017.

149 *Alexander Korol had spent:* Korol, author interview, 2015.

149 *It was past 9:00 p.m:* The time of the convoy's departure is given by Esaulov as ten
at night (*City That Doesn't Exist,* 27) and confirmed by Valery Slutsky, bus driver,
author interview, Pripyat, February 2006.

149 *Two red Ikarus buses:* Esaulov would recall later that although there weren't too
many passengers during this first trip—twenty-four people who could ride up-
right (plus two others who couldn't and were transported by ambulance)—he requested a spare Ikarus bus just in case, for fear that one might break down along
the way. Esaulov, *City That Doesn't Exist,* 26–27; Shcherbak, "Report on First Anniversary of Chernobyl Accident," trans. JPRS, pt. 1, 31.

149 *Sheets of plastic:* Leonenko, author interview, 2016.

149 *"The rods came halfway, and then stopped":* Korol, author interview, 2015.

150 *30 percent burns:* Shcherbak, *Chernobyl,* 51.

150 *In their big corner apartment:* Viktor and Valentina Brukhanov, author interview,
2015.

150 *When Veniamin Prianichnikov got through:* Prianichnikov, author interview, 2006.

150 *In the small hours:* Esaulov writes that the convoy arrived at Borispol at three thirty
in the morning (*City That Doesn't Exist,* 28–29).

151 *"People are resting":* Handwritten log of events for April 26–27, in File on Special
Measures in Pripyat Zone, Internal Affairs Department of the Kiev Oblast Party
Committee, archive of Chernobyl Museum, 13.

9. SUNDAY, APRIL 27, PRIPYAT

152 *The first of the transport helicopters:* Nikolai Antoshkin, author interview, Moscow,
October 2015.

152 *Dispatched from the central command post:* The expert was Colonel Anatoly Kush-
nin. See his account of events in Kiselyov, "Inside the Beast," 50. Additional details: Lubomir Mimka, author interview, Kiev, February 2016.

152 *As soon as he arrived:* He reported first to Ivanov, head of civil defense, and the
commander of the Soviet chemical troops Vladimir Pikalov, who finally arrived
at eleven thirty on Saturday night, according to Voznyak and Troitsky, *Chernobyl:
It Was Like This,* 214.

152 *"We need helicopters":* Antoshkin, author interview, 2015.

152 *Using a telephone:* Ibid.; Mimka, author interview, 2016; Colonel Boris Nesterov,
author interview, Dnipro, Ukraine, December 2016; Major A. Zhilin, "No such
thing as someone else's grief" [Чужого горя не бывает], *Aviatsiya i Kosmonavtika,* no. 8 (August 1986): 10.

153 *Inside the hotel:* Prushinsky, "This Can't Be—But It Happened," 318.

153 *Legasov estimated:* Legasov, "My duty is to tell about this," in Mould, *Chernobyl
Record,* 292.

153 *The intense heat could soon melt:* Legasov Tapes, cassette One, 8.

153 *A rate of around one tonne an hour:* Legasov, "My duty is to tell about this," in Mould, *Chernobyl Record,* 292; Legasov Tapes, cassette One, 8. For Legasov's report to the Politburo about his analysis, see Maleyev, *Chernobyl: Days and Years:* "Meeting of the Politburo of the CPSU Central Committee: Protocol No. 3" [Заседание Политбюро ЦК КПСС 5 мая 1986 года: Протокол № 3], 249–52.

153 *The blaze could roar on for more than two months:* Legasov's guess for the total weight of graphite in Unit Four, both before and after the explosion, significantly exceeded most others. But even the lower estimates for how much graphite remained inside the core after the accident—such as the 1,500 tonnes cited by a KGB memo on May 11, 1986—made for around two months of continuous burning. See the KGB memo in Danilyuk, ed., "Chernobyl Tragedy," *Zarkhiviv,* document no. 31: *Special Report of the UkSSR OG KGB Chief in the Town of Chernobyl to the UkSSR KGB Chairman.*

153 *Ordinary firefighting techniques:* V. Bar'yakhtar, V. Poyarkov, V. Kholosha, and N. Shteinberg, "The Accident: Chronology, Causes and Releases," in G. J. Vargo, ed., *The Chornobyl Accident: A Comprehensive Risk Assessment* (Columbus, OH: Battelle Press, 2000), 13.

153 *The graphite and nuclear fuel were burning:* Legasov Tapes, cassette One, p. 8; Grigori Medvedev, *The Truth about Chernobyl,* 176; Zhores Medvedev, *The Legacy of Chernobyl,* 43.

154 *Colossal fields of gamma radiation:* Bar'yakhtar et al., "The Accident: Chronology, Causes and Releases," 13.

154 *One physicist, unable to find the answer:* Evgeny Ignatenko, ed., *Chernobyl: Events and Lessons* [Чернобыль: события и уроки] (Moscow: Politizdat, 1989), 128.

154 *In his office at the Kurchatov Institute:* Legasov Tapes, cassette One, 9; Grigori Medvedev, *Truth About Chernobyl,* 176.

154 *Meanwhile, the team:* Armen Abagyan, account in Voznyak and Troitsky, *Chernobyl: It Was Like This,* 220.

154 *At 2:00 a.m., Scherbina telephoned:* Vladimir Dolgikh, interview transcript, June 1990, 2RR archive file no. 1/3/5, 4. The fact that Scherbina's mind was not yet made up at 2:30 a.m. is attested to by a senior Kiev transportation official who, at around that time, arrived near Pripyat with the column of buses and came to the White House, where he reported to Scherbina. The chairman asked him, "And who sent you?" V. M. Reva, first vice president of the Ukrainian State Automobile Transport Corporation, testimony at the 46th session of the Supreme Rada, December 11, 1991, transcript online at http://rada.gov.ua/meeting/stenogr/show /4642.html.

154 *By the time the scientists finally crawled into their beds:* Drach, author interview, 2017; Nesterov, author interview, 2016.

155 *"I've made my decision":* Ivanov's diary entry, reproduced in "Chernobyl, Part 3: Evacuation" [Часть 3: Эвакуация], *Voennye Znaniya* 40, no. 3 (1988): 38.

155 *Ivanov gave him the radiation report:* Ibid.; Leonenko, author interview, 2016. A different view is presented by Leonid Drach (author interview, 2017), who recalls that between 1:00 a.m. and 2:00 a.m. on Sunday, Pikalov was among those who told Scherbina he had no choice but to evacuate the city.

155 *But he still held back from giving the order:* According to the handwritten notes in the log kept at the Pripyat *militsia* headquarters, at 6:54 a.m., the first secretary of Kiev Oblast Party Committee, G. I. Revenko, reported that "the decision on evacuation will be made after 9:00 a.m." The KGB confirmed this forecast at 7:45 a.m. Pripyat *militsia*, File on Special Measures in the Pripyat Zone (Chernobyl Museum), 12–13.

155 *Soon after 8:00 a.m.:* Time of the flight is given by Antoshkin in *Regarding Chernobyl* (unpublished memoir) as 8:12 a.m.

155 *They were joined by Generals Pikalov and Antoshkin:* Nesterov, author interview, 2016; Zhilin, "No such thing as someone else's grief," 10.

156 *To even the most recalcitrant Soviet eye:* Legasov Tapes, cassette One, p. 6; Mould, *Chernobyl Record*, 291; Margarita Legasova, *Academician Valery A. Legasov*, 119.

156 *As the helicopter headed back to Pripyat:* Legasov in Mould, *Chernobyl Record*, 290.

156 *At ten o'clock:* Vladimir Pikalov, "Interview with Commander of Chemical Troops," interview by A. Gorokhov, *Pravda* (December 25, 1986), translated in JPRS, *Chernobyl Nuclear Accident Documents*, 92; Ivanov, "Chernobyl, Part 3: Evacuation," 38.

156 *At 1:10 p.m.:* The time of the broadcast is given as 1:10 p.m. in Voznyak and Troitsky, *Chernobyl: It Was Like This*, 223. Others recall this happened on or before noon: Drach, author interview, 2017.

156 *In a strident, confident voice:* For the original text of the announcement, see Andrei Sidorchik, "Deadly Experiment. Chronology of the Chernobyl NPP Catastrophe" [Смертельный эксперимент. Хронология катастрофы на Чернобыльской АЭС], *Argumenty i fakty*, April 26, 2016, www.aif.ru/society/history/smertelnyy _eksperiment_hronologiya_katastrofy_na_chernobylskoy_aes. A recording of the announcement can be found at www.youtube.com/watch?v=1l3g3m8Vrgs.

156 *Drafted that morning:* Leonid Drach (author interview, 2017), said that he worked on a draft of the announcement with Nikolai Nikolayev, deputy chairman of the Ukraine Council of Ministers. Sklyarov recalls working on it as well, together with Ivan Plyushch, deputy chairman of the Kiev region *ispolkom* (Sklyarov, author interview, 2016).

157 *The emergency proclamation:* Esaulov, *City That Doesn't Exist*, 45. Vitali Sklyarov explained that the proclamation was designed not only to preclude panic but also to discourage the citizens from filling the available transport with heavy luggage and personal possessions. Sklyarov, author interview, Kiev, February 2016.

157 *They should close their windows:* Lyubov Kovalevskaya, quoted in Shcherbak, "Report on First Anniversary of Chernobyl," trans. JPRS, pt. 1, 41.

157 *Earlier that morning:* Natalia Yuvchenko, author interview, 2015.

158 *Almost a month's salary:* As a schoolteacher, Natalia was earning 120 rubles a month.

158 *On the second floor of the White House:* Protsenko, author interview, 2016.

159 *In all, there were some 51,300:* These numbers are from the handwritten log of emergency response measures kept by the local *militsia* major general, who noted later that 47,000 people had been evacuated and that up to 1,800 plant operators and 2,500 construction workers remained behind. Also staying in Pripyat were 600 to 700 staff of the Department of Internal Affairs and armed forces, in addition to city administrators and civil defense personnel (Pripyat *militsia*, File on Special Measures in the Pripyat Zone, April 27, 1986, 29). However, a large proportion of the population had left the city by other means long before the evacuation began, although the estimated numbers of those who did so vary considerably (see detailed note below).

159 *To get all the families out safely:* Protsenko, author interviews, 2015 and 2016; Ukrainian Ministry of Internal Affairs, Report no. 287c/Gd [287c/Гд], April 27, 1986 (confidential, signed by Interior Minister Ivan Gladush), archive of the Chernobyl Museum.

159 *At the same time, in Kiev:* "Report of the Ministry of Transport of the Ukrainian SSR to the Ukraine Central Committee of the Communist Party," April 27, 1986 (no. 382c, confidential, signed by Minister Volkov), archive of the Chernobyl Museum.

159 *By 3:50 a.m.:* Pripyat *militsia*, File on Special Measures in the Pripyat Zone, 10–13.

159 *The bus stops of Kiev were crowded with frustrated passengers:* Shcherbak, "Report on First Anniversary of Chernobyl," trans. JPRS, pt. 1, 42–43.

159 *Boiled potatoes, bread, lard:* Natalia Khodemchuk, author interview, Kiev, 2017.

159 *Despite warnings to stay inside:* Anelia Perkovskaya (the secretary of the Komsomol at Pripyat *gorkom*), account in Shcherbak, "Report on First Anniversary of Chernobyl," trans. JPRS, pt. 1, 40 and 43.

159 *Some families set off:* Boris Nesterov, *Heaven and Earth: Memories and Reflections of a Military Pilot* [Небо и земля: Воспоминания и размышления военного летчика] (Kherson, 2016), 240.

160 *At the same time, the crews of two helicopters:* Antoshkin would later insist that "bombing" of the reactor was forbidden before the evacuation was complete (author interview, 2017), but it seems likely this is wishful thinking, with the benefit of hindsight, and is contradicted by other accounts. For example, Boris Nesterov, who flew these first missions, said he began dropping material into the reactor at around 3:00 p.m. and could see the evacuation unfold from his cockpit (author interview, 2016).

160 *The operation, approved by Boris Scherbina:* A. A. Dyachenko, ed., *Chernobyl. Duty and Courage* [Чернобыль. Долг и мужество], vol. 1 (Moscow: Voenizdat, 2001), 233.

160 *A complex cocktail of substances:* Legasov Tapes, cassette One, 10; Shasharin, "The Chernobyl Tragedy," 91.

160 *Lead, in particular:* Sklyarov, *Chernobyl Was . . . Tomorrow,* 61 and 69.

161 *In the meantime, Scherbina sent General Antoshkin:* Shasharin, testimony in Grigori Medvedev, *Truth About Chernobyl,* 192; Protsenko, author interview, 2015; Mimka, author interview, 2016; Antoshkin, author interview, 2017. In the interview, Antoshkin disputed Shasharin's suggestion that during this episode the general was filling sandbags while still wearing his full dress uniform.

161 *The quantities required were enormous:* Dyachenko, ed., *Chernobyl. Duty and Courage,* 234.

161 *Eventually between 100 and 150 men and women:* Mimka, author interview, 2016; Logachev, author interview, 2017.

161 *Scherbina remained implacable:* Gennadi Shasharin and Anatoly Zagats (chief engineer of Yuzhatomenergomontazh), testimonies in Grigori Medvedev, *Truth About Chernobyl,* 192–93.

162 *If he was aware of the rising level of contamination:* Sklyarov, *Chernobyl Was . . . Tomorrow,* 52.

162 *Early on Sunday afternoon, the first ten bags of sand:* Shasharin, testimony in Grigori Medvedev, *Truth About Chernobyl,* 193; Mimka, author interview, 2016; Nesterov, author interview, 2016.

162 *There were 1,225 buses in all:* Report of Ukraine Ministry of Transport of the Ukrainian SSR to the Ukraine Central Committee of the Communist Party of Ukraine on April 27, 1986 (no. 382c), archive of the Chernobyl Museum; Protsenko, author interview, 2016; Natalia Yuvchenko, author interview, 2016.

162 *At 2:00 p.m.:* The time is given as one thirty in the report of the Ministry of Transport to the Ukraine Central Committee, archive of the Chernobyl Museum; but as 2:00 p.m. in the handwritten chronology of events in the Pripyat *militsia* log of Chernobyl accident response measures (File on Special Measures in the Pripyat Zone, Chernobyl Museum, 29–30). A Kiev transport official who oversaw the buses also states 2:00 p.m. (Reva, testimony at the Supreme Rada, December 11, 1991).

162 *Maria Protsenko was waiting:* Protsenko, author interview, 2015.

163 *Outside the 540 separate entrances:* Ivanov, "Chernobyl. Part 3: Evacuation," 38; Voznyak and Troitsky, *Chernobyl: It Was Like This,* 223.

163 *At around 3:00 p.m., Colonel Boris Nesterov:* Nesterov, author interview, 2016; Nesterov, *Heaven and Earth,* 236–43.

164 *At 5:00 p.m., Maria Protsenko folded her map:* Protsenko, author interview, 2015. A large number of residents left by their own means, either before learning of the accident or after. The local branch of the Ukrainian Interior Ministry estimated this number to be 8,800 people: "Situation report as of 8 p.m., April 28, 1986," in Pripyat *militsia* log of Special Measures in the Pripyat Zone, Chernobyl Museum, 30. Other sources estimate this number to be as high as 20,000: Baranovska, ed., *Chernobyl Tragedy,* document no. 59: "Memorandum of the Department of Science and Education of the Central Committee of Communist Party of Ukraine on Immediate Measures Pertaining to the Accident at Chernobyl NPP," April 29, 1986.

This memorandum states that only 27,500 people were evacuated using buses and other transport specially provided by officials.

165 *As the multicolored convoy of buses wound its way:* Natalia Yuvchenko, author interview, 2015.

165 *Part of the convoy was well beyond:* Logachev, author interview, 2017.

165 *One member of the power station staff:* Glukhov, author interview, 2015.

165 *Viktor Brukhanov's wife, Valentina, wept:* Viktor and Valentina Brukhanov, author interview, 2016.

165 *The passengers whispered anxiously:* Natalia Yuvchenko, author interview, 2015.

165 *On the third floor of the White House:* Protsenko, author interview, 2015.

Part 2. Death of an Empire

10. THE CLOUD

169 *Unleashed in the violence of the explosion:* World Health Organization (WHO), "Chernobyl Reactor Accident: Report of a Consultation," Regional Office for Europe, report no. ICP/CEH 129, May 6, 1986 (provisional), 4.

169 *The cloud carried gaseous xenon 133:* Helen ApSimon and Julian Wilson, "Tracking the Cloud from Chernobyl," *New Scientist*, no. 1517 (July 17, 1986): 42–43; Zhores Medvedev, *Legacy of Chernobyl*, 89–90.

169 *At its heart, it pulsed:* ApSimon and Wilson, "Tracking the Cloud from Chernobyl," 45; Zhores Medvedev, *Legacy of Chernobyl*, 195.

169 *By the time Soviet scientists finally began regular airborne monitoring:* At this point, the cloud initially released by the explosion had already crossed into Poland and Finland: Zhores Medvedev, *Legacy of Chernobyl*, 195.

169 *Within twenty-four hours:* WHO, "Chernobyl Reactor Accident: Report of a Consultation," 4.

169 *At midday on Sunday, an automatic monitoring device:* Zhores Medvedev, *Legacy of Chernobyl*, 196–97.

169 *Late that night, the plume encountered rain clouds:* ApSimon and Wilson, "Tracking the Cloud," 42 and 44; Zhores Medvedev, *Legacy of Chernobyl*, 197.

170 *Shortly before seven o'clock on Monday morning:* Cliff Robinson, author interview by telephone, March 2016.

170 *They were building a large underground repository:* The facility was finished in 1988. See "This is where Sweden keeps its radioactive operational waste," Swedish Nuclear Fuel and Waste Management Company (SKB), November 2016, www.skb.com/our-operations/sfr.

170 *The reactor was only six years old:* Erik K. Stern, *Crisis Decisionmaking: A Cognitive Institutional Approach* (Stockholm: Swedish National Defence College, 2003), 130.

171 *At 9:30 a.m., the plant manager, Karl Erik Sandstedt:* Stern, *Crisis Decisionmaking*, 131–32; Nigel Hawkes et al., *The Worst Accident in the World: Chernobyl, the End of the Nuclear Dream* (London: William Heinemann and Pan Books, 1988), 116.

171 *Thirty minutes later:* Robinson, author interview, 2016.

171 *But by then, state nuclear and defense agencies:* Stern, *Crisis Decisionmaking*, 134–36.

172 *At around eleven in the morning Moscow time, Heydar Aliyev:* Heydar Aliyev, interview transcript, 2RR archive file no. 3/1/6, 14–15.

172 *One of the most powerful men in the Soviet Union:* Aliyev headed the KGB in Azerbaijan from 1967 to 1969: "Heydar Aliyev, President of the Republic of Azerbaijan" [Гейдар Алиев, президент Азербайджанской Республики], interview by Mikhail Gusman, TASS, September 26, 2011, http://tass.ru/arhiv/554855.

172 *Authorities in Kiev, without prompting from Moscow:* Angus Roxburgh, *The Second Russian Revolution: The Struggle for Power in the Kremlin* (New York: Pharos Books, 1992), 41–42.

172 *Aliyev realized:* Aliyev, interview transcript, 2RR, 14–15.

172 *The dozen men:* List of participants: minutes from the Politburo meeting (April 28, 1986), in Maleyev, *Chernobyl. Days and Years*, 241; Gorbachev's office: Aliyev, interview transcript, 2RR, 14–15.

172 *In spite of recent renovations:* Valery Boldin, *Ten Years That Shook the World: The Gorbachev Era as Witnessed by His Chief of Staff* (New York: Basic Books, 1994), 162–63.

172 *"What happened?":* Alexander Yakovlev, interview transcript, 2RR archive file no. 3/10/7, 5.

172 *Vladimir Dolgikh, the Central Committee secretary:* Dolgikh, interview transcript, 2RR archive file no. 1/3/5, 4.

172 *He described an explosion:* Working record of the April 28, 1986, Politburo meeting reproduced in Rudolf G. Pikhoya, *Soviet Union: The History of Power 1945–1991* [*Советский Союз: История власти. 1945–1991*] (Novosibirsk: Sibirsky Khronograf, 2000), 429–30.

173 *Information was still scant, and conflicting:* Yakovlev, interview transcript, 2RR, 5. Some Party elders struggled to grasp the significance of even the few hard facts they did receive. In a copy of one of the first KGB reports on the accident delivered to the Ukrainian Central Committee in Kiev on April 28, someone underlined the recorded radiation figures and scribbled in the margin, "What does this mean?" See the second page of the document titled "On the Explosion at the NPP" [О взрыве на АЭС], April 28, 1986, archival material of the State Security Service of Ukraine, f. 16, op. 11-A [ф. 16, оп. 11-A], www.archives.gov.ua/Sections/Chornobyl _30/GDA_SBU/index.php?2.

173 *Nothing more than a slogan:* Kotkin, *Armageddon Averted*, 67.

173 *"We can't procrastinate":* Politburo meeting minutes (April 28, 1986), in Pikhoya, *Soviet Union*, 431.

173 *Gorbachev's grip on power remained tenuous:* Committed reformers in the Politburo at the time were in a minority of four: Yeltsin, Yakovlev, Shevardnadze, and Gorbachev himself. Ligachev was a hardliner and Ryzhkov a moderate conservative. Remnick, *Lenin's Tomb*, 48.

173 *The official record of the meeting:* The working record quotes Ligachev (who, by most other accounts, opposed information sharing) as saying, "The people have

been accommodated well. We should make a statement about the incident without delay." Politburo meeting minutes (April 28, 1986), in Pikhoya, *Soviet Union*, 431.

173 *The second most powerful man in the Kremlin:* Jonathan Harris, "Ligachev, Egor Kuzmich," in Joseph Wieczynski, ed., *The Gorbachev Encyclopedia* (Salt Lake City: Schlacks, 1993), 246.

173 *"Come off it!":* Heydar Aliyev, testimony in the documentary *The Second Russian Revolution* (1991), "Episode Two: The Battle for Glasnost," online at www.youtube .com/watch?v=5PafRkPMFWI; Aliyev, interview transcripts, 2RR, archive file no. 3/1/6 and 1/4/2.

173 *Others at the table:* Yakovlev, interview transcript, 2RR, 6.

173 *"The statement should be formulated":* Politburo meeting minutes (April 28, 1986), in Pikhoya, *Soviet Union*, 431.

174 *Ligachev had apparently prevailed:* Aliyev, interview transcript 1/4/2, 2RR, 9. Leonid Dobrokhotov, Central Committee spokesman, says during an interview in the second episode of *The Second Russian Revolution:* "The instructions were traditional—that is to say, we had to play down the catastrophe, to prevent panic among the people, and to fight against what was then called bourgeois falsification, bourgeois propaganda, and inventions."

174 *By 2:00 p.m. in Stockholm, Swedish state authorities:* Stern, *Crisis Decisionmaking*, 136.

174 *Back in the town of Chernobyl:* Sklyarov, *Chernobyl Was . . . Tomorrow*, 70.

174 *That afternoon in Moscow:* Stern, *Crisis Decisionmaking*, 137–38.

175 *The official told Örn:* Hawkes et al., *Worst Accident in the World*, 117.

175 *"One of the atomic reactors has been damaged":* Text of the announcement is from the official summary of the April 28 Politburo meeting available in RGANI, Opis 53, Reel 1.1007, File 1: "Excerpts from the protocol of meeting no. 8 of the CPSU Politburo" [Выписка из протокола № 8 заседания Политбюро ЦК КПСС от 28 апреля 1986 года]. The time of the broadcast is specified in Alexander Amerisov, "A Chronology of Soviet Media Coverage," *Bulletin of the Atomic Scientists* 42, no. 7 (August/September 1986): 38. For Western coverage of the announcement, see William J. Eaton, "Soviets Report Nuclear Accident: Radiation Cloud Sweeps Northern Europe; Termed Not Threatening," *Los Angeles Times*, April 29, 1986; and Serge Schmemann, "Soviet Announces Nuclear Accident at Electric Plant," *New York Times*, April 29, 1986.

175 *At 9:25 p.m. Moscow time,* Vremya: BBC Summary of World Broadcasts, "Accident at Chernobyl Nuclear Power Station," SU/8246/I, April 30, 1986 (Wednesday). The video of the *Vremya* news segment is available at "The announcement of the *Vremya* program about Chernobyl of 04.28.1986" [Сообщение программы Время о Чернобыле от 28-04-1986], published April 2011 and accessed May 2018: www.youtube.com/watch?v=VG6eIuAfLoM.

175 *Their editors did their best to keep it quiet:* Marples, *Chernobyl and Nuclear Power in the USSR*, 3.

175 *The second extraordinary meeting of the Politburo in two days:* V. I. Vorotnikov, *This Is*

How It Went . . . From the Diary of a Member of the Politburo of the Central Committee of the Communist Party of the Soviet Union [А было это так . . . Из дневника члена Политбюро ЦК КПСС] (Moscow: Soyuz Veteranov Knigoizdaniya SI–MAR, 1995), 96–97.

176 *Vladimir Dolgikh gave his colleagues the latest news:* Politburo meeting minutes (April 29, 1986), from Russian Government Archives Fond 3, Opis 120, Document 65, reproduced in Maleyev, *Chernobyl. Days and Years,* 245. A different version of events is proposed by Pikhoya, whose summary of meeting proceedings suggests that Dolgikh described deteriorating conditions at the plant: Pikhoya, *Soviet Union,* 432.

176 *They were facing disaster:* Politburo meeting minutes (April 29, 1986), in Maleyev, *Chernobyl. Days and Years,* 246. Vorotnikov maintains that it was only at this second meeting that the reports made the emerging scale of the accident clear (*This Is How It Went,* 96–97).

176 *"The more honest we are, the better":* Politburo meeting minutes (April 29, 1986), reproduced in Maleyev, *Chernobyl. Days and Years,* 247 and 249.

176 *They agreed to cable statements:* "Resolution of the CPSU Central Committee: On Additional Measures Related to the Liquidation of the Accident at the Chernobyl Nuclear Power Plant" [О дополнительных мерах, связанных с ликвидацией аварии на Чернобыльской АЭС], top secret, April 29, 1986, in RGANI, Opis 53, Reel 1.1007, File 2.

177 *"Should we give information to our people?":* Politburo meeting minutes (April 29, 1986), in Maleyev, *Chernobyl. Days and Years,* 248.

177 Vremya *broadcast a new statement:* Amerisov, "A Chronology of Soviet Media Coverage," 38; Marples, *Chernobyl and Nuclear Power in the USSR,* 4; Mickiewicz, *Split Signals,* 61–62.

177 *Luther Whittington of the wire service:* Nicholas Daniloff, *Of Spies and Spokesmen: My Life as a Cold War Correspondent* (Columbia: University of Missouri Press, 2008), 343. In their book on the history of UPI, Gregory Gordon and Ronald Cohen suggested that Whittington was the victim of a deliberate attempt to discredit Western reporting, orchestrated by the KGB. Gregory Gordon and Ronald E. Cohen, *Down to the Wire: UPI's Fight for Survival* (New York: McGraw-Hill, 1990), 340–41.

177 *"'2,000 DIE' IN NUKEMARE'":* Luther Whittington, "'2,000 Die' in Nukemare; Soviets Appeal for Help as N-plant Burns out of Control," *New York Post,* April 29, 1986; "'2000 Dead' in Atom Horror: Reports in Russia Danger Zone Tell of Hospitals Packed with Radiation Accident Victims," *Daily Mail,* April 29, 1986.

178 *That night, the lurid death toll:* Hawkes et al., *Worst Accident in the World,* 126.

178 *A secret intelligence assessment:* "Estimate of Fatalities at Chernobyl Reactor Accident," cable from Morton I. Abramovitz to George Shultz, Secret, May 2, 1986, CREST record CIA-RDP88G01117R000401020003-1, approved for release December 29, 2011.

178 *Meanwhile, the radioactive cloud:* ApSimon and Wilson, "Tracking the Cloud from Chernobyl," 44.

178 *The West German and Swedish governments lodged furious complaints:* William J. Eaton and Willion Tuohy, "Soviets Seek Advice on A-Plant Fire 'Disaster': Bonn, Stockholm Help Sought, but Moscow Says Only 2 Died," *Los Angeles Times,* April 30, 1986; Karen DeYoung, "Stockholm, Bonn Ask for Details of Chernobyl Mishap: Soviets Seek West's Help to Cope With Nuclear Disaster," *Washington Post,* April 30, 1986; Stern, *Crisis Decisionmaking,* 230.

178 *In Denmark, pharmacies:* Stern, *Crisis Decisionmaking,* 147; DeYoung, "Stockholm, Bonn Ask for Details."

178 *In Communist Poland:* Murray Campbell, "Soviet A-leak 'world's worst': 10,000 lung cancer deaths, harm to food cycle feared," *Globe and Mail,* April 30, 1986.

178 *"The world has no idea of the catastrophe":* Hawkes et al., *Worst Accident in the World,* 127.

178 *Soviet spokesmen dismissed these stories:* Marples, *Chernobyl and Nuclear Power in the USSR,* 6–7.

179 *Chebrikov notified his superiors:* V. Chebrikov, "On the reaction of foreign diplomats and correspondents to the announcement of an accident at Chernobyl NPP" [О реакции иностранных дипломатов икорреспондентов на сообщение об аварии на Чернобыльской АЭС], KGB memo to Central Committee of the CPSU, April 20, 1986, in RGANI, Opis 53, Reel 1.1007, File 3.

179 *Apparently attempting to sever his remaining communications:* Daniloff, *Of Spies and Spokesmen,* 344; Daniloff, author interview, 2017.

179 *Fifteen thousand people had been killed:* Guy Hawtin, "Report: 15,000 Buried in Nuke Disposal Site," *New York Post,* May 2, 1986.

179 *Using just three helicopters:* Antoshkin, *Regarding Chernobyl,* 2.

180 *"Like hitting an elephant with a BB gun!":* Antoshkin, author interview, 2017. In his unpublished memoir, *Regarding Chernobyl,* Antoshkin recalls slightly different figures: fifty-five tonnes of sand, and ten of boron. Piers Paul Read reports that Scherbina at first told the chemical warfare generals, Ivanov and Pikalov, that Antoshkin was simply incompetent (*Ablaze,* 123–24).

180 *Heavy-lift helicopters:* Nesterov, *Heaven and Earth,* 245. For a description of Mi-26, see "Russia's airborne 'cow,'" BBC News Online, August 20, 2002.

180 *But it proved almost impossible:* Nesterov, *Heaven and Earth,* 247.

180 *Most crews flew a total of ten to fifteen sorties:* Antoshkin, author interview, 2015, and 11–13 in his unpublished memoir *The Role of Aviation in Localizing the Consequences of the Catastrophe at Chernobyl* [Роль авиации в локализации последствий катастрофы на Чернобыльской АЭС].

180 *The temperature dropped from more than 1,000 degrees:* Dolgikh, testimony to the Politburo on April 29, 1986, in meeting minutes reproduced in Maleyev, *Chernobyl. Days and Years,* 245. See also 258 for Legasov's testimony to the Politburo on May 5, 1986.

180 *The government commission was forced to withdraw:* Shasharin, "The Chernobyl Tragedy," 96.

181 *The territory immediately surrounding the plant:* Areas around the plant would

soon be classified as three concentric circles, the innermost of which measured about 1.5 kilometers across: Mary Mycio, *Wormwood Forest: A Natural History of Chernobyl* (Washington, DC: Joseph Henry Press, 2005), 23. The term *osobaya zona* can be found, for example, in a KGB memo from December 1986, although there it denotes a larger area, about nine kilometers in diameter: Danilyuk, ed., "Chernobyl Tragedy," *Z arkhiviv*, document no. 73: "Special report of the UkSSR KGB to the USSR KGB 6th Department concerning the radioactive situation and the progress in works on the cleaning up operation after the accident at the Chernobyl' NPS," December 31, 1986.

181   *The first meeting of the Politburo Operations Group:* Baranovska, ed., *The Chernobyl Tragedy*, document no. 60: "Protocol of the first meeting of the Politburo Task Force on liquidating the consequences of the Chernobyl NPP accident," April 29, 1986, 80–81.

181   *Legasov was afraid to ask:* Legasov Tapes, cassette One, 14; Nikolai Ryzhkov, *Ten Years of Great Shocks* [*Десять лет великих потрясений*] (Moscow: Kniga-Prosveshchenie-Miloserdie, 1995), 167.

181   *The first 2,500 tonnes arrived:* Lyashko, *Weight of Memory*, 362.

181   *By the time the light failed:* Antoshkin, *Regarding Chernobyl*, 3.

181   *A scientific report:* Baranovska, ed., *The Chernobyl Tragedy*, document no. 59: "Memorandum of the Department of Science and Education of the Central Committee of the Communist Party of Ukraine on immediate measures connected to the accident at the Chernobyl NPP," April 29, 1986.

181   *The regional civil defense leaders had made preparations:* Ivanov, "Chernobyl, Part 3: Evacuation," 39. The number 10,000 is specified by Lyashko, *Weight of Memory*, 355.

181   *A shipment of parachutes was delivered:* Mimka, author interview, 2016; Antoshkin, author interview, 2017; Nikolai Antoshkin, "Helicopters over Chernobyl" [Вертолеты над Чернобылем], interview by Sergei Lelekov, *Nezavisimaya Gazeta*, April 28, 2006, http://nvo.ng.ru/history/2006-04-28/1_chernobil.html.

181   *Each chute could carry as much as 1.5 tonnes:* Antoshkin, author interview, 2017.

181   *When the general delivered his report that night:* Nikolai Antoshkin, handwritten testimony, archive of the Chernobyl Museum.

182   *This time it turned almost due south:* Zhores Medvedev, *Legacy of Chernobyl*, 158–59.

182   *At exactly one o'clock that afternoon:* Y. Izrael, ed., *Chernobyl: Radioactive Contamination of the Environment* [*Чернобыль: Радиоактивное загрязнение природных сред*] (Leningrad: Gidrometeoizdat, 1990), 56. The accepted measure in the USSR for normal background radiation was between 4 and 20 microroentgen per hour. *Radiation Safety Norms–76* [*Нормы радиационной безопасности–76*], Moscow: Atomizdat, 1978, cited in "For Reference" [Справочно], undated, archive of the Chernobyl Museum. In his memoir, Kiev district radiation reconnaissance officer Alexander Logachev states that he regarded the normal background in Ukraine as 11 microroentgen per hour (*The Truth*).

182 *The progress of the radioactive cloud had been tracked:* Alla Yaroshinskaya, *Chernobyl: Crime Without Punishment,* 73–75.

182 *Inside the Ukrainian Ministry of Health:* Iurii Shcherbak, account in Zhores Medvedev, *Legacy of Chernobyl,* 160; Shcherbak, interview transcript (June 12, 1990), 2RR archive file no. 3/8/5, 2.

182 *Familiar with the dangers of radiation:* Zgursky had previously led the S. P. Korolev Manufacturing Company (later renamed Meridian), which produced specialized electronics, including gamma measurement devices. See "More than 60 years in the market of detection equipment and appliances" [Более 60 лет на рынке измерительной и бытовой техники], Meridian, http://www.merydian.kiev.ua/.

182 *He tried to convince Scherbitsky:* Alexander Kitral, "Gorbachev to Scherbitsky: 'Fail to hold the parade, and I'll leave you to rot!'" [Горбачев—Щербицкому: «Не проведешь парад—сгною!»], *Komsomolskaya Pravda v Ukraine,* April 26, 2011, https://kp.ua/life/277409-horbachev-scherbytskomu-ne-provedesh-parad-shnoui.

183 *With only ten minutes to go, he was nowhere to be seen:* Lyashko confirms that Scherbitsky arrived late and that he spent some time conferring "in an undertone" with E. V. Kachalovsky, who headed the Ukrainian government's task force on responding to the Chernobyl accident: Lyashko, *Weight of Memory,* 356. See also the interview with Vitali Korotich, then a prominent magazine editor in Moscow, in *The Second Russian Revolution,* "Episode Two: The Battle For Glasnost" (BBC, 1991).

183 *"I told him":* Kitral, "Gorbachev to Scherbitsky"; Serhii Plokhy, *The Gates of Europe: A History of Ukraine* (New York: Basic Books, 2015), 310. Scherbitsky's wife, Rada, confirmed the story about the Party card in an interview in 2006: Rada Scherbitskaya, interview by Yelena Sheremeta, "After Chernobyl, Gorbachev told Vladimir Vasiliyevich, 'If you don't hold the parade, say goodbye to the Party" [Рада Щербицкая: «После Чернобыля Горбачев сказал Владимиру Васильевичу: «Если не проведешь первомайскую демонстрацию, то можешь распрощаться с партией»], *Fakty i kommentarii,* February 17, 2006: http://fakty.ua/43896-rada-csherbickaya-quot-posle-chernobylya-gorbachev-skazal-vladimiru-vasilevichu-quot-esli-ne-provedesh-pervomajskuyu-demonstraciyu-to-mozhesh-rasprocshatsya-s-partiej-quot.

183 *"To hell with it":* Kitral, "Gorbachev to Scherbitsky"; Plokhy, *Gates of Europe,* 310–11. In 1991, as the Soviet Union was finally collapsing, author and member of the Supreme Soviet Iurii Shcherbak would say it had already become impossible to establish who truly issued the order for the parade to go ahead, because everything was discussed on the telephone and no written instructions were issued by anyone involved. Afterward, Scherbitsky's people insisted it was Moscow's directive; in the Kremlin, they blamed the Ukrainians (Shcherbak, interview transcript no. 3/8/5, 2RR, 7). For example, Nikolai Ryzhkov contests the Ukrainian account, insisting that authority over the parade resided with Scherbitsky alone. (See Ryzhkov, interview by Interfax, April 23, 2016: www.interfax.ru/world/505124.) Ryzhkov declined to be interviewed for this book.

183 *There were seas of red banners:* Video footage of the parade is featured in *The Second Russian Revolution*, Episode 2: The Battle for Glasnost: www.youtube.com /watch?v=tyW6wbHft2M.

184 *Some concessions to the perils of the fallout:* Kitral, "Gorbachev to Scherbitsky."

184 *Some on the rostrum that morning:* Sklyarov, *Chernobyl Was . . . Tomorrow*, 146.

184 *Later, when the wind changed direction again:* Alan Flowers, author interview by telephone, February 2016; Justin Sparks, "Russia Diverted Chernobyl Rain, Says Scientist," *Sunday Times*, August 8, 2004; Richard Gray, "How We Made Chernobyl Rain," *Sunday Telegraph*, April 22, 2007. Moscow has repeatedly denied that cloud seeding took place after the accident, but two of the pilots involved in the operation—one of whom was later awarded a medal for involvement in the operation—described their efforts in the 2007 BBC documentary *The Science of Superstorms*.

184 *The May Day procession swept through Red Square:* UPI, "Tens of Thousands in March: Nuclear Disaster Ignored at Soviet May Day Parade," *Los Angeles Times*, May 1, 1986. During the celebrations, two cosmonauts orbiting Earth on the Soviet space station *Mir* contributed a live message from space.

184 *But afterward, Prime Minister Ryzhkov convened:* Velikhov, *Strawberries from Chernobyl*, 245. Velikhov, interview transcript (June 12, 1990), 2RR archive file no. 1/1/14, 1.

184 *The group confronted the emergencies:* "Protocol no. 3 of the meeting of the Politburo Operations Group of the CPSU Central Committee on problems related to the aftermath of the Chernobyl NPP accident" [Протокол № 3 заседания Оперативной группы Политбюро ЦК КПСС по вопросам связанным с ликвидацией последствий аварии на Чернобыльской АЭС], May 1, 1986, in RGANI, Opis 51, Reel 1.1006, File 19.

185 *This new team would be led by Ivan Silayev:* Ibid. In November 1985 Silayev had been appointed deputy chairman of the USSR Council of Ministers, deputy prime minister and chairman of the council's Bureau for Machine-Building.

185 *Ryzhkov went to see Gorbachev in his office:* Ryzhkov, *Ten Years of Great Shocks*, 170–71.

185 *Flew to Kiev without him:* Nikolai Ryzhkov, interview by Elena Novoselova, "The Chronicle of Silence" [Хроника молчания], *Rossiiskaya Gazeta*, April 25, 2016, https://rg.ru/2016/04/25/tridcat-let-nazad-proizoshla-avariia-na-chernobyl skoj-aes.html.

185 *Accompanied by Scherbitsky, the Ukrainian first secretary:* Ryzhkov, *Ten Years of Great Shocks*, 170–72. Ryzhkov describes the map he was using in the interview by Novoselova, *Rossiiskaya Gazeta*, 2016.

186 *At 2:00 p.m.:* Ivanov, "Chernobyl, Part 3: Evacuation," 39.

186 *All top secret:* Sklyarov, *Chernobyl Was . . . Tomorrow*, 89.

11. THE CHINA SYNDROME

187 *From high up on the roof of the Hotel Polesia:* Mimka, author interview, 2016; author visit to the Hotel Polesia, Pripyat, April 25, 2016.

188 *Taking off in a continuous carousel:* Footage of the helicopters lifting and transporting these loads can be see at around 1:06 in *Chernobyl: A Warning* [Чернобыль: Предупреждение], a 1987 Russian state television documentary, available online at www.youtube.com/watch?v=mwxbS_ChNNk (accessed May 2018).

188 *Although many routinely underreported it:* Antoshkin, handwritten testimony, Chernobyl Museum.

188 *Bitter potassium iodide tablets:* Mimka, author interview, 2016.

188 *"If you want to be a dad":* A. N. Semenov, "For the 10th Anniversary of the Chernobyl Catastrophe," in Semenov, ed., *Chernobyl: Ten Years On*, 22.

189 *Physicists were hauled into their offices:* Alexander Borovoi (head of the neutrino laboratory at Kurchatov Institute at the time of the accident), account in Alexander Kupny, *Memories of Lives Given: Memories of Liquidators* [Живы, пока нас помнят: Воспоминания ликвидаторов] (Kharkiv: Zoloty Storynki, 2011), 6–7.

189 *Five or six times every day:* E. P. Ryazantsev, "It Was in May 1986," in Viktor A. Sidorenko, ed., *The Contribution of Kurchatov Institute Staff to the Liquidation of the Accident at the Chernobyl NPP* [Вклад Курчатовцев в ликвидацию последствий аварии на чернобыльской АЭС] (Moscow: Kurchatov Institute, 2012), 85.

189 *They watched as the pilots' payloads:* V. M. Fedulenko, "Some Things Have Not Been Forgotten," in Sidorenko, ed., *The Contribution of Kurchatov Institute Staff*, 79.

189 *A beautiful crimson halo:* Ryazantsev, "It Was in May 1986," 86.

189 *The volcanoes of Kamchatka:* Mimka, author interview, 2016.

189 *At the very start, one member of the Kurchatov group:* Fedulenko, "Some Things Have Not Been Forgotten," 82; Read, *Ablaze*, 132–33.

190 *Day after day, the volume of material:* These statistics, which differ from those recalled by Antoshkin, are drawn from data in helicopter pilot logs recorded at the time, provided by Alexander Borovoi to Alexander Sich ("The Chornobyl Accident Revisited," 241).

190 *They began dropping lead:* Shasharin, "Chernobyl Tragedy," 107.

190 *A hastily created new formation:* Vladimir Gudov, *Special Battalion no. 731* [731 спецбатальон] (Kiev: Kyivskyi Universitet Publishing Center, 2010), trans. Tamara Abramenkova as *731 Special Battalion: Documentary Story* (Kiev: N. Veselicka, 2012). See 54 in the original Russian edition or 80 in the English translation.

190 *The hot weather and the rotor wash:* Piotr Zborovsky, interview by Sergei Babakov, "I'm still there today, in the Chernobyl zone" [Я и сегодня там, в Чернобыльской зоне], *Zerkalo nedeli Ukraina*, September 18, 1998: http://gazeta.zn.ua/SOCIETY /ya_i_segodnya_tam,_v_chernobylskoy_zone.html, translated in Gudov, *731 Special Battalion*, 101. Also see pp. 124–25 for N. Bosy, "Open Letter of a Commander of a Radiological Protection Battalion 731 [. . .] to Battalion Staff."

190 *1,200 tonnes of lead, sand, and other materials:* Antoshkin, handwritten testimony, Chernobyl Museum. Antoshkin states that he deliberately underreported these volumes so that Scherbina did not set an even higher target for the next day. The actual total dropped on May 1 attributed to the pilot logs cited by Sich is 1,900 tonnes ("Chornobyl Accident Revisited," 241).

190 *Some members of the government commission got to their feet:* Antoshkin, handwritten testimony, Chernobyl Museum.

191 *Instead of continuing to fall:* International Atomic Energy Agency, International Nuclear Safety Advisory Group, "Summary Report on the Post-Accident Review Meeting on the Chernobyl Accident," Safety series no. 75–INSAG-1, 1986, 35; Sich, "Chornobyl Accident Revisited," 241–42, fig. 4.1 and fig. 4.4.

191 *Approaching 1,700 degrees centigrade:* Legasov's report to the Politburo, in minutes of the May 5, 1986, meeting reproduced in Maleyev, *Chernobyl. Days and Years*, 258. The transcript cites "20 degrees," but this seems likely to be an erroneous transcription of 2,000 degrees Centigrade, since Legasov adds that the temperature has been rising by about 135 degrees per day since Saturday, April 26, when it measured 1,100 degrees. Based on these calculations, by the evening of Thursday, May 4, the reactor would be at around 1,595 degrees. In *Legasov Tapes* (cassette One, 20), he similarly mentions 2,000 degrees Centigrade as "approximately the highest temperature we have observed." In reality, all of these figures could be little more than educated guesses, since the scientists did not yet have any way of taking accurate readings from inside the reactor space.

191 *The academicians now feared:* Sich, "Chornobyl Accident Revisited," 241 and 257–58.

191 *If the temperature of the molten fuel:* Ryzhkov, statement to Politburo on May 5: minutes reproduced in Maleyev, *Chernobyl. Days and Years*, 252. Sich ("Chornobyl Accident Revisited," 242) specifies the temperature necessary for liquefaction as between 2,300 and 2,900 degrees centigrade.

191 *A whole range of toxic radionuclides:* P. A. Polad-Zade (deputy minister of water of the USSR), "Too Bad It Took a Tragedy" [Жаль, что для этого нужна трагедия], in Semenov, ed., *Chernobyl: Ten Years On*, 195.

191 *But the second threat:* Karpan, *Chernobyl to Fukushima*, 68; Vitali Masol (head of the Ukraine State Planning Committee and deputy chairman of the Ukraine Council of Ministers at the time of the accident), interview by Elena Sheremeta, "We were quietly preparing to evacuate Kiev" [Виталий Масол: «Мы тихонечко готовились к эвакуации Киева»], *Fakty i kommentarii*, April 26, 2006: http://fakty.ua/45679-vitalij-masol-quot-my-tihonechko-gotovilis-k-evakuacii-kieva-quot.

192 *On Friday, May 2, the new team:* The resolution to send in a duplicate team was reached by the Kremlin's task force on Chernobyl on May 1, 1986: "Protocol no. 3 of the meeting of the Politburo Operations Group," in RGANI.

192 *Thoroughly irradiated:* Drach, author interview, 2017; Kopchinsky, recollections in Kopchinsky and Steinberg, *Chernobyl*, 53.

192 *The commission members had not been given iodine tablets:* Sklyarov, *Chernobyl Was . . . Tomorrow*, 52; Shasharin writes that there were no dosimeters available to the commission members at first and that "a subsequent analysis showed that the exposure dose ranged from 60 to 100 rem (without the internal radiation)" ("Chernobyl Tragedy," 99).

192 *Now their eyes and throats were red and raw:* Evgeny P. Velikhov, *My Journey: I Shall Travel Back to 1935 in Felt Boots* [Мой путь. Я на валенках поеду в 35-й год] (Moscow: AST, 2016), translated by Andrei Chakhovskoi as *Strawberries from Chernobyl: My Seventy-Five Years in the Heart of a Turbulent Russia*, 253. Also see Abagayn, account in Voznyak and Troitsky, *Chernobyl: It Was Like This*, 216.

192 *Others felt sick:* Sklyarov, *Chernobyl Was . . . Tomorrow*, 141.

192 *They surrendered their clothes and expensive foreign-made watches:* Sklyarov, *Chernobyl Was . . . Tomorrow*, 83; Drach, author interview, 2017.

193 *Legasov chose to stay behind:* Vladimir Gubarev, testimony in Margarita Legasova, *Academician Valery A. Legasov*, 343.

193 *Velikhov had no direct experience:* Velikhov, *Strawberries from Chernobyl*, 245–46.

193 *His manner didn't impress the generals:* Read, *Ablaze*, 138–39.

193 *But Velikhov:* Bolshov, author interview, 2017; Vladimir Gubarev (science editor at *Pravda*), memorandum to the USSR Central Committee, summarized by Nicholas Daniloff, "Chernobyl and Its Political Fallout: A Reassessment," *Demokratizatsiya: The Journal of Post-Soviet Democratization* 12, no. 1 (Winter 2004): 123. Alexander Borovoi describes Gorbachev's personal animosity toward Legasov in Alla Astakhova, interview with Alexander Borovoi, "The Liquidator" [Ликвидатор], *Itogi* 828, no. 17 (April 23, 2012), www.itogi.ru/obsh-spetzproekt /2012/17/177051.html.

193 *Now, in addition to their different personalities:* Rafael V. Arutyunyan, "The China Syndrome" [Китайский синдром], *Priroda*, no. 11 (November 1990): 77–83. In his taped recollections of the events, Legasov mentions that Velikhov had seen the film recently: Legasov Tapes, cassette One, 19.

193 *The chances of a full meltdown:* Shasharin, "Chernobyl Tragedy," 100; Legasov Tapes, cassette One, 20.

193 *A 50 percent margin of error:* International Atomic Energy Agency, INSAG–1, 35.

193 *They knew nothing:* A. A. Borovoi and E. P. Velikhov, *The Chernobyl Experience: Part 1, Work on the "Shelter" Structure* [Опыт Чернобыля: Часть 1, работы на объекте «Укрытие»] (Moscow: Kurchatov Institute, 2012), 28.

193 *An enclosed body of water:* Shasharin, "Chernobyl Tragedy," 100.

193 *In the West, scientists had been simulating:* Arutyunyan, "China Syndrome," 77–83.

194 *Velikhov contacted the head of his research lab:* Bolshov, author interview, 2017.

194 *The temperature inside Reactor Number Four continued to rise:* Legasov, statement at the Politburo meeting on May 5, 1986, in Maleyev, *Chernobyl. Days and Years*, 259.

194 *Velikhov called Gorbachev:* Velikhov, *My Journey*, 274.

194 *Less volatile than Boris Scherbina:* Velikhov, *Strawberries from Chernobyl*, 251.

194 *But he faced an even more dire situation:* Velikhov, interview transcript, 2RR, 1; *Chernobyl: A Warning* (Soviet documentary, 1986); Read, *Ablaze*, 137–38.

195 *Many slept for only:* BBC Summary of World Broadcasts, "Velikhov and Silayev: 'Situation No Longer Poses Major Threat' " (text of a *Vesti* video report from Chernobyl on May 11, 1986), translated May 13, 1986.

195 *He summoned subway construction engineers:* The head of the Kiev Metro construction company (Kievmetrostroi) arrived on site May 3, according to the account of Nikolai Belous, a senior surveyor, in Shcherbak, *Chernobyl*, 172.

195 *Five thousand cubic meters:* Ryzhkov, statement at the Politburo meeting on May 5, 1986, in Maleyev, *Chernobyl. Days and Years*, 252.

195 *In the meantime:* Mimka, author interview, 2016. A Ukrainian KGB memo of May 5, 1986, records plans to drop another thousand tonnes of loads on the reactor the following day (Danilyuk, ed., "Chernobyl Tragedy," *Z arkhiviv*, document no. 28: *Report of the UkSSR KGB 6th Department to the USSR KGB Concerning the Radioactive Situation and Progress in Investigating the Accident at the Chernobyl NPS*).

195 *At 1:00 a.m. on May 3:* Zborovsky, interview by Babakov, *Zerkalo nedeli*, 1998. Zborovsky recalls that this incident took place at one o'clock on the night of May 1–2, but Silayev was not scheduled to fly into Chernobyl until the morning of May 2 at the earliest. (His appointment and impending travel to Chernobyl were discussed during the Politburo meeting on the afternoon of May 1: "Protocol no. 3 of the meeting of the Politburo Operations Group," in RGANI. Velikhov's presence was also recorded at that meeting.) So it seems likely that Zborovsky meant instead the night of May 2–3.

196 *The steam suppression pools:* Shasharin, "Chernobyl Tragedy," 100; Sich, "Chernobyl Accident Revisited," 254 and 257. See photographs of the pools in Borovoi and Velikhov, *The Chernobyl Experience: Part 1*, 123 and 142.

196 *But on April 26 the condensation system:* Karpan, *Chernobyl to Fukushima*, 68–69; Alexey Ananenko, a senior engineer at Unit Two reactor shop, recollections [Воспоминания старшего инженера-механика реакторного цеха №2 Алексея Ананенка], *Soyuz Chernobyl*, undated (before September 2013), www.souzchernobyl.org/?id=2440.

197 *It was still the small hours:* Zborovsky, interview by Babakov, *Zerkalo nedeli*, 1998.

198 *Gradually, the water level rose:* Zborovsky, testimony in Gudov, *731 Special Battalion*, 112. Karpan explains that the intake point was located in the stairway compartment 05/1 of the Auxiliary Reactor Equipment block, beneath Unit Three (*Chernobyl to Fukushima*, 69).

198 *Back in Moscow, Evgeny Velikhov's team:* Bolshov, author interview, 2017.

198 *They sent samples to Kiev:* Borovoi and Velikhov, *Chernobyl Experience Part 1*, 29–30.

198 *They quickly confirmed:* Arutyunyan, "China Syndrome," 78–81.

198 *But they also found:* Bolshov, author interview, 2017.

198 *In Chernobyl, the commission remained:* Zborovsky, testimony in Gudov, *731 Special Battalion*, 103–9.

199 *The plant physicists, consumed with fear:* Prianichnikov, author interview, 2006.

199 *By 6:00 p.m. on Sunday, Legasov's readings:* Legasov, statement at the Politburo meeting on May 5, 1986, in Maleyev, *Chernobyl. Days and Years*, 258.

12. THE BATTLE OF CHERNOBYL

200 *Shortly after 8:00 p.m.*: The White House, "Presidential Movements" and "The Daily Diary of President Ronald Reagan," April and May 1986, Ronald Reagan Presidential Library and Museum, online at www.reaganlibrary.gov/sites/default /files/digitallibrary/dailydiary/1986-05.pdf; Paul Lewis, "Seven Nations Seeking Stable Currency," *New York Times*, May 6, 1986.

200 *The first reports of the radiation*: Ronald Reagan, diary entry, Wednesday, April 30, 1986, in Douglas Brinkley, ed., *Reagan Diaries*, vol. 2: *November 1985–January 1989* (New York: HarperCollins, 2009), 408; George P. Shultz, *Turmoil and Triumph: My Years as Secretary of State* (New York: Charles Scribner's Sons, 1993), 714.

200 *From high-resolution spy satellite photographs:* Laurin Dodd (RBMK reactor expert in Nuclear Systems and Concepts Department, Pacific Northwest National Laboratory, March 1986 to May 1994), author interview by telephone, May 2018.

200 *And officials at the US Nuclear Regulatory Commission:* Stephen Engelberg, "2D Soviet Reactor Worries U.S. Aides," *New York Times*, May 5, 1986.

200 *American nuclear experts could only speculate:* Dodd, author interview, 2018.

200 *In a classified report:* Eduard Shevardnadze, "Memorandum, CPSU Central Committee, no. 623/GS" [ЦК КПСС № 623/ГС], classified, May 3, 1986, in RGANI, opis 53, reel 1.1007, file 3.

201 *President Reagan broadcast:* Ronald Reagan, "Radio Address to the Nation on the President's Trip to Indonesia and Japan," May 4, 1986, *The American Presidency Project* (collaboration of Gerhard Peters and John T. Woolley), www.presidency .ucsb.edu/ws/?pid=37208.

201 *Radioactive rain fell on Japan:* P. Klages, "Atom Rain over U.S.," *Telegraph*, May 6, 1986; D. Moore, "UN Nuclear Experts Go to USSR," *Telegraph*, May 6, 1986.

201 *The following afternoon:* Moore, "UN Nuclear Experts Go to USSR."

201 *In the hours before their arrival:* "Draft minutes, the meeting of the Politburo of the CPSU Central Committee on May 5, 1986" [Рабочая запись, Заседание Политбюро ЦК КПСС 5 мая 1986 г.] (Russian Government Archives collection 3, opis 120, document 65, 1–18), reproduced in Maleyev, *Chernobyl. Days and Years*, 249–64.

202 *"I can only imagine":* Politburo meeting minutes (May 5, 1986), in Maleyev, *Chernobyl. Days and Years*, 253.

203 *"a nuclear explosion":* Ibid., 252.

204 *The republican authorities had begun:* Masol, "We were quietly preparing to evacuate Kiev"; Vitali Masol, author interview, Kiev, June 2017.

204 *"We've got to pick up our pace":* "Minutes from the Politburo meeting (May 5, 1986), in Maleyev, *Chernobyl. Days and Years*, 249–64.

204 *Zborovsky had set out:* Zborovsky, testimony in Gudov, *731 Special Battalion*, 108.

204 *When they arrived on the scene:* Vladimir Trinos, interview by Irina Rybinskaya, "Fireman Vladimir Trinos, one of the first to arrive at Chernobyl after the explosion: 'It was inconvenient to wear gloves, so the guys worked with their bare hands, crawling on their knees through radioactive water'" [Пожарный Владимир

Тринос, одним из первых попавший на ЧАЭС после взрыва: «в рукавицах было неудобно, поэтому ребята работали голыми руками, ползая на коленях по радиоактивной воде»], *Fakty i kommentarii*, April 26, 2001: http://fakty.ua /95948-pozharnyj-vladimir-trinos-odnim-iz-pervyh-popavshij-na-chaes-posle -vzryva-quot-v-rukavicah-bylo-neudobno-poetomu-rebyata-rabotali-golymi -rukami-polzaya-na-kolenyah-po-radioaktivnoj-vode-quot. Abandoned fire trucks are also mentioned by Nikolai Steinberg in his recollections from arriving on the scene on May 7: Kopchinsky and Steinberg, *Chernobyl*, 56.

204 *They drilled again and again:* Read, *Ablaze*, 135.

204 *At first, Captain Zborovsky wasn't afraid:* Zborovsky, testimony in Gudov, *731 Special Battalion*, 111.

205 *The specialists and management from the plant:* Kopchinsky and Steinberg, *Chernobyl*, 57–59.

205 *Since the final evacuation of Pripyat:* Glukhov, author interview, 2015.

205 *Decorated with whimsical sculptures:* Photographs of the camp can be found at https://www.facebook.com/pg/skazochny/photos/?tab=album&album_id=163 1999203712325 and http://chornobyl.in.ua/chernobyl-pamiatnik.html.

205 *Now the woods and the fields nearby:* Kopchinsky and Steinberg, *Chernobyl*, 55–56.

205 *First, the subway engineers:* V. Kiselev, deputy chief engineer of the Ministry of Transport department for special projects (known as Department 157 and responsible for building the Moscow Metro), account in Dyachenko, ed., *Chernobyl: Duty and Courage*, vol. 1, 38–40; Belous, account in Shcherbak, *Chernobyl*, 172.

206 *At the same time, technicians:* Steinberg, author interview, 2015; Kopchinsky and Steinberg, *Chernobyl*, 67.

206 *Silayev's government commission sent out instructions:* Read, *Ablaze*, 139–40; Steinberg, author interview, 2015; Kopchinsky and Steinberg, *Chernobyl*, 67.

206 *Carried by a pair of Antoshkin's:* Mimka, author interview, 2016.

206 *"Find the nitrogen":* Read, *Ablaze*, 140.

207 *It was eight in the evening on Tuesday:* Zborovsky, testimony in Gudov, *731 Special Battalion*, 107–9.

207 *The men stopped the trucks:* Karpan, *Chernobyl to Fukushima*, 69.

207 *They ran out the hoses:* Trinos, interview by Rybinskaya, *Fakty i kommentarii*, 2001.

207 *Leaving the engines running:* Zborovsky, testimony in Gudov, *731 Special Battalion*, 109–10.

207 *At last, the water:* Read, *Ablaze*, 136.

207 *Every few hours, three men:* Trinos, interview by Rybinskaya, *Fakty i kommentarii*, 2001; Zborovsky, interview by Babakov, *Zerkalo nedeli*, 1998.

207 *At three in the morning:* Trinos, interview by Rybinskaya, *Fakty i kommentarii*, 2001; Read, *Ablaze*, 136–37.

207 *One truck's engine coughed:* Trinos, interview by Rybinskaya, *Fakty i kommentarii*, 2001.

207 *The men were all frightened:* Read, *Ablaze*, 136.

208 *Another began ranting:* Trinos, interview by Rybinskaya, *Fakty i kommentarii*, 2001.

208 *"Don't bring out the beast in me":* Zborovsky, interview by Babakov, *Zerkalo nedeli*, 1998.

208 *Details of what had happened at the plant began seeping:* Yuri Shcherbak, author interview, Kiev, February 2016. News spread by word of mouth as many of the 47,000 former residents of Pripyat were distributed across Ukraine, and rumors filled the information vacuum left by the state. Kopchinsky and Steinberg, *Chernobyl*, 39–40.

208 *The surveillance department of the Ministry of Internal Affairs:* The Seventh Directorate of the Ministry of Internal Affairs (MVD) of the Ukrainian SSR, *Report on Results of Public Opinion Monitoring with Regard to the Accident at Chernobyl NPP* [Докладная записка о результатах изучения общественного мнения в связи с аварией на Чернобыльской АЭС], classified, addressed to the Minister of Internal Affairs of Ukraine I. Gladush, April 30, 1986, archive of the Chernobyl Museum.

208 *But the streets of Kiev:* Zhores Medvedev, *The Legacy of Chernobyl*, 161.

208 *Radiation dose rates in the city had increased:* Department of Science of the Central Committee of the Communist Party of Ukraine, "On several urgent measures to prevent health harm to Kiev's population from the accident at Chernobyl NPP" [О некоторых неотложных мерах по предотвращению ущерба здоровью населения г. Киева вследствие аварии на Чернобыльской АЭС], May 4, 1986, archive of the Chernobyl Museum.

208 *The head of the Ukrainian KGB warned:* Stepan Mukha, statement at the Ukrainian Politburo meeting, in Baranovska, ed., *The Chernobyl Tragedy*, Document no. 73: "Transcript of the meeting of the Operational Group of the Politburo of the Communist Party of Ukraine," May 3, 1986.

209 *By then, word had gone around:* When this news reached the Politburo, Gorbachev and Ligachev discussed taking steps to remove Scherbitsky from his position at the head of the Republic. Kopchinsky and Steinberg, *Chernobyl*, 45–46.

209 *Days earlier, at a central Kiev pharmacy:* Shcherbak, interview transcript, 2RR, p. 4; Shcherbak, author interview, 2016.

209 *Worse still, rumors had also:* Shcherbak, author interview, 2016; Shcherbak, *Chernobyl*, 157–59; Boris Kachura (member of the Ukrainian Politburo, 1980–90), transcript of interview conducted by Tatyana Saenko on July 19, 1996, *The Collapse of the Soviet Union: The Oral History of Independent Ukraine, 1988–1991*, http://oralhistory.org.ua/interview-ua/360.

209 *That evening, crowds gathered:* Read, *Ablaze*, 185–86; Gary Lee, "More Evacuated in USSR: Indications Seen of Fuel Melting Through Chernobyl Reactor Four," *Washington Post*, May 9, 1986.

209 *The Soviet internal passport system:* Read, *Ablaze*, 185–86.

209 *Fleets of orange street-cleaning trucks:* Zhores Medvedev writes that water trucks did not begin regular washing in Kiev until May 6 or 7 (*Legacy of Chernobyl*, 161).

Orange trucks are also mentioned in Serge Schmemann, "The Talk of Kiev," *New York Times*, May 31, 1986.

209 *"There is no truth to the rumor"*: Interview with Deputy Minister of Health of the Ukrainian SSR A. M. Kasianenko, *Pravda Ukrainy*, May 11, 1986, cited in Marples, *Chernobyl and Nuclear Power in the USSR*, 149.

209 *Crowds of frantic passengers*: Shcherbak, *Chernobyl*, 152; Grigori Medvedev, "Chernobyl Notebook," trans. JPRS, 61.

209 *At the railway station*: Yuri Kozyrev, author interview, Kiev, 2017.

210 *Twenty thousand people left by car or bus*: Plokhy, *Chernobyl*, 212.

210 *Western reporters witnessed*: Felicity Barringer, "On Moscow Trains, Children of Kiev," *New York Times*, May 9, 1986.

210 *Fearing mass panic*: Lyashko, *Weight of Memory*, 372–73.

210 *"Tell him that our outhouse"*: Velikhov, *My Journey*, 277–78.

210 *It wasn't until around four in the morning*: Trinos, interview by Rybinskaya, *Fakty i kommentarii*, 2001.

210 *Deputy Minister Silayev insisted*: Shasharin, "Chernobyl Tragedy," 102; Ananenko, recollections on *Soyuz Chernobyl*.

211 *Three men from the plant staff*: Shasharin, "The Chernobyl Tragedy," 102.

211 *Clutching wrenches and flashlights*: Ananenko, recollections on *Soyuz Chernobyl*.

211 *Baranov kept watch*: Ibid.

212 *Inside he found 1,000 rubles*: Zborovsky, testimony in Gudov, *731 Special Battalion*, 113–14.

212 *The academicians' relief*: Kopchinsky and Steinberg, *Chernobyl*, 68.

212 *Some estimates now suggested*: E. Ignatenko, *Two Years of Liquidating the Consequences of the Chernobyl Disaster* [Два года ликвидации последствий Чернобыльской катастрофы] (Moscow, Energoatomizdat, 1997), 62, cited in Karpan, *Chernobyl to Fukushima*, 72.

212 *They were repeatedly stopped*: Belous, account in Shcherbak, *Chernobyl*, 175–76.

212 *At the same time*: Bolshov, author interview, 2017; "Protocol No. 8 of the Meeting of the Politburo CPSU Operations Group on Problems Related to the Aftermath of the Chernobyl Nuclear Accident" [Протокол № 8 заседания Оперативной группы Политбюро ЦК КПСС по вопросам, связанным с ликвидацией последствий аварии на Чернобыльской АЭС], May 7, 1986, in RGANI, opis 51, reel 1.1006, file 20.

212 *The most desperate measures yet*: William J. Eaton, "Soviets Tunneling Beneath Reactor; Official Hints at Meltdown into Earth; Number of Evacuees Reaches 84,000," *Los Angeles Times*, May 9, 1986.

212 *In their lab outside Moscow*: Arutyunyan, "The China Syndrome," 79; Bolshov, author interview, 2017.

213 *They were aghast*: Bolshov, author interview, 2017; Arutyunyan, "China Syndrome," 81.

213 *The scientists no longer saw themselves*: Bolshov, author interview, 2017.

214 *They were met at the airport*: Velikhov, *My Journey*, 278–79.

214 *Their green overalls:* TV footage of their landing is in *Two Colors of Time*, Pt. 1, mark 3.55, https://www.youtube.com/watch?v=ax54gzlzDpg.

214 *What the academician did not share:* Velikhov, *Strawberries from Chernobyl*, 251.

214 *The graphite fire:* International Atomic Energy Agency, INSAG-1; Borovoi and Velikhov, *The Chernobyl Experience: Part 1*, 3.

214 *The temperature on the surface:* "Protocol no. 9 of the meeting of the Politburo CPSU Operations Group on problems related to the aftermath of the Chernobyl nuclear accident" [Протокол № 9 заседания Оперативной группы Политбюро ЦК КПСС по вопросам, связанным с ликвидацией последствий аварии на Чернобыльской АЭС], May 8, 1986, RGANI, opis 51, reel 1.1006, file 21. A KGB report of May 11, 1986, attributes the fall in the temperature to the injection of gaseous nitrogen on May 7 and 8, but this conclusion remains questionable at best. Danilyuk, ed., "Chernobyl Tragedy," *Z arkhiviv*, document no. 31: "Special Report of the UkSSR OG KGB chief in the town of Chernobyl to the UkSSR KGB Chairman."

215 *"I can see perfectly":* Velikhov, *My Journey*, 279.

215 *At a press conference:* BBC Summary of World Broadcasts, "IAEA Delegation Gives Press Conference in Moscow" (report published by TASS in English and broadcast by Moscow World Service on May 9, 1986), translated May 12, 1986.

215 *That Sunday, May 11:* BBC Summary of World Broadcasts, "Velikhov and Silayev: 'Situation No Longer Poses Major Threat,'" May 11, 1986; and Serge Schmemann, "Kremlin Asserts 'Danger Is Over,'" *New York Times*, May 12, 1986. Some video footage from this report is contained in the 1987 Soviet documentary *Chernobyl: A Warning* at 35:30.

215 *Back in Moscow:* Bolshov, author interview, 2017.

216 *Five meters high and thirty meters square:* Kozlova, *The Battle with Uncertainty*, 77.

216 *"Build it":* Bolshov, author interview, 2017.

13. INSIDE HOSPITAL NUMBER SIX

217 *"Two steps back!":* Esaulov, *The City That Doesn't Exist*, 39–41; Svetlana Kirichenko, author interview, Kiev, April 2016.

217 *By the time night fell:* Baranovska, ed., *The Chernobyl Tragedy*, document no. 58: "Update from the Ukrainian SSR Interior Ministry to the Central Committee of the Communist Party of Ukraine on the Evacuation From the Accident Zone," April 28, 1986. Undated handwritten list on p. 28 in Pripyat *militsia* File on Special Measures in the Pripyat Zone (archive of the Chernobyl Museum).

218 *"Clean . . . Contaminated":* Esaulov, *City That Doesn't Exist*, 40.

218 *Valentina, a trained engineer:* Viktor and Valentina Brukhanov, author interview, 2015; Andrey V. Illesh, *Chernobyl: A Russian Journalist's Eyewitness Account* (New York: Richardson & Steirman, 1987), 62–63.

218 *But Valentina had been separated:* Viktor and Valentina Brukhanov, author interview, 2015.

218 *Thirty kilometers away, Natalia Yuvchenko:* Natalia Yuvchenko, author interviews, 2015 and 2016.

218 *By Wednesday, the official news blackout:* Nikolai Steinberg writes that on April 30 he and the other senior staff of the Balakovo nuclear power station knew only that there had been an accident of some kind. They judged its severity by taking dosimetry readings from the sandals of a woman who had been visiting Pripyat and had left on the evening of April 26 without learning the true scale of what had happened. Kopchinsky and Steinberg, *Chernobyl*, 10–12.

219 *Nine stories of austere brown brick:* Description of building and its surroundings from Gale and Hauser, *Final Warning*, 51, and author visit to the Institute of Biophysics, Moscow, October 15, 2016.

220 *The first patients from the plant:* Anzhelika Barabanova (burns specialist in the radiation medicine department at Hospital Number Six), author interview, Moscow, October 2016; Angelina Guskova and Igor Gusev, "Medical Aspects of the Accident at Chernobyl," in Gusev et al., eds., *Medical Management of Radiation Accidents*, 199, table 12.1.

220 *They had been met:* Smagin, account in Chernousenko, *Insight from the Inside*, 66–67. Smagin left Kiev at noon on Sunday on a second special flight to Moscow, and said they were driven around the airport for an hour before being released from the plane.

220 *A six-hundred-bed facility:* Barabanova, author interview, 2016.

220 *Some were still:* Ibid., 2016; H. Jack Geiger, MD, "The Accident at Chernobyl and the Medical Response," *Journal of the American Medical Association* (*JAMA*) 256, no. 5 (August 1, 1986): 610.

220 *The aircraft that delivered:* Barabanova, author interview, 2016; Alexander Borovoi, author interview, October 2016.

220 *By evening on Sunday:* Angelina Guskova, *The Country's Nuclear Industry Through the Eyes of a Doctor* [Атомная отрасль страны глазами врача] (Moscow: Real Time, 2004), 141–42. Other sources note slightly different figures for the number of Chernobyl victims admitted to Hospital Number Six. The figure of 202 is cited in Alexander Baranov, Robert Peter Gale, Angelina Guskova et al., "Bone Marrow Transplantation After the Chernobyl Nuclear Accident," *New England Journal of Medicine* 321, no. 4 (July 27, 1989), 207. Dr. Anzhelika Barabanova (author interview, 2016) puts the number at slightly more than 200.

220 *Ten had received:* Barabanova, author interview, 2016.

220 *The head of the Clinical Department:* L. A. Ilyin and A. V. Barabanova, "Obituary: Angelina Konstantinova Guskova," *Journal of Radiological Protection* 35 (2015): 733.

221 *When Guskova failed to return:* Guskova's younger sister ensured that the letters were never sent: Guskova, interview by Gubarev, *Nauka i zhizn*, 2007.

221 *In Mayak:* Vladislav Larin, *"Mayak" Kombinat: A Problem for the Ages* [Комбинат "Маяк"—проблема на века], 2nd edition (Moscow: Ecopresscenter, 2001), 199–200; Brown, *Plutopia*, 172.

221 *Later, the young women:* Brown, *Plutopia*, 173–75.

221 *That same year, at the age of thirty-three:* Date of birth (March 29, 1924): "Angelina

Konstantinovna Guskova: Biography" [Гуськова Ангелина Константиновна: биография], Rosatom; Guskova, interview by Gubarev, *Nauka i zhizn*, 2007.

222 *The price of progress:* This was true, for example, of the survivors of the K-19 submarine accident in 1961. Six of the worst-affected patients were sent to Hospital Number Six, according to Barabanova, and afterward told to lie to their doctors about the cause of their complaints. Matt Bivens, "Horror of Soviet Nuclear Sub's '61 Tragedy Told," *Los Angeles Times*, January 3, 1994; Barabanova, author interview, 2016.

222 *Alarmed by the refusal:* Guskova, *The Country's Nuclear Industry Through the Eyes of a Doctor*, 141.

222 *The following year:* A. K. Guskova and G. D. Baysogolov, *Radiation Sickness in Man* [Лучевая болезнь человека] (Moscow: Meditsina, 1971); Ilyin and Barabanova, "Obituary: Angelina K. Guskova," 733.

222 *By 1986, Guskova:* Ilyin and Barabanova, "Obituary: Angelina K. Guskova."

222 *She had treated more than a thousand:* Mould, *Chernobyl Record*, 92.

222 *It took only a few moments:* Natalia Yuvchenko, author interview, 2015.

223 *The hospital was dim:* Robert Gale, author interview by telephone, June 2016; Richard Champlin, author interview by telephone, September 2016.

223 *When the elevator:* Barabanova, author interview, 2016.

223 *By the time they awoke:* Gunnar Bergdahl, *The Voice of Ludmilla*, trans. Alexander Keiller (Goteborg: Goteborg Film Festival, 2002), 43–45.

223 *Some felt so well:* Barabanova, author interview, 2016; Alexander Nazarkovsky, author interview, Kiev, February 2006; Uskov, account in Shcherbak, *Chernobyl*, 129–30.

224 *Others noticed a reddening:* Read, *Ablaze*, 144. The nature of radiation skin injuries is detailed in Fred A. Mettler Jr., "Assessment and Management of Local Radiation Injury," in Fred A. Mettler Jr., Charles A. Kelsey, Robert C. Ricks, eds., *Medical Management of Radiation Accidents*, 1st ed. (Boca Raton, FL : CRC Press, 1990), 127–49.

224 *Yuvchenko's head had been shaved:* Barabanova, author interview, 2016.

224 *The radioactivity:* Uskov, account in Shcherbak, *Chernobyl*, 130.

224 *"Let's go and have a smoke":* Natalia Yuvchenko, author interview, 2015.

224 *As befits a disease:* Barabanova, author interview, 2016.

225 *A few meters here or there:* Dr. Richard Champlin, "With the Chernobyl Victims: An American Doctor's Inside Report From Moscow's Hospital No. 6," *Los Angeles Times*, July 6, 1986.

226 *In the chaos that followed:* Leonid Khamyanov, account in Kopchinsky and Steinberg, *Chernobyl*, 80–81.

226 *But Guskova's decades of work:* Barabanova, author interview, 2016.

226 *It was a laborious process:* Champlin, "With the Chernobyl Victims."

227 *"Within the first three weeks":* Natalia Yuvchenko, author interview, 2016.

227 *When a doctor came:* Piotr Khmel, author interview, 2015. Despite clear public statements to the contrary made at the time by Soviet health officials, includ-

ing Dr. Guskova herself, the belief that alcohol could cleanse the human body of radioactive poisons persisted in the USSR long after the accident. In fact, in the laboratory, ethanol has been demonstrated to have a mild radioprotective effect at the cellular level, although it's unlikely a human being could drink enough alcohol to combat the effects of a lethal dose of radiation. However, at least one study demonstrates that mosquitoes are protected from the effects of radiation by drinking beer: S. D. Rodriguez, R. K. Brar, L. L. Drake et al. "The effect of the radio-protective agents ethanol, trimethylglycine, and beer on survival of X-ray-sterilized male *Aedes aegypti*," *Parasites & Vectors* 6, no. 1 (July 2013): 211, doi:10.1186/1756-3305-6-211.

227   *By now, the relatives:* Bergdahl, *The Voice of Ludmilla*, 46.

227   *From his bed, Pravik sent:* Letter quoted in Voznyak and Troitsky, *Chernobyl*, 196.

228   *On Tuesday they received a telegram:* Telegram from Leonid Toptunov to Vera Toptunova, April 29, 1986, archive of the Chernobyl Museum.

228   *When they arrived:* Date of arrival given as April 30 in letter from Toptunov's parents Vera and Fyodor, reproduced in Shcherbak, *Chernobyl*, 362.

228   *"Everything is fine!":* Vera Toptunova, author interview, 2015.

228   *Dr. Robert Gale was a man of regular habits:* Details drawn from Gale and Hauser, *Final Warning*, 33–36; Robert Gale, "Witness to Disaster: An American Doctor at Chernobyl," *Life*, August 1986; Gale, author interview by telephone, 2016; Sabine Jacobs (assistant to Robert Gale), author interview, Los Angeles, September 2016.

229   *Gale knew:* Gale and Hauser, *Final Warning*, 36–37.

229   *In Moscow, he met Lenin:* Hammer would recast his first trip to Moscow as part of a voluntary humanitarian mission to help save the lives of Soviet children from typhus; in reality he had traveled to the USSR after his father had been imprisoned for carrying out an illegal abortion that had killed both mother and child—an operation that had actually been conducted by Armand, who would never fully qualify as a doctor. Once in the USSR, Hammer was set up by the Communist government as the owner of a useless asbestos mine and a pencil factory, which functioned as fronts through which the Cheka—the forerunner of the KGB—could finance a network of spies in the United States. The details of Hammer's double life, which would only be fully revealed after his death in 1990 and the fall of the Soviet Union, are described in Edward Jay Epstein, *Dossier: The Secret History of Armand Hammer* (New York: Random House, 1996).

229   *"An almost unique bridge":* Gale and Hauser, *Final Warning*, 38.

230   *By Thursday afternoon:* "Top U.S. Doc Races Death," *New York Post*, May 2, 1986.

230   *On the landings of Hospital Number Six:* Natalia Yuvchenko, author interview, 2016.

230   *Those trained in nuclear engineering:* Read, *Ablaze*, 156.

230   *At their son's bedside:* Vera Toptunova, author interview, 2015.

230   *On the morning of Thursday:* Bergdahl, *Voice of Ludmilla*, 48–50.

231   *Identifying marrow donors:* Gale and Hauser, *Final Warning*, 57.

231 *For those relatives shown by the tests:* Details of procedure from Gale and Hauser, *Final Warning*, 34 and 56; and Champlin, "With the Chernobyl Victims."

231 *When Vasily Ignatenko heard:* Bergdahl, *Voice of Ludmilla*, 48–49.

232 *By the end of the first week:* According to Barabanova's records, Toptunov received transplants on the second and seventh days after the accident (April 27 and May 2) and Akimov on the fourth day after the accident (April 29).

232 *But three more patients:* Gale and Hauser, *Final Warning*, 54–55.

232 *This treatment stood even less chance:* Champlin, "With the Chernobyl Victims"; Barabanova, author interview, 2016.

232 *By then, the limitations:* Champlin, "With the Chernobyl Victims."

232 *Yet this analysis:* Guskova and Gusev, "Medical Aspects of the Accident at Chernobyl," 200; Barabanova, author interview, 2016.

232 *And as the visible signs:* Barabanova recalled that beta burns began to manifest on day six or seven: author interview, 2016.

232 *On May 2 Dr. Baranov estimated:* Read, *Ablaze*, 145.

232 *Their families had high hopes:* Elvira Sitnikova, testimony in Shcherbak, *Chernobyl*, 281.

232 *After checking in:* Gale and Hauser, *Final Warning*, 47–50 and 161; Barabanova, author interview, 2016; Read, *Ablaze*, 143–44.

233 *The two men were driven:* Read, *Ablaze*, 152.

233 *On the eighth floor:* Gale writes in *Final Warning* that the sterile unit was on the fifth floor, but in subsequent testimonies, several witnesses—including Arkady Uskov and Ludmilla Ignatenko—agree that it was on the eighth.

233 *Up here was the hospital's sterile:* Herbert L. Abrams, "How Radiation Victims Suffer," *Bulletin of Atomic Scientists* 42, no. 7 (1986): 16; Barabanova, author interview, 2016.

233 *In the sterile unit:* Gale and Hauser, *Final Warning*, 52–53; Barabanova, author interview, 2016.

233 *Young soldiers now came:* By May 2, a detachment of soldiers with special chemical protection uniforms and equipment had arrived in Hospital Number Six and pitched tents outside on the lawn. Bergdahl, *Voice of Ludmilla*, 51; and Yuri Grigoriev, interview by Alina Kharaz, "It was like being at the front" [Там было как на фронте], *Vzgliad*, April 26, 2010, www.vz.ru/society/2010/4/26/396742.html.

233 *Some of the staff, especially:* Sitnikova, testimony in Shcherbak, *Chernobyl*, 281.

234 *The dozens of pills:* In his hospital diary, Arkady Uskov noted having to take "some 30 pills a day" on his second week of treatment. Uskov, account in Shcherbak, *Chernobyl*, 131.

234 *His hair began to fall out:* Bergdahl, *Voice of Ludmilla*, 49–53.

234 *The worst-affected patients:* Mould, *Chernobyl Record*, 81–82; Gale and Hauser, *Final Warning*, 62–63.

234 *Unlike thermal burns:* Barabanova, author interview, 2016.

235 *Within the first twelve days:* Read, *Ablaze*, 152–53; Gale and Hauser, *Final Warning*, 79; Adriana Petryna, *Life Exposed: Biological Citizens after Chernobyl* (Princeton,

NJ: Princeton University Press, 2013), 45; Champlin, "With the Chernobyl Victims."

235 *But the doctors knew:* Geiger, "The Accident at Chernobyl and the Medical Response," 610.

235 *His contaminated overalls alone:* Barabanova, author interview, 2016.

235 *"Don't worry":* Read, *Ablaze*, 157.

235 *He told a friend:* Davletbayev, "The Final Shift," 382.

235 *"I'll never go back":* Read, *Ablaze*, 156.

235 *By the time Sergei Yankovsky:* Sergei Yankovsky, author interview, Kiev, February 7, 2016; Barabanova, author interview, 2016.

236 *On May 6:* Davletbayev, "The Last Shift," 382.

236 *The patients watched from the windows:* Uskov, diary entry, quoted in Shcherbak, *Chernobyl*, 131.

236 *Had begun to shed his skin:* Bergdahl, *Voice of Ludmilla*, 52.

236 *Lying alone in his room:* Khmel, author interview, 2016.

236 *The deaths began:* Dates of all deaths are provided in "List of Fatalities in the Accident at Chernobyl Nuclear Power Plant," Chernobyl and Pripyat electronic archive.

236 *Grotesque rumors:* Zakharov, interview by Taras Shumeyko, 2006.

236 *His eyes open, his skin black:* Luba Akimov, testimony in Grigori Medvedev, *Truth About Chernobyl*, 253–54.

236 *Dr. Guskova now forbade:* Uskov, account in Shcherbak, *Chernobyl*, 131–34.

237 *While the first of his comrades:* Parry, "How I Survived Chernobyl."

237 *Yuvchenko moved into intensive care:* Ibid.; Natalia Yuvchenko, author interview, 2015; and Barabanova, author interview, 2016. Although Natalia Yuvchenko is firm on this point, Barabanova, her husband's doctor, insists that she never considered the need to amputate.

237 *On Tuesday, May 13:* Bergdahl, *Voice of Ludmilla*, 56–58.

238 *With 90 percent of his skin covered:* Barabanova, author interview, 2016; Vera Toptunova, author interview, 2015; Toptunov's medical records, in Barabanova's personal archive.

238 *Viktor Proskuryakov:* Uskov, diary entry, quoted in Shcherbak, *Chernobyl*, 131–33.

238 *By the end of the third week in May:* Natalia Yuvchenko, author interview, 2016; Alexander Yuvchenko, interview by Bond, *New Scientist*, 2004.

238 *Alone in their rooms:* Ibid., 133.

14. THE LIQUIDATORS

239 *On Wednesday, May 14, 1986:* Marples, *Chernobyl and Nuclear Power in the USSR*, 32. For full text of the speech, see "M. S. Gorbachev's address on Soviet television (Chernobyl)" [Выступление М. С. Горбачева посоветскому телевидению (Чернобыль)], May 14, 1986, Gorbachev Foundation, www.gorby.ru/userfiles /file/chernobyl_pril_6.pdf.

239 *Reading from a prepared statement:* Don Kirk, "Gorbachev Tries Public Approach," *USA Today*, May 15, 1986.

239 *The accident at Chernobyl:* Celestine Bohlen, "Gorbachev Says 9 Died from Nuclear Accident; Extends Soviet Test Ban," *Washington Post,* May 15, 1986.

239 *Gorbachev railed against the "mountain of lies":* BBC Summary of World Broadcasts, "Television Address by Gorbachev," text of broadcast, Soviet television 1700 GMT, May 14, 1986, translated May 16, 1986.

239 *Forty-eight hours earlier:* Maleyev, *Chernobyl: Days and Years,* 51.

240 *A military task force:* Mikhail Revchuk, account in Gudov, *731 Special Battalion,* 92; Marples, *Social Impact,* 184; Danilyuk, ed., "Chernobyl Tragedy," *Z arkhiviv,* document no. 51: "Report of the UkSSR OG KGBM and the USSR KGB on the town of Chernobyl to the USSR KGB concerning the radioactive situation and the progress in works on the cleaning up operation after the accident at the Chernobyl NPS," July 4, 1986.

240 *Now Marshal Sokolov:* Maleyev, *Chernobyl: Days and Years,* 54.

240 *From every republic of the USSR:* Kozlova, *The Battle with Uncertainty,* 67 and 378; V. Lukyanenko and S. Ryabov, "USSR Cities Rush to Send Critical Cargo," *Pravda Ukrainy,* May 17, 1986, translated in JPRS, *Chernobyl Nuclear Accident Documents.*

240 *The spirit of patriotic mass mobilization:* Andrey Illesh, "Survivors Write about Night of April 26," *Izvestia,* May 19, 1986; and V. Gubarev and M. Odinets, "Communists in the Front Ranks: The Chernobyl AES—Days of Heroism," *Pravda,* May 16, 1986, both translated in JPRS, *Chernobyl Nuclear Accident Documents.*

241 *"Had temporarily got out of control":* Eduard Pershin, "They Were the First to Enter the Fire," *Literaturna Ukraina,* May 22, 1986, translated in JPRS, *Chernobyl Nuclear Accident Documents.*

241 *The residents of the evacuated territory:* V. Prokopchuk, "We Report the Details: Above and Around No. 4," *Trud,* May 22, 1986, translated in JPRS, *Chernobyl Nuclear Accident Documents.*

241 *The first cleanup efforts:* Discussions on isolating the radiation and decontaminating the zone were already taking place on May 3, according to a KGB memo filed the next day. Danilyuk, ed., "Chernobyl Tragedy," *Z arkhiviv,* document no. 26: "Report of the UkSSR KGB 6th Department to the USSR KGB concerning the radioactive situation and progress in investigating the accident at the Chernobyl NPS," May 4, 1986.

241 *No formal plans:* Positions of Pikalov and the Health Ministry are outlined in Dyachenko, ed., *Chernobyl: Duty and Courage,* vol. 1, 89–91. Maleyev, *Chernobyl: Days and Years,* 61, gives the date of this decree as May 24. Kopchinsky and Steinberg mention that the 25 rem limit was imposed by Ministry of Energy Order No. 254, dated May 12, 1986 (*Chernobyl,* 59).

241 *Horrified by the lack of preparation:* Nikolai Istomin, head of the occupational health and safety department at Chernobyl, account in Kopchinsky and Steinberg, *Chernobyl,* 83–85. See also Evgeny Akimov, testimony in Chernousenko, *Chernobyl: Insight from the Inside,* 120–21.

242 *No comprehensive survey:* M. A. Klochkov, testimony in Dyachenko, ed., *Chernobyl: Duty and Courage,* vol. 1, 70.

242 *There was a chronic shortage:* Kopchinsky and Steinberg, *Chernobyl*, 88; Valery Koldin, author interview, Moscow, April 2017; Kiselev, testimony in Dyachenko, ed., *Chernobyl: Duty and Courage*, vol. 1, 39.

242 *The task of clearing:* Klochkov, testimony in Dyachenko, ed., *Chernobyl: Duty and Courage*, vol. 1, 71.

242 *By May 4, the first two colossal:* Ibid., 70–71. Interrogated by a lieutenant general and a Soviet minister on the reasons for the failure, the composure of the officer in charge of the operation eventually snapped: "Why?" he shouted. "Why? I don't know! Go and see for yourself!" At this, the bosses' technical curiosity abruptly evaporated.

243 *Abandoned in a nearby field:* Zhores Medvedev, *Legacy of Chernobyl*, 101.

243 *While the Ministry of Energy urgently:* "Protocol no. 8 of the meeting of the Politburo Operations Group," May 7, 1986, in RGANI. Concrete covering: Danilyuk, ed., "Chernobyl Tragedy," *Z arkhiviv*, document no. 33: "Report of the UkSSR KGB 6th Department concerning the radioactive situation and progress in investigating the accident," May 13, 1986. See also document no. 31, which mentions readiness to begin on May 11: "Special report of the UkSSR OG KGB chief in Chernobyl to the UkSSR KGB Chairman," May 11, 1986.

243 *Construction teams fed the gray slurry:* Kopchinsky and Steinberg, *Chernobyl*, 93.

243 *Special Battalion 731 began work:* Gudov, *731 Special Battalion*, 126, Kopchinsky and Steinberg, *Chernobyl*, 93.

243 *Their shifts lasted as little:* Revchuk, account in Gudov, *731 Special Battalion*, 92–93.

243 *Called to help clear chunks of reactor graphite:* Kiselev, testimony in Dyachenko, ed., *Chernobyl: Duty and Courage*, vol. 1, 40; Yuri Kolyada, testimony in Shcherbak, *Chernobyl*, 199.

243 *Such tasks exposed:* Petryna, *Life Exposed*, xix.

244 *Meanwhile, underground, the battle:* Prianichnikov, author interview, 2006.

244 *600 degrees centigrade:* Danilyuk, ed., "Chernobyl Tragedy," *Z arkhiviv*, document no. 34: "Report of the UkSSR OG KGBM and the USSR KGB in the town of Chernobyl to the USSR KGB concerning the radioactive situation and progress in investigating the accident," May 15, 1986.

244 *Using a plasma torch:* Vladimir Demchenko, account in Gudov, *731 Special Battalion*, 90.

245 *Four hundred miners:* These workers included 234 miners from the Donbas region in Ukraine and 154 from the Moscow Coal Basin: Borovoi and Velikhov, *Chernobyl Experience: Part 1*, 32.

245 *Once again, the deadline:* Orders came down for the miners to begin work on May 16 and complete all digging work by June 22. By July 2, the cooling pipe network was planned to be ready. Dmitriyev, account in Kozlova, *Battle with Uncertainty*, 64–66.

245 *The miners began tunneling:* Reikhtman, author interview, 2015.

245 *Digging with hand tools:* Yuri Tamoykin, account in Kozlova, *Battle with Uncertainty*, 71.

245 *When the chamber was complete:* Ibid., 68–72.

246 *The forty-kilogram graphite blocks:* Dmitriyev (66) and Tamoykin (72–73), accounts in Kozlova, *Battle with Uncertainty*, 66.

246 *Final assembly began:* Tamoykin, account in Kozlova, *Battle with Uncertainty*, 72.

246 *But long before the project was completed:* Prianichnikov, author interview, 2006; Kozlova, *Battle with Uncertainty*, 75–77.

246 *General of the Army Valentin Varennikov:* Steinberg, recollections in Kopchinsky and Steinberg, *Chernobyl*, 101. For Varennikov's biography, see "Gen. Valentin Varennikov Dies at 85; Director of the Soviet War in Afghanistan," Associated Press, May 6, 2009.

246 *When the general arrived:* Dyachenko, ed., *Chernobyl, Duty and Courage*, vol. 1, 43. Minenergo construction workers: "Protocol no. 8 of the meeting of the Politburo Operations Group," May 7, 1986, in RGANI.

246 *The Politburo now recognized:* Vladimir Maleyev, author interview, Moscow, April 2017. Details of alcoholism and drug use in the Soviet Armed Forces in Murray Feshbach and Alfred Friendly Jr., *Ecocide in the USSR: Health and Nature under Siege* (New York: Basic Books, 1992), 165–66.

246 *A decree unprecedented in peacetime:* See relevant parts of the decree (no. 634-188) in Vladimir Maleyev, "Chernobyl: The Symbol of Courage" [Чернобыль: символ мужества], *Krasnaya Zvezda*, April 25, 2017, archive.redstar.ru/index.php/2011-07-25-15-55-35/item/33010-chernobyl-simvol-muzhestva.

247 *They were told they were required:* Colonel Valery Koldin, author interview, Moscow, April 2017.

247 *By the beginning of July:* Danilyuk, ed., "Chernobyl Tragedy," *Z arkhiviv*, document no. 51: "Report of the UkSSR OG KGBM and the USSR KGB," July 4, 1986.

247 *The journey was long and hot:* V. Filatov, "Chernobyl AES—Test of Courage," *Krasnaya Zvezda*, May 24, 1986, translated in JPRS, *Chernobyl Nuclear Accident Documents.*

247 *Whether rising from:* Yuri Kozyrev (senior physicist at the Ukrainian Institute of Physics), author interview, Kiev, April 2016.

247 *Those who understood the threat:* In one interview he gave after his release from prison, former deputy chief engineer Dyatlov, for example, demonstrated this habit. See Michael Dobbs, "Chernobyl's 'Shameless Lies,'" *Washington Post*, April 27, 1992.

247 *But others remained unaware:* Kozyrev, author interview, 2016.

248 *Only the fate of the crows:* Klochkov, testimony in Dyachenko, ed., *Chernobyl: Duty and Courage*, vol. 1, 73.

248 *Daily radiation surveys:* Zhores Medvedev, *Legacy of Chernobyl*, 77–78.

248 *The threat to the population:* International Atomic Energy Agency, "Cleanup of Large Areas Contaminated as a Result of a Nuclear Accident," IAEA Technical Reports Series No. 330 (IAEA, Vienna, 1989), Annex A: The Cleanup After the Accident at the Chernobyl Power Plant, 104–8.

248 *Wind and weather:* Legasov, "My duty is to tell about this," in Mould, *Chernobyl Record*, 294n9.

248 *The work of decontaminating:* IAEA, "Cleanup of Large Areas," 109.

248 *The word "liquidation":* Brown, *Plutopia*, 234.

249 *No one in the USSR—or, indeed, anywhere else:* International Atomic Energy Agency, INSAG-1, 40.

249 *When General Pikalov:* Read, *Ablaze*, 102 and 130–31. During the May 5 Politburo meeting (minutes reproduced in Maleyev, *Chernobyl. Days and Years*, 255), Ryzhkov likewise described a longer cleanup time than estimated by Pikalov, although in Ryzhkov's report, this estimate was between one and two years. "This is unacceptable," Ryzhkov concluded.

249 *On their return to Moscow:* "Protocol no. 10 of the meeting of the Politburo Operations Group of the CPSU Central Committee on problems related to the aftermath of the Chernobyl NPP accident" [Протокол № 10 заседания Оперативной группы Политбюро ЦК КПСС по вопросам, связанным с ликвидацией последствий аварии на Чернобыльской АЭС], May 10, 1986, in RGANI, opis 51, reel 1.1006, file 22.

249 *Its construction teams were overwhelmed:* Igor Belyaev, author interview, Moscow, April 2017.

250 *The USSR had been the first:* Gorbachev to the Politburo, May 15, 1986, quoted by Volkogonov and Shukman, *Autopsy for an Empire*, 480.

250 *Sredmash chief Efim Slavsky arrived the next day:* Belyaev, author interview, April 2017; I. Belyaev, *Chernobyl: Death Watch* [Чернобыль: Вахта смерти], 2nd ed. (IPK Pareto-Print, 2009), 7. The date of Slavsky's arrival, May 21, fell a day after the formation of Construction Supervisory Agency no. 605 (Kozlova, *Battle with Uncertainty*, 217).

250 *"Lads, you'll have to take the risk":* Belyaev, author interview, 2017.

250 *The following afternoon:* Ibid.; Read, *Ablaze*, 208; BBC Summary of World Broadcasts, "Other Reports; Work at Reactor and in Chernobyl: Interviews with Silayev and Ministers," select Soviet TV and radio programming on May 18 and 19, translated May 20, 1986.

250 *In public, the Soviet government:* On May 8, *Izvestia* acknowledged that there was some surface contamination beyond the zone but emphasized that it posed no threat to human health: Zhores Medvedev, *Legacy of Chernobyl*, 158.

250 *But in its secret sessions:* "Protocol no. 10 of the Politburo Operations Group meeting," May 10, 1986, in RGANI.

251 *Civil defense troops threw:* Nikolai Tarakanov, author interview, Moscow, October 2016; Tarakanov, *The Bitter Truth of Chernobyl* [Горькая правда Чернобыля] (Moscow: Center for Social Support of Chernobyl's Invalids, 2011), 5–6.

251 *On May 12 it banned:* "Resolution of a selective meeting of the executive committee, the Soviet of People's Deputies of the Kiev region" [Решение суженного заседания исполкома Киевского областного Совета народных депутатов], May 12, 1986, archive of the Chernobyl Museum.

251 *On every approach to Kiev:* Lyashko, *Weight of Memory*, 372.

251 *The Kremlin's chief scientists:* Read, *Ablaze,* 187–88; Lyashko, *Weight of Memory,* 373–75.

252 *Romanenko, appeared once more on TV:* A. Y. Romanenko, "Ukrainian Minister of Health: School Year to End by 15th May," transcript of TV appearance on May 8, 1986, translated by BBC Summary of World Broadcasts on May 12, 1986; Read, *Ablaze,* 189.

252 *The evacuation began five days later:* Lyashko, *Weight of Memory,* 376–78; Alexander Sirota, author interview, Ivankov, 2017.

253 *On May 22 Scherbitsky put his signature:* "On the activities of local Soviets of people's deputies of the Kiev region in relation to the accident at Chernobyl" [О работе местных Советов народных депутатов Киевской области в связи с аварией на Чернобыльской АЭС], May 21, 1986, archive of the Chernobyl Museum; V.Scherbitsky, "Information on ongoing work pertaining to the accident at Chernobyl NPP" [Информация о проводимой работе в связи с аварией на Чернобыльской АЭС], report no. I/50 to Central Committee of the CPSU, May 22, 1986, archive of the Chernobyl Museum.

253 *But for all this apparent care:* Oleg Schepin (deputy minister of health of the USSR), "VCh-gram from Moscow" [ВЧ-грамма из Москвы], May 21, 1986, archive of the Chernobyl Museum; Petryna, *Life Exposed,* 43 and 226n18).

254 *Back inside the zone:* Baranovska, ed., *Chernobyl Tragedy,* document no. 91: "Materials of the Ukrainian SSR State Agroindustrial Committee on the state of the industry in the wake of the accident at Chernobyl NPP," May 6, 1986; and document no. 135: "Proposal from the Ministry of Internal Affairs of the Ukrainian SSR on the organization of hunting squads for clearing the 30-kilometer zone of dead and stray animals," May 23, 1986.

254 *Twenty thousand agricultural and domestic animals:* IAEA, "Environmental Consequences of the Chernobyl Accident and Their Remediation: Twenty Years of Experience," Report of the Chernobyl Forum Expert Group "Environment" no. STI/PUB/1239, April 2006, 75.

254 *Doza or Rentgen:* Dyachenko, ed., *Chernobyl: Duty and Courage,* vol. 1, 78.

254 *Soviet contingency plans:* Zhores Medvedev cites Leonid Ilyin, then the vice president of the USSR Academy of Medical Sciences, as saying that the Soviet response strategy involved a onetime ejection of radionuclides into the atmosphere: *Legacy of Chernobyl,* 76 and 326n6. See also Anatoly Dyachenko, "The Experience of Employing Security Agencies in the Liquidation of the Catastrophe at the Chernobyl Nuclear Power Plant" [Опыт применения силовых структур при ликвидации последствий катастрофы на Чернобыльской АЭС], *Voyennaya mysl,* no. 4 (2003): 77–79.

255 *Radiation experts were summoned:* Natalia Manzurova and Cathy Sullivan, *Hard Duty: A Woman's Experience at Chernobyl* (Tesuque, NM: Natalia Manzurova and Cathy Sullivan, 2006), 19.

255 *At first, the chemical troops:* IAEA, "Cleanup of Large Areas," 116.

255 *Some materials proved more recalcitrant:* Wolfgang Spyra and Michael Katzsch, eds., *Environmental Security and Public Safety: Problems and Needs in Conversion Policy and Research after 15 Years of Conversion in Central and Eastern Europe*, NATO Security through Science Series (New York: Springer, 2007): 181.

255 *In yards and gardens:* IAEA, "Cleanup of Large Areas," 124.

255 *Soviet technicians tried:* Klochkov, testimony in Dyachenko, ed., *Chernobyl: Duty and Courage*, vol. 1, 74.

255 *They sprayed the shoulders:* Irina Simanovskaya, account in Kupny, *Memories of Lives Given*, 39.

255 *Specialists from NIKIMT:* Elena Kozlova, author interview, Moscow, April 2017.

256 *Meanwhile, the threat posed:* Polad-Zade, "Too Bad It Took a Tragedy," 198–99; L. I. Malyshev and M. N. Rozin (both senior water engineers with the Ministry of Energy at time of accident), "In the Fight for Clean Water," in Semenov, ed., *Chernobyl: Ten Years On*, 238.

256 *Close to Pripyat:* IAEA, "Present and Future Environmental Impact of the Chernobyl Accident," report no. IAEA-TECDOC-1240, August 2001, 65.

256 *Within ten days, the dense stands of pine:* Nikolai Steinberg writes that the trees were already an unusual color, but not yet red, on May 7: Kopchinsky and Steinberg, *Chernobyl*, 56.

256 *The soldiers and scientists:* Dyachenko, ed., *Chernobyl: Duty and Courage*, vol. 1, 79.

256 *In the fields of the collective farms:* Zhores Medvedev, *The Legacy of Chernobyl*, 90–91; Manzurova and Sullivan, *Hard Duty*, 31.

257 *The specialists' optimistic forecast:* IAEA, "Cleanup of Large Areas," 114.

257 *But in a place:* The total amount of soil removed in the process of decontamination was about 500,000 cubic meters. Zhores Medvedev, *Legacy of Chernobyl*, 102.

257 *Encircled by a besieging army:* By the end of 1986, more than 70,000 men and 111 military units would have served in the zone, according to Boris Scherbina's memo to the Central Committee on October 15, 1987: "Memorandum, CPSU Central Committee, no. Shch-2882s" [ЦК КПСС № Щ-2882с], classified, in RGANI, opis 53, reel 1.1007, file 74.

257 *The detritus of combat:* See *Chernobyl: Chronicle of Difficult Weeks*, shot by the first documentary film crew permitted access to the zone, for footage of the plant and its surroundings during this period.

257 *The banished citizens:* Esaulov, *City That Doesn't Exist*, 53–55.

257 *On June 6 alone:* Baranovska, ed., *Chernobyl Tragedy*, document no. 177: "Report of the Ukrainian MVD on maintaining public order within the 30-kilometer zone and in locations housing the evacuated population," June 7, 1986.

258 *Attempts to make Pripyat habitable:* Esaulov, *City That Doesn't Exist*, 51.

258 *The members of the Pripyat city council:* Protsenko, author interview, 2016.

258 *On June 10, engineering troops:* "The creation of the protective barrier in the Chernobyl NPP zone during efforts to liquidate the 1986 accident's consequences" [Создание рубежа охраны в зоне Чернобыльской АЭС при ликвидации последствий катастрофы в 1986 году], Interregional Non-Governmental Move-

ment of Nuclear Power and Industry Veterans, Soyuz Chernobyl, May 6, 2013, www.veteranrosatom.ru/articles/articles_173.html.

258 *A centralized electronic alarm system:* "Evgeny Trofimovich Mishin" [Мишин Евгений Трофимович], Interregional Non-Governmental Movement of Nuclear Power and Industry Veterans, undated, www.veteranrosatom.ru/heroes/heroes _86.html.

258 *Around the edge of the 30-kilometer zone:* Dmitry Bisin, account in Kozlova, *Battle with Uncertainty,* 202.

259 *By June 24, they had completed:* Maleyev, *Chernobyl. Days and Years,* 68–69.

259 *A twelve-person committee met:* Esaulov, *City That Doesn't Exist,* 53–54.

259 *A five-month campaign:* Decontamination efforts in Pripyat continued until October 2, 1986. Belyaev, *Chernobyl: Death Watch,* 158.

260 *A dedicated force of 160,000 men:* Kozyrev, author interview, 2016.

260 *"Forget it":* Protsenko, author interview, 2015.

15. THE INVESTIGATION

261 *When Sergei Yankovsky arrived:* Sergei Yankovsky, author interviews, Kiev, February 2016 and May 2017.

261 *A strictly capitalist problem:* For crime statistics during the later years of the Soviet Union, see Wieczynski, ed., *Gorbachev Encyclopedia,* 90–92.

261 *It had been two in the morning:* Ibid.

263 *The investigation into the causes:* Yankovsky, author interview, 2017. The deputy prosecutor general was Oleg Soroka, and the head of the Second Department was Nikolai Voskovtsev.

263 *That same evening:* Karpan, *Chernobyl to Fukushima,* 113; Kopchinsky and Steinberg, *Chernobyl,* 47.

263 *A pair of experts on RBMK reactors:* The scientists were Alexander Kalugin and Konstantin Fedulenko. See Read, *Ablaze,* 123; Fedulenko, "Some Things Have Not Been Forgotten," 74–75.

264 *"Cause of accident unruly and uncontrollable":* Read, *Ablaze,* 126.

264 *By the end of the first week in May:* Valentin Zhiltsov (laboratory director at VNI-IAES, the Soviet atomic plants research institute), account in Shcherbak, *Chernobyl,* 182–83 and 186.

265 *Back at the station:* Steinberg, recollections in Kopchinsky and Steinberg, *Chernobyl,* 56–57; Viktor and Valentina Brukhanov, author interview, 2015; Steinberg, author interview, 2017; Read, *Ablaze,* 201.

265 *"Damn it":* Yankovsky, author interview, 2017.

265 *"You don't look well":* Grigori Medvedev, *Truth About Chernobyl,* 225–26, and "Chernobyl Notebook" [Чернобыльская тетрадь], *Novy Mir,* no. 6 (June 1989), available online at http://lib.ru/MEMUARY/CHERNOBYL/medvedev.txt.

265 *Two weeks later:* Read, *Ablaze,* 201.

266 *"The cause lies":* "A Top Soviet Aide Details Situation at Stricken Plant," Associated Press, May 3, 1986. In a June 1990 interview for the British documentary series *The*

Second Russian Revolution, Vladimir Dolgikh, the Central Committee secretary who oversaw the energy industry, stated that Yeltsin called this press conference on his own initiative. Dolgikh, interview transcript, 2RR, 5.

266 *"The accident was caused"*: Andranik Petrosyants, " 'Highly Improbable Factors' Caused Chemical Explosion," *Los Angeles Times*, May 9, 1986.

266 *Hardliners inside the Ministry*: Read, *Ablaze*, 198.

266 *The academician had returned home*: Margarita Legasova, "Defenceless Victor: From the Recollections of Academician V. Legasov's Widow" [Беззащитный победитель: Из воспоминаний вдовы акад. В. Легасова], *Trud*, June 1996, translated in Mould, *Chernobyl Record*, 304–5; Margarita Legasova, *Academician Valery A. Legasov*, 381.

267 *Nevertheless, Legasov threw himself*: Inga Legasov, author interview, 2017.

267 *Meanwhile, behind closed doors*: The document was also known as the "Act on the investigation of the causes of the accident at Unit no. 4 of Chernobyl NPP" [Акт расследования причин аварии на энергоблоке No. 4 Чернобыльской АЭС]. Karpan, *Chernobyl to Fukushima*, 113 and 146–47.

268 *In response, Aleksandrov convened*: Kopchinsky and Steinberg, *Chernobyl*, 48.

268 *The meetings went on for hours*: Karpan, *Chernobyl to Fukushima*, 113–15; Shasharin, "Chernobyl Tragedy," 105; Gennadi Shasharin, "Letter to Gorbachev (draft)" [Письмо М. С. Горбачеву (черновик)], May 1986, available online at http://accidont.ru/letter.html and in translation in Karpan, *Chernobyl to Fukushima*, 214–17.

268 *But Gennadi Shasharin*: Shasharin, "Letter to Gorbachev," in Karpan, *Chernobyl to Fukushima*, 215–16.

269 *Viktor Brukhanov returned*: The new director was Erik Pozdishev. Viktor Brukhanov, interview by Sergei Babakov, *Zerkalo nedeli*, 1999. The date of Pozdishev's arrival was May 27, 1986, according to Steinberg (recollections in Kopchinsky and Steinberg, *Chernobyl*, 61). Also see Read, *Ablaze*, 202.

269 *"What are we going to do"*: The new chief engineer was Nikolai Steinberg. Steinberg, author interview, 2017; Viktor and Valentina Brukhanov, author interview, 2016.

269 *Inside the headquarters of the Second Department*: Yankovsky, author interviews, 2016 and 2017. The other two Ukrainian nuclear power plants, at Rovno and at Khmelnitsky, used VVER reactors.

270 *On Wednesday, July 2*: Read, *Ablaze*, 201. (Read gives the date as June 2, but this is incorrect: the Politburo meeting held the next day was July 3.)

270 *By now, the deposed director*: Viktor Brukhanov, interview by Maria Vasyl, *Fakty i kommentarii*, 2000.

270 *At precisely eleven o'clock*: Michael Dobbs, *Down with Big Brother: The Fall of the Soviet Empire* (New York: Vintage Books, 1998), 163.

270 *"The accident was the result"*: Politburo meeting minutes (top secret, single copy), July 3, 1986, reproduced in Yaroshinskaya, *Chernobyl: Crime Without Punishment*, 272–73. In his recorded summary of the meeting, Vorotnikov confirms Scherbi-

na's discussion of the shortfalls of the RBMK and its designers' failure to understand and eliminate them: Vorotnikov, *This Is How It Went*, 104.

271 *By the time Scherbina had finished:* Dobbs, *Down with Big Brother*, 163–64. See also meeting minutes excerpted form Gorbachev Foundation archives in Mikhail S. Gorbachev, *Collected Works* [*Собрание сочинений*] (Moscow: Ves Mir, 2008), vol. 4, 276–77.

271 *The meeting blazed on:* Read, *Ablaze*, 202; Yaroshinskaya, *Chernobyl: Crime Without Punishment*, 274.

271 *Slavsky continued to blame:* Politburo meeting minutes, July 3, 1986, reproduced in Anatoly Chernyaev, A. Veber, and Vadim Medvedev, eds., *In the Politburo of the Central Committee of the Communist Party of the Soviet Union . . . From the notes of Anatoly Chernyaev, Vadim Medvedev, Georgi Shakhnazarov (1985–1991)* [*В Политбюро ЦК КПСС. . . По записям Анатолия Черняева, Вадима Медведева, Георгия Шахназарова (1985–1991)*], 2nd ed. (Moscow: Alpina Business Books, 2008), 57–62. Also see "The meeting of the Politburo of the CPSU Central Committee on July 3, 1986: On Chernobyl" [Заседание Политбюро ЦК КПСС 3 июля 1986 года: О Чернобыле], Gorbachev Foundation, http://www.gorby.ru /userfiles/file/chernobyl_pril_5.pdf.

271 *The representatives of the Ministry of Energy:* Vorotnikov, *This Is How It Went*, 104; "On Chernobyl" [О Чернобыле], excerpt from the July 3, 1986, Politburo meeting in a compilation of Politburo protocols published by the Gorbachev Foundation, www.gorby.ru/userfiles/protokoly_politbyuro.pdf.

271 *Meshkov unwisely insisted:* Chernyaev, Veber, and Medvedev, eds., *In the Politburo*, 58. Also see "On Chernobyl," Gorbachev Foundation.

272 *"It is our fault":* Yaroshinskaya, *Chernobyl: Crime Without Punishment*, 279.

272 *"The accident was inevitable":* Dobbs, *Down with Big Brother*, 164–65; Chernyaev, Veber, and Medvedev, eds., *In the Politburo*, 59–60.

272 *These were drafted into a resolution:* "Resolution of the Central Committee of the CPSU: On the results of investigation of the mistakes that caused the Chernobyl nuclear accident, on measures to address its aftermath, and on the safety of the atomic power industry" [Постановление ЦК КПСС: О результатах расследования причин аварии на Чернобыльской АЭС и мерах по ликвидации ее последствий, обеспечению безопасности атомной энергетики], top secret, July 7, 1986, in RGANI, opis 53, reel 1.1007, file 12. The document was ratified by a unanimous vote of the Politburo on July 14, 1986, according to a signed voting sheet.

273 *"Openness is of huge benefit":* Gorbachev, *Collected Works*, vol. 4, 279.

273 *Not everyone agreed:* "Catalogue of information pertaining to the accident at block no. 4 of the Chernobyl NPP that is subject to classification" [Перечень сведений, подлежащих засекречиванию по вопросам, связанным с аварией на блоке № 4 Чернобыльской АЭС (ЧАЭС)], July 8, 1986, archive of the State Security Service of Ukraine fond 11, file 992, online at the Ukrainian Liberation Movement electronic archive: http://avr.org.ua/index.php/viewDoc/24475.

273 *On his arrival in Kiev:* Read, *Ablaze*, 202; Brukhanov, interview by Sergei Babakov, *Zerkalo nedeli*, 1999.

274 *On the evening of Saturday:* Associated Press, "Text of the Politburo Statement About Chernobyl," *New York Times*, July 21, 1986; Lawrence Martin, "Negligence Cited in Chernobyl Report," *Globe and Mail* (Canada), July 21, 1986.

274 *At home in Tashkent:* Viktor and Valentina Brukhanov, author interview, 2015.

274 *Handed down its own verdicts:* "Punishment for Chernobyl Officials," *Radynska Ukraina*, July 27, 1986, translated in the BBC Summary of World Broadcasts, August 2, 1986.

274 *Eleven river cruise ships:* Lyashko, *Weight of Memory*, 369.

274 *On August 12 the deputy chief engineer:* Brukhanov, interview by Sergei Babakov, *Zerkalo nedeli*, 1999; Viktor and Valentina Brukhanov, author interview, 2015.

275 *Two weeks later, on August 25:* Walter C. Patterson, "Chernobyl—The Official Story," *Bulletin of the Atomic Scientists* 42, no. 9 (November 1986): 34–36. For archival footage of Legasov's IAEA appearance, see the documentary film *The Mystery of Academician Legasov's Death* [Тайна смерти академика Легасова], directed by Yuliya Shamal and Sergei Marmeladov (Moscow: Afis-TV for Channel Rossiya, 2004).

275 *Legasov had spent much of the summer:* Read, *Ablaze*, 196.

275 *And yet, glasnost or not:* Alexander Kalugin, interview in *The Mystery of Academician Legasov's Death*. Kalugin provides a similar summary of this note in his 1990 article "Today's understanding of the accident" [Сегодняшнее понимание аварии], *Priroda*, available online at https://scepsis.net/library/id_698.html.

276 *While his notions:* Read, *Ablaze*, 196–97.

276 *Legasov's delivery:* Steinberg, account in Kopchinsky and Steinberg, *Chernobyl*, 148–49; *The Mystery of Academician Legasov's Death*. For specific parsing of Legasov's language—and particularly the use of the word "drawbacks" rather than "defects"—see Walt Patterson, "Futures: Why a kind of hush fell over the Chernobyl conference / Western atomic agencies' attitude to the Soviet nuclear accident," *The Guardian*, October 4, 1986.

276 *"About half":* "Soviets: Half of Chernobyl-Type Reactors Shut," *Chicago Tribune*, August 26, 1986. Fourteen RBMK units remained in operation at the time, according to Dodd, *Industrial Decision-Making*, Appendix D.

276 *By the time they departed:* Patterson, "Chernobyl—The Official Story," 36. Alexander Borovoi, author interview, Moscow, October 2015. Interview with Alexander Borovoi, "The Liquidator."

277 *In the middle of the conference:* Richard Wilson, author interview, Cambridge, MA, August 2016; Alexander Shlyakhter and Richard Wilson, "Chernobyl: The Inevitable Results of Secrecy," *Public Understanding of Science* 1, no. 3 (July 1992): 255; Zhores Medvedev, *Legacy of Chernobyl*, 99.

277 *"I did not lie in Vienna":* As recalled by Andrei Sakharov, according to Shlyakhter and Wilson, "Chernobyl: The Inevitable Results of Secrecy," 254.

16. THE SARCOPHAGUS

278 *In the darkened room:* Tarakanov, author interview, 2016; Nikolai Tarakanov, *The Bitter Truth of Chernobyl* [Горькая правда Чернобыля] (Moscow: Center for Social Support of Chernobyl's Invalids, 2011). For documentary footage, see "Chernobyl. Cleaning the Roofs. Soldiers (Reservists)," a segment of the documentary series *Chernobyl 1986.04.26 P. S.* [Чернобыль. 1986.04.26 P. S.], narrated by Valery Starodumov (Kiev: Telecon, 2016), online at www.youtube.com/watch?v=ti-Wd TF2Q. Also see *Chernobyl 3828* [Чернобыль 3828], directed by Sergei Zabolotny (Kiev: Telecon, 2011).

278 *According to their height:* Tarakanov, *Bitter Truth of Chernobyl*, 142.

278 *He named each area:* Tarakanov, author interview, 2016. Radiation levels: Starodumov, commentary in *Chernobyl 3828*. Starodumov worked as a radiation scout at the time of this operation.

279 *As much as 10,000 roentgen:* Yuri Samoilenko, interview by Igor Osipchuk, "When it became obvious that clearing the NPP roofs of radioactive debris would have to be done by hand by thousands of people, the Government Commission sent soldiers there" [Когда стало ясно, что очищать крыши ЧАЭС от радиоактивных завалов придется вручную силами тысяч человек, правительственная ком иссия послала туда солдат], *Fakty i Kommentarii*, April 25, 2003, http://fakty .ua/75759-kogda-stalo-yasno-chto-ochicshat-kryshi-chaes-ot-radioaktivnyh -zavalov-pridetsya-vruchnuyu-silami-tysyach-chelovek-pravitelstvennaya -komissiya-poslala-tuda-soldat.

280 *The final concept:* Lev Bocharov (chief engineer, US-605 team three), author interview, Moscow, April 2017; V. Kurnosov et al., report no. IAEA-CN-48/253: "Experience of Entombing the Damaged Fourth Power Unit of the Chernobyl Nuclear Power Plant" [Опыт захоронения аварийного четвертого энергоблока Чернобыльской АЭС], in IAEA, *Nuclear Power Performance and Safety*, proceedings of the IAEA conference in Vienna (September 28 to October 2, 1987), vol. 5, 1988, 170. Other tallies of design proposals on the short list have also been reported. Y. Yurchenko notes twenty-eight blueprints (Kozlova, *Battle with Uncertainty*, 205). Nikolai Steinberg cites more than a hundred (author interview, 2006).

280 *Hollow lead balls:* Kopchinsky and Steinberg, *Chernobyl*, 128; Kozlova, *Battle with Uncertainty*, 209.

280 *Others had proposed:* Kozlova, *Battle with Uncertainty*, 209; Steinberg, author interview, 2006.

280 *During the first meetings:* Kopchinsky and Steinberg, *Chernobyl*, 128.

281 *Among the proposed architectural solutions:* Blueprints in the archive of Lev Bocharov (author interview, 2017).

281 *But the technical challenges:* Belyaev, author interview, 2017.

281 *So the engineers planned:* Kozlova, *Battle with Uncertainty*, 206–7.

281 *And time was short:* Baranovska, ed., *The Chernobyl Tragedy*, document no. 172: "Resolution of the Central Committee of the CPSU and the USSR Council of Ministers 'On Measures to Conserve Chernobyl NPP Objects Pertaining to the Acci-

dent at Energy Block No. 4, and to Prevent Water Runoff from Plant Territory,' " June 5, 1986.

282 *To limit their overall:* Viktor Sheyanov (chief engineer, US-605 team one), account in Kozlova, *Battle with Uncertainty,* 217.

282 *Create the infrastructure necessary:* General Y. Savinov, testimony in I. A. Belyaev, *Sredmash Brand Concrete* [Бетон марки "Средмаш"] (Moscow: Izdat, 1996), 39.

282 *But the often middle-aged:* Savinov explains that the reservists were forty-five to fifty years old and that he regarded them as amateur soldiers who approached their tasks in the same improvised way as the partisans of World War II. Belyaev, *Sredmash Brand Concrete,* 39.

282 *The first shift's most important task:* Sheyanov, account in Kozlova, *Battle with Uncertainty,* 218.

282 *Before building work began:* Bocharov, author interview, 2017; Belyaev, author interview, 2017.

282 *Began to lay siege to the reactor:* Kozlova, *Battle with Uncertainty,* 260.

283 *More than 6 meters high:* Ibid., 220 and 229; Belyaev, author interview, 2017.

283 *The surface of the ground around them:* Kozlova, *Battle with Uncertainty,* 226.

283 *The work was relentless:* Lev Bocharov, account in Kozlova, *Battle with Uncertainty,* 290. Construction supervisor Valentin Mozhnov recalls that the maximum daily volume of concrete reached 5,600 cubic meters (261).

283 *It was rushed to the remains:* Bocharov and Nikifor Strashevsky (senior engineer), accounts in Kozlova, *Battle with Uncertainty,* 290 and 326.

283 *In July and August:* L. Krivoshein, account in Kozlova, *Battle with Uncertainty,* 96; Tarakanov, *Bitter Truth,* 142.

283 *Capable of lifting:* Kozlova, *Battle with Uncertainty,* 243.

284 *If brought too close:* Yurchenko, account in Kozlova, *Battle with Uncertainty,* 245 and 252.

284 *The steel forms of the Cascade Wall:* A. V. Shevchenko (senior construction engineer, US-605 team two), account in Kozlova, *Battle with Uncertainty,* 251.

284 *While the Sredmash engineers were at work:* Borovoi, author interview, 2015; Alexander Borovoi, *My Chernobyl* [Мой Чернобыль] (Moscow: Izdat, 1996), 54.

285 *His rival, Velikhov:* Semenov, "For the 10th Anniversary of the Catastrophe at Chernobyl NPP," 41.

285 *Yet their initial efforts:* K. P. Checherov, "The Unpeaceful Atom of Chernobyl" [Немирный атом Чернобыля], nos. 6–7 (2006–2007), online at http://vivovoco .astronet.ru/VV/PAPERS/MEN/CHERNOBYL.HTM. For the location of this compartment, see 3D diagrams in Sich, "Chornobyl Accident Revisited," 288 and 296–98.

285 *Alexander Borovoi, a thickset forty-nine-year-old:* Borovoi, author interview, 2015; Borovoi, *My Chernobyl,* 39–40.

286 *The Politburo had publicly promised:* In a TV interview from Chernobyl on June 1, Vladimir Voronin, deputy chairman of the Council of Ministers and the third chief of the government commission, said he was "fully confident" that Units One

and Two would be restarted by winter "in accordance with the timescale planned by the government." BBC Summary of World Broadcasts, "1st June TV Report of Work at AES: Statement by Voronin," summary of television programming on June 1, 1986 (translated June 3, 1986).

286 *But now that the truth:* Kopchinsky and Steinberg, *Chernobyl*, 98 and 108–12. The plastic coverings remain on floors and staircases throughout the plant today.

286 *The entire ventilation system:* Ibid., 102–7.

287 *A five-tonne concrete panel:* Tarakanov, author interview, 2016.

287 *The government commission turned once again:* Elena Kozlova, author interview, Moscow, April 2017; Kozlova, *Battle with Uncertainty*, 190–92.

287 *Technicians planned to clear the debris using robots:* For more details on this effort, see Y. Yurchenko, report no. IAEA-CN-48/256: "Assessment of the Effectiveness of Mechanical Decontamination Technologies and Technical Devices Used at the Damaged Unit of the Chernobyl Nuclear Power Plant" [Оценка эффективности технологий и технических средств механической дезактивации аварийного блока Чернобыльской АЭС], in IAEA, *Nuclear Power Performance and Safety*, 1988, 164–65.

288 *On September 16 General Tarakanov received:* Tarakanov, author interview, 2016; recollections of Nikolai Tarakanov (*Bitter Truth of Chernobyl*, 144–45), translated in Chernousenko, *Insight from the Inside*, 151. Description of Samoilenko's appearance at the time is drawn from documentary footage in *Chernobyl 3828*.

288 *Using a sketch plan:* The sketch map is described by Tarakanov in *Bitter Truth of Chernobyl*, 141, and reproduced in Karpan, *Chernobyl to Fukushima*, 14.

288 *Tarakanov's soldiers launched:* Tarakanov, author interview, 2016; Tarakanov, *Bitter Truth of Chernobyl*, 151.

289 *"I'm asking any one of you":* Tarakanov, *Bitter Truth of Chernobyl*, 170.

289 *Many were young:* Alexander Fedotov (former liquidator), interviewed in *The Battle of Chernobyl*, dir. Thomas Johnson (France: Play Film, 2006). Although the majority of the men were *partizans*, the platforms of the vent stack were cleared by cadets from the firefighting school in Kharkov who had volunteered for the task, young men barely out of their teens. The cadets proved especially dedicated and, in some cases, stayed out longer than their allowed time to do extra work.

289 *Years afterward, the general would insist:* Tarakanov, interview in *The Battle of Chernobyl*.

290 *Their eyes hurt:* Igor Kostin and Alexander Fedotov, interviews in *The Battle of Chernobyl*. Kostin's biography and photographs taken on the plant's roof are included in Igor Kostin, *Chernobyl: Confessions of a Reporter* (New York: Umbrage Editions, 2006), 76–81 and 225–37.

290 *Recorded in a ledger:* Tarakanov, author interview, 2016; "List of personnel of army units and subdivisions of the USSR Ministry of Defense that took part in the operation to remove nuclear fuel, highly radioactive graphite and other products of the explosion from the roof of energy block no. 3, machine hall and vent supports of the Chernobyl NPP in the period from September 19 to October 1, 1986" [Список

личного состава воинских частей и подразделений МО СССР, принимавших участие в операции по удалению ядерного топлива, высокорадиоактивного зараженного графита и других продуктов взрыва с крыш 3го энергоблока, машзала и трубных площадок ЧАЭС в период с 19 сентября по 1 октября 1986 года], personal archive of Nikolai Tarakanov.

290 *For twelve days:* Starodumov, narration in *Chernobyl 3828*; Tarakanov, author interview, 2016.

290 *At a quarter to five that afternoon:* Kopchinsky and Steinberg, *Chernobyl*, 115.

290 *On the roof of Unit Three:* See documentary footage in *Chernobyl 3828*, narrated by Starodumov, who was one of the scouts raising the flag. Kostin's photograph is reproduced in his book *Chernobyl: Confessions of a Reporter*, 95.

291 *Tarakanov was climbing into his car:* Tarakanov, author interview, 2016.

291 *Splashed on the front page:* An image of the article titled "The Taming of the Reactor" [Укрощение реактора], is reproduced in Kozlova, *Battle with Uncertainty*, 284.

291 *The chief engineer of the shift:* Bocharov, author interview, 2017; Josephson, *Red Atom*, 69; IAEA, *Nuclear Applications for Steam and Hot Water Supply*, report no. TECDOC-615, July 1991, 73; Stefan Guth, "Picturing Atomic-Powered Communism," paper given at the international conference Picturing Power. Photography in Socialist Societies, University of Bremen, December 9–12, 2015.

291 *Had already fallen behind schedule:* An October 5 KGB memo specified that the Sarcophagus had missed the original Sredmash deadline but that the roofing operation was expected to commence on October 11, and Unit Two was scheduled to come online on October 20 (Danilyuk, ed., "Chernobyl Tragedy," *Z arkhiviv*, document no. 65, *Report of the USSR OG KGB and UkSSR KGB to the USSR KGB Concerning the Radioactive Situation and the Progress in Works on the Cleaning Up Operation After the Accident at the Chernobyl NPS*, October 5, 1986).

292 *A massive steel house of cards:* Kozlova, *Battle with Uncertainty*, 324.

292 *Colossal and ungainly:* Ibid., 358–59; Belyaev, *Chernobyl: Death Watch*, 145.

292 *Bocharov and his engineers set up:* Bocharov, author interview, 2017; Kozlova, *Battle with Uncertainty*, 270.

293 *The* batiskaf: The NIKIMT technicians made several versions of the bathyscaphe, each slightly different. Pictures and a description are provided in Alexander Khodeyev's account in Kozlova, *Battle with Uncertainty*, 161–62, and by Pavel Safronov, account, 380.

293 *The chief designer's plan:* Bocharov, author interview, 2017.

294 *They would have to start the Sarcophagus again:* Bocharov, account in Kozlova, *Battle with Uncertainty*, 382.

294 *Before leaving, some were presented with awards:* Koldin, author interview, 2017.

294 *Regardless of the triumphant:* Raab, *All Shook Up*, 172–73.

294 *Some bribed the draft officer:* Marples, *The Social Impact of the Chernobyl Disaster*, 191.

294 *One group of two hundred:* James M. Markham, "Estonians Resist Chernobyl Duty, Paper Says," *New York Times*, August 27, 1986.

294 *Military police patrols in Kiev:* Logachev, Taras Shumeyko interview, 2017.

294 *High wages:* Wages paid in the zone were calculated at a multiple of individual salaries, according to Maria Protsenko (author interview, 2016). Additionally, at the end of May, the Politburo approved a schedule of special one-off payments for those who distinguished themselves during the liquidation. See Baranovska, ed., *The Chernobyl Tragedy*, document no. 154: "Resolution of the Central Committee of the CPSU and the USSR Council of Ministers 'On conducting decontamination work in Ukrainian SSR and Belarusian SSR regions affected by radioactive pollution after the accident at Chernobyl NPP,'" May 29, 1986.

294 *Vladimir Usatenko was thirty-six years old:* Vladimir Usatenko, author interview, Kiev, December 2016.

296 *Twenty-eight missions:* Ibid. For his work, Usatenko earned five times his usual salary as an electrical engineer, plus a 100-ruble bonus as a noncommissioned officer, earning 1,400 rubles in total.

296 *Led by a physicist:* Bocharov, author interview, 2017; Bocharov, account in Kozlova, *Battle with Uncertainty*, 361–78; Belyaev, *Chernobyl: Death Watch*, 144–45.

296 *At ten in the evening:* Belyaev, *Chernobyl: Death Watch*, 146 and 149.

297 *There was still no sign:* Borovoi, author interview, 2015; Astakhova, "The Liquidator."

297 *The roof and windows:* Kozlova, *Battle with Uncertainty*, 515.

297 *The engineers boasted:* Belyaev, *Chernobyl: Death Watch*, 165. These figures, cited frequently in Soviet reports, do not survive close scrutiny. In his thesis, Alexander Sich shows that cramming this volume of concrete into a building the size of the Sarcophagus is a geometric impossibility (Sich, "Chornobyl Accident Revisited," 26n12).

297 *The costs had risen:* Kozlova, *Battle with Uncertainty*, 518.

297 *As he gazed up:* Belyaev, *Chernobyl: Death Watch*, 162.

297 *It would be Slavsky's final achievement:* Belyaev, author interview, 2017.

298 *"Kandahar?":* Bocharov, author interview, 2017.

17. THE FORBIDDEN ZONE

299 *By the beginning of August 1986:* Gary Lee, "Chernobyl's Victims Lie Under Stark Marble, Far From Ukraine," *Washington Post*, July 2, 1986; Carol J. Williams, "Chernobyl Victims Buried at Memorial Site," Associated Press, June 24, 1986; Thom Shanker, "2 Graves Lift Chernobyl Toll to 30," *Chicago Tribune*, August 3, 1986. Description of cemetery layout: Grigori Medvedev, *Truth About Chernobyl*, 262.

299 *In September, Dr. Angelina Guskova:* "'No Significant Increase in Cancer Sufferers Foreseen' After Accident," excerpts from interviews with Guskova and L. A. Ilyin (vice president of the Soviet Academy of Medical Sciences and director of the

USSR Ministry of Health Institute of Biophysics), published in *Izvestia* on September 19, 1987, and translated by BBC Summary of World Broadcasts on September 27, 1986; Reuters, "Chernobyl Costs Reach $3.9 Billion," *Globe and Mail* (Canada), September 20, 1986.

299 *The body of pump operator:* Shcherbak, *Chernobyl,* 340.

299 *Since then, twenty-nine more:* International Atomic Energy Agency, INSAG-1, 64. One death, of a woman, was attributed to a brain hemorrhage also apparently a consequence of ARS: Gusev et al., eds., *Medical Management of Radiation Accidents,* 201.

299 *Of the thirteen patients who had been treated:* International Atomic Energy Agency, INSAG-1, 64–65; Zhores Medvedev, *Legacy of Chernobyl,* 140.

300 *Deputy Chief Engineer Anatoly Dyatlov:* Dyatlov, *Chernobyl: How It Was,* 54 and 109.

300 *Major Leonid Telyatnikov:* Felicity Barringer, "One Year After Chernobyl, a Tense Tale of Survival," *New York Times,* April 6, 1987.

300 *The doctors regarded the survival:* Barabanova, author interview, 2016. Also see Davletbayev, "Last Shift," 373.

300 *Alexander Yuvchenko:* Natalia Yuvchenko, author interview, 2016. Despite her flinty attitude to her staff, Angelina Guskova apparently displayed a special warmth for her favorite patients. According to Natalia, the veteran radiation specialist appeared at Yuvchenko's bedside, twittering with pet names and reassurances, like a devoted grandmother. "Sashenka!" she said. "Everything is going to be fine! Why are you worried?"

301 *It was June:* Ibid.

301 *Just outside the Exclusion Zone, in the town of Polesskoye:* Esaulov, *City That Doesn't Exist,* 50.

302 *In May the Soviet Red Cross Society:* Ibid., 69.

302 *On July 25 they received an answer:* Ibid., 14 and 55; Natalia Yuvchenko, author interview, 2016.

302 *The refugees were allowed:* Esaulov, *City That Doesn't Exist,* 55–56.

303 *Some found it hard:* A diary entry of a checkpoint worker tasked with accompanying former residents on their apartment visits, published in *Komsomolskaya Pravda* in October 1986 and reproduced in David R. Marples, *The Social Impact of the Chernobyl Disaster* (New York: St. Martin's Press, 1988), 173.

303 *It was September:* Natalia Yuvchenko, author interview, 2016.

303 *Other residents retrieved:* Valery Slutsky, author interview, Pripyat, 2006.

303 *Valentina Brukhanov:* Viktor and Valentina Brukhanov, author interview, 2016. Svetlana Samodelova in "The private catastrophe of Chernobyl's director" [Личная катастрофа директора Чернобыля], *Moskovsky komsomolets,* April 22, 2011, www.mk.ru/politics/russia/2011/04/21/583211-lichnaya-katastrofa-direktora-chernobyilya.html.

303 *It was often late at night:* Esaulov, *City That Doesn't Exist,* 56.

303 *They manned the barriers:* Diary entries of MEPhI checkpoint workers, *Komsomolskaya pravda*, October 1986, reproduced in Marples, *Social Impact of the Chernobyl Disaster*, 172–77.

304 *The visits to the deserted city:* Esaulov, *City That Doesn't Exist*, 56.

304 *The town council planned:* Ibid., 67–68.

304 *A charity rock concert:* BBC Summary of World Broadcasts, " 'Highlights' of Rock Concert for Chernobyl Victims Shown on TV," summary of Soviet television programming on July 11, 1986 (translated July 15).

305 *Account No. 904 had already received:* BBC Summary of World Broadcasts, "Contributions to Chernobyl aid fund," summary of TASS news report on August 11, 1986 (translated August 15, 1986).

305 *In June the Politburo passed a resolution:* Baranovska, ed., *The Chernobyl Tragedy*, document no. 173: "Resolution of the Central Committee of the CPSU and the USSR Council of Ministers 'On providing homes and social amenities to the population evacuated from the Chernobyl zone," June 5, 1986.

305 *Fifty thousand men:* "New Homes for Evacuees: AES Workers' Township," *Pravda*, July 23, 1986, translated by BBC Summary of World Broadcasts on July 28, 1986.

305 *The first settlement:* Marples, *Social Impact of the Chernobyl Disaster*, 197.

305 *Each home was reportedly:* Ibid., 198.

305 *11,500 new single-family houses:* Lyashko, *Weight of Memory*, 370.

305 *But the Politburo task force:* Ibid., 371–72; Valentin Kupny, author interview, Slavutych, Ukraine, February 2016; Natalia Khodemchuk, author interview, Kiev, May 2017.

305 *Mysteriously came to a halt:* Esaulov, *City That Doesn't Exist*, 58–59.

306 *They were shunned:* Natalia Khodemchuk, author interview, 2017.

306 *At school:* Samodelova, "The private catastrophe of Chernobyl's director."

306 *The radiation readings in the stairwells:* G. K. Zlobin and V. Y. Pinchuk, eds., *Chernobyl: Post-Accident Construction Program* [Чорнобиль: Післяаварійна програма будівництва], Kiev Construction Academy (Kiev: Fedorov, 1998), 311.

306 *With the first reactor:* E. N. Pozdishev, interview by *Pravda* correspondents, "Chernobyl AES: Chronicle of Events—In Test Mode," *Pravda*, October 10, 1986, translated in "Aftermath of Chernobyl Nuclear Power Plant Accident—Part II," Foreign Broadcast Information Service, USSR Report: Political and Sociological Affairs, January 22, 1987.

306 *Unit Three remained:* Kopchinsky and Steinberg, *Chernobyl*, 125; Danilyuk, ed., "Chernobyl Tragedy," *Z arkhiviv*, document no. 73: "Special report of the USSR KGB and UkSSR KGB 6th Department concerning the radioactive situation and the progress in works on the cleaning up operation after the accident at the Chernobyl NPS," December 31, 1986.

306 *The commission even issued orders:* Kopchinsky and Steinberg, *Chernobyl*, 117.

306 *In the meantime*, Pravda *reported:* O. Ignatyev and M. Odinets, "House Warming at Zelenyy Mys," *Pravda*, October 20, 1986, translated in "Aftermath of Chernobyl—

Part II," Foreign Broadcast Information Service; Marples, *Social Impact of the Chernobyl Disaster,* 225–26.

307 *"Tens upon tens of millions":* BBC Summary of World Broadcasts, "Gromyko's Presentation of Awards to 'Heroes' of Chernobyl," summary of TASS news report on January 14, 1987 (translated January 16, 1987).

307 *The few awards:* Grigori Medvedev, *Truth About Chernobyl,* 264.

307 *At one point:* Natalia Yuvchenko, author interview, 2015.

307 *Instead of recognition:* Yankovsky, author interview, 2017. The date of this notification is given as November 28, 1986, by Samodelova in "The private catastrophe of Chernobyl's director."

308 *He was permitted:* Viktor and Valentina Brukhanov, author interview, 2016.

308 *Occasionally Brukhanov had:* Brukhanov, interview by Maria Vasyl, *Fakty i kommentarii,* 2000.

308 *For a while:* Viktor and Valentina Brukhanov, author interview, 2015.

308 *At first, Brukhanov refused:* Brukhanov, interview by Maria Vasyl, *Fakty i kommentarii,* 2000.

308 *But his wife:* Yuri Sorokin (Viktor Brukhanov's attorney in court), author interview, Moscow, October 2016.

308 *That same month:* Yankovsky, author interview, 2017.

308 *The director found a letter:* The expert, once more, was Vladimir Volkov (see chapter 4), who this time wrote a letter of protest to Gorbachev himself.

309 *They sent a total of forty-eight:* Sorokin, author interview, 2016. The investigator Yankovsky recalls that there were 57 volumes of material, including KGB surveillance recordings of telephone conversations and data from the plant (Yankovsky, author interview, 2017).

309 *Four other senior members of the plant staff:* The sixth man on trial, Yuri Laushkin, the inspector of the state nuclear industry regulator (Gosatomenergonadzor) based at the plant, was the only one not accused of a crime under Article 220 regarding an "explosion-prone facility." He was tried under Article 167, for negligence. Schmid, *Producing Power,* 4–5 and 206n29 and 206n30; and A. Rekunkov, Prosecutor General of the USSR, "On the completion of the criminal investigation into the accident at Chernobyl NPP" [О завершении расследования по уголовному делу об аварии на Чернобыльской АЭС], memo to the Central Committee of the CPSU, in RGANI, opis 53, reel 1.1007, file 56.

309 *An inventive legal gambit:* Karpan, *Chernobyl to Fukushima,* 125; Schmid, *Producing Power,* 4. The legal category of "explosion-prone" facilities was normally reserved for plants and storage spaces housing large volumes of hot oil, fertilizer, acid, and other chemicals. See A. G. Smirnov and L. B. Godgelf, *The classification of explosive areas in national and international standards and regulations* [Классификация взрывоопасных зон в национальных и международных стандартах, правилах] (Moscow: Tiazhpromelectroproyekt, 1992), online at http://aquagroup.ru/norm docs/1232.

309 *To bolster the case:* Voznyak and Troitsky, *Chernobyl*, 249; Karpan, *Chernobyl to Fukushima*, 126.

309 *Too mentally unstable:* A. Smagin, testimony in Grigori Medvedev, *Truth About Chernobyl*, 256–57.

309 *While the wretched technician:* Voznyak and Troitsky, *Chernobyl*, 246. For details of the orginal trial date, see "On the criminal trial related to the accident at Chernobyl NPP" [О судебном разбирательстве уголовного дела, связанного с аварией на Чернобыльской АЭС], February 27, 1987, memo to the Central Committee of the CPSU, RGANI, opis 53, reel 1.1007, file 58. A follow-up memo from two months later suggests an alternative reason for the postponement was to avoid the trial coinciding with the first anniversary of the disaster: "On the criminal trial related to the accident at Chernobyl NPP" [О судебном разбирательстве уголовного дела, связанного с аварией на Чернобыльской АЭС], April 10, 1987, memo to the Central Committee of the CPSU, RGANI, opis 4, reel 1.989, file 22.

309 *Maria Protsenko returned:* Protsenko, author interview, 2015.

310 *On April 18:* Protsenko, author interview, 2016. The precise language of the standard diagnosis is reported in Chernousenko, *Insight from the Inside*, 163.

310 *Back in Pripyat:* L. Kaybysheva, "News panorama" from Chernobyl, *Izvestia*, March 13, 1987, translated by BBC Summary of World Broadcasts on March 26, 1987; Alexander Sich, "Truth Was an Early Casualty," *Bulletin of Atomic Scientists* 52, no. 3 (1996): 41.

310 *Now sole masters:* Felicity Barringer, "A Reporter's Notebook: A Haunted Chernobyl," *New York Times*, June 24, 1987.

310 *Tens of thousands of liquidators:* During the course of 1987, approximately 120,000 military personnel were rotated through the Exclusion Zone as part of the liquidation effort. Yuriy Skaletsky and Oleg Nasvit (National Security and Defense Council of Ukraine), "Military liquidators in liquidation of the consequences of Chornobyl NPP accident: myths and realities," in T. Imanaka, ed., *Multi-side Approach to the Realities of the Chernobyl NPP Accident* (Kyoto University Press, 2008), 92.

310 *Dust from highly contaminated areas:* Danilyuk, ed., "Chernobyl Tragedy," *Z arkhiviv*, document no. 82: "Special report of the UkSSR KGBM on Kiev and Kiev region to the UkSSR KGB 6th Department concerning the radioactive situation and the progress in works on the cleaning up operation after the accident at the Chernobyl NPS," May 19, 1987.

311 *The Kombinat leaders:* V. Gubarev and M. Odinets, "Chernobyl: Two years on, the echo of the 'zone,'" and commentary by V. A. Masol (chairman of Ukraine's Council of Ministers) in *Pravda*, April 24, 1988, translated by the BBC Summary of World Broadcasts on April 29, 1988.

311 *In the meantime, looting:* Ivan Gladush (interior minister of Ukraine at time of accident), interview by Dmitry Kiyansky, "Let our museum be the only and the last" [Пусть наш музей будет единственным и последним], *Zerkalo nedeli Ukraina*,

April 28, 2000, https://zn.ua/society/pust_nash_muzey_budet_edinstvennym_i _poslednim.html.

311 *Logachev watched in amazement:* Alexander Logachev, interview by Taras Shumeyko, 2017.

311 *The cars and motorcycles left behind:* Esaulov, *City That Doesn't Exist*, 65; Maria Protsenko, author interview, 2016.

311 *As the first anniversary of the disaster approached:* L. Kravchenko, list of proposed print, TV, and radio stories, in "Plan of essential propaganda measures to commemorate the first anniversary of the Chernobyl nuclear accident, approved by the Central Committee" [План основных пропагандистских мероприятий в связи с годовщиной аварии на Чернобыльской АЭС], April 10, 1987, in RGANI, opis 53, reel 1.1007, file 27.

311 *The official Soviet report:* "Annex 7: Medical–Biological Problems," in "USSR State Committee Report on Chernobyl," Vienna, August 1986.

312 *Western specialists:* David R. Marples, "Phobia or not, people are ill at Chernobyl," *Globe and Mail* (Canada), September 15, 1987; Felicity Barringer, "Fear of Chernobyl Radiation Lingers for the People of Kiev," *New York Times*, May 23, 1988.

312 *Robert Gale told the press:* Stuart Diamond, "Chernobyl's Toll in Future at Issue," *New York Times*, August 29, 1986.

312 *Here, more than a year after the accident:* Valeri Slutsky, author interview, 2006; Felicity Barringer, "Pripyat Journal: Crows and Sensors Watch Over Dead City," *New York Times*, June 22, 1987; Sue Fox, "Young Guardian: Memories of Chernobyl— Some of the things Dr. Robert Gale remembers from the aftermath of the world's worst nuclear disaster," *The Guardian*, May 18, 1988; Celestine Bohlen, "Chernobyl's Slow Recovery; Plant Open, but Pripyat Still a Ghost Town," *Washington Post*, June 21, 1987; Thom Shanker, "As Reactors Hum, 'Life Goes On' at Mammoth Tomb," *Chicago Tribune*, June 15, 1987.

313 *And one morning would be awoken:* Viktor Haynes and Marco Bojcun, *The Chernobyl Disaster* (London: Hogarth, 1988), 98. When that day might come, no one was any longer prepared to say. "I can't predict the future," a spokesman for Kombinat explained. "Maybe in ten or fifteen years."

18. THE TRIAL

314 *The trial of Viktor Brukhanov:* Voznyak and Troitsky, *Chernobyl*, 244–50.

314 *A handful of representatives:* Martin Walker, "Three Go on Trial After World's Worst Atomic Disaster," *Guardian*, July 7, 1987, cited by Schmid, *Producing Power*, 205, fn.13

315 *"I think I'm not guilty":* Voznyak and Troitsky, *Chernobyl*, 253.

315 *Wearing a suit jacket:* For photos from the trial, see "Chernobyl trial" [Черно быльский суд], Chernobyl and Pripyat electronic archive, December 18, 2010, http://pripyat-city.ru/main/36-chernobylskiy-sud.html.

315 *He gave an account:* Voznyak and Troitsky, *Chernobyl*, 254–55.

315 *And yet he told the court:* Brukhanov, court testimony in Karpan, *Chernobyl to Fukushima*, 130–33.

316 *"The answer to this question":* Ibid., 137.

316 *"Who do you think is guilty?":* Ibid., 173.

316 *Chief engineer Nikolai Fomin:* Voznyak and Troitsky, *Chernobyl*, 252.

316 *His face pale and glistening:* See video footage of court proceedings at "The Chernobyl Trial" [Чернобыльский суд], online at www.youtube.com/watch?v =BrH2lmP5Wao (accessed May 2018).

316 *He explained how he had been incapacitated:* Voznyak and Troitsky, *Chernobyl*, 259; Karpan, *Chernobyl to Fukushima*, 144.

317 *Unaware of the scale:* Fomin, court testimony in Karpan, *Chernobyl to Fukushima*, 151.

317 *"Dyatlov and Akimov":* Fomin, court testimony in ibid., 143.

317 *Of all the defendants:* Voznyak and Troitsky, *Chernobyl*, 252; Karpan, *Chernobyl to Fukushima*, 162.

317 *He said responsibility:* Voznyak and Troitsky, *Chernobyl*, 259.

317 *Despite being contradicted:* Dyatlov, court testimony in Karpan, *Chernobyl to Fukushima*, 155 and 164. He would later admit to giving the mission to the trainees, in his memoir *How It Was*, 49.

317 *Although none of the accused:* Read, *Ablaze*, 231.

317 *Reporters were told:* Voznyak and Troitsky, *Chernobyl*, 270.

318 *Yet many of the expert witnesses:* Karpan, *Chernobyl to Fukushima*, 205–6; Voznyak and Troitsky, *Chernobyl*, 261.

318 *The court stifled:* Read, *Ablaze*, 231–32.

318 *On July 23:* Ibid., 231; Voznyak and Troitsky, *Chernobyl*, 262–63.

318 *"There is no reason to believe":* Ibid.

319 *They both recognized that:* Sorokin, author interview, 2016.

319 *Fomin accepted his guilt:* Voznyak and Troitsky, *Chernobyl*, 264–68.

319 Sarcophagus: A Tragedy: William J. Eaton, "Candor Stressed in Stage Account; Soviet Drama Spotlights Chernobyl Incompetence," *Los Angeles Times*, September 17, 1986; Martin Walker, "Moscow Play Pans Nuclear Farce: Piece on Chernobyl Accident to Tour Soviet Cities," *Guardian*, September 18, 1986.

319 *"Of course, they should be punished":* Thom Shanker, "Life Resumes at Chernobyl as Trials Begin," *Chicago Tribune*, June 16, 1987.

319 *During an adjournment:* Read, *Ablaze*, 233.

319 *Judge Brize delivered his verdict:* Voznyak and Troitsky, *Chernobyl*, 271.

320 *Valentina Brukhanov fainted:* Viktor and Valentina Brukhanov, author interview, 2016; Samodelova, "The private catastrophe of Chernobyl's director."

320 *Driven from the Palace:* Viktor and Valentina Brukhanov, author interview, 2015; Voznyak and Troitsky, *Chernobyl*, 271. "*Stolypin* cars" was a common name for the railroad cattle wagons used to transport convicts, and pickled herring was given to prisoners to discourage hunger.

320 *When he finally arrived:* Samodelova, "Private catastrophe of Chernobyl's direc-

tor"; Viktor Brukhanov, interview, "The Incomprehensible Atom" [Непонятый атом], *Profil*, April 24, 2006, www.profile.ru/obshchestvo/item/50192-items_18 814; Viktor and Valentina Brukhanov, author interview, 2016.

320 *As the end of 1987 approached:* Marples, *Social Impact of the Chernobyl Disaster*, 226–27 and 235; Baranovska, ed., *Chernobyl Tragedy*, document no. 372: "Information from the Central Committee of the Communist Party of Ukraine to the Central Committee of the CPSU on the status of construction of the city of Slavutych," August 5, 1987; and document no. 373: "The letter of V. Scherbitsky to the USSR Council of Ministers about construction shortfalls in the city of Slavutych," September 21, 1987. The city would eventually welcome its first five hundred residents in April 1988 (Reuters, "New Town Opens to Workers from Chernobyl Power Plant," *New York Times*, April 19, 1988).

320 *A radiation survey of Slavutych:* Baranovska, ed., *Chernobyl Tragedy*, document no. 374: "Report of the Joint Commission of USSR ministries and agencies on the radioactive situation in the city of Slavutych," September 21, 1987.

321 *The last of the three surviving reactors:* BBC Summary of World Broadcasts, "Chernobyl Nuclear Station Third Set Restart," summary of Soviet television programming, December 4, 1987 (translated December 11, 1987).

321 *Unit Three, although now separated:* Kopchinsky and Steinberg, *Chernobyl*, 119–20. Even in 1990, there were still fuel pills scattered on the roof of Unit Three (Karpan, *Chernobyl to Fukushima*, 13). Attempts to rectify the problem in the autumn of 1987 are detailed in Borovoi and Velikhov, *The Chernobyl Experience, Part 1*, 114–16.

321 *A tacit admission:* Schmidt, *Producing Power*, 153 and 271n86.

321 *The authorities revised:* Ibid., 152. In an interview with a West German environmental magazine at the end of the year, Legasov said that the safety modifications would cost the equivalent of between $3 million and $5 million at each station. BBC Summary of World Broadcasts, "Better safeguards for nuclear stations," West German Press Agency, November 22, 1987 (translated on December 4, 1987).

321 *Little had really changed:* The report noted that 320 equipment failures occurred at Soviet nuclear power plants since the accident at Chernobyl, and that 160 of them led to emergency reactor shutdowns: Memorandum to the CPSU Central Committee by I. Yastrebov (head of the Department of Heavy Industry and Power Engineering) and O. Belyakov (head of the Department of Defense Industry), "On the work of the USSR Ministry of Atomic Energy and the Ministry of Medium Machine Building on securing operational safety of nuclear power plants as a result of implementing the CPSU Central Committee resolution of July 14, 1986" [О работе Министерства атомной энергетики СССР и Министерства среднего машиностроения СССР по обеспечению безопасности эксплуатации атомных электростанций в свете постановления ЦК КПСС от 14 июля 1986 года о результатах расследования причин аварии на Чернобыльской АЭС], May 29, 1987, in RGANI, opis 53, reel 1.1007, file 61.

321 *Demoralized by the way:* Danilyuk, ed., *Z arkhiviv*, document no. 82: "Special report of the UkSSR KGBM," May 19, 1987.

322 *In public, Valery Legasov:* Legasov, writing in *Pravda* in June 1986, cited in Mould, *Chernobyl Record*, 299n12.

322 *But privately:* Legasov Tapes, cassette Three, 11–14.

322 *He made repeated visits:* Margarita Legasova, "Defenceless Victor," in Mould, *Chernobyl Record*, 304.

322 *He proposed that:* Read, *Ablaze*, 254.

323 *Even his role at Chernobyl:* Vladimir S. Gubarev, "On the Death of V. Legasov," excerpts from *The Agony of Sredmash* [*Агония Средмаша*] (Moscow: Adademkniga, 2006), reproduced in Margarita Legasova, *Academician Valery A. Legasov*, 343.

323 *A perestroika of its own:* Ibid., 340.

323 *Legasov, citing his poor health:* Read, *Ablaze*, 256.

324 *He began reading the Bible:* Legasova, "Defenceless Victor," in Mould, *Chernobyl Record*, 305.

324 *Using a new Japanese Dictaphone:* Margarita Legasova, *Academician Valery A. Legasov*, 382; Read, *Ablaze*, 257.

324 *Afterward, Gubarev attempted:* Read, *Ablaze*, 257–58; Gubarev, "On the Death of V. Legasov," 346.

324 *In a separate interview with Yunost:* Shcherbak, "Report on First Anniversary of Chernobyl," trans. JPRS, pt. 2, 20–21.

325 *By the beginning of 1988:* Read, *Ablaze*, 259–60.

325 *That afternoon, Legasov's daughter:* Inga Legasov, author interview, 2017.

325 *At lunchtime:* Ibid.; the time of the discovery is given in *The Mystery of Academician Legasov's Death.*

325 *When a colleague:* Borovoi, author interview, 2015.

326 *"Why did he abandon me?":* Read, *Ablaze*, 261.

326 *The opening address:* E. I. Chazov, USSR Minister of Health, "Opening speech," in the proceedings of the event published by the IAEA: "Medical Aspects of the Chernobyl Accident: Proceedings of An All-Union Conference Organized by the USSR Ministry of Health and the All-Union Scientific Centre of Radiation Medicine, USSR Academy of Medical Sciences, and Held in Kiev, 11–13 May 1988," report no. IAEA-TECDOC-516, 1989, 9–10. The number of adults and children living in affected areas is given in G. M. Avetisov et al., "Protective Measures to Reduce Population Exposure Doses and Effectiveness of These Measures," 151.

326 *In Kiev, even two years:* Felicity Barringer, "Fear of Chernobyl Radiation Lingers for the People of Kiev," *New York Times*, May 23, 1988.

326 *But the mandarins:* Kopchinsky and Steinberg, *Chernobyl*, 41.

326 *"They inflict great damages":* Leonid Ilyn, quoted in Barringer, "Fear of Chernobyl Radiation Lingers for the People of Kiev."

327 *And the general secretary's own realization:* Taubman, *Gorbachev*, 235–43.

327 *What began with more open reporting:* Kotkin, *Armageddon Averted*, 68.

327 *An edited extract:* "'My Duty Is to Tell about This': From Academician V. Legasov's Notes" [«Мой долг рассказать об этом» Из записок академика В. Легасова], *Pravda*, May 20, 1988, translated in Mould, *Chernobyl Record*, 300.

328 *Two new nuclear plants:* The Minsk station was hastily converted to a natural-gas-powered plant. The other construction project, near Krasnodar, was abandoned. Quentin Peel, "Work Abandoned on Soviet Reactor," *Financial Post* (Toronto), September 9, 1988; Sich, "Chornobyl Accident Revisited," 165.

328 *In spite of glasnost:* Grigori Medvedev, interview transcript, June 1990, 2RR. The full story, published in June 1989, was preceded by extracts published in March by the *Kommunist* magazine.

328 *A letter he personally addressed:* Sakharov's message (dated November 1988) is enclosed in the Central Committee memorandum "On Academician A. D. Sakharov's letter" [О письме академика А. Д. Сахарова], signed by the head of the committee's ideology department, January 23, 1989, in RGANI, opis 53, reel 1.1007, file 81.

328 *"Everything that pertains":* Grigori Medvedev, "Chernobyl Notebook," trans. JPRS, 1.

328 *Even larger than that within:* See maps of contamination disclosed in March 1989 in Zhores Medvedev, *Legacy of Chernobyl*, 86–88.

328 *"Glasnost wins after all"*: Charles Mitchell, "New Chernobyl Contamination Charges," UPI, February 2, 1989.

329 *The land was so poisoned:* Francis X. Clines, "Soviet Villages Voice Fears on Chernobyl," *New York Times*, July 31, 1989.

329 *Traveled to the scene:* Gerald Nadler, "Gorbachev Visits Chernobyl," UPI, February 24, 1989; Bill Keller, "Gorbachev, at Chernobyl, Urges Environment Plan," *New York Times*, February 24, 1989.

329 *Zelenyi Svit:* "Ukrainian Ecological Association 'Green World': About UEA" [Українська екологічна асоціація «Зелений світ»: Про УЕА], www.zelenysvit .org.ua/?page=about.

329 *The crowd veered off script:* John F. Burns, "A Rude Dose of Reality for Gorbachev," *New York Times*, February 21, 1989.

329 *As the third anniversary:* Nadler, "Gorbachev Visits Chernobyl"; Remnick, *Lenin's Tomb*, 245; Zhores Medvedev, *Legacy of Chernobyl*, 87.

329 *One member of a team:* BBC Summary of World Broadcasts, "'Sanctuary' Designated Around Chernobyl Plant and Animal Mutations Appearing," summary of TASS news reports on May 19, 1989 (in English) and July 31, 1989 (in Russian), translated August 26, 1989.

330 *Secretly mixed into sausages:* David Remnick, "Chernobyl's Coffin Bonus," *Washington Post*, November 24, 1989; Josephson, *Red Atom*, 165–66. The Politburo report cited a controversy in Yaroslavl, a city whose meat processing plant was revealed to be supplied with contaminated meat. Local officials insisted they had been acting with the approval of the Soviet Sanitation Service despite their own earlier denials that any Chernobyl meat was shipped to the area. "On the Radio Report from Yaroslavl Region" [О радиосообщении из Ярославской области], memo by the head of the agrarian department of the CPSU Central Committee, December 29, 1989, in RGANI, opis 53, reel 1.1007, file 87.

330 *Strange new phenomena:* BBC Summary of World Broadcasts, "'Sanctuary' Desig-

nated Around Chernobyl Plant" and "An international research centre is to be set up at the Chernobyl AES," summary of a TASS news report on September 15, 1989 (translated September 16, 1989).

330 *The price for the construction:* V. Kholosha and V. Poyarkov, "Economy: Chernobyl Accident Losses," in Vargo, ed., *Chornobyl Accident*, 215.

330 *One estimate put the eventual bill:* Kholosha and Poyarkov estimate that $128 billion was the sum of all direct and indirect costs borne by Ukraine alone between 1986 and 1997, noting that Ukraine has assumed most of the ongoing expenses in the post-Soviet period. The official report of the USSR Finance Ministry in 1990 put the direct cost attributed to the accident at $12.6 billion for the USSR as a whole, and Ukraine's share was about 30 percent (Kholosha and Poyarkov, "Economy: Chernobyl Accident Losses," 220). The Soviet defense budget was disclosed by Gorbachev in 1989, revising the lower official figure of about $32 billion per year ("Soviet Military Budget: $128 Billion Bombshell," *New York Times*, May 31, 1989).

331 *In Lithuania, six thousand:* Bill Keller, "Public Mistrust Curbs Soviet Nuclear Efforts," *New York Times*, October 13, 1988.

331 *In Minsk, a reported:* Reports by AFP (October 1, 1989) and *Sovetskaya Kultura* (October 6, 1989), summarized in BBC Summary of World Broadcasts, "The Chernobyl Situation: Other reports, Nuclear Power and Test Sites," October 30, 1989.

331 *The disaster unleashed:* Ben A. Franklin, "Report Calls Mistrust a Threat to Atom Power," *New York Times*, March 8, 1987.

331 *The United States faced:* Serge Schmemann, "Chernobyl and the Europeans: Radiation and Doubts Linger," *New York Times*, June 12, 1988.

332 *A focal point of regional opposition:* Dodd, *Industrial Decision Making and High-Risk Technology*, 129–30.

332 *Eight hundred waste disposal sites:* V. Kukhar', V. Poyarkov, and V. Kholosha, "Radioactive Waste: Storage and Disposal Sites," in Vargo, ed., *Chornobyl Accident*, 85.

332 *Even offered twice the average:* Yuri Risovanny, interview by David R. Maples, "Revelations of a Chernobyl Insider," *Bulletin of the Atomic Scientists* 46, no. 10 (1990): 18; Antoshkin, *The Role of Aviation*, 1.

332 *As many as six hundred thousand:* Burton Bennett, Michael Repacholi, and Zhanat Carr, eds., "Health Effects of the Chernobyl Accident and Special Care Programmes," Report of the UN Chernobyl Forum Expert Group "Health," World Health Organization, 2006, 2.

333 *A dedicated clinic:* Chernousenko, *Chernobyl: Insight from the Inside*, 160.

333 *Reluctant to connect their symptoms:* Ibid., 163. According to an instruction sent by the Soviet Ministry of Defense to recruiting centers throughout the USSR, military doctors were forbidden from mentioning Chernobyl work on medical certificates they issued to liquidators. Radiation doses below the level that caused Acute Radiation Syndrome were also to be omitted ("Explanation by the Central

Military Medical Commission of the USSR Ministry of Defense," no. 205 [July 8, 1987], cited in Yaroshinskaya, *Chernobyl: Crime Without Punishment*, 47).

334 *Captain Sergei Volodin:* Volodin, author interview, 2006.

334 *Some died of heart disease:* Gusev, Guskova, and Mettler, eds., *Medical Management of Radiation Accidents*, 204–5t12.4.

334 *Major Telyatnikov:* "Late Chernobyl Fireman's Blood Tests to Be Disclosed," *Japan Times*, April 19, 2006; Anna Korolevska, author interview, 2015.

334 *For others, the psychological burden:* Guskova, *The Country's Nuclear Industry Through the Eyes of a Doctor*, 156; Barabanova, author interview, 2016.

334 *In a desolate cafe:* Prianichnikov, author interview, 2006.

334 *"The invisible enemy":* Antoshkin, interview in *Battle of Chernobyl*, 2006.

335 *When I visited:* Alexander Yuvchenko, author interview, 2006; Natalia Yuvchenko, author interview, 2015.

335 *Yet when he began:* Natalia Yuvchenko, author interview, 2016.

19. THE ELEPHANT'S FOOT

336 *The afternoon of Monday, April 25:* Author visit to Pripyat, April 25, 2016; Mycio, *Wormwood Forest*, 5.

336 *Open to the elements:* Mycio, *Wormwood Forest*, 5–6 and 239.

338 *One arm outstretched in alarm:* The handprints of the figure in bronze were cast from those of Khodemchuk's widow, Natalia (Natalia Khodemchuk, author interview, 2017).

337 *When I first visited the Chernobyl station:* Author visit, February 10, 2016.

338 *From the very start:* Borovoi, "My Chernobyl," 45–48.

339 *Although fully aware:* Borovoi, author interview, 2015.

339 *In the autumn of 1986:* Ibid.; Borovoi, "My Chernobyl," 86–87.

340 *Unable to find:* Borovoi, "My Chernobyl," 90–92.

340 *Sixteen thousand tonnes:* Sich, "Chornobyl Accident Revisited," 241.

340 *The sample revealed:* Borovoi and Velikhov, *Chernobyl Experience: Part 1*, 118–19.

340 *But it contained no trace:* Borovoi, author interview, 2015; Sich, "Chornobyl Accident Revisited," 326n.

340 *By measuring:* Borovoi, "My Chernobyl," 52 and 99–100.

340 *At the beginning of 1988:* Borovoi and Velikhov, *Chernobyl Experience: Part 1*, 66–71.

340 *By the late spring:* Borovoi, "My Chernobyl," 104–9; Borovoi, author interview, 2015. See also documentary footage in *Inside Chernobyl's Sarcophagus*, directed by Edward Briffa (United Kingdom: BBC Horizon, 1991) (subsequently rereleased in 1996).

341 *The smallest trace:* Only 0.01 percent of the lead dropped from the helicopters was found in the corium (Sich, "Chornobyl Accident Revisited," 331).

341 *Mounds up to fifteen meters high:* Spartak T. Belyayev, Alexandr A. Borovoy, and I. P. Bouzouloukov, "Technical Management on the Chernobyl Site: Status and Future of the 'Sarcophagus,'" in European Nuclear Society (ENS), *Nuclear Accidents and the Future of Energy: Lessons Learned from Chernobyl*, Proceedings of the

ENS International Conference in Paris, France, April 15–17, 1991, 27, cited in Sich, "Chornobyl Accident Revisited," 248n34.

341 *Some lead ingots:* Checherov, "Unpeaceful Atom of Chernobyl."

341 *Simply burned itself out:* Sich, "Chornobyl Accident Revisited," 331.

341 *Almost entirely pointless:* This issue is explored in detail in ibid., 243–50.

341 *But the Complex Expedition also revealed:* Borovoi and Velikhov, *Chernobyl Experience: Part 1*, 118; Borovoi, author interview, 2015; Sich, "Chornobyl Accident Revisited," 332.

342 *The zirconium cladding:* Alexander Sich estimates that 71 percent of the 190.3 tonnes of uranium fuel flowed downward from the reactor shaft ("Chornobyl Accident Revisited," 288). The weight of the lower biological shield is noted on 195 and 409.

342 *It burned clean through:* Ibid., 293n; Borovoi and Velikhov, *Chernobyl Experience, Part 1*, 30–31.

342 *Spreading out to the south and east:* Sich provides a map of the four flow routes: "Chornobyl Accident Revisited," 322.

342 *It had burned and seethed:* Borovoi, author interview, 2015; Sich, "Chornobyl Accident Revisited," 322.

342 *Puddles fifteen centimeters deep:* Sich, "Chornobyl Accident Revisited," 308.

342 *When the lava dropped:* Ibid., 323. Elsewhere, according to Sich, the heat of radioactive decay ensured that the solidified corium remained hot even in 1991, five years after the accident had taken place. Also see p. 245, which diagrams a cross section of the damaged Unit Four.

343 *"For the time being":* Conclusions of the expert group: S. T. Belyaev, A. A. Borovoi, V. G. Volkov et al., "Technical Validation of the Nuclear Safety of the Shelter" [Техническое обоснование ядерной безопасности объекта Укрытие], report on the scientific research work conducted by the Complex Expedition, 1990, cited in Borovoi and Velikhov, *Chernobyl Experience: Part 1*, 147–48. The monitoring system (named "Finish"): Ibid., 148–49.

343 *Increasingly forgotten:* Borovoi, author interview, 2015.

343 *Eventually the men ran short:* Documentary footage in *Inside Chernobyl's Sarcophagus*, 1991; Borovoi, "My Chernobyl," 110.

343 *So fascinating and important:* Borovoi, "My Chernobyl," 30 and 34.

344 *They granted individual nicknames:* Borovoi and Velikhov, *Chernobyl Experience: Part 1*, 119, 134, and 141.

344 *Chernobylite:* Borovoi, author interview, 2015; Sich, author interview, 2018; Valery Soyfer, "Chernobylite: Technogenic Mineral," *Khimiya i zhizn*, November 1990, translated in JPRS report JPRS-UCH-91-004: "Science and Technology: USSR Chemistry," March 27, 1991.

344 *Could soon collapse:* Borovoi, "My Chernobyl," 37.

344 *Former plant director:* "Information on the criminal case against V. P. Brukhanov" [Справка по уголовному делу в отношении Брюханова В. П.], personal archive of Yuri Sorokin.

344 *The good Czech overcoat:* Viktor and Valentina Brukhanov, author interview, 2016.

345 *A two-line letter:* S. B. Romazin (president of the Collegium on Criminal Cases of the USSR Supreme Court), Letter no. 02DC-36-87, addressed to Y. G. Sorokin, December 26, 1991, personal archive of Yuri Sorokin.

345 *Longed to return to Pripyat:* Viktor and Valentina Brukhanov, author interview, 2016. A 2011 newspaper profile of Brukhanov reports that he returned to work at Chernobyl after his release as head of the technical department and was greeted warmly by the staff (Samodelova, "The private catastrophe of Chernobyl's director"). But his wife, Valentina, stated in an interview with the author that Brukhanov's first job after prison was in Kiev and involved administrative assistance for a former colleague.

345 *Eventually Vitali Sklyarov:* Viktor and Valentina Brukhanov, author interview, 2016; Vitali Sklyarov, author interview, 2016; Viktor Brukhanov, interview by Babakov, *Zerkalo nedeli*, 1999.

345 *The fallen director:* Read, *Ablaze*, 336.

345 *The former nuclear safety inspector:* Samodelova, "The private catastrophe of Chernobyl's director."

345 *Granted early release:* Read, *Ablaze*, 336. Date of Fomin's release (September 26, 1988): "Information on the criminal case against V. P. Brukhanov," personal archive of Yuri Sorokin.

345 *Had spent his years of incarceration:* Anatoly Dyatlov, "Why INSAG Has Still Got It Wrong," *Nuclear Engineering International* 40, no. 494 (September 1995): 17; Anatoly Dyatlov, letter to Leonid Toptunov's parents, Vera and Fyodor, June 1, 1989, personal archive of Vera Toptunova.

346 *Also granted early release:* Date of Dyatlov's release (October 1, 1990): "Information on the criminal case against V. P. Brukhanov," personal archive of Yuri Sorokin.

346 *In the face of opposition from NIKIET:* Steinberg, recollections in Kopchinsky and Steinberg, *Chernobyl*, 149–51.

346 *One senior member:* Armen Abagyan, interviews by *Asahi Shimbun*, July 17 and August 31, 1990, cited in Kopchinsky and Steinberg, *Chernobyl*, 151.

347 *In May 1991:* Kopchinsky and Steinberg, *Chernobyl*, 152; Read, *Ablaze*, 324.

347 *"scientific, technological":* Steinberg, quoted in Read, *Ablaze*, 324.

347 *"Under those circumstances":* Ibid.

347 *"Thus the Chernobyl accident":* Ibid., 324–25.

348 *"Those who hang a rifle":* Read, *Ablaze*, 325.

348 *But the barons:* Kopchinsky and Steinberg, *Chernobyl*, 152.

348 *"New information":* International Atomic Energy Agency, INSAG-7, 16.

349 *"In many respects":* Ibid., 22.

349 *Attracted little attention:* Alexander Sich, author interview, Steubenville, OH, April 2018.

349 *Until his own death:* "Brief biography of A. S. Dyatlov" [Краткая биография Дятлова А. С.], preface to Dyatlov, *How It Was*, 3.

349 *Ukrainian Orders for Courage:* Karpan, *Chernobyl to Fukushima,* 24–25; Decree of the President of Ukraine No.1156/2008 at the official website of the President of Ukraine: https://www.president.gov.ua/documents/11562008-8322.

20. A TOMB FOR VALERY KHODEMCHUK

350 *It wasn't until the end of that year:* Natalia Yuvchenko, author interviews, 2015 and 2016.

352 *Almost twenty-five years after the explosion:* Author visit to Red Forest, February 5, 2011.

353 *The territory of the Exclusion Zone had expanded repeatedly:* Mycio, *Wormwood Forest,* 68–69; Sergiy Paskevych and Denis Vishnevsky, *Chernobyl: Real World* [Чернобыль. Реальный мир] (Moscow: Eksmo, 2011). Also see Mikhail D. Bondarkov et al., "Environmental Radiation Monitoring in the Chernobyl Exclusion Zone—History and Results 25 Years After," *Health Physics* 101, no. 4 (October 2011): 442–85.

354 *In Britain, restrictions on the sale of sheep:* Liam O'Brien, "After 26 Years, Farms Emerge from the Cloud of Chernobyl," *Independent,* June 1, 2012.

354 *Subsequent studies found:* "Wild Boars Roam Czech Forests—and Some of Them Are Radioactive," Reuters, February 22, 2017.

354 *The first evidence for the phenomenon:* Sergei Gaschak, deputy director for science, Chornobyl International Radioecology Laboratory, author interview, Chernobyl exclusion zone, February 2011.

354 *After the breakup of the USSR:* Adam Higginbotham, "Is Chernobyl a Wild Kingdom or a Radioactive Den of Decay?," *Wired,* April 2011; Gaschak, author interview, 2011.

355 *The idea of the miracle of the zone:* For example, Mycio, *Wormwood Forest,* 99–116; *Radioactive Wolves,* documentary film, directed by Klaus Feichtenberger (PBS: ORF/epo-film, 2011).

355 *Yet scientific evidence:* The continuing controversy over this area of research was addressed by Mary Mycio in "Do Animals in Chernobyl's Fallout Zone Glow?," *Slate,* January 21, 2013.

355 *Winter wheat seeds taken from the Exclusion Zone:* Dmitry Grodzinsky, head of the Department of Biophysics and Radiobiology of the Institute of Cell Biology and Genetic Engineering of the National Academy of Sciences of Ukraine, author interview, Kiev, February 2011.

355 Stephanie Pappas, "How Plants Survived Chernobyl," *Science,* May 15, 2009.

355 *The World Health Organization asserted:* WHO/IAEA/UNDP, "Chernobyl: The True Scale of the Accident," joint press release, September 5, 2005, cited in Petryna, *Life Exposed,* xx.

355 *This bore out decades of earlier research:* Jorgensen, *Strange Glow,* 226–30.

355 *But some researchers insisted:* Grodzinsky, author interview, 2011. See also Anders Pape Møller and Timothy Alexander Mousseau, "Biological Consequences of Chernobyl: 20 Years On," *Trends in Ecology & Evolution* 21, no. 4 (April 2006): 200–220.

356 *"That's what we want to know":* Moller, author interview, 2011. By 2017, scientists affiliated with the US National Cancer Institute had begun a genome study examining the long-term effect of radiation on a small sample of the population affected by the accident. Dr. Kiyohiko Mabuchi, head of Chernobyl Research Unit, National Cancer Institute, author interview, September 2018.

356 *"The Chernobyl zone is not as scary":* Andrew Osborn, "Chernobyl: The Toxic Tourist Attraction," *Telegraph*, March 6, 2011.

356 *The authorities had already tolerated:* These so-called "squatters" began to find their way back into the forests around the plant almost immediately after the first forced evacuations from the thirty-kilometer zone, along the same trails many had used to evade the Nazis during the Great Patriotic War. In 1988, the MVD reported that 980 people had already returned to their homes; 113 had never left in the first place, according to an MVD report reproduced in Anton Borodavka, *Faces of Chernobyl*, 2013, 19. Borodavka attributes the term "aborigines of the nuclear reservation" to the noted Ukrainian poet Lina Kostenko (*Faces of Chernobyl*, 12).

356 *The first contract to build:* The contract was for the Vogtle nuclear power plant in Georgia. Terry Macalister, "Westinghouse Wins First US Nuclear Deal in 30 Years," *Guardian*, April 9, 2008.

356 *At the beginning of March 2011:* The new reactors were Units Three and Four, which the Ukrainian government planned to add to the Khmelnitsky nuclear power plant. "Construction cost of blocks 3 and 4 of Khmelnitsky NPP will be about $4.2 billion" [Стоимость строительства 3 и 4 блоков Хмельницкой АЭС составит около $4,2 млрд], Interfax, March 3, 2011.

357 *Yet nuclear power endured:* List of nuclear power reactors: US Nuclear Regulatory Commission, "Operating Nuclear Power Reactors (by Location or Name)," updated April 4, 2018, www.nrc.gov/info-finder/reactors.

357 *France continued to generate:* "Nuclear Power in France," World Nuclear Association, updated June 2018, www.world-nuclear.org/information-library/country -profiles/countries-a-f/france.aspx; "Nuclear Power in China," World Nuclear Association, updated May 2018, www.world-nuclear.org/information-library/country -profiles/countries-a-f/china-nuclear-power.aspx.

357 *Humanity was predicted to double the amount of energy it uses by 2050:* These forecasts vary depending on the prediction models employed. A more recent estimate suggests that demand will double by 2060. "World Energy Scenarios 2016: Executive Summary," World Energy Council, https://www.worldenergy.org/wp-content /uploads/2016/10/World-Energy-Scenarios-2016_Executive-Summary-1.pdf.

357 *The fine particulates from fossil fuel plants:* United States: "The Toll from Coal: An Updated Assessment of Death and Disease from America's Dirtiest Energy Source," Clean Air Task Force, September 2010, 4.

357 *Even to begin to head off climate change:* Barry W. Brook et al., "Why Nuclear Energy Is Sustainable and Has to Be Part of the Energy Mix," *Sustainable Materials and Technologies*, volumes 1–2 (December 2014): 8–16.

358 *Statistically safer than every competing energy industry:* "Mortality Rate World-

wide in 2018, by Energy Source (in Deaths Per Terawatt Hours)," Statista.com, www.statista.com/statistics/494425/death-rate-worldwide-by-energy-source; Phil McKenna, "Fossil Fuels Are Far Deadlier Than Nuclear Power," *New Scientist*, March 23, 2011.

358 *In principle, these fourth-generation reactors:* This thesis is examined in detail in Gwyneth Cravens, *Power to Save the World: The Truth About Nuclear Energy* (New York: Vintage Books, 2008); and the documentary film *Pandora's Promise*, directed by Robert Stone (Impact Partners, 2013).

358 *The liquid fluoride thorium reactor:* Robert Hargraves and Ralph Moir, "Liquid Fuel Nuclear Reactors," *Physics and Society* (a newsletter of the American Physical Society), January 2011.

358 *In 2015 Microsoft founder Bill Gates:* Gates was one of the financial backers of Terra-Power, funding research into a fourth-generation "traveling wave" reactor. See Richard Martin, "China Details Next-Gen Nuclear Reactor Program," *MIT Technology Review*, October 16, 2015; Richard Martin, "China Could Have a Meltdown-Proof Nuclear Reactor Next Year," *MIT Technology Review*, February 11, 2016.

358 *"The problem of coal has become clear":* Stephen Chen, "Chinese Scientists Urged to Develop New Thorium Nuclear Reactors by 2024," *South China Morning Post*, March 18, 2014.

359 *"were not nearly as substantial":* WHO/IAEA/UNDP, "Chernobyl: The True Scale of the Accident." http://www.who.int/mediacentre/news/releases/2005/pr38/en/.

359 *The Chernobyl Forum:* WHO, "Health Effects of the Chernobyl Accident: An Overview," April 2006, www.who.int/ionizing_radiation/chernobyl/backgrounder/en; Elisabeth Cardis et al., "Estimates of the Cancer Burden in Europe from Radioactive Fallout from the Chernobyl Accident," *International Journal of Cancer* 119, no. 6 (2006): 1224–35.

359 *"paralyzing fatalism":* WHO/IAEA/UNDP, "Chernobyl: The True Scale of the Accident," quoted in Petryna, *Life Exposed*, xv.

359 *In a follow-up report:* "1986–2016: Chernobyl at 30—An Update," WHO press release, April 25, 2016.

360 *"Basically nothing happened here":* Adriana Petryna, "Nuclear Payouts: Knowledge and Compensation in the Chernobyl Aftermath," *Anthropology Now*, November 19, 2009.

360 *Yet these conclusions:* Petryna, *Life Exposed*, xix–xx.

361 *She lived alone with six cats:* Protsenko, author interview, 2015.

361 *"It still stinks of radiation":* Ibid., 2016.

361 *I found Viktor Brukhanov:* Viktor and Valentina Brukhanov, author interview, 2015.

362 *"The director has primary responsibility":* Ibid., Viktor and Valentina Brukhanov, author interview, 2016.

363 *By the morning of April 26, 2016:* Author attendance at ceremony marking the thirtieth anniversary of the Chernobyl catastrophe, Chernobyl nuclear power plant, April 26, 2016.

363 *"Satan sleeps beside the Pripyat"*: The poem is "Satan Sleeps Beside the Pripyat" [На березі Прип'яті спить сатана] by Lina Kostenko, translated here by Tetiana Vodianytska.

364 *He spoke of the accident's catalytic role*: Petro Poroshenko, "The President's address at the ceremony marking the 30th anniversary of the Chernobyl catastrophe" [Виступ Президента під час заходів у зв'язку з 30-ми роковинами Чорнобильської катастрофи], speech on the New Safe Confinement site, April 26, 2016, online at the President of Ukraine website: www.president.gov.ua/news /vistup-prezidenta-pid-chas-zahodiv-u-zvyazku-z-30-mi-rokovin-37042.

364 *The official cost of its completion*: Nicolas Caille (Novarka project director), speech at the dedication ceremony for the New Safe Confinement, Chernobyl NPP, November 29, 2016; "Unique Engineering Feat Concluded as Chernobyl Arch Has Reached Resting Place," EBRD press release, November 29, 2016; Laurin Dodd, author interview, telephone, May 2018.

365 *"Ukrainians are a strong people"*: Poroshenko, "The President's address marking the 30th anniversary of the Chernobyl catastrophe."

365 *Six months later*: Author attendance at New Safe Confinement dedication ceremony, Chernobyl NPP, November 29, 2016.

366 *Back in Moscow, the original architects*: Bocharov, author interview, 2017; Belyaev, author interview, 2017.

366 *"We have closed a wound"*: Hans Blix, speech at the dedication ceremony, Chernobyl NPP, November 29, 2016.

366 *Neither man nor machine*: Laurin Dodd, author interview, May 2018; Artur Korneyev, quoted in Henry Fountain, "Chernobyl: Capping a Catastrophe," *New York Times*, April 27, 2014.

EPILOGUE

367 *Anatoly Aleksandrov*: "Former Academy President Aleksandrov on Chernobyl, Sakharov," *Ogonek* no. 35, August 1990, 6–10, translated by JPRS.

367 *Major General Nikolai Antoshkin*: author interview; "Nikolai Timofeyevich Antoshkin," [Антошкин Николай Тимофеевич] *Geroi Strany*, www.warheroes.ru /hero/hero.asp?Hero_id=1011.

368 *Alexander Borovoi*: Alla Astakhova, interview with Alexander Borovoi, "The Liquidator," [Ликвидатор] *Itogi* 828, no. 17, April 23, 2012, www.itogi.ru/obsh-spetz proekt /2012/17/177051.html.

368 *Following his fall from power*: Taubman, *Gorbachev*, 650–663; Mikhail Gorbachev, "Turning Point at Chernobyl," *Project Syndicate*, April 14, 2006.

369 *Dr. Angelina Guskova*: Guskova, *The Country's Nuclear Industry Through the Eyes of a Doctor*, 156.

370 *Prime Minister Nikolai Ryzhkov*: US Department of the Treasury, "Treasury Sanctions Russian Officials, Members of the Russian Leadership's Inner Circle, and an Entity for Involvement in the Situation in Ukraine," March 20, 2014.

370 *Boris Scherbina:* Andriyanov and Chirskov. *Boris Scherbina*, 386–88; Drach, author interview.

370 *Vladimir Scherbitsky:* Rada Scherbitskaya, interview with Sheremeta, "After Chernobyl, Gorbachev told Vladimir Vasiliyevich," 2006; Baranovka, ed., *The Chernobyl Tragedy*, document no. 482: "Resolution on the termination of the criminal case opened February 11, 1992, with regard to the conduct of officials and state and public institutions after the Chernobyl NPP accident," April 24, 1993.

371 *Ukrainian Energy Minister:* Vitali Sklyarov, interview by Natalia Yatsenko, "Vitali Sklyarov, energy advisor to the Ukrainian prime minister: 'What's happening in our energy sector is self-suffocation,'" [Советник премьер-министра Украины по вопросам энергетики Виталий Скляров: Самоудушение—вот что происходит с нашей энергетикой], *Zerkalo nedeli Ukraina*, October 7, 1994, https://zn.ua/ECONOMICS/sovetnik_premier-ministra_ukrainy_po_voprosam _energetiki_vitaliy_sklyarov_samoudushenie_-_vot_chto_p.html.

371 *Following treatment:* Ekaterina Sazhneva, "The living hero of a dead city" [Живой герой мертвого города], *Kultura*, February 2, 2016, http://portal-kultura.ru /articles/history/129184-zhivoy-geroy-mertvogo-gorodasazh; Tamara Stadnychenko-Cornelison, "Military engineer denounces handling of Chernobyl Accident," *The Ukrainian Weekly*, April 26, 1992.

371 *Vladimir Usatenko:* Vladimir Usatenko, interview with Oleksandr Hrebet, "A Chernobyl liquidator talks of the most dangerous nuclear waste repository in Ukraine," [Ліквідатор аварії на ЧАЕС розповів про найнебезпечніше сховище ядерних відходів в У країні] *Zerkalo nedeli*, December 14, 2016, https://dt.ua/UKRAINE /likvidator-avariyi-na-chaes-rozpoviv-pronaynebezpechnishomu-shovische -yadernih-vidhodiv-v -ukrayini-227461_.html.

372 *Detective Sergei Yankovski:* Sergei Yankovski, author interview, Kiev, 2017.

372 *After pumping out:* Zborovsky, interview with Babakov, in Gudov, *Special Battalion no. 731*, 36, 78; John Daniszewski, "Reluctant Ukraine to Shut Last Reactor at Chernobyl," *Los Angeles Times*, December 14, 2000.

Bibliography

Archival Collections and Materials

The American Presidency Project, hosted by the University of California, Santa Barbara, www.presidency.ucsb.edu/index.php.

Archive of the State Security Service of Ukraine [Галузевий державний архів Служби безпеки України (СБУ)]. Ukraine State Security Service Central Office. Kiev, Ukraine.

Barabanova, Dr. Angelika. Personal archive.

BBC Monitoring Service, Summary of World Broadcasts, daily reports on the Soviet Union and Eastern Europe.

"Chernobyl, Pripyat: A Bit About Everything" [Чернобыль, Припять: обо всем понемногу]. Chernobyl and Pripyat electronic archive maintained by Alexander Kamayev. Online at http://pripyat-city.ru.

"The Chernobyl Tragedy: The Crime of the Soviet Government" [Чорнобильська трагедія—злочин радянської влади]. Ukrainian Liberation Movement electronic archive: Joint project of Ukrainian Center for Research on the Liberation Movement, Ivan Franko National University of Lviv, and National Museum-Memorial of Victims of Occupation Regimes (Prison on Łącki). Online at http://avr.org.ua/index.php/ROZDILY_RES?idUpCat=867.

CIA Records Search Tool (CREST), the Central Intelligence Agency. Online at www.cia.gov/library/readingroom/collection/crest-25-year-program-archive.

Gorbachev Foundation Electronic Archive: Virtual Museum "M. S. Gorbachev. Life and Reforms" [Виртуальный музей "М.С.Горбачев. Жизнь и реформы"]. Online at www.gorby.ru/en/archival/archive_library.

Joint Publication Research Service (JPRS), *Chernobyl Nuclear Accident Documents*, foreign media monitoring reports produced by the US government's National Technical Information Service (including CIA, Department of Defense, Department of Energy, Congressional, GAO, and Foreign Press Monitoring Files), released March 2011.

Legasov, Valery. "On the Accident at Chernobyl AES" [Об аварии на Чернобыльской АЭС] aka "Legasov Tapes." Transcript of five cassette tapes dictated by V. A. Legasov, online at http://lib.web-malina.com/getbook.php?bid=2755.

Ronald Reagan Presidential Library and Museum Electronic Archive: "Daily Diary" (www.reaganlibrary.gov/digital-library/daily-diary) and "Public Papers of the President" (www.reaganlibrary.gov/sspeeches).

Russian State Archive of Contemporary History (RGANI) [Российский госуда рственный архив новейшей истории (РГАНИ)]. Hoover Institution. Microfilm, Fond 89: Communist Party of the Soviet Union on Trial, 1919–1992. Copies held at the Lamont Library: Archive of Contemporary History. Harvard University, Cambridge, MA.

The Second Russian Revolution (2RR) Collection. Material relating to the documentary film series *The Second Russian Revolution* (tapes and interview transcripts). Reference no. GB 97 2RR. LSE Library Archive, London.

Sorokin, Yuri. Personal archive.

State Film Archive of Ukraine, Kiev, Ukraine.

Tarakanov, General Nikolai. Personal archive.

Toptunova, Vera. Personal archive.

Ukrainian National Chernobyl Museum [Национальный музей "Чернобыль"]. Archival documents and materials. Kiev, Ukraine.

Veterans of Rosatom website. www.veteranrosatom.ru.

Wilson Center Digital Archive. http://digitalarchive.wilsoncenter.org.

Author Interviews

Job title given at time of accident unless otherwise specified.

Antoshkin, Nikolai. Major general, chief of staff of the Soviet Air Defense Forces' Seventeenth Airborne Army, Kiev military district. Moscow. October 21 and 23, 2015; October 13, 2016.

Barabanova, Anzhelika. Chief of radiation burn surgery, clinical department, Hospital Number Six. Moscow, October 14, 2016.

Belyaev, Igor. Head of the Main Directorate of the Ministry of Medium Machine Building of the USSR; deputy head of Sredmash US-605, June–November 1986. Moscow, April 17, 2017.

Bocharov, Lev. Deputy chief engineer of the Main Directorate for Design and Capital Construction of the Ministry of Medium Machine Building. Chief engineer of Shift Three, Sredmash US-605, September–December 1986. Moscow, April 14, 2017.

Bolshov, Leonid. Research physicist in the Troitsk branch of the Kurchatov Institute of Atomic Energy; director of the Nuclear Safety Institute of the Russian Academy of Sciences, 1988. Moscow, April 15, 2017.

Borovoi, Alexander. Head of the neutrino physics research group, Kurchatov Institute of Atomic Energy; chief scientist of the Chernobyl Complex Expedition, 1988–2003. Moscow, October 15, 2015.

Breus, Alexey. Senior reactor control engineer, Second shift, Unit Four of Chernobyl nuclear power plant. Kiev, July 11, 2015.

Brukhanov, Viktor (director of the Chernobyl nuclear power plant), and Valentina

Brukhanov (Chernobyl plant heat treatment specialist). Kiev, September 6, 2015, and February 14, 2016.

Champlin, Richard. Chief of bone marrow transplant surgery at UCLA Medical Center. Telephone, September 21, 2016.

Daniloff, Nicholas. Moscow bureau chief, *U.S. News & World Report*. Telephone interview, September 26, 2017.

Dodd, Laurin. RBMK reactor expert in Nuclear Systems and Concepts Department, Battelle, Pacific Northwest National Laboratory, Richland, WA, March 1986–May 1994; managing director, Shelter Implementation Plan Project Management Unit, Chernobyl nuclear power plant, April 2006–March 2014. Telephone, May 4, 2018.

Drach, Leonid. Head of the Nuclear Energy Sector of the USSR Council of Ministers; member of the first government commission investigating the causes of the Chernobyl accident. Moscow, April 19, 2017.

Esaulov, Alexander. Deputy chairman of the Pripyat municipal executive committee, or *ispolkom*. Irpin, Ukraine, July 2015.

Flowers, Alan. Nuclear physicist and lecturer at Kingston Polytechnic, London; subsequently radiation protection officer at Kingston University; honorary doctor of radioecology with the International Sakharov Environmental Institute of Belarusian State University, Minsk. Telephone, February 25, 2016.

Gale, Robert. Professor of medicine at UCLA Medical Center; chairman of the Scientific Advisory Committee of the International Bone Marrow Transplant Registry. June 22, 2016 (telephone) and August 11, 2016 (Big Sky, MT).

Gaschak, Sergei. In 2018, deputy director for Science, Chornobyl International Radioecology Laboratory. Chernobyl Exclusion Zone, February 7, 2011.

Glukhov, Andrei. Senior reactor control engineer, Unit One, Chernobyl nuclear power plant, April 1981–August 1984; head of Operational Support Group, Department of Nuclear Safety, Chernobyl nuclear power plant, 1984–1986. Slavutych, Ukraine, July 12, 2015; Chernobyl nuclear power plant, Ukraine, February 11 2016; telephone, July 3, 2018.

Goldston, Robert. Head of physics research team, Tokamak Test Reactor, Princeton Plasma Physics Laboratory. Princeton, NJ, February 15, 2018.

Grebeniuk, Vladimir. Colonel; commander of 427th Red Banner Mechanized Regiment of the Civil Defense Forces stationed in Kiev. Kiev, February 9, 2015.

Grodzinsky, Dmitri. Head of the Department of Biophysics and Radiobiology of the Institute of Cell Biology and Genetic Engineering of the National Academy of Sciences of Ukraine. Kiev, February 8, 2011, and July 13, 2015.

Gubarev, Vladimir. Science editor, *Pravda*. Moscow, October 23, 2015.

Ignatenko, Sergey. Director of the Electrical Grid for the Right Bank of the Dnieper for the Kiev region, Ukrainian Ministry of Energy and Electrification. Kiev, April 22, 2016.

Jacob, Sabine. In 2016, assistant to Dr. Robert Gale. Los Angeles, September 23, 2016.

Jorgensen, Timothy. In 2018, associate professor in the Department of Radiation Medicine at Georgetown University. Telephone, June 19, 2016.

Khmel, Piotr. Lieutenant, Paramilitary Fire Brigade Number Two. Kiev, February 16, 2006, and July 14, 2015.

Khodemchuk, Natalia. Engineer in Pripyat water pumping station Number Two and wife of Valery Khodemchuk. Kiev, May 28, 2017.

Kirichenko, Svetlana. Chief economist of the Pripyat *ispolkom*. Kiev, April 23, 2016.

Kizima, Vasily. Director of construction, Chernobyl nuclear power plant. Kiev, February 7, 2016.

Klochko, Viktor. Colonel; head of the Pripyat department of the KGB. (Interview by Taras Shumeyko), Kiev, September 7, 2015.

Koldin, Valery. Colonel; deputy head of the Department of Military Construction, Third Shift, Sredmash US-605 (October–December 1986). Moscow, April 18, 2017.

Koliadin, Anatoly. Electrician in thermal automation and measurement workshop, Unit Four, Chernobyl nuclear power plant; editor of *Chernobyl Post*, 2003–2017, Kiev. July 10, 2015.

Kopchinsky, Georgi. Senior advisor on nuclear power, Communist Party Central Committee, Moscow. Kiev, November 28, 2016.

Korneyev, Yuri. Turbine operator, Fifth Shift, Unit Four, Chernobyl nuclear power plant. Kiev, September 8 2015.

Korol, Alexander. Trainee reactor control engineer, Unit Four, Chernobyl nuclear power plant. Kiev, September 9, 2015 (interview by author) and April 17, 2018 (interview by Taras Shumeyko).

Korolevska, Anna. In 2018, deputy director for science of the Chernobyl Museum. Kiev, July 10, 2015; February 8, 2016.

Kovtutsky, Viktor. Chief accountant at Chernobyl nuclear power plant construction department. Kiev, April 23, 2016.

Kozlova, Elena. Materials research technician, NIKIMT. Moscow, April 17, 2017.

Kozyrev, Yuri. Senior scientist at the Gaseous Electronics Department of the Institute of Physics in Kiev. Kiev, April 21, 2017.

Kryat, Anatoly. Chief of the Nuclear Physics Laboratory in the Department of Nuclear Safety, Chernobyl nuclear power plant. Kiev, February 15, 2016.

Kupny, Valentin. Director of Zaporizhia NPP; director of Chernobyl Shelter, 1995–2002. Slavutych, Ukraine, February 12, 2016.

Legasov, Inga. Daughter of Academician Valery Legasov. Moscow, April 18, 2017.

Leonenko, Vitali. Director of Medical-Sanitary Center Number 126, Pripyat. Vepryk, Ukraine, December 3, 2016.

Lisovenko, Vasily; Major, head of the Third Division of the Sixth Department, Ukrainian KGB. Vyshenski, Ukraine, September 10, 2015.

Logachev, Alexander. Senior lieutenant; chief radiation scout of the 427th Red Banner Mechanized Regiment of the Civil Defense Forces stationed in Kiev. Kiev, June 1, 2017 (conducted by the author) and June 3, 2017 (conducted by Taras Shumeyko).

Mabuchi, Dr. Kiyohiko. Head of Chernobyl Research Unit and senior scientist, National Cancer Institute, Division of Epidemiology and Genetics. Telephone, September 13, 2018.

Maleyev, Vladimir. In 1987, lieutenant colonel commanding 14th Radiation and Chemical Reconnaissance regiment of the Soviet Chemical Warfare Forces. Moscow, April 16, 2017.

Masol, Vitali. Head of the Ukrainian State Committee for Material and Technical Supply (GOSSNAB); prime minister of Ukraine, 1994–1995. Kiev, June 1, 2017.

McNeil, Oscar. Managing director of Chernobyl Shelter Implementation Plan, 2015–present, Telephone, June 11, 2018.

Mimka, Lubomir. Colonel; deputy chief of staff of the Soviet Air Defense Forces' Seventeenth Airborne Army, Kiev military district. Kiev, February 13, 2016.

Moller, Anders. Research director; in 2018, Ecology, Systematics and Evolution Laboratory, University of Paris-Sud. Paris, France, January 26, 2011.

Mousseau, Timothy. In 2018, professor of biological sciences, University of South Carolina, Columbia. Columbia, SC, January 2011.

Nazarkovsky, Alexander. Senior electromechanical engineer in charge of construction and installation quality control, Chernobyl nuclear power plant. Kiev, February 16, 2006.

Nesterov, Boris. Colonel; deputy commander of the Air Forces of the Kiev Military District. Dnipro, Ukraine, December 2, 2016.

Nosko, Valeri. Major, Third Division of the Sixth Department, Ukrainian KGB. Kiev, September 9, 2015.

Parashyn, Serhiy. Communist Party secretary, Chernobyl nuclear power plant; director, Chernobyl nuclear power plant, 1994–1998. Kiev, November 30, 2016.

Petrovsky, Alexander. Sergeant, Third Watch, Paramilitary Fire Brigade Number Two. Bohdany, Ukraine, November 30, 2016.

Prianichnikov, Veniamin. Director of plant technical training programs, Chernobyl nuclear power plant. Kiev, February 13, 2006.

Protsenko, Maria. Chief architect for the city of Pripyat. Kiev, September 5, 2015; April 24, 2016; and May 28, 2017.

Reikhtman, Georgi. Deputy head of shift, Unit One, Chernobyl nuclear power plant. Kiev, September 9, 2015.

Robinson, Cliff. Radiochemistry laboratory technician, Forsmark nuclear power station, Sweden. Telephone, March 2, 2016.

Sevastianov, Alexander. In 2016, shift engineer of Main Building Department, Chernobyl nuclear power plant. Chernobyl nuclear power plant, February 10, 2016.

Shcherbak, Iurii. Author; research professor in the epidemiological department of the Ukrainian Ministry of Health, Kiev; special correspondent for *Literaturnaya Gazeta*, Moscow, from May 1986 onward; founding member of Green World; delegate to the Supreme Soviet of the USSR, 1989–1991. Kiev, February 9, 2016.

Shyrokov, Sergei. Chief of the Atomic Energy Department, Ministry of Energy and Electrification, Ukrainian Soviet Socialist Republic. Kiev, December 1, 2016.

Sich, Alexander. MIT nuclear engineering PhD candidate and member of the Chernobyl Complex Expedition under the Kurchatov Institute and the Ukrainian Academy of Sciences, November 1990–April 1992. Telephone, December 21, 2016, and Steubenville, Ohio, April 20–21, 2018.

Sirota, Alexander. Student attending School Number One, Pripyat. Ivankov, Ukraine, June 4, 2017.

Sklyarov, Vitali. Minister of energy and electrification, Ukrainian Soviet Socialist Republic. Kiev, February 6, 2016, and May 30, 2017.

Slutsky, Valery. Bus driver in the city of Pripyat. Pripyat, Ukraine, February 17, 2006.

Sorokin, Yuri. Defense attorney for Viktor Brukhanov. Moscow, October 13, 2016.

Steinberg, Nikolai. Engineer at the Chernobyl nuclear power plant beginning in 1971, and leaving in 1983 as chief turbine engineer, Units Three and Four. In April 1986, deputy chief engineer of Balakovo Nuclear Power Plant. Kiev, February 14, 2006; September 4, 2015, and May 28, 2017.

Stolyarchuk, Boris. Senior unit engineer (SUIB), Fifth Shift, Unit Four, Chernobyl nuclear power plant. Kiev, July 14, 2015 and December 5, 2016.

Svetetsky, Anatoly. Head of technological safety systems, reactor and turbine department of Units Three and Four, Chernobyl nuclear power plant. Kiev, May 28, 2017 (interview by Taras Shumeyko).

Tarakanov, Nikolai. Major general; deputy chief of staff, Civil Defense Forces of the USSR. Moscow, October 22, 2015.

Toptunova, Vera. Mother of Leonid Toptunov. Kiev, September 7 and September 10, 2015.

Usatenko, Vladimir. Chief electrical engineer of the Dnieper Science Institute, Kharkov; drafted in October 1986 and posted to the Chernobyl zone as a sergeant in military repair company 73413. Kiev, December 1, 2016.

Volodin, Sergei. Captain; helicopter pilot with the 225th Composite Air Squadron, Kiev military district. Kiev, February 15, 2006, and July 12, 2015.

Von Hippel, Frank. Chairman of the Federation of American Scientists, 1983–1991. Princeton, NJ, February 15, 2018.

Wilson, Richard. Mallinckrodt Professor of Physics at Harvard University; chairman of the American Physical Society Study Group on Severe Reactor Accidents. Cambridge, MA, August 11, 2016.

Yankovsky, Sergei. Chief investigator of the Kiev region prosecutor's office. Kiev, February 7, 2016, and May 31, 2017.

Young, Martin. In 2018, director, policy and risk, World Energy Council. Telephone, August 3, 2018.

Yuvchenko, Alexander. Senior mechanical engineer, Fifth Shift, Unit Four, Chernobyl nuclear power plant. Moscow, February 12, 2006.

Yuvchenko, Natalia. Schoolteacher in Pripyat School Number Four; wife of Alexander Yuvchenko. Moscow, October 22, 2015, and October 11, 2016.

Zakharov, Anatoly. Driver, Third Watch, Paramilitary Fire Brigade Number Two. Kiev, February 15, 2006, and February 8, 2016.

Books and Memoirs

Akhromeyev, Sergei, and Georgi Korniyenko. *Through the Eyes of a Marshal and a Diplomat: A Critical Look at USSR Foreign Policy Before and After 1985 [Глазами* маршала и дипломата: *Критический взгляд на внешнюю политику СССР до и после 1985 года*]. Moscow: Mezhdunarodnye otnosheniya, 1992.

Albats, Yevgenia. *The State Within a State: The KGB and Its Hold on Russia—Past, Present, and Future*. Translated by Catherine A. Fitzpatrick. New York: Farrar, Straus and Giroux, 1994.

Alexievich, Svetlana. *Voices From Chernobyl*. Translated by Keith Gessen. London: Dalkey Archive Press, 2005.

Andriyanov, V., and V. Chirskov. *Boris Scherbina* [*Борис Щербина*]. Moscow: Molodaya Gvardiya, 2009.

Antoshkin, Nikolai. *Regarding Chernobyl* [*По Чернобылю*]. Unpublished memoir.

———. *The Role of Aviation in Localizing the Consequences of the Catastrophe at Chernobyl* [*Роль авиации в локализации последствий катастрофы на Чернобыльской АЭС*]. Unpublished memoir.

Arnold, Lorna. *Windscale, 1957: Anatomy of a Nuclear Accident*. New York: St. Martin's Press, 1992.

Baranovska, N., ed. *The Chernobyl Tragedy: Documents and Materials* [*Чорнобильська Трагедія: Документи і матеріали*]. Kiev: Naukova Dumka, 1996.

Belyaev, Igor. *Chernobyl: Death Watch* [*Чернобыль: Вахта Смерти*]. 2nd ed. IPK Pareto-Print, 2009.

Belyaev, Igor A. *Sredmash Brand Concrete* [*Бетон марки "Средмаш"*]. Moscow: Izdat, 1996, http://elib.biblioatom.ru/text/belyaev_beton-marki-sredmash_1996.

Bergdahl, Gunnar. *The Voice of Ludmilla*. Translated by Alexander Keiller. Goteborg: Goteborg Film Festival, 2002.

Boldin, Valery. *Ten Years That Shook the World: The Gorbachev Era as Witnessed by His Chief of Staff*. New York: Basic Books, 1994.

Borodavka, Anton. *Faces of Chernobyl: Exclusion Zone* [*Лица Чернобыля: Зона отчуждения*]. Kiev: Statsky Sovetnik, 2013.

Borovoi, A. A., and E. P. Velikhov. *The Chernobyl Experience: Part 1 (Work on the "Shelter" Structure)* [*Опыт Чернобыля: Часть 1 (работы на объекте «Укрытие»)*]. Moscow: Kurchatov Institute, 2012, www.nrcki.ru/files/pdf/1464174600.pdf.

Brown, Kate. *Plutopia: Nuclear Families, Atomic Cities, and the Great Soviet and American Plutonium Disasters*. Oxford: Oxford University Press, 2015.

Chernousenko, Vladimir M. *Chernobyl: Insight from the Inside*. New York: Springer, 1991.

Chernyaev, Anatoly, A. Veber, and Vadim Medvedev, eds. *In the Politburo of the Central Committee of the Communist Party of the Soviet Union . . . From the Notes of Anatoly Chernyaev, Vadim Medvedev, Georgi Shakhnazarov (1985–1991)* [*В Политбюро ЦК КПСС . . . По записям Анатолия Черняева, Вадима Медведева, Георгия Шахназарова (1985–1991)*]. 2nd ed. Moscow: Alpina Business Books, 2008.

Chernyaev, Anatoly S. *My Six Years With Gorbachev*. University Park, PA: Pennsylvania State University Press, 2000.

Cooke, Stephanie. *In Mortal Hands: A Cautionary History of the Nuclear Age*. New York: Bloomsbury, 2010.

Cravens, Gwyneth. *Power to Save the World: The Truth about Nuclear Energy.* New York: Vintage Books, 2008.

Crawley, Gerard M., ed. *Energy from the Nucleus: The Science and Engineering of Fission and Fusion.* Hackensack, NJ: World Scientific Publishing, 2017.

Daniloff, Nicholas. *Of Spies and Spokesmen: My Life as a Cold War Correspondent.* Columbia: University of Missouri Press, 2008.

Danilyuk, Yuri, ed. "Chernobyl Tragedy in Documents and Materials" [Чорнобильська трагедія в документах та матеріалах]. Special issue, *Z arkhiviv VUChK–GPU–NKVD–KGB* 1, no. 16, 2001.

Denisevich, K. B., et al. *Book 4: The Development of Atomic Power and Unified Electricity Systems* [Книга 4: Развитие атомной энергетики и объединенных энергосистем]. Kiev: Energetika, 2011, http://energetika.in.ua/ru/books/book-4/section-2/section-3.

Dobbs, Michael. *Down with Big Brother: The Fall of the Soviet Empire.* New York: Vintage Books, 1998.

Dodd, Charles K. *Industrial Decision-Making and High-Risk Technology: Siting Nuclear Power Facilities in the USSR.* Lanham, MD: Rowman & Littlefield, 1994.

Donets, Natalia, et al. *25 Years of the National Olympic and Sports Committee of the Republic of Moldova* [25 de ani ai Comitetului Naţional Olimpic şi Sportiv din Republica Moldova]. Chisinau: Elan Poligraph, 2016.

Dyachenko, A. A., ed. *Chernobyl: Duty and Courage* [Чернобыль. Долг и мужество]. Vols. 1 and 2. Moscow: Voenizdat, 2001.

Dyatlov, Anatoly. *Chernobyl: How It Was* [Чернобыль: Как это было]. Moscow: Nauchtekhlitizdat, 2003. Online at http://pripyat-city.ru/books/25-chernobyl-kak-yeto-bylo.html.

Epstein, Edward Jay. *Dossier: The Secret History of Armand Hammer.* New York: Random House, 1996.

Esaulov, Alexander. *The City That Doesn't Exist* [Город, которого нет]. Vinnytsia: Teza, 2013.

Fischer, David. *History of the Atomic Energy Agency: The First Forty Years.* Vienna: IAEA, 1997.

Gale, Robert Peter, and Thomas Hauser. *Final Warning: The Legacy of Chernobyl.* New York: Warner Books, 1989.

Gale, Robert Peter, and Eric Lax. *Radiation: What It Is, What You Need to Know.* New York: Vintage Books, 2013.

Gorbachev, Mikhail S. *Collected Works* [Собрание сочинений]. Moscow: Ves Mir, 2008.

Gubarev, Vladimir. *Sarcophagus: A Tragedy.* New York: Vintage Books, 1987.

Gudov, Vladimir. *Special Battalion no. 731* [731 спецбатальон]. Kiev: Kyivskyi Universitet Publishing Center, 2010. Translated by Tamara Abramenkova as *731 Special Battalion: Documentary Story* (Kiev: N. Veselicka, 2012).

Gusev, Igor A., Angelina K. Guskova, and Fred A. Mettler Jr., eds. *Medical Management of Radiation Accidents.* 2nd ed. Boca Raton, FL: CRC Press, 2001.

Guskova, A. K., and G. D. Baysogolov. *Radiation Sickness in Man* [*Лучевая болезнь человека*]. Moscow: Meditsina, 1971.

Guskova, Angelina. *The Country's Nuclear Industry Through the Eyes of a Doctor* [*Атомная отрасль страны глазами врача*]. Moscow: Real Time, 2004. Online at http://elib.biblioatom.ru/text/guskova_atomnaya-otrasl-glazami-vracha _2004.

Ham, Paul. *Hiroshima Nagasaki: The Real Story of the Atomic Bombings and Their Aftermath*. New York: Thomas Dunne Books/St. Martin's Press, 2014.

Hansen, Chuck. *U.S. Nuclear Weapons: The Secret History*. Arlington, TX: Aerofax, 1988.

Hawkes, Nigel, et al. *The Worst Accident in the World: Chernobyl, the End of the Nuclear Dream*. London: William Heinemann and Pan Books, 1988.

Haynes, Viktor, and Marco Bojcun. *The Chernobyl Disaster*. London: Hogarth Press, 1988.

Holloway, David. *Stalin and the Bomb: The Soviet Union and Atomic Energy, 1939–1956*. New Haven: Yale University Press, 1994.

Howes, Ruth H., and Caroline L. Herzenberg. *Their Day in the Sun: Women of the Manhattan Project*. Philadelphia: Temple University Press, 1999.

Ignatenko, Evgeny, ed. *Chernobyl: Events and Lessons* [*Чернобыль: события и уроки*]. Moscow: Politizdat, 1989.

Illesh, Andrey V. *Chernobyl: A Russian Journalist's Eyewitness Account*. New York: Richardson & Steirman, 1987.

Izrael, Y., ed. *Chernobyl: Radioactive Contamination of the Environment* [*Чернобыль: Радиоактивное загрязнение природных сред*]. Leningrad: Gidrometeoizdat, 1990.

Jorgensen, Timothy J. *Strange Glow: The Story of Radiation*. Princeton, NJ: Princeton University Press, 2017.

Josephson, Paul R. *Red Atom: Russia's Nuclear Power Program from Stalin to Today*. Pittsburgh, PA: University of Pittsburgh Press, 2005.

———. "Rockets, Reactors and Soviet Culture." In Loren Graham, ed. *Science and the Soviet Social Order*. Cambridge, MA: Harvard University Press, 1990.

Karpan, Nikolay. "First days of the Chernobyl accident. Private experience." Memoir, 2008. Retrieved from the Kyoto University Research Reactor Institute website, at www.rri.kyoto-u.ac.jp/NSRG/en/Karpan2008English.pdf.

———. *From Chernobyl to Fukushima* [*От Чернобыля до Фукусимы*]. Kiev: S. Podgornov, 2011. Translated by Andrey Arkhipets. Kiev: S. Podgornov, 2012.

———. *Chernobyl: Revenge of the Peaceful Atom* [*Чернобыль. Месть мирного атома*]. Kiev: CHP Country Life, 2005.

Kopchinsky, Georgi, and Nikolai Steinberg, *Chernobyl: On the Past, Present and Future* [*Чернобыль: О прошлом, настоящем и будущем*] (Kiev: Osnova, 2011)

Kostin, Igor. *Chernobyl: Confessions of a Reporter*. New York: Umbrage Editions, 2006.

Kotkin, Stephen. *Armageddon Averted: The Soviet Collapse, 1970–2000*. 2nd ed. New York: Oxford University Press, 2008.

Kozlova, Elena. *The Battle with Uncertainty* [*Схватка с неизвестностью*]. Moscow:

Izdat, 2011. Online at http://elib.biblioatom.ru/text/kozlova_shvatka-s-neizvest nostyu_2011.

Kruglov, A. K. *The History of the Soviet Atomic Industry*. New York: Taylor & Francis, 2002.

Kupny, Alexander. *Memories of Lives Given: Memories of Liquidators* [Живы, пока нас помнят: Воспоминания ликвидаторов]. Kharkiv: Zoloty Storynki, 2011.

Larin, Vladislav. *"Mayak" Kombinat: A Problem for the Ages* [Комбинат "Маяк"— проблема на века], 2nd ed. Moscow: Ecopresscenter, 2001.

Legasova, Margarita. *Academician Valery Alekseyevich Legasov* [Академик Валерий Алексеевич Легасов]. Moscow: Spektr, 2014.

——. "Defenceless Victor: From the Recollections of Academician V. Legasov's Widow" [Беззащитный победитель: Из воспоминаний вдовы акад. В. Легасова], *Trud*, June 1996. Translated in Mould, *Chernobyl Record*.

Lewin, Moshe. *The Soviet Century*. Reprint. Edited by Gregory Elliott. New York: Verso, 2016.

Ligachev, Yegor. *Inside Gorbachev's Kremlin: The Memoirs of Yegor Ligachev*. Translated by Catherine A. Fitzpatrick, Michele A. Berdy, and Dobrochna Dyrcz-Freeman. Introduction by Stephen F. Cohen. New York: Pantheon, 1993.

Logachev, Alexander. *The Truth* [Истина]. Unpublished memoir, 2005.

Lyashko, Alexander. *The Weight of Memory: On the Rungs of Power* [Груз памяти: На ступенях власти]. Volume 2 in a trilogy. Kiev: Delovaya Ukraina, 2001.

Lyerly, Ray L., and Walter Mitchell III. *Nuclear Power Plants*. "Understanding the Atom" series. Rev. ed. Washington, DC: Atomic Energy Commission, 1973.

Mahaffey, James. *Atomic Accidents: A History of Nuclear Meltdowns and Disasters: From the Ozark Mountains to Fukushima*. New York: Pegasus Books, 2014.

——. *Atomic Awakening: A New Look at the History and Future of Nuclear Power*. New York: Pegasus, 2009.

Maleyev, Vladimir. *Chernobyl. Days and Years: The Chronicle of the Chernobyl Campaign* [Чернобыль. Дни и годы: летопись Чернобыльской кампании]. Moscow: Kuna, 2010.

Manzurova, Natalia, and Cathy Sullivan. *Hard Duty: A Woman's Experience at Chernobyl*. Tesuque, NM: Natalia Manzurova and Cathy Sullivan, 2006.

Marples, David R. *Chernobyl and Nuclear Power in the USSR*. New York: St. Martin's Press, 1986.

——. *The Social Impact of the Chernobyl Disaster*. New York: St. Martin's Press, 1988.

Medvedev, Grigori. *Chernobyl Chronicle* [Чернобыльская хроника]. Moscow: Sovermennik, 1989. Translated by Evelyn Rossiter as *The Truth About Chernobyl* (New York: Basic Books, 1991).

Medvedev, Roy A., and Zhores A. Medvedev. *The Unknown Stalin*. Translated by Ellen Dahrendorf. New York: I. B. Tauris, 2003.

Medvedev, Zhores A. *The Legacy of Chernobyl*. New York: Norton, 1990.

Mettler, Fred A., Jr., Charles A. Kelsey, and Robert C. Ricks, eds. *Medical Management of Radiation Accidents*. 1st ed., Boca Raton, FL: CRC Press, 1990.

Mickiewicz, Ellen Propper. *Split Signals: Television and Politics in the Soviet Union.* New York: Oxford University Press, 1990.

Mould, Richard F. *Chernobyl Record: The Definitive History of the Chernobyl Catastrophe.* Boca Raton, FL: CRC Press, 2000.

Mycio, Mary. *Wormwood Forest: A Natural History of Chernobyl.* Washington, DC: Joseph Henry Press, 2005.

Nesterov, Boris. *Heaven and Earth: Memories and Reflections of a Military Pilot* [*Небо и земля: Воспоминания и размышления военного летчика*]. Kherson, 2016.

Nove, Alec. *The Soviet Economy: An Introduction.* 2nd rev. ed. New York: Praeger, 1969.

Paskevych, Sergiy, and Denis Vishnevsky. *Chernobyl: Real World* [*Чернобыль. Реальный мир*]. Moscow: Eksmo, 2011.

Petryna, Adriana. *Life Exposed: Biological Citizens after Chernobyl.* Princeton, NJ: Princeton University Press, 2013.

Pikhoya, R. G. *Soviet Union: The History of Power. 1945–1991* [*Советский Союз: История власти. 1945–1991*]. Novosibirsk: Sibirsky Khronograf, 2000.

Pikhoya, R. G., and A. K. Sokolov. *The History of Modern Russia: Late 1970s to 1991* [*История современной России: Конец 1970х–1991гг*]. Moscow: Fond Pervogo Prezidenta Rossii B. N. Yeltsina, 2008.

Plokhy, Serhii. *Chernobyl: The History of a Nuclear Catastrophe.* New York: Basic Books, 2018.

———. *The Gates of Europe: A History of Ukraine.* New York: Basic Books, 2015.

Popov, Fedor. *Arzamas–16: Seven Years with Andrei Sakharov* [*Арзамас-16: семь лет с Андреем Сахаровым*]. Moscow: Institut, 1998.

Raab, Nigel. *All Shook Up: The Shifting Soviet Response to Catastrophes, 1917–1991.* Montreal: McGill-Queen's University Press, 2017.

Read, Piers Paul. *Ablaze: The Story of the Heroes and Victims of Chernobyl.* New York: Random House, 1993.

Remnick, David. *Lenin's Tomb: The Last Days of the Soviet Empire.* New York: Vintage Books, 1994.

Roth-Ey, Kristin. *Moscow Prime Time: How the Soviet Union Built the Media Empire That Lost the Cultural Cold War.* Ithaca: Cornell University Press, 2014.

Roxburgh, Angus. *Moscow Calling: Memoirs of a Foreign Correspondent.* Berlin: Birlinn, 2017.

———. *The Second Russian Revolution: The Struggle for Power in the Kremlin.* New York: Pharos Books, 1992.

Russian Ministry of Emergency Situations. "The aftermath of the man-made radiation exposure and the challenge of rehabilitating the Ural region" [Последствия техногенного радиационного воздействия и проблемы реабилитации Уральского региона]. Moscow, 2002, http://chernobyl-mchs.ru/upload/program_rus/program_rus_1993-2010/Posledstviy_Ural.pdf.

Ryzhkov, Nikolai. *Ten Years of Great Shocks* [*Десять лет великих потрясений*]. Moscow: Kniga-Prosveshchenie-Miloserdie, 1995.

Schmid, Sonja D. *Producing Power: The Pre-Chernobyl History of the Soviet Nuclear Industry.* Cambridge, MA: MIT Press, 2015.

Sebag Montefiore, Simon. *Stalin: The Court of the Red Tsar.* New York: Knopf, 2004.

Semenov, A. N., ed. *Chernobyl: Ten Years On: Inevitability or Accident?* [Чернобыль. Десять лет спустя. Неизбежность или случайность?]. Moscow: Energoatomizdat, 1995.

Service, Robert. *A History of Modern Russia: From Tsarism to the Twenty-First Century.* Cambridge, MA: Harvard University Press, 2010.

Shcherbak, Iurii. *Chernobyl* [Чернобыль]. Moscow: Sovietsky Pisatel, 1991.

———. *Chernobyl: A Documentary Story.* Translated by Ian Press. Foreword by David R. Marples. London: Macmillan, 1989.

Shlapentokh, Vladimir. *A Normal Totalitarian Society: How the Soviet Union Functioned and How It Collapsed.* Armonk, NY: M.E. Sharpe, 2001.

Shultz, George P. *Turmoil and Triumph: My Years as Secretary of State.* New York: Charles Scribner's Sons, 1993.

Sich, Alexander. "The Chornobyl Accident Revisited: Source Term Analysis and Reconstruction of Events During the Active Phase." PhD diss., Massachusetts Institute of Technology, 1994.

Sidorenko, V. A., ed. *The History of Atomic Energy in the Soviet Union and Russia* [История атомной энергетики Советского Союза и России]. Moscow: Kurchatov Institute/Izdat, 2001, http://elib.biblioatom.ru/text/istoriya-atomnoy-energe tiki_v1_2001.

Sidorenko, Viktor A., ed. *The Contribution of Kurchatov Institute Staff to the Liquidation of the Accident at the Chernobyl NPP* [Вклад Курчатовцев в ликвидацию последствий аварии на чернобыльской АЭС]. Moscow: Kurchatov Institute, 2012, www.nrcki.ru/files/pdf/1464174688.pdf.

Sklyarov, V. F. *Sublimation of Time* [Сублимация времени]. Kiev: Kvic, 2015.

———. *Tomorrow Was Chernobyl: A Documentary Account* [Завтра был Чернобыль: Документальная повесть]. Kiev: Osvita, 1991. Translated by Victor Batachov as *Chernobyl Was . . . Tomorrow: A Shocking Firsthand Account* (Montreal: Presses d'Amérique, 1993).

Smirnov, A. G., and L. B. Godgelf. *The Classification of Explosive Areas in National and International Standards and Regulations* [Классификация взрывоопасных зон в национальных и международных стандартах, правилах]. Moscow: Tiazhprom-electroproyekt, 1992, online at http://aquagroup.ru/normdocs/1232.

Smirnov, Y. N., ed. *Igor Vasilyevich Kurchatov in Recollections and Documents* [Игорь Васильевич Курчатов в воспоминаниях и документах]. 2nd ed. Moscow: Kurchatov Institute/Izdat, 2004, http://elib.biblioatom.ru/text/kurchatov-v-vospom inaniyah-i-dokumentah_2004.

Stern, Eric K. *Crisis Decisionmaking: A Cognitive Institutional Approach.* Stockholm: Swedish National Defence College, 2003.

Tarakanov, Nikolai. *The Bitter Truth of Chernobyl* [Горькая правда Чернобыля]. Moscow: Center for Social Support of Chernobyl's Invalids, 2011.

Taubman, William. *Gorbachev: His Life and Times.* New York: Simon & Schuster, 2017.

Vargo, G. J., ed. *The Chornobyl Accident: A Comprehensive Risk Assessment.* Columbus, OH: Battelle Press, 2000.

Velikhov, Evgeny P. *My Journey: I Shall Travel Back to 1935 in Felt Boots* [*Мой путь. Я на валенках поеду в 35-й год*]. E-Book. Moscow: AST, 2016. Translated by Andrei Chakhovskoi as *Strawberries from Chernobyl: My Seventy-Five Years in the Heart of a Turbulent Russia* (CreateSpace Independent Publishing Platform, 2012).

Volkogonov, Dmitri, and Harold Shukman. *Autopsy for an Empire: The Seven Leaders Who Built the Soviet Regime.* New York: Free Press, 1999.

Vorotnikov, V. I. *This Is How It Went . . . From the Diary of a Member of the Politburo of the Central Committee of the Communist Party of the Soviet Union* [*А было это так . . . Из дневника члена Политбюро ЦК КПСС*]. Moscow: Soyuz Veteranov Knigoizdaniya SI–MAR, 1995.

Voznyak, Vasily, and Stanislav Troitsky. *Chernobyl: It Was Like This—The View from the Inside* [*Чернобыль: Так это было—взгляд изнутри*]. Moscow: Libris, 1993.

Weeks, Theodore R. *Across the Revolutionary Divide: Russia and the USSR, 1861–1945.* Chichester, UK: Wiley-Blackwell, 2010.

Wieczynski, Joseph, ed. *The Gorbachev Encyclopedia.* Salt Lake City, UT: Schlacks, 1993.

Yaroshinskaya, Alla A. *Chernobyl: The Big Lie* [*Чернобыль: Большая ложь*]. Moscow: Vremya, 2011. Translated by Sergei Roy as *Chernobyl: Crime Without Punishment.* Edited by Rosalie Bertell and Lynn Howard Ehrle. New Brunswick, NJ: Transaction Publishers, 2011).

———. *Chernobyl: The Forbidden Truth.* Translated by Michèle Kahn and Julia Sallabank. Foreword by David R. Marples. Lincoln: University of Nebraska Press, 1995.

Yelinskaya, T. N., ed. *Chernobyl: Labor and Heroism. Dedicated to Krasnoyarsk Liquidators of the Chernobyl Accident* [*Чернобыль: Труд и подвиг. Красноярским ликвидаторам Чернобыльской аварии посвящается*]. Krasnoyarsk: Polykor, 2011, www.krskstate.ru/dat/bin/atlas_let_attach/1008_book_1_131.pdf.

Yevsiukov, Yuri. *Pripyat* [*Припять*]. Kiev: Mystetstvo, 1986.

Zemtsov, Ilya. *Lexicon of Soviet Political Terms.* Edited by Gay M. Hammerman. Fairfax, VA: Hero Books, 1985.

Zickel, Raymond E., ed. *Soviet Union: A Country Study.* 2nd ed. Washington, DC: US Government Printing Office, 1991.

Zlobin, G. K., and V. Y. Pinchuk, eds. *Chernobyl: Post-Accident Construction Program* [*Чорнобиль: Післяаварійна програма будівництва*], Kiev Construction Academy. Kiev: Fedorov, 1998.

Articles and Reports

Abakumov, Vitali. "Analyzing the causes and circumstances of the 1975 accident on Unit One of Leningrad NPP (perspective of an engineer-physicist, participant and witness to the events)" [Анализ причин и обстоятельств аварии 1975 года на 1-м блоке ЛАЭС (комментарий инженера-физика, участника и очевидца событий)]. April 10, 2013, http://accidont.ru/Accid75.html.

Abrams, Herbert L. "How Radiation Victims Suffer." *Bulletin of Atomic Scientists* 42, no. 7 (1986). (August/September 1986): 13–17.

Amerisov, Alexander. "A Chronology of Soviet Media Coverage." *Bulletin of the Atomic Scientists* 42, no. 7 (August/September 1986): 38–39.

Ananenko, Alexei. "Recollections of a Senior Machine Engineer of the 2nd Reactor Shop" [Воспоминания старшего инженера-механика реакторного цеха №2 Алексея Ананенка]. Interview. Soyuz Chernobyl. April 4, 2013, www.souzcher nobyl.org/?id=2440.

"Angelina Konstantinovna Guskova: Biography" [Гуськова Ангелина Константиновна: биография], Rosatom, www.biblioatom.ru/founders/guskova_angelina _konstantinovna.

ApSimon, Helen, and Julian Wilson. "Tracking the Cloud from Chernobyl." *New Scientist*, no. 1517 (July 17, 1986): 42–45.

Arutunyan, Rafael V. "The China Syndrome" [Китайский синдром]. *Priroda*, no. 11 (November 1990): 77–83.

Associated Press. "Text of the Politburo Statement About Chernobyl." *New York Times*, July 21, 1986.

———. "A Top Soviet Aide Details Situation at Stricken Plant." May 3, 1986.

Astakhova, Alla. Interview with Alexander Borovoi. "The Liquidator" [Ликвидатор], *Itogi* 828, no. 17 (April 23, 2012), www.itogi.ru/obsh-spetzproekt/2012/17 /177051.html.

Babakov, Sergei. Interview with Piotr Zborovsky. "I'm still there today, in the Chernobyl zone" [. . . Я и сегодня там, в Чернобыльской зоне]. *Zerkalo nedeli Ukraina*, September 18, 1998, http://gazeta.zn.ua/SOCIETY/ya_i_segodnya _tam,_v_chernobylskoy_zone.html. Translated as ". . . I am still there, in nowadays Chernobyl zone," in Gudov, *731 Special Battalion: Documentary Story* (Kiev: N. Veselicka, 2012).

———. Interview with Viktor Brukhanov. "I don't accept the charges against me . . ." [С предъявленными мне обвинениями не согласен . . .]. *Zerkalo nedeli*, August 27, 1999, https://zn.ua/society/c_predyavlennymi_mne_obvineniyami_ne _soglasen.html.

Baranov, Alexandr, Robert Peter Gale, Angelina Guskova et al. "Bone Marrow Transplantation after the Chernobyl Nuclear Accident." *New England Journal of Medicine* 321, no. 4.

Barré, Bertrand. "Fundamentals of Nuclear Fission." In Gerard M. Crawley, ed., *Energy from the Nucleus: The Science and Engineering of Fission and Fusion*. Hackensack, NJ: World Scientific Publishing, 2016.

Barringer, Felicity. "Fear of Chernobyl Radiation Lingers for the People of Kiev." *New York Times*, May 23, 1988.

———. "One Year After Chernobyl, a Tense Tale of Survival." *New York Times*, April 6, 1987.

———. "On Moscow Trains, Children of Kiev." *New York Times*, May 9, 1986.

———. "Pripyat Journal: Crows and Sensors Watch Over Dead City." *New York Times*, June 22, 1987.

Batorshin, G. Sh., and Y. G. Mokrov. "Experience in Eliminating the Consequences of the 1957 Accident at the Mayak Production Association." International Experts' Meeting on Decommissioning and Remediation After a Nuclear Accident, IAEA, Vienna, Austria, January 28 to February 1, 2013, www-pub.iaea.org/iaeameetings /IEM4/Session2/Mokrov.pdf.

Baumann, Paul. "NL Man Was First Victim of Atomic Experiments." *The Day*, August 6, 1985.

Bennett, Burton, Michael Repacholi, and Zhanat Carr, eds. "Health Effects of the Chernobyl Accident and Special Care Programmes." Report of the UN Chernobyl Forum Expert Group on Health, World Health Organization, 2006.

Bivens, Matt. "Horror of Soviet Nuclear Sub's '61 Tragedy Told." *Los Angeles Times*, January 3, 1994.

Bohlen, Celestine. "Chernobyl's Slow Recovery; Plant Open, but Pripyat Still a Ghost Town." *Washington Post*, June 21, 1987.

———. "Gorbachev Says 9 Died From Nuclear Accident; Extends Soviet Test Ban." *Washington Post*, May 15, 1986.

Bolyasny, Alexander. "The First 'Orderly' of the First Zone" [Первый «санитар» первой зоны]. *Vestnik* 320 no. 9 (April 2003), www.vestnik.com/issues/2003 /0430/koi/bolyasny.htm.

Bond, Michael. Interview with Alexander Yuvchenko. "Cheating Chernobyl." *New Scientist*, August 21, 2004.

Bondarkov, Mikhail D., et al. "Environmental Radiation Monitoring in the Chernobyl Exclusion Zone—History and Results 25 Years After." *Health Physics* 101, no. 4 (October 2011).

Borovoi, A. A. "My Chernobyl" [Мой Чернобыль]. *Novy Mir*, no. 3 (1996).

Brook, Barry W., et al. "Why Nuclear Energy Is Sustainable and Has to Be Part of the Energy Mix." *Sustainable Materials and Technologies* 1–2 (December 2014): 8–16.

Brukhanov, Viktor. Interview. "The Incomprehensible Atom" [Непонятый атом]. *Profil*, April 24, 2006, www.profile.ru/obshchestvo/item/50192-items_18814.

Bukharin, Oleg A. "The Cold War Atomic Intelligence Game, 1945–70." *Studies in Intelligence* 48, no. 2.

Burns, John F. "A Rude Dose of Reality for Gorbachev." *New York Times*, February 21, 1989.

Bushmelev, V. Y. "For Efim Pavlovich Slavsky's 115th Birthday" [К 115-летию Ефима Павловича Славского], Interregional Non-Governmental Movement of Nuclear Power and Industry Veterans, October 26, 2013, www.veteranrosatom.ru/articles /articles_276.html.

Campbell, Murray. "Soviet A-Leak 'World's Worst': 10,000 Lung Cancer Deaths, Harm to Food Cycle Feared." *Globe and Mail*, April 30, 1986.

Cardis, Elisabeth, et al. "Estimates of the Cancer Burden in Europe from Radioactive Fallout from the Chernobyl Accident." *International Journal of Cancer* 119, no. 6 (2006): 1224–35.

Champlin, Dr. Richard. "With the Chernobyl Victims: An American Doctor's Inside Report from Moscow's Hospital No. 6." *Los Angeles Times*, July 6, 1986.

Checherov, Konstantin P. "Evolving accounts of the causes and processes behind the Block 4 accident at the Chernobyl NPP on 26 April 1986" [Развитие представлений о причинах и процессах аварии на 4-м блоке ЧАЭС 26 апреля 1986 г.]. In *Problems of Chornobyl*, Issue 5: Materials of International Scientific and Practical Conference "Shelter-98," 1999: 176–78, www.iaea.org/inis /collection/NCLCollectionStore/_Public/32/020/32020472.pdf.

———. "On the physical nature of the explosion on ChNPP energy block no. 4." [О физической природе взрыва на 4-м энергоблоке ЧАЭС]. *Energia*, no. 6, 2002, http://portalus.ru/modules/ecology/rus_readme.php?subaction=showfull&id =1096468666.

———. "The Unpeaceful Atom of Chernobyl" [Немирный атом Чернобыль]. *Chelovek* no. 6, 2006. Online at http://vivovoco.astronet.ru/VV/PAPERS/MEN /CHERNOBYL.HTM.

Cherkasov, Vitali. "On the 15th anniversary of the atomic catastrophe: Chernobyl's sores" [К 15-летию атомной катастрофы: язвы Чернобыля]. *Pravda*, April 25, 2011, www.pravda.ru/politics/25-04-2001/817996-0.

Chernobyl NPP website. "Materials: Liquidation Heroes" [Материалы: Герои-ликвидаторы]. http://chnpp.gov.ua/ru/component/content/article?id=82.

Clean Air Task Force, United States. "The Toll from Coal: An Updated Assessment of Death and Disease from America's Dirtiest Energy Source." September 2010.

Clines, Francis X. "Soviet Villages Voice Fears on Chernobyl." *New York Times*, July 31, 1989.

Daily Mail. "'2000 Dead' in Atom Horror: Reports in Russia Danger Zone Tell of Hospitals Packed with Radiation Accident Victims." April 29, 1986.

Daniloff, Nicholas. "Chernobyl and Its Political Fallout: A Reassessment." *Demokratizatsiya: The Journal of Post-Soviet Democratization* 12, no. 1 (Winter 2004): 117–32.

Daniszewski, John. "Reluctant Ukraine to Shut Last Reactor at Chernobyl," *Los Angeles Times*, December 14, 2000.

Davletbayev, Razim. "The Last Shift" [Последняя смена]. In Semenov, ed., *Chernobyl: Ten Years On*, 366–83.

DeYoung, Karen. "Stockholm, Bonn Ask for Details of Chernobyl Mishap: Soviets Seek West's Help to Cope With Nuclear Disaster." *Washington Post*, April 30, 1986.

Diamond, Stuart. "Chernobyl's Toll in Future at Issue." *New York Times*, August 29, 1986.

Dobbs, Michael. "Chernobyl's 'Shameless Lies.'" *Washington Post*, April 27, 1992.

Drozdov, Sergei. "Aerial Battle over Chernobyl" [Воздушная битва при Чернобыле]. *Aviatsiya i vremya* 2 (2011), online at www.xliby.ru/transport_i_aviacija/aviacija _i_vremja_2011_02/p6.php.

Dyachenko, Anatoly. "The Experience of Employing Security Agencies in the Liquidation of the Catastrophe at the Chernobyl Nuclear Power Plant" [Опыт применения силовых структур при ликвидации последствий катастрофы на Чернобыльской АЭС]. *Military Thought* [*Военная мысль*], no. 4 (2003): 77–80.

Dyatlov, Anatoly S. "Why INSAG has still got it wrong." *Nuclear Engineering International* 40, no. 494 (September 1995): 219–23.

Eaton, William J. "Candor Stressed in Stage Account; Soviet Drama Spotlights Chernobyl Incompetence." *Los Angeles Times*, September 17, 1986.

———. "Soviets Report Nuclear Accident: Radiation Cloud Sweeps Northern Europe; Termed Not Threatening." *Los Angeles Times*, April 29, 1986.

———. "Soviets Tunneling Beneath Reactor; Official Hints at Meltdown into Earth; Number of Evacuees Reaches 84,000." *Los Angeles Times*, May 9, 1986.

Eaton, William J., and William Tuohy. "Soviets Seek Advice on A-Plant Fire 'Disaster': Bonn, Stockholm Help Sought, but Moscow Says Only 2 Died." *Los Angeles Times*, April 30, 1986.

Engelberg, Stephen. "2D Soviet Reactor Worries U.S. Aides." *New York Times*, May 5, 1986.

Fedulenko, Valentin. "Some Things Have Not Been Forgotten." In Sidorenko, ed., *The Contribution of Kurchatov Institute Staff.*

———. "Perspectives on the Accident: Memoirs of a Participant and Expert Opinion. Part 3" [Версии аварии: мемуары участника и мнение эксперта. Часть 3]. Chernobyl.by, September 19, 2008, www.chernobyl.by/accident/28-versii-avarii -memuary-uchastnika-i-mnenie.html.

Fountain, Henry. "Chernobyl: Capping a Catastrophe." *New York Times*, April 27, 2014.

Fox, Sue. "Young Guardian: Memories of Chernobyl—Some of the Things Dr. Robert Gale Remembers from the Aftermath of the World's Worst Nuclear Disaster." *Guardian*, May 18, 1988.

Franklin, Ben A. "Report Calls Mistrust a Threat to Atom Power." *New York Times*, March 8, 1987.

Geiger, H. Jack. "The Accident at Chernobyl and the Medical Response." *Journal of the American Medical Association (JAMA)* 256, no. 5 (August 1, 1986).

Geist, Edward. "Political Fallout: The Failure of Emergency Management at Chernobyl." *Slavic Review* 74, no. 1 (Spring 2015): 104–26.

Ger, E. "Reactors Were Evolving Faster Than the Culture of Safety" [Реакторы развивались быстрее, чем культура безопасности]. Interview with Viktor Sidorenko. *Rosatom's Living History*, 2015, http://memory.biblioatom.ru/persona /sidorenko__v_a/sidorenko__v_a.

Gorbachev, Mikhail. "Turning Point at Chernobyl." *Project Syndicate*, April 14, 2006.

Gray, Richard. "How We Made Chernobyl Rain." *Sunday Telegraph*, April 22, 2007.

Grogan, David. "An Eyewitness to Disaster, Soviet Fireman Leonid Telyatnikov Recounts the Horror of Chernobyl." *People*, October 5, 1987.

Gubarev, Vladimir. Interview with Angelina Guskova. "On the Edge of the Atomic Sword" [На лезвии атомного меча]. *Nauka i zhizn*, no. 4 (2007), www.nkj.ru/archive /articles/9759/.

———. "On the Death of V. Legasov." Excerpts from *The Agony of Sredmash* [Агония Средмаша] (Moscow: Adademkniga, 2006), reproduced in Margarita Legasova, *Academician Valery A. Legasov.*

Guskova, Angelina, and Igor Gusev. "Medical Aspects of the Accident at Chernobyl." In Igor A. Gusev, Angelina K. Guskova, and Fred A. Mettler, eds., *Medical Management of Radiation Accidents* (New York: CRC Press, 2001), 195–210.

Gusman, Mikhail. "Geidar Aliyev, President of the Republic of Azerbaijan" [Гейдар Алиев, президент Азербайджанской Республики]. TASS, September 26, 2011, http://tass.ru/arhiv/554855.

Guth, Stefan. "Picturing Atomic-Powered Communism." Paper presented at the international conference Picturing Power: Photography in Socialist Societies, University of Bremen, December 9–12, 2015.

Harding, Graham. "Sovetskoe Shampanskoye—Stalin's 'Plebeian Luxury,'" *Wine As Was*, blog, August 26, 2014.

Hargraves, Robert, and Ralph Moir. "Liquid Fuel Nuclear Reactors." *Physics and Society* (a newsletter of the American Physical Society), January 2011.

Harrison, John, et al. "The Polonium-210 poisoning of Mr Alexander Litvinenko." *Journal of Radiological Protection* 371, no. 1 (February 28, 2017), 266–278.

Hawtin, Guy. "Report: 15,000 Buried in Nuke Disposal Site." *New York Post*, May 2, 1986.

Higginbotham, Adam. "Chernobyl 20 Years On." *Guardian*, March 25, 2006.

———. "Is Chernobyl a Wild Kingdom or a Radioactive Den of Decay?" *Wired*, April 2011.

Hrebet, Oleksandr. Interview with Vladimir Usatenko, "A Chernobyl liquidator talks of the most dangerous nuclear waste repository in Ukraine" [Ліквідатор аварії на ЧАЕС розповів про найнебезпечніше сховище ядерних відходів в Україні], *Zerkalo nedeli*, December 14, 2016, https://dt.ua/UKRAINE/likvidator-avariyi -na-chaes-rozpoviv-pro-naynebezpechnishomu-shovische-yadernih-vidhodiv-v -ukrayini-227461_.html.

Ilyin, L. A., and A. V. Barabanova. "Obituary: Angelina K. Guskova." *Journal of Radiological Protection* 35, No. 33 (September 7, 2015), 733–34.

Interfax. "Construction Cost of Blocks 3 and 4 of Khmelnitsky NPP Will Be About $4.2 Billion" [Стоимость строительства 3 и 4 блоков Хмельницкой АЭС составит около $4,2 млрд]. Interfax, March 3, 2011.

International Atomic Energy Agency. "Cleanup of Large Areas Contaminated as a Result of a Nuclear Accident." IAEA Technical Reports Series No. 330. IAEA, Vienna, 1998.

———. "Environmental Consequences of the Chernobyl Accident and Their Remediation: Twenty Years of Experience." Report of the Chernobyl Forum Expert Group "Environment" no. STI/PUB/1239, April 2006.

———. "Medical Aspects of the Chernobyl Accident: Proceedings of An All-Union Conference Organized by the USSR Ministry of Health and the All-Union Scientific Centre of Radiation Medicine, USSR Academy of Medical Sciences, and held in Kiev, 11–13 May 1988," report no. IAEA-TECDOC-516, 1989.

———. "Nuclear Applications for Steam and Hot Water Supply." Report no. TEC-DOC-615, July 1991.

———. "Present and Future Environmental Impact of the Chernobyl Accident." Report no. IAEA–TECDOC-1240, August 2001.

International Nuclear Safety Advisory Group. "The Chernobyl Accident: Updating of INSAG-1." Safety series no. 75–INSAG-7. Vienna: International Atomic Energy Agency, 1992.

———. "Summary Report of the Post-accident Review Meeting on the Chernobyl Accident." Safety Series no. 75–INSAG-1. Vienna, Austria, 1986.

"Ivan Stepanovich Silaev" [Иван Степанович Силаев]. Biography on the website of the Republic of Chuvashia, http://gov.cap.ru/SiteMap.aspx?gov_id=15&id=36079, accessed on 12 November 2017.

Ivanov, Boris. "Chernobyl." *Voennye znaniya* 40, nos. 1–4 (January–April 1988).

Jacquot, Jeremy. "Numbers: Nuclear Weapons, From Making a Bomb to Making a Stockpile to Making Peace." *Discover*, October 23, 2010.

Japan Times. "Late Chernobyl fireman's blood tests to be disclosed." April 19, 2006.

Keller, Bill. "Gorbachev, at Chernobyl, Urges Environment Plan." *New York Times*, February 24, 1989.

———. "Public Mistrust Curbs Soviet Nuclear Efforts." *New York Times*, October 13, 1988.

Kharaz, Alina. "It was like being at the front" [Там было как на фронте]. Interview with Yuri Grigoriev, *Vzgliad*, April 26, 2010.

Kholosha, V., and V. Poyarkov. "Economy: Chernobyl Accident Losses." In Vargo, ed., *The Chornobyl Accident*.

Kirk, Don. "Gorbachev Tries Public Approach." *USA Today*, May 15, 1986.

Kiselyov, Sergei. "Inside the Beast." Translated by Viktoria Tripolskaya-Mitlyng. *Bulletin of the Atomic Scientists* 52, no. 3 (May–June 1996): 43–51.

Kitral, Alexander. "Gorbachev to Scherbitsky: 'Fail to hold the parade, and I'll leave you to rot!'" [Горбачев—Щербицкому: «Не проведешь парад—сгною!»]. *Komsomolskaya Pravda v Ukraine*, April 26, 2011, https://kp.ua/life/277409-horbachev -scherbytskomu-ne-provedesh-parad-shnoui.

Kiyansky, Dmitry. "Let our museum be the only and the last" [Пусть наш музей будет единственным и последним], interview with Ivan Gladush (interior minister of Ukraine at time of accident), *Zerkalo nedeli Ukraina*, April 28, 2000, https://zn.ua /society/pust_nash_muzey_budet_edinstvennym_i_poslednim.html.

Klages, P. "Atom Rain over U.S." *Telegraph*, May 6, 1986.

Kovalevska, Lyubov. "Not a Private Matter" [Не приватна справа]. *Literaturna Ukraina*, March 27, 1986, www.myslenedrevo.com.ua/uk/Sci/HistSources/Chor nobyl/Prolog/NePryvatnaSprava.html.

Kruchik, Igor. "Mother of the Atomgrad" [Мати Атомограда]. *Tizhden*, September 5, 2008, http://tyzhden.ua/Publication/3758.

Kukhar', V., V. Poyarkov, and V. Kholosha. "Radioactive Waste: Storage and Disposal Sites." In Vargo, ed., *The Chornobyl Accident*.

Kurnosov, V., et al. Report no. IAEA-CN-48/253: "Experience of Entombing the Damaged Fourth Power Unit of the Chernobyl Nuclear Power Plant" [Опыт захоронения аварийного четвертого энергоблока Чернобыльской АЭС]. In IAEA, *Nuclear Power Performance and Safety*, proceedings of the IAEA conference in Vienna (September 28 to October 2, 1987), vol. 5, 1988.

Kuzina, Svetlana. "Kurchatov wanted to know what stars were made of—and created bombs" [Курчатов хотел узнать, из чего состоят звезды. И создал бомбы]. *Komsomolskaya Pravda*, January 10, 2013, www.kp.ru/daily/26012.4/2936276.

Lee, Gary. "Chernobyl's Victims Lie Under Stark Marble, Far From Ukraine." *Washington Post*, July 2, 1986.

———. "More Evacuated in USSR: Indications Seen of Fuel Melting Through Chernobyl Reactor Four." *Washington Post*, May 9, 1986.

Legasov, Valery. "My Duty Is to Tell about This." In Mould (ed.) *Chernobyl Record*, 2000: 287–306.

Legasov, Valery; V. A. Sidorenko; M. S. Babayev; and I. I. Kuzmin. "Safety Issues at Atomic Power Stations" [Проблемы безопасности на атомных электростанциях]. *Priroda* no. 6 (1980).

Macalister, Terry. "Westinghouse wins first US nuclear deal in 30 years." *The Guardian*, April 9, 2008.

Maleyev, Vladimir. "Chernobyl: The Symbol of Courage" [Чернобыль: символ мужества]. *Krasnaya Zvezda*, April 25, 2017, archive.redstar.ru/index.php/2011-07-25-15-55-35/item/33010-chernobyl-simvol-muzhestva.

Malyshev, L. I., and M. N. Rozin. "In the Fight for Clean Water" [В борьбе за чистую воду]. In Semenov, ed., *Chernobyl: Ten Years On*, 231–44.

Marin, V. V. "On the Activities of the Task Force of the Politburo of the CPSU Central Committee at the Chernobyl NPP" [О деятельности оперативной группы Политбюро ЦК КПСС на Чернобыльской АЭС]. In Semenov, ed., *Chernobyl: Ten Years On*.

Markham, James M. "Estonians Resist Chernobyl Duty, Paper Says." *New York Times*, August 27, 1986.

Marples, David R. "Phobia or not, people are ill at Chernobyl." *Globe and Mail* (Canada), September 15, 1987.

———. "Revelations of a Chernobyl Insider." *Bulletin of the Atomic Scientists* 46, no. 10, (December) 1990.

Martin, Lawrence. "Negligence cited in Chernobyl report." *Globe and Mail* (Canada), July 21, 1986.

Martin, Richard. "China Could Have a Meltdown-Proof Nuclear Reactor Next Year." *MIT Technology Review*, February 11, 2016.

———. "China Details Next-Gen Nuclear Reactor Program." *MIT Technology Review*, October 16, 2015.

Masharovsky, Maj. Gen. M. "Operation of Helicopters During the Chernobyl Accident." *Current Aeromedical Issues in Rotary Wing Operations: Papers Presented at the RTO Human Factors and Medicine Panel (HFM)*. Symposium held in San Diego, CA, October 19–21, 1998, RTO/NATO, 1999.

McKenna, Phil. "Fossil Fuels Are Far Deadlier Than Nuclear Power." *New Scientist*, March 23, 2011.

Medvedev, Grigori. "Chernobyl Notebook" [Чернобыльская тетрадь]. *Novy Mir*, no. 6 (June 1989). Translated by JPRS Economic Affairs (Report no. JPRS-UEA-89-034, October 23, 1989).

Mervine, Evelyn. "Nature's Nuclear Reactors: The 2-Billion-Year-Old Natural Fission Reactors in Gabon, Western Africa." *Scientific American* blog, July 13, 2011.

Mettler, Fred A., Jr., and Charles A. Kelsey. "Fundamentals of Radiation Accidents." Gusev, Guskova, and Mettler, eds., *Medical Management of Radiation Accidents*, 2001.

Ministry of Energy and Electrification of the USSR. "Chernobyl NPP: Master Plan of the Settlement" [Чернобыльская АЭС: Генеральный план поселка]. Moscow, Gidroproekt, 1971.

Mitchell, Charles. "New Chernobyl Contamination Charges." UPI, February 2, 1989.

Moore, D. "U.N. Nuclear Experts to Go to USSR." *Daily Telegraph*, May 5, 1986.

Morelle, Rebecca. "Windscale Fallout Underestimated." October 6, 2007, BBC News.

Mycio, Mary. "Do Animals in Chernobyl's Fallout Zone Glow?" *Slate*, January 21, 2013.

Nadler, Gerald. "Gorbachev Visits Chernobyl." UPI, February 24, 1989.

National Nuclear Laboratory. "Boiling Water Reactor Technology: International Status and UK Experience." Position paper, National Nuclear Laboratory, 2013.

National Research Council. *Health Risks from Exposure to Low Levels of Ionizing Radiation: BEIR VII Phase 2*. Washington, DC: The National Academies Press, 2006.

New York Times. "Atomic Bomb Worker Died 'From Burns.'" September 21, 1945.

———. "Soviet Military Budget: $128 Billion Bombshell." May 31, 1989.

Nikolaevich, Oleg. Interview with Viktor Brukhanov. "Stories about Tashkent Natives: True and Sometimes Unknown. Part 1" [Истории о ташкентцах правдивые и не всем известные. Часть 1]. *Letters about Tashkent*, April 29, 2016, http://my tashkent.uz/2016/04/29/istorii-o-tashkenttsah-pravdivye-i-ne-vsem-izvestnye -chast-1.

Novoselova, Elena. Interview with Nikolai Ryzhkov. "The Chronicle of Silence" [Хроника молчания]. *Rossiiskaya Gazeta*, April 25, 2016, https://rg.ru/2016/04 /25/tridcat-let-nazad-proizoshla-avariia-na-chernobylskoj-aes.html.

O'Brien, Liam. "After 26 years, farms emerge from the cloud of Chernobyl." *The Independent*, June 1, 2012.

Osipchuk, Igor. "The legendary Academician Aleksandrov fought with the White Guard in his youth" [Легендарный академик Александров в юности был белогвардейцем], *Fakty i kommentarii*, February 4, 2014, http://fakty.ua/176084 -legendarnyj-prezident-sovetskoj-akademii-nauk-v-yunosti-byl-belogvardejcem.

———. "When it became obvious that cleaning the NPP roofs of radioactive debris would have to be done by hand by thousands of people, the Government Commission sent soldiers there" [Когда стало ясно, что очищать крыши ЧАЭС от радиоактивных завалов придется вручную силами тысяч человек, правительственная комиссия послала туда солдат], interview with Yuri Samoilenko, *Fakty i Kommentarii*, April 25, 2003, http://fakty.ua/75759-kogda-stalo -yasno-chto-ochicshat-kryshi-chaes-ot-radioaktivnyh-zavalov-pridetsya-vruch nuyu-silami-tysyach-chelovek-pravitelstvennaya-komissiya-poslala-tuda-soldat.

Oskolkov, B. Y. "Treatment of radioactive waste in the initial period of liquidating the consequences of the Chernobyl NPP accident. Overview and analysis" [Обращение с радиоактивными отходами первоначальный период ликвидации последствий аварии на ЧАЭС. Обзор и анализ]. Chornobyl Center for Nuclear Safety, January 2014.

Pappas, Stephanie. "How Plants Survived Chernobyl." *Science*, May 15, 2009.

Parry, Vivienne. "How I Survived Chernobyl." *Guardian*, August 24, 2004.

Patterson, Walt. "Futures: Why a Kind of Hush Fell Over the Chernobyl Conference / Western Atomic Agencies' Attitude to the Soviet Nuclear Accident." *Guardian*, October 4, 1986.

Peel, Quentin. "Work abandoned on Soviet reactor." *Financial Post* (Toronto), September 9, 1988.

Petrosyants, Andranik. "'Highly Improbable Factors' Caused Chemical Explosion." *Los Angeles Times*, May 9, 1986.

Petryna, Adriana. "Nuclear Payouts: Knowledge and Compensation in the Chernobyl Aftermath." *Anthropology Now*, November 19, 2009.

Petty, Dwayne Keith. "Inside Dawson Forest: A History of the Georgia Nuclear Aircraft Laboratory." *Pickens County Progress*, January 2, 2007, online at http://archive.li /GMnGk.

Polad-Zade, P. A. "Too Bad It Took a Tragedy" [Жаль, что для этого нужна трагедия]. In Semenov, ed., *Chernobyl. Ten Years On*, 195–200.

Poroshenko, Petro. "The President's address at the ceremony marking the 30th anniversary of the Chernobyl catastrophe" [Виступ Президента під час заходів у зв'язку з 30-ми роковинами Чорнобильської катастрофи]. Speech on the New Safe Confinement site, April 26, 2016, online at: www.president.gov.ua/news/vist up-prezidenta-pid-chas-zahodiv-u-zvyazku-z-30-mi-rokovin-37042.

Potter, William C. "Soviet Decision-Making for Chernobyl: An Analysis of System Performance and Policy Change." Report to the National Council for Soviet and East European Research, March 1990.

Prushinsky, B. Y. "This Can't Be—But It Happened" [Этого не может быть—но это случилось]. In Semenov, ed., *Chernobyl. Ten Years On*, 308–24.

Remnick, David. "Chernobyl's Coffin Bonus." *Washington Post*, November 24, 1989.

———. "Echo in the Dark." *New Yorker*, September 22, 2008.

Reuters. "Chernobyl Costs Reach $3.9 Billion." *Globe and Mail* (Canada), September 20, 1986.

———. "New Town Opens to Workers from Chernobyl Power Plant." *New York Times*, April 19, 1988.

———. "Wild Boars Roam Czech Forests—and Some of Them Are Radioactive." February 22, 2017.

Reva, V. M. Testimony at the 46th session of the Supreme Rada, December 11, 1991. Kiev, Ukraine. Transcript online at http://rada.gov.ua/meeting/stenogr/show /4642.html.

Rogovin, Mitchell, and George T. Frampton Jr. (NRC Special Inquiry Group). *Three Mile Island: A Report to the Commissioners and to the Public*. Washington, DC: Government Printing Office, 1980.

Rybinskaya, Irina. Interview with Vladimir Trinos. "Fireman Vladimir Trinos, one of the first to arrive at Chernobyl after the explosion: 'It was inconvenient to wear gloves, so the guys worked with their bare hands, crawling on their knees through radioactive water . . .'" [Пожарный Владимир Тринос, одним из первых попавший на ЧАЭС после взрыва: «в рукавицах было неудобно, поэтому ребята работали голыми руками, ползая на коленях по радиоактивной воде . . . »]. *Fakty i kommentarii*, April 26, 2001, http://fakty.ua/95948-pozharnyj-vladimir-trinos-odnim-iz-pervyh-popav shij-na-chaes-posle-vzryva-quot-v-rukavicah-bylo-neudobno-poetomu-rebyata -rabotali-golymi-rukami-polzaya-na-kolenyah-po-radioaktivnoj-vode-quot.

Rylskii, Maksim. Interview with Vitali Sklyarov. "The Nuclear Power Industry in the Ukraine." *Soviet Life* 353, no. 2.

Samarin, Anton. Interview with Viktor Brukhanov. "Chernobyl hasn't taught anyone anything" [Чернобыль никого и ничему не научил]. *Odnako*, April 26, 2010, www.odnako.org/magazine/material/chernobil-nikogo-i-nichemu-ne-nau chil-1/.

Samodelova, Svetlana. "The private catastrophe of Chernobyl's director" [Личная катастрофа директора Чернобыля]. *Moskovsky Komsomolets*, April 22, 2011, www.mk.ru/politics/russia/2011/04/21/583211-lichnaya-katastrofa-direktora -chernobyilya.html.

Sazhneva, Ekaterina. "The living hero of a dead city" [Живой герой мертвого города]. *Kultura*, February 2, 2016, http://portal-kultura.ru/articles/history/129184-zhivoy -geroy-mertvogo-gorodasazh.

Schmemann, Serge. "Chernobyl and the Europeans: Radiation and Doubts Linger." *New York Times*, June 12, 1988.

——. "Kremlin Asserts 'Danger Is Over.'" *New York Times*, May 12, 1986.

——. "Soviet Announces Nuclear Accident at Electric Plant." *New York Times*, April 29, 1986.

——. "The Talk of Kiev." *New York Times*, May 31, 1986.

Semenov, A. N. "For the 10th Anniversary of the Chernobyl Catastrophe." In Semenov, ed., *Chernobyl: Ten Years On*, 7–74.

Shanker, Thom. "As Reactors Hum, 'Life Goes On' at Mammoth Tomb." *Chicago Tribune*, June 15, 1987.

——. "Life Resumes at Chernobyl as Trials Begin." *Chicago Tribune*, June 16, 1987.

——. "2 Graves Lift Chernobyl Toll to 30." *Chicago Tribune*, August 3, 1986.

Shasharin, G. "Chernobyl Tragedy" [Чернобыльская трагедия]. *Novy Mir* 797, no. 9 (1991): 164–79. Republished with the same title in Semenov, ed., *Chernobyl: Ten Years On*, 75–132.

Shcherbak, Iurii. "Chernobyl: A Documentary Tale" [Чернобыль: Документальная повесть]. *Yunost*, nos. 6–7 (1987). Translated by JPRS Soviet Union Political Affairs as "Fictionalized Report on First Anniversary of Chernobyl Accident" (Report no. JPRS-UPA-87-029, September 15, 1987).

Sheremeta, Yelena. Interview with Rada Scherbitskaya. "After Chernobyl, Gorbachev told Vladimir Vasiliyevich, 'If you don't hold the parade, say good-bye to the party.'" [После Чернобыля Горбачев сказал Владимиру Васильевичу: «Если не проведешь первомайскую демонстрацию, то можешь распрощаться с партией»]. *Fakty i kommentarii*, February 17, 2006, http://fakty.ua/43896-rada-csherbickaya-quot-posle-chernobylya-gorbachev-skazal-vladimiru-vasilevichu-quot-esli-ne-provedesh-pervomajskuyu-demonstraciyu-to-mozhesh-rasprocshatsya-s-partiej-quot.

———. Interview with Vitali Masol. "Vitali Masol: 'We were quietly preparing to evacuate Kiev'" [Виталий Масол: «Мы тихонечко готовились к эвакуации Киева»]. *Fakty i kommentarii*, April 26, 2006, http://fakty.ua/45679-vitalij-masol-quot-my-tihonechko-gotovilis-k-evakuacii-kieva-quot.

Shunevich, Vladimir. "Former ChNPP Director Brukhanov: 'When after the accident my mother learned that I'd been expelled from the party, her heart broke . . .'" [Бывший директор ЧАЭС Виктор Брюханов: «Когда после взрыва реактора моя мама узнала, что меня исключили из партии, у нее разорвалось сердце . . .»]. *Fakty i kommentarii*, December 1, 2010, http://fakty.ua/123508-byvshij-direktor-chaes-viktor-bryuhanov-kogda-v-1986-godu-posle-vzryva-reaktora-moya-mama-uznala-chto-menya-isklyuchili-iz-partii-u-nee-razorvalos-serdce.

———. Interview with Viktor Brukhanov. "Former director of the Chernobyl Atomic Power Station Viktor Brukhanov: 'At night, driving by Unit Four, I saw that the structure above the reactor is . . . Gone!'" [Бывший директор Чернобыльской Атомной Электростанции Виктор Брюханов: «Ночью, проезжая мимо четвертого блока, увидел, что верхнего строения над реактором . . . Нету!»]. *Fakty i kommentarii*, April 28, 2006, http://fakty.ua/45760-byvshij-direktor-chernobylskoj-atomnoj-elektrostancii-viktor-bryuhanov-quot-nochyu-proezzhaya-mimo-chetvertogo-bloka-uvidel-chto-verhnego-stroeniya-nad-reaktorom-netu-quot.

Sich, Alexander. "Truth Was an Early Casualty." *Bulletin of Atomic Scientists* 52, no. 3 (May–June 1996).

Sidorchik, Andrei. "Deadly experiment. Chronology of the Chernobyl NPP catastrophe" [Смертельный эксперимент. Хронология катастрофы на Чернобыльской АЭС]. *Argumenty i fakty*, April 26, 2016, www.aif.ru/society/history/smertelnyy_eksperiment_hronologiya_katastrofy_na_chernobylskoy_aes.

Skaletsky, Yuriy, and Oleg Nasvit (National Security and Defense Council of Ukraine). "Military liquidators in liquidation of the consequences of Chornobyl NPP accident: myths and realities." In T. Imanaka, ed., *Multi-side Approach to the Realities of the Chernobyl NPP Accident* (Kyoto: Kyoto University, 2008).

"Slavsky Efim Pavlovich" [Славский Ефим Павлович]. Family history, www.famhist.ru/famhist/ap/001ef29d.htm.

Sparks, Justin. "Russia Diverted Chernobyl Rain, Says Scientist." *Sunday Times*, August 8, 2004.

Sports.ru. "Soccer in Pripyat: The History of the 'Builder' Soccer Club" [Футбол в Припяти. История футбольного клуба «Строитель»]. April 27, 2014, www.sports.ru/tribuna/blogs/golden_ball/605515.html.

Stadnychenko-Cornelison, Tamara. "Military engineer denounces handling of Chernobyl Accident." *The Ukrainian Weekly*, April 26, 1992.

Stastita. "Mortality Rate Worldwide in 2018, by Energy Source (in Deaths Per Terawatt Hours)." Statista.com, www.statista.com/statistics/494425/death-rate-worldwide-by-energy-source.

Stein, George. "Chernobyl's Flaws Exposed in March by Ukraine Paper." *Los Angeles Times*, May 2, 1986.

Strasser, Emily. "The Weight of a Butterfly." *Bulletin of the Atomic Scientists*, February 25, 2015, https://thebulletin.org/2015/02/the-weight-of-a-butterfly/.

Telyatnikov, Leonid. "Firefight at Chernobyl." Transcript of the address at the Fourth Great American Firehouse Exposition and Muster, Baltimore, MD, September 17, 1987, online at Fire Files Digital Library, https://fire.omeka.net/items/show/625.

Tolstikov, V. S., and V. N. Kuznetsov. "The 1957 Radiation Accident in Southern Urals: Truth and Speculation" [Южно-уральская радиационная авария 1957 года: Правда и домыслы]. *Vremya* 32, no. 8 (August 2017): 13.

United Press International. "Tens of Thousands in March: Nuclear Disaster Ignored at Soviet May Day Parade." *Los Angeles Times*, May 1, 1986.

US Department of the Treasury. "Treasury Sanctions Russian Officials, Members of the Russian Leadership's Inner Circle, and an Entity for Involvement in the Situation in Ukraine." March 20, 2014, https://www.treasury.gov/press-center/press-releases/Pages/jl23331.aspx.

US Nuclear Regulatory Commission. "Operating Nuclear Power Reactors (by Location or Name)." Updated April 4, 2018, www.nrc.gov/info-finder/reactors.

———. "Report on the Accident at the Chernobyl Nuclear Power Station (NUREG–1250)." US Department of Energy, January 1987.

USSR State Committee on the Utilization of Atomic Energy. "The Accident at the Chernobyl Nuclear Power Plant and Its Consequences" [also known as "Vienna report"]. Information compiled for the IAEA Experts' Meeting, Vienna, 25–29 August, 1986, Parts I and II (August 1986).

Vasyl, Maria. Interview with Viktor Brukhanov. "Former ChNPP director Brukhanov: 'Had they found legal grounds to have me shot, they would have done so.'" [Бывший директор ЧАЭС Виктор Брюханов: «Если бы нашли для меня расстрельную статью, то, думаю, расстреляли бы.»]. *Fakty i kommentarii*, October 18, 2000, http://fakty.ua/104690-byvshij-direktor-chaes-viktor-bryuhanov-quot-esli-by-nashli-dlya-menya-rasstrelnuyu-statyu-to-dumayu-rasstrelyali-by-quot.

Veklicheva, R. "A Soviet Way of Life: The Test" [Образ жизни—Советский. Испытание]. *Vperiod* (official newspaper of Obninsk Communist Party Committee), June 17, 1986.

Voloshko, Vladimir. "The Town That Died at the Age of Sixteen" [Город, погибший в 16 лет]. http://pripyat.com/people-and-fates/gorod-pogibshii-v-16-let.html.

Von Hippel, Frank N., and Matthew Bunn. "Saga of the Siberian Plutonium-Production

Reactors." Federation of American Scientists Public Interest Report, 53 (November/ December 2000), https://fas.org/faspir/v53n6.htm.

Vorobyev, A. "Chernobyl Catastrophe Five Years On" [Чернобыльская катастрофа пять лет спустя]. *Novy Mir* 797, no. 9 (1991): 179–83.

Walker, Martin. "Moscow Play Pans Nuclear Farce: Piece on Chernobyl Accident to Tour Soviet Cities." *Guardian*, September 18, 1986.

Weber, E. T., et al. "Chernobyl Lessons Learned: Review of N Reactor." Report WHC-SP-0257, prepared for the US Department of Energy Assistant Secretary for Defense Programs. October 1987.

Wellock, Thomas. "'Too Cheap to Meter': A History of the Phrase." *United States Nuclear Regulatory Commission* (blog), June 3, 2016.

Whittington, Luther. "'2,000 Die' in Nukemare; Soviets Appeal for Help as N-plant Burns out of Control." *New York Post*, April 29, 1986.

Williams, Carol J. "Chernobyl Victims Buried at Memorial Site." Associated Press. June 24, 1986.

World Energy Council. World Energy Scenarios 2016, Executive Summary. World Energy Council report, https://www.worldenergy.org/wp-content/uploads/2016/10/World-Energy-Scenarios-2016_Executive-Summary-1.pdf.

World Health Organization (WHO), Regional Office For Europe. "Chernobyl Reactor Accident: Report of a Consultation." Report no. ICP/CEH 129, May 6, 1986 (provisional).

WHO. "Health Effects of the Chernobyl Accident: An Overview." April 2006, www.who.int/ionizing_radiation/chernobyl/backgrounder/en/.

———. "1986–2016: Chernobyl at 30—An update." Press release, April 25, 2016.

WHO/IAEA/UNDP. "Chernobyl: The true scale of the accident." Joint press release, September 5, 2005.

World Nuclear Association. "Nuclear Power in France." Updated June 2018, www.world-nuclear.org/information-library/country-profiles/countries-a-f/france.aspx.

———. "Nuclear Power in China," World Nuclear Association, updated May 2018, www.world-nuclear.org/information-library/country-profiles/countries-a-f/china-nuclear-power.aspx.

Yadrihinsky, A. A. "Atomic Accident at Unit Four of Chernobyl NPP and Nuclear Safety of RBMK Reactors" [Ядерная авария на 4 блоке Чернобыльской АЭС и ядерная безопасность реакторов РБМК]. Gosatomenergonadzor Inspectorate at the Kursk Nuclear Power Station, 1989.

Yatsenko, Natalia. Interview with Vitali Sklyarov. "Vitali Sklyarov, energy advisor to the Ukrainian prime minister: 'What's happening in our energy sector is self-suffocation.'" [Советник премьер-министра Украины по вопросам энергетики Виталий Скляров: «Самоудушение—вот что происходит с нашей энергетикой»]. *Zerkalo nedeli Ukraina*, October 7, 1994, https://zn.ua/ECONOMICS/sovetnik_premier-ministra_ukrainy_po_voprosam_energetiki_vitaliy_sklyarov_samoudushenie_-_vot_chto_p.html.

Yurchenko, Y. Report no. IAEA-CN-48/256: *Assessment of the Effectiveness of Mechanical Decontamination Technologies and Technical Devices Used at the Damaged Unit of the Chernobyl Nuclear Power Plant* [Оценка эффективности технологий и технических средств механической дезактивации аварийного блока Чернобыльской АЭС.]. In IAEA, *Nuclear Power Performance and Safety*, 1988.

Zhilin, A. "No such thing as someone else's grief" [Чужого горя не бывает]. *Aviatsiya i Kosmonavtika* no. 8 (August 1986).

Film and TV

The Atom Joins the Grid. Documentary film. London: British Pathé, October 1956.

The Battle of Chernobyl. Documentary film. Directed by Thomas Johnson. France: Play Film, 2006.

The Bells of Chernobyl (Ten Years Later). Documentary film. Directed by Kurt Langbein et al. NTU/BTRC/TVP/RTL/Tele Images/Strix TV/TV Asahi, 1996.

Chernobyl: A Warning [Чернобыль: Предупреждение]. Documentary film. Narrated by Lev Nikolayev. Directed by Vladimir Osminin. Moscow: Channel One, 1987.

Chernobyl: Chronicle of Difficult Weeks. Documentary film. Directed by Vladimir Shevchenko. Kiev: Ukrainian News and Documentary Studio, 1986.

Chernobyl: Two Colors of Time [Чернобыль: Два цвета времени]. Three-part documentary. Directed by I. Kobrin. Kiev: Ukrtelefim, 1989.

Chernobyl 1986.04.26 Post Scriptum [Чернобыль. 1986.04.26 Post Scriptum]. Documentary series narrated by Valery Starodumov. Kiev: Telecon, 2016.

Chernobyl 3828 [Чернобыль 3828]. Documentary film. Narrated by Valery Starodumov. Directed by Sergei Zabolotny. Kiev: Telecon, 2011.

The Construction of the Chernobyl Nuclear Power Plant [Будівництво Чорнобилсської АЕС]. Documentary film. Kiev: Ukrainian Studio of Documentary Chronicle Films, 1974. State Film Archive of Ukraine.

Inside Chernobyl's Sarcophagus. Documentary film. Directed by Edward Briffa. United Kingdom: BBC Horizon, 1991 (rereleased 1996).

The Mystery of Academician Legasov's Death [Тайна смерти академика Легасова]. Documentary film. Directed by Yuliya Shamal and Sergei Marmeladov. Moscow: Afis-TV for Channel Rossiya, 2004.

Pandora's Promise. Documentary film. Directed by Robert Stone. Impact Partners, 2013.

Radioactive Wolves. Documentary film. Directed by Klaus Feichtenberger. Epo-Film, ORF/Universum and Thirteen in association with BBC, NDR, and WNET New York Public Media, 2011.

The Science of Superstorms. Documentary series. Produced by Michael Mosley. London: BBC, 2007.

The Second Russian Revolution. Documentary series. Produced by Norma Percy and directed by Mark Anderson. London: BBC, 1991.

Windscale 1957: Britain's Biggest Nuclear Disaster. Documentary film. Directed by Sarah Aspinall. BBC, 2007.

Zero Hour: Disaster at Chernobyl. Documentary and dramatic reconstruction. Directed by Renny Bartlett. Discovery, 2004.

Photo Credits

Index

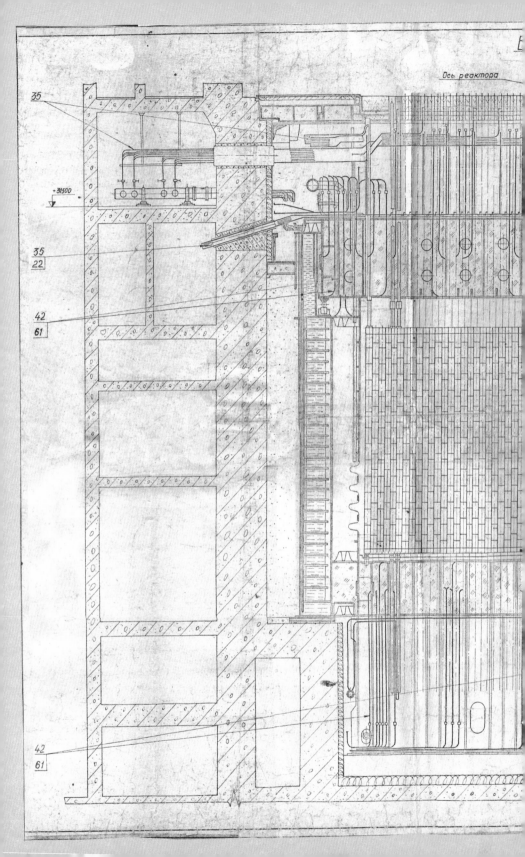

Ось реактора

35

+ 31500

35
22

42
61

42
61